MW00605000

Nanotechnology and Tissue Engineering

The Scaffold

Nanotechnology and Tissue Engineering

The Scaffold

Cato T. Laurencin, M.D., Ph.D.

Lakshmi S. Nair, Ph.D.

CRC Press is an imprint of the
Taylor & Francis Group, an **informa** business

CRC Press
Taylor & Francis Group
6000 Broken Sound Parkway NW, Suite 300
Boca Raton, FL 33487-2742

Printed in the United States of America on acid-free paper
10 9 8 7 6 5 4 3 2 1

International Standard Book Number-13: 978-1-4200-5182-7 (Hardcover)

Library of Congress Cataloging-in-Publication Data

Nanotechnology and tissue engineering: the scaffold / [edited by] Cato Laurencin and Lakshmi Nair.
p.; cm.
Includes bibliographical references and index.
ISBN 978-1-4200-5182-7 (hardcover: alk. paper) 1.Tissue engineering. 2. Nanotechnology. 3. Biomedical materials. I. Laurencin, Cato T. II. Nair, Lakshmi. III. Title.
[DNLM: 1. Tissue Engineering--methods. 2. Nanostructures. 3. Nanotechnology. 4. Tissue Scaffolds. QT 37 N187 2008]

R857.T55N36 2008
610.284--dc22 2008013040

Visit the Taylor & Francis Web site at
http://www.taylorandfrancis.com

and the CRC Press Web site at
http://www.crcpress.com

Dedication

To my wife Cynthia, and my children,
Ti, Michaela, and Victoria

To my husband Prem and my son Bharath

Contents

SECTION III Promising Nanofabrication Techniques to Develop Scaffolds for Tissue Engineering

SECTION IV Applications of Nanostructured Scaffolds in Biology and Medicine

Foreword

I am delighted to write the foreword for *Nanotechnology and Tissue Engineering: The Scaffold* edited by Professor Laurencin and Dr. Nair. This book will be extremely useful as a reference source for all those working in the area of biomaterials, tissue engineering, and bio-nanotechnology. Today, science and technology at the nanoscale is capable of providing unprecedented understanding, control, and manipulation of matter at the atomic and molecular level. Nanotechnology has already had a significant impact on modern medicine through the development of novel targeted therapies, diagnostic and imaging techniques, as well as by providing novel strategies to repair and regenerate tissues. Nanofabrication techniques have opened the door toward developing unique structures, which would be of great value in regenerative medicine, by giving us the ability to mimic biological structures with molecular level precision, and thereby controlling or modulating cellular functions. The present work is timely and provides a fine summary of the present status of bio-nanotechnology aimed toward developing biomimetic scaffolds for tissue regeneration. The attraction of the book lies in the judicious combination of concise and comprehensive chapters covering the fundamentals of ideal scaffolds for tissue engineering, cellular behavior toward nanostructures, and state-of-the-art nanofabrication techniques for developing biomimetic nanostructures for tissue engineering. The collection of 14 chapters written by experts in their fields from different parts of the world presents an excellent overview of the subject for a wide audience. In my opinion, the book will provide a valuable resource to the field of bio-nanotechnology and tissue engineering.

C.N.R. Rao
Honorary President of Jawaharlal Nehru Centre for Advanced Scientific Research, India

Distinguished Visiting Professor, Department of Materials, University of California, Santa Barbara

Foreword

Preface

The aim of *Nanotechnology and Tissue Engineering: The Scaffold* is to provide a state-of-the-art, comprehensive account of research in the rapidly emerging areas of tissue engineering and nanotechnology. The tremendous advances in bio-nanotechnology during the past decade have significantly impacted the area of tissue engineering, opening up new avenues to realize the dream of regenerative medicine. This book is a natural presentation of the marriage of nanotechnology and tissue engineering, an emerging technology which has the potential to revolutionize medicine in the near future. The content of each chapter is written with a detailed background as the book covers a multidisciplinary area with audiences from various fields of professions ranging from engineers, clinicians, and scientists to graduate students and senior undergraduate students. The chapters also provide an extensive bibliography for the reader who wants to explore the subject to a greater depth.

The book is organized into 14 chapters, carrying the reader through the fundamentals of tissue engineering, comprehensive analysis of the unique cellular responses toward nanostructured materials, emerging nanofabrication techniques, and state-of-the-art reviews on the exciting breakthroughs using nanostructures in engineering three major tissues of the human body: neural, vascular, and musculoskeletal. Each of the 14 chapters is subindexed and titled so that the book can easily be used as a reference source.

We have chosen to present the materials under four sections: Section I clearly emphasizes the importance of scaffolds in tissue engineering. This section includes four chapters. Chapter 1 presents a broad overview of the area of tissue engineering and sets the stage for the rest of the book. Chapter 2 vividly presents the structure and functions of the extracellular matrix, the structure which tissue engineers are attempting to recreate using novel technologies. Chapter 3 discusses the functions and requirements of synthetic scaffolds for engineering tissues and Chapter 4 reviews the various microfabrication techniques currently being investigated for developing tissue engineering scaffolds. Section II is meant to emphasize the effect of nanostructures on cellular responses and tissue regeneration. Chapter 5 presents an in-depth discussion of the current literature on cellular responses toward nanostructured materials. Chapter 6 provides an overview of the various nanoscale biological surface modifications of biomaterials to improve cellular responses. Section III presents an overview of some of the most promising nanofabrication techniques to develop scaffolds for tissue engineering. Chapters 7 through 9 give a comprehensive account of the process of electrospinning and its versatility as a fabrication technique to form tissue engineering scaffolds. Chapter 10 discusses the various lithographic techniques toward developing nanostructured scaffolds for tissue engineering. Chapter 11 presents self-assembly as a unique fabrication method for developing biologically active scaffolds for accelerated tissue regeneration. Section IV provides an overview of some of the applications of nanostructured scaffolds in biology and medicine. Chapter 12 discusses the applications of nanostructured materials in neural tissue engineering, Chapter 13 in cardiovascular tissue engineering, and Chapter 14 in musculoskeletal tissue engineering.

It is our hope that all the chapters, written by eminent experts in the field, will provide a platform to better understand the impact of nanotechnology in the area of tissue engineering and regenerative medicine.

Cato T. Laurencin
Lakshmi S. Nair

Editors

Cato T. Laurencin is a professor of biomedical engineering and chemical engineering at the University of Virginia. He is also the Lillian T. Pratt distinguished professor and chairman of the Department of Orthopaedic Surgery at the University of Virginia, and orthopaedic surgeon-in-chief of the University of Virginia Health System.

Dr. Laurencin earned his BSE in chemical engineering from Princeton University, his MD from Harvard Medical School where he graduated magna cum laude, and his PhD in biochemical engineering/biotechnology from the Massachusetts Institute of Technology where he was a Hugh Hampton Young Scholar. Dr. Laurencin completed a residency in orthopaedic surgery at Harvard University where he was chief resident at the Beth Israel Hospital, Harvard Medical School. He also completed a clinical fellowship in shoulder surgery and sports medicine at the Hospital for Special Surgery, Cornell Medical College in New York. Clinically, Dr. Laurencin is board certified in orthopaedic surgery. He is a fellow of the American College of Surgeons, a fellow of the American Surgical Association, and a fellow of the American Academy of Orthopaedic Surgeons. For his clinical work, Dr. Laurencin was named one of the top 101 doctors in America by *Black Enterprise*, and has been named to America's Top Doctors and America's Top Surgeons.

Dr. Laurencin's academic interests are in the areas of tissue engineering, biomaterials, drug delivery, and nanotechnology. Dr. Laurencin was honored at the White House where he received the Presidential Faculty Fellowship Award from President William Jefferson Clinton in recognition of his research work bridging medicine and engineering. Dr. Laurencin is an international fellow in biomaterials science and engineering. He is the recipient of the William Grimes award from the American Institute of Chemical Engineers, the Nicolas Andry Award from the Association of Bone and Joint Surgeons for Orthopaedic Research, the Clemson Award from the Society for Biomaterials for contributions to the biomaterials literature, and the Leadership in Technology Award from the New Millennium Foundation.

Dr. Laurencin serves on the editorial board of 12 journals including *the Journal of Biomedical Materials Research*, *Biomaterials*, and *the Journal of Biomedical Nanotechnology*. He is also an assistant editor for *Clinical Orthopaedics and Related Research*.

In public policy, Dr. Laurencin is a member of the National Science Foundation's Directorate of Engineering Advisory Committee (ADCOM), and was a member of the National Science Advisory Board of the U.S. Food and Drug Administration (FDA). He recently also served in the leadership of the National Medical Association as speaker of the house of delegates and was a member of the NIH National Advisory Council for Musculoskeletal and Skin Diseases (NIAMS).

Dr. Laurencin is an elected member of the Institute of Medicine of the National Academy of Sciences.

Lakshmi Nair is an assistant professor at the Department of Orthopedic Surgery, University of Virginia. She received her MSc in analytical chemistry and MPhil in chemistry from the University of Kerala, India. Dr. Nair received her PhD in polymer chemistry on surface modification of polymers from Sree Chitra Tirunal Institute for Medical Sciences and Technology, India. She did postdoctoral training in the areas of biomaterials and tissue engineering at Drexel University and University of Virginia before joining the current position. Her research interests include hydrogels, nanomaterials, and tissue engineering. She has more than 60 publications in the area of biomaterials, hydrogels, drug delivery, and tissue engineering.

Editors

Contributors

C. Mauli Agrawal
Department of Biomedical Engineering
The University of Texas, San Antonio
San Antonio, Texas

Ravi Bellamkonda
WHC Department of Biomedical Engineering
Georgia Institute of Technology
Atlanta, Georgia

and

Emory University
Atlanta, Georgia

Christopher J. Bettinger
Department of Materials Science and Engineering
Massachusetts Institute of Technology
Cambridge, Massachusetts

Mina J. Bissell
Life Sciences Division
Lawrence Berkeley National Laboratory
Berkeley, California

Jeffrey T. Borenstein
Biomedical Engineering Center
Charles Stark Draper Laboratory
Cambridge, Massachusetts

Edward A. Botchwey
Departments of Biomedical Engineering
and Orthopaedic Surgery
University of Virginia
Charlottesville, Virginia

Casey K. Chan
Division of Bioengineering
Faculty of Engineering
National University of Singapore
Singapore

and

Department of Orthopaedic Surgery
Yong Loo Lin School of Medicine
National University of Singapore
Singapore

Michael Cho
Department of Bioengineering
University of Illinois
Chicago, Illinois

Batur Ercan
Division of Engineering
Brown University
Providence, Rhode Island

Akihiro Horii
Massachusetts Institute of Technology
Cambridge, Massachusetts

and

Olympus Corporation
Hachiouji-shi, Tokyo, Japan

Tejas Shyam Karande
Department of Biomedical Engineering
The University of Texas at Austin
Austin, Texas

and

Department of Orthopaedics
The University of Texas Health Science Center
at San Antonio
San Antonio, Texas

Jeffrey M. Karp
Department of Medicine
Brigham and Women's Hospital
Harvard Medical School
Cambridge, Massachusetts

and

Harvard-MIT Division of Health
Science and Technology
Massachusetts Institute of Technology
Cambridge, Massachusetts

Sangamesh G. Kumbar
Department of Orthopaedic Surgery
University of Virginia
Charlottesville, Virginia

Robert Langer
Departments of Chemical Engineering and Bioengineering
Department of Mechanical Engineering
Massachusetts Institute of Technology
Cambridge, Massachusetts
and
Harvard-MIT Division of Health Science and Technology
Massachusetts Institute of Technology
Cambridge, Massachusetts

Cato T. Laurencin
Department of Orthopaedic Surgery, Biomedical Engineering, and Chemical Engineering
University of Virginia
Charlottesville, Virginia

Duron A. Lee
Department of Biomedical Engineering
Drexel University
Philadelphia, Pennsylvania

Susan Liao
Division of Bioengineering
Faculty of Engineering
National University of Singapore
Singapore
and
Department of Orthopaedic Surgery
Yong Loo Lin School of Medicine
National University of Singapore
Singapore

Chwee Teck Lim
Division of Bioengineering
Faculty of Engineering
National University of Singapore
Singapore
and
Department of Mechanical Engineering
Faculty of Engineering
National University of Singapore
Singapore

Gregory S. McCarty
Department of Biomedical Engineering
North Carolina State University
Raleigh, North Carolina
and
University of North Carolina at Chapel Hill
Chapel Hill, North Carolina

Constantine M. Megaridis
Department of Mechanical and Industrial Engineering
University of Illinois
Chicago, Illinois

Jonathan G. Merrell
Department of Biomedical Engineering
University of Virginia
Charlottesville, Virginia

Benjamin Moody
Department of Biomedical Engineering
North Carolina State University
Raleigh, North Carolina
and
University of North Carolina at Chapel Hill
Chapel Hill, North Carolina

Michael J. Moore
Department of Biomedical Engineering
Tulane University
New Orleans, Louisiana

Vivek Mukhatyar
WHC Department of Biomedical Engineering
Georgia Institute of Technology
Atlanta, Georgia
and
Emory University
Atlanta, Georgia

William L. Neeley
Department of Chemical Engineering
Massachusetts Institute of Technology
Cambridge, Massachusetts

Syam Prasad Nukavarapu
Department of Orthopaedic Surgery
University of Virginia
Charlottesville, Virginia

Seeram Ramakrishna
Division of Bioengineering
Faculty of Engineering
National University of Singapore
Singapore

and

Department of Mechanical Engineering
Faculty of Engineering
National University of Singapore
Singapore

Anita Shukla
Department of Chemical Engineering
Massachusetts Institute of Technology
Cambridge, Massachusetts

Eunice Phay Shing Tan
Division of Bioengineering
Faculty of Engineering
National University of Singapore
Singapore

Eva A. Turley
Departments of Oncology and Biochemistry
University of Western Ontario
Ontario, Canada

and

London Regional Cancer Program
London Health Sciences Center
London, Ontario, Canada

Mandana Veiseh
Life Sciences Division
Lawrence Berkeley National Laboratory
Berkeley, California

Xiumei Wang
Center for Biomedical Engineering
Massachusetts Institute of Technology
Cambridge, Massachusetts

Thomas Webster
Division of Engineering and Department of Orthopaedics
Brown University
Providence, Rhode Island

Kristen A. Wieghaus
Department of Biomedical Engineering
University of Virginia
Charlottesville, Virginia

Joel K. Wise
Department of Bioengineering
University of Illinois
Chicago, Illinois

Fan Yang
Department of Chemical Engineering
Massachusetts Institute of Technology
Cambridge, Massachusetts

Alexander L. Yarin
Department of Mechanical and Industrial Engineering
University of Illinois
Chicago, Illinois

Julie Yeh
WHC Department of Biomedical Engineering
Georgia Institute of Technology
Atlanta, Georgia

and

Emory University
Atlanta, Georgia

Shuguang Zhang
Center for Biomedical Engineering
Massachusetts Institute of Technology
Cambridge, Massachusetts

Eyal Zussman
Faculty of Mechanical Engineering
Technion-Israel Institute of Technology
Haifa, Israel

Section I

Importance of Scaffolds in Tissue Engineering

1 Tissue Engineering: The Therapeutic Strategy of the Twenty-First Century

Fan Yang, William L. Neeley, Michael J. Moore, Jeffrey M. Karp, Anita Shukla, and Robert Langer

CONTENTS

1.1 INTRODUCTION

The loss or failure of an organ or tissue is one of the most frequent, devastating, and costly problems in health care. Tissue engineering was born largely of the need for investigators to turn to multidisciplinary approaches to solve this long-standing problem in medicine. Advances in medicine have been paralleled by increased interactions among multiple disciplines such as biology, material sciences, and engineering, which led to progress in diagnostics, monitoring, and emergence of implanted devices and tissue grafts. Moreover, as medicine continued to advance, and the survivability of major disorders and injuries increased, so did the number of patients receiving and awaiting these critical treatments, and the need for alternative therapies became clearly apparent.

Clinicians have been a powerful driving force for innovation in medicine. The origin of tissue engineering stems from the demands by surgeons in regenerating functionally active tissue to replace those lost due to trauma, congenital malformations, or various disease processes. Current methods for organ and tissue replacement mainly utilize autografts, allografts, or metallic devices. Effective as these approaches are, they are associated with clear limitations including donor site morbidity, shortages in supply, poor integration, and potential immunologic reactions. These limitations further emphasize the importance of a timely development and successful translation of therapies based on tissue engineering principles. Internists also have historically turned to more complex therapies, from pharmacological administration of small molecules, to use of proteins, DNA, and other macromolecules, to extracorporeal devices for the replacement of lost cell or tissue function. Cell therapies became attractive for their ability to carry out numerous complex biochemical functions. Thus, these difficult problems in clinical medicine have continually inspired scientists and physicians in their quest to uncover biological mechanisms for exploitation at the bedside.

Our research in collaboration with Vacanti's group started seeking an alternative for patients awaiting liver transplants. Together, we sought ways to expand the cell seeding concept to three dimensions as an effort toward whole organ replacement. Our collaboration led to a publication describing the use of synthetic, resorbable, polymeric meshes for cell transplantation (Vacanti et al. 1988). This approach was adopted by a number of chemical engineers and others working with synthetic polymers, influencing many to employ similar techniques with degradable polymers.

Many turned their skills in biology or engineering toward tissue engineering, and the excitement felt in the academic sector was closely paralleled by that in the private sector. Due in part to federal agencies' early predilection toward funding hypothesis-driven research and in part to a contemporaneous flurry of corporate investment in biotechnology, tissue engineering research enjoyed a large influx in private funding. From the mid-1980s through the end of the millennium, over $3.5 billion was invested worldwide in research and development, and over 90% of those funds were supplied by the private sector. At the end of 2000, over 70 companies were participating in tissue engineering research and development and/or manufacturing. They were spending an estimated $600 million annually and employed about 3300 full-time equivalent scientists and support staff, all while only two products had received FDA approval (Lysaght and Reyes 2001).

In the first decade of the twenty-first century, scientific advances continue at a steady pace. Federal agencies have been increasingly stalwart in their nurturing of the field, both in the United States and abroad, not only by increasing funding but also by sponsoring workshops and studies and helping to define its future (McIntire 2003; Viola et al. 2003). Most importantly, as the complexities and challenges of engineering living tissue have become more fully understood, research has

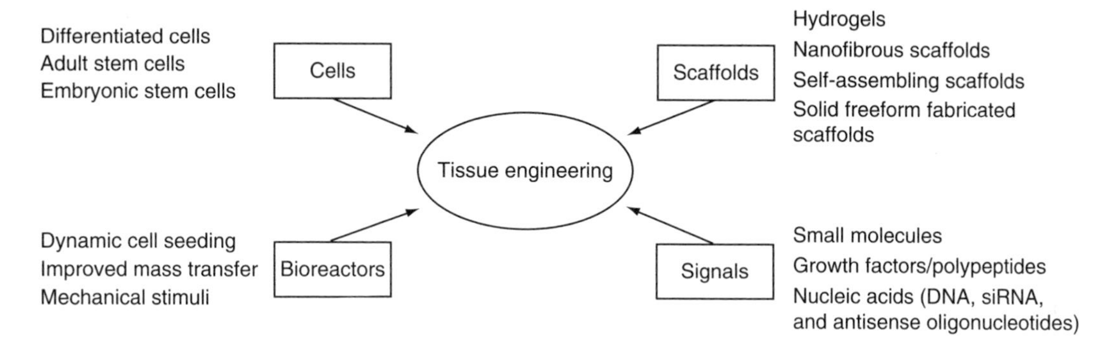

FIGURE 1.1 Key components of tissue engineering.

plumbed ever greater depths of innovation and technology. Tissue engineers have become increasingly drawn toward tangential fields, particularly stem cell and developmental biology and nanotechnology (Ingber et al. 2006; Vunjak-Novakovic and Kaplan 2006). So, even as first-generation products continue to come to market, the scientific foundation is being laid for engineering ever more complex and functional tissues. In this chapter, we briefly review a number of important advances in the field from its inception to the present day, including cells, scaffolds, tissue-inducing factors, and bioreactors (Figure 1.1).

1.2 CELLS

1.2.1 Cell Sources for Tissue Engineering

Cells are the building blocks of tissue, and cells present in the grafted tissue are believed to play a critical role in promoting tissue healing and regeneration. Therefore, most tissue engineering approaches involve isolating and expanding cells in vitro. Cell source is an important parameter to consider when applying tissue engineering strategies to restore lost tissue and functions. One of the major obstacles in engineering tissue constructs for clinical use is the shortage of available human cells. Conventional approaches usually utilize the fully differentiated adult cell types that make up the target organ or tissue. This often requires harvesting tissue such as autogenous or allogeneic tissue, enzymatically digesting the tissue to release cells, and culturing the dissociated cells in tissue culture flasks to initiate cell expansion. For example, autologous chondrocyte transplantation (ACT) is a cell-based procedure for cartilage repair that involves obtaining chondrocytes from the patient, expanding the cells in vitro, and transplanting the cells back into the same patient (Brittberg et al. 1994; Peterson et al. 2000). However, tissue engineering approaches need large numbers of cells, whereas the proliferation capability of fully differentiated cells is very limited. Furthermore, fully differentiated adult cells tend to lose their phenotype or dedifferentiate during in vitro expansion (Schnabel et al. 2002). Given the limitations of the fully differentiated cells, scientists and clinicians have collaborated to harness the potential of stem cells, which many believe hold the key to unlocking the secrets of tissue regeneration. In this chapter, we will mainly focus on advances in the stem cell field that have generated significant excitement in the past decade.

1.2.2 Potential of Stem Cells for Tissue Engineering Applications

Stem cells provide alternative cell sources for tissue engineering, such as craniofacial repair (Bruder et al. 1994; Aubin 1998; Shamblott et al. 1998; Thomson et al. 1998; Pittenger et al. 1999; Sottile et al. 2003; Cowan et al. 2004; Kim et al. 2005a). Unlike other types of cells in the body, stem cells are unspecialized cells that are capable of self-renewal for long periods yet maintain their capacity to differentiate into multiple specialized cell types upon exposure to specific induction cues. Development of techniques for culturing and regulating human stem cells could lead to unprecedented regenerative treatments and cures for diseases that cannot presently be treated via other means. It has been estimated that approximately 3000 people die every day in the United States from diseases that could have been treated with stem cell–derived tissues (Lanza et al. 2001). In addition to the generation of tissues and organs to treat cancer, trauma, inflammation, or age-related tissue deterioration, stem cells are also potentially useful for treatment of numerous diseases including Parkinson's disease, Alzheimer's disease, osteoporosis, and heart disease. Stem cells are currently being tested therapeutically for the treatment of liver diseases, coronary diseases, autoimmune and metabolic disorders, chronic inflammatory diseases, and other advanced cancers. Stem cells may be xenogenic, allogeneic, or autologous, where autologous cells are preferred as they will not evoke an immunologic response, and thus the harmful side effects of immunosuppressive agents can be avoided. Autologous stem and progenitor cells may be derived postnatally in adulthood or early in life from umbilical cord blood (Cetrulo 2006) or tissue (Baksh et al. 2007). Autologous-like cells may also be generated using therapeutic cloning or somatic cell nuclear transfer (SCNT), the process through which Dolly the sheep was cloned in 1997 (Hwang et al. 2004). Studies to date have demonstrated that cells derived from SCNT can be expanded in culture and can organize into tissue structures after transplantation in vivo in combination with biodegradable scaffolds (Lanza et al. 1999).

1.2.3 Stem Cell Source

Depending on the development stage of the tissues from which the stem cells are isolated, stem cells can be broadly divided into two categories: adult stem cells and embryonic stem cells (Shamblott et al. 1998; Thomson et al. 1998; Pittenger et al. 1999). Adult stem cells can be found in many adult tissue types including bone marrow, peripheral blood, adipose tissue, nervous tissue, muscle, dermis, etc. (Table 1.1). Adult stem cells are considered to be multipotent, which can give rise to several other cell types. Among the adults stem cells, bone marrow–derived stem cells (MSCs) have been shown to have the capability of differentiating into multiple tissue types, including bone, cartilage, muscle, tendon, etc., and hold great potential for autologous cell-based therapy (Pittenger et al. 1999). Another important characteristic of MSCs for regenerative medicine is their potential allogenic use without immunosuppressive therapy (Le Blanc et al. 2003; Maitra et al. 2004; Aggarwal and Pittenger 2005). In addition to the adult tissues mentioned above, stem cells have

TABLE 1.1
Types and Sources of Human Stem Cells

Origin	Types of Stem Cells	Sources of Isolation
Adult	Mesenchymal stem cells	Bone marrow
	Hemopoietic stem cells	Bone marrow and peripheral blood
	Neural stem cells	Neural tissue
	Adipose-derived stem cells	Adipose tissue
	Muscle-derived stem cells	Muscle
	Epidermal-derived stem cells	Skin, hair
	Umbilical cord blood stem cells	Umbilical cord blood
	Umbilical cord matrix stem cells	Wharton's jelly
Embryonic	Embryonic stem cells	Inner cell mass of 5–7 day blastocyst
	Embryonic germ cells	Gonadal ridge of 6–11 week fetus

also been identified in fetal tissues such as umbilical cord blood and Wharton's jelly. Although it was originally believed that stem cells derived from a particular tissue could only regenerate that specific tissue, numerous studies have disproved this idea (Macpherson et al. 2005). For example, both bone marrow and adipose tissue–derived mesenchymal stem cells may differentiate into cells and tissues of mesodermal origin including adipocytes, chondrocytes, osteoblasts, and skeletal myocytes and can be used to generate respective tissues including fat, cartilage, bone, and muscle (Caplan and Bruder 2001; Zuk et al. 2001; Baksh et al. 2003; Izadpanah et al. 2006). Unlike isolates of bone marrow, which typically require multiple punctures with a large bore needle, subcutaneous adipose tissue can be obtained through surgical removal with scalpels or through liposuction, which some patients may view as advantageous. However, despite their ability to differentiate into multiple cell types, adult stem cells are generally considered to give rise to only a limited range of differentiated cell types in comparison to embryonic stem cells.

Compared with MSCs, which can only be expanded in an undifferentiated state for limited passages, embryonic stem (ES) cells or embryonic germ (EG) cells can self-renew without differentiation for much longer. This property makes them attractive candidates as cell sources for tissue engineering, where large cell numbers are often needed. ES cells are derived from the inner cell mass of blastocysts, and EG cells are isolated from developing gonadal ridge. Since these cells are isolated from embryonic stage, they are considered to be pluripotent and can develop into any of the three germ layers: endoderm (interior stomach lining, gastrointestinal tract, the lungs), mesoderm (muscle, bone, blood, urogenital), or ectoderm (epidermal tissues and nervous system). It was only in 1998 that the political and ethical controversy surrounding stem cells erupted with the creation of human ES cells derived from discarded human embryos (Thomson et al. 1998). In addition to direct therapeutic use, ES cells represent an attractive cell source for the study of developmental biology, for drug/toxin screening studies, and for the development of therapeutic agents to aid in tissue or organ replacement therapies. Although ES cells may hold the secret to multiple cures and groundbreaking advancements in the field of regenerative medicine, they raise significant ethical concerns because they are harvested from embryos.

1.2.4 Pure Stem Cell-Based Therapies

Although stem cell research is still in its infancy, there are some remarkable success stories including blood transfusions and bone marrow transplantation that have been used in thousands of patients to successfully treat low blood volume and diseases of the blood and bone marrow such as lymphoma. Bone marrow transplantation represents the most common clinically approved method of stem cell-based therapy. Here growth factors such as granulocyte colony-stimulating factor (G-CSF) are initially administered to amplify and mobilize hematopoietic stem cells into the peripheral circulation where they can easily be collected using leukapheresis techniques. The transplantation of bone marrow which has been used since the 1960s involves infusion of the stem cells into the recipients' peripheral circulation through an intravenous catheter. The stem cells home to the bone marrow where they proliferate and start to produce blood cells. Remarkably, even a single hematopoietic stem cell can be used to fully reconstitute the lymphohematopoietic system (Osawa et al. 1996). Numerous pure stem cell-based therapies are currently in clinical trials. For example, Osiris Therapeutics Inc. is working on a product called Prochymal, which is a treatment for a life-threatening disease called acute graft versus host disease (AGHD) which attacks the gastrointestinal tract, skin, and liver. AGHD affects half of all patients who receive a bone marrow transplant for anemia and other diseases. A trial is also currently under way using Prochymal to assess its ability to reduce the symptoms of moderate-to-severe Crohn's disease. Australia's adult stem cell company, Mesoblast Ltd., recently commenced a phase 2 clinical trial of its allogeneic adult stem cell therapy for patients with heart attacks. This therapy involves injection of stem cells via catheter into damaged heart muscle and aims to improve heart function and reduce congestive heart failure.

1.2.5 Scaffold-Based Stem Cell Therapies

Unlike pure cell-based therapies where stem cells are injected directly into the peripheral circulation or a specific tissue, many applications require a cell carrier to transport and/or arrange the stem cells within a 3D configuration, or to isolate them within a particular location in the body. Moreover, certain applications require differentiation of the cells down particular lineages prior to transplantation. These approaches are common in the field of tissue engineering where stem cells are combined with engineered matrices either to build transplantable tissues ex vivo or to inject or implant viable constructs that are programmed to promote or initiate regeneration. Whereas fundamental research centers on developing an understanding of the mechanisms that regulate stem cells, applied fields such as tissue engineering aim to harness this knowledge to initiate tissue-specific regeneration. The field of tissue engineering has promoted the transition of standard two-dimensional (2D) stem cell culture systems to 3D platforms in an attempt to mimic the in vivo 3D culture environment which may be more conducive for regulating stem cell function.

Previous research on ES cells mainly focused on the understanding of stem cell biology, and tissue engineering applications using ES cells or EG cells in combination with scaffolds is still at its infancy (Elisseeff et al. 2006; Hwang et al. 2006). Scaffolds can promote cell and tissue development by providing a 3D environment in which cells can proliferate, attach, and deposit extracellular matrix (ECM) (Peppas and Langer 1994; Hubbell 1995; Langer and Tirrell 2004; Lutolf and Hubbell 2005). Various biological signals such as growth factors or peptides can also be incorporated into scaffolds to promote the desired differentiation (Hubbell 1999; Healy et al. 1999). For example, it has been shown that bone MSCs can undergo osteogenesis in a 3D hydrogel scaffold, and incorporation of cell adhesion peptide YRGDS into the scaffold promotes the osteogenesis of MSCs in a dosage-dependent manner, with 2.5 mM being the optimal concentration (Yang et al. 2005b).

One of the major challenges with scaffold-based transplantation of cells is lack of engraftment that typically results within damaged avascular tissue due to a deficiency in mass transport of oxygen and nutrients, a requirement for cell survival and for proper cell function. Maximal reported rates of angiogenesis are $\sim$1 mm/day (Folkman 1971; Li et al. 2000) and cells need to be within $\sim$100–200 μm of the nearest blood vessel (Muschler et al. 2004). Thus, transplanted cells within the core of large defects (>1–2 cm) do not survive long enough to contribute to the healing process. Specifically, it may take many weeks or months for complete vascularization of the defect (Mooney et al. 1994; Sanders et al. 2002) leading to tissue ischemia and necrosis (Helmlinger et al. 1997) (cell and tissue death) in graft sites as small as 1–2 mm (Muschler et al. 2004). This significantly reduces the capacity for an exogenous cell source to contribute to the regenerative process. Furthermore, most scaffold-based tissue engineering strategies passively permit filling of the scaffold pores with blood clot (Karp et al. 2004), which represents a static and potentially harsh environment for the transplanted cells. Although hematomas contain factors such as vascular endothelial growth factor (VEGF) that induce neovascularization (Street et al. 2000), hematomas are acidic and hypoxic and exhibit elevated levels of phosphorous, potassium, and lactic acid, which are cytotoxic to multiple cell types (Wray 1970). Therefore, transplanted cells are susceptible to death given their distance from host vasculature, and their position within the static and relatively harsh environment of the blood clot. This has significantly limited advancement in the field of tissue and organ replacement. After three decades of substantial research in this area, the potential to provide tissues and organs to millions of patients suffering from trauma, congenital defects, and chronic diseases has yet to be fully realized (Mikos et al. 2006). Although this is partly due to uncertainty and difficulties with clinical markets, typical results in preclinical animal models remain highly variable with poor rates of success in larger defects and in higher animal species likely due to poor survival of the transplanted cells (Petite et al. 2000; Muschler et al. 2004). Although it is not surprising that the effectiveness of cell-based therapies relies on the retention of cell viability after implantation (Wilson et al. 2002; Kruyt et al. 2003), little attention has been focused on this issue.

Recently, an advanced cell-instructive tissue engineering approach was successfully employed that utilized (1) high density arginine, glycine, aspartic acid (RGD)-containing cell adhesion ligands, (2) an exogenous differentiated myoblast cell source, and (3) growth factors to enhance the regenerative capacity of the transplanted cells through promoting their survival, preventing their terminal differentiation, and promoting their outward migration (Hill et al. 2006). Specifically, cells were delivered on porous alginate/calcium sulfate scaffolds that contained both hepatocyte growth factor (HGF) and fibroblast growth factor-2 (FGF-2), which were employed to maintain the cells in an activated, proliferating, but nondifferentiated state. Whereas control groups had only a modest effect on muscle regeneration, a combination strategy employing controlled release of HGF and FGF-2 in combination with scaffolds and cells dramatically enhanced the participation of transplanted cells leading to significant tissue regeneration. Despite the relatively small size of the scaffolds employed here (50 mm^3) and the uncertainties in translating this strategy into larger clinically relevant defects, the work demonstrates a proof of concept for cell-based therapies that can be designed to direct tissue regeneration.

1.3 SCAFFOLDS AND FABRICATION

1.3.1 Importance of Scaffolds to Promote Tissue Formation

Three common strategies employed in tissue regeneration are infusion of isolated cells, treatment with tissue-inducing substances, and implantation of a cell–scaffold composite (Langer and Vacanti 1993). Of the three strategies, the use of cell–scaffold composites generally leads to a more successful outcome. These scaffolds are often critical, both in vitro as well as in vivo, to recapitulating the normal tissue development process and allowing cells to formulate their own microenvironments. In contrast to using cells alone, a scaffold provides a 3D matrix on which the cells can proliferate and migrate, produce matrix, and form a functional tissue with a desired shape. The scaffold also provides structural stability for developing tissue and allows incorporation of biological or mechanical signals to enhance tissue formation. The biological and mechanical properties of scaffolds may vary depending on the application, and can be designed to provide an environment with appropriate signals that stimulate cells to proliferate and/or differentiate.

The importance of the ECM scaffold in cell development should not be underestimated. Nearly 30 years ago, Bissell proposed dynamic reciprocity, which states that a tissue achieves a specific function in part through interactions of the cells with the ECM (Bissell et al. 1982). Subsequent work demonstrated that gene expression can be mediated by the ECM binding to ECM receptors on the cell surface, which provide a link to the cytoskeleton and eventually the nuclear matrix (Nickerson 2001). The inclusion of neighboring cell interactions and soluble signals originating systemically or from cells in the immediate or distant vicinity provides a more complete model of tissue environment (Nelson and Bissell 2006).

Much attention has been given to the simulation of the extracellular environment. Of particular interest is the creation of scaffolds as substitutes for the ECM. Scaffolds for tissue regeneration occupy a fundamental role in tissue development since they must support the proliferation and differentiation of cells as they mature into a functional tissue. Regeneration to the native state necessitates removal of the artificial scaffold, most commonly by bioabsorption. To this end, numerous natural and synthetic materials have been proposed for use in tissue scaffolds (Nair and Laurencin 2006; Velema and Kaplan 2006). Drawbacks do exist for the use of existing materials in particular tissue engineering applications; however, 3D scaffold fabrication and incorporation of biofactors into the scaffold comprise the central challenges in the field today. Furthermore, although biomaterials received much attention during the 1990s, the current emphasis on hybrid living–artificial systems requires continued development of fabrication methods.

As a substitute for the ECM, a scaffold in general should impart a 3D geometry, have appropriate mechanical properties, enable cell attachment, and facilitate the development of a functional tissue. At the microscopic level, a highly porous structure is needed for diffusion of nutrients and waste products through the scaffold. The optimal pore size should be tailored to the specific cell type and be large enough to allow for cell migration and ECM formation yet not be so small that pore occlusion occurs. The scaffold surface architecture and chemistry should facilitate cell migration through the scaffold, provide developmental signals to the cells, and promote cell recruitment from the surrounding tissue. Additionally, in most cases the scaffold should be constructed from a degradable nontoxic material (Leong et al. 2003). Recent advancements in scaffold fabrication technologies are discussed below.

1.3.2 Scaffold Fabrication

1.3.2.1 Conventional Methods and Limitations

The formation of a porous structure constitutes a central goal of scaffold fabrication and a number of techniques were developed to achieve this aim including phase separation (Lo et al. 1995) (nonsolvent-induced phase separation and thermally induced phase separation), gas foaming (Mooney et al. 1996), solvent casting/particulate leaching (Wald et al. 1993), and freeze drying (Dagalakis et al. 1980). Because of the relative ease in using these techniques to fabricate scaffolds, they are still commonly used. A core limitation of these technologies is the lack of precise control over scaffold specifications such as pore size, shape, distribution, and interconnectivity as well as the overall scaffold shape. Numerous studies note the importance of pore size in the ability of cells to adhere and proliferate on a scaffold (Hulbert et al. 1970), but recent work with scaffolds produced using solid freeform fabrication (SFF) techniques where the pore size is precisely controlled suggests that eliminating the variability in the pore size and structure decouples the dependence of cell adhesion and proliferation on pore characteristics (Itala et al. 2001; Hollister 2005) (Figure 1.2). However, the porosity of the material, which is defined as the proportion of void space in a solid, is still a critical factor (Karageorgiou and Kaplan 2005). The fabrication of hierarchical porous structures, which consist of both a nano- or microscopic pore structure and a macroscopic pore structure, is more readily realized using SFF methods. These techniques allow the reproducible fabrication of scaffolds directly from a computer-aided design (CAD) file. The ability to translate an electronic data set into a scaffold opens up the possibility for patient-specific scaffolds based on computed tomography (CT) or MRI data (Mankovich et al. 1990; Hollister et al. 2000; Wettergreen et al. 2005).

1.3.2.2 Solid Freeform Fabrication Methods

1.3.2.2.1 Fused Deposition Modeling

Fused deposition modeling (FDM) (Crump 1992) is a process whereby a molten material is extruded through a nozzle and deposited as a layer on a surface. At the completion of the layer deposition, the sample stage is lowered and a new layer is deposited. In this fashion, the technique fabricates a 3D structure. A benefit of this method is the absence of organic solvents in the fabrication process. The process is computer controlled, which allows the use of CAD data in the design of the scaffold. The technique has been used to prepare porous scaffolds from polymers such as PCL (Hutmacher et al. 2001), PEG–PCL–PLA (Hoque et al. 2005), and HA/PCL composite (Sun et al. 2007). The requirement of a melt feed limits the range of materials that can be used and excludes sensitive molecules such as proteins from being directly incorporated into the scaffold.

1.3.2.2.2 3D Printing

The technique of 3D printing (Sachs et al. 1993) consists of applying a layer of powder onto a surface and using an ink-jet printer head to spray the surface precisely with a binder to join the powder particles. The process is repeated after spreading a new layer of powder on top of the previous layer, which results in the creation of a 3D structure. In the past, organic solvents

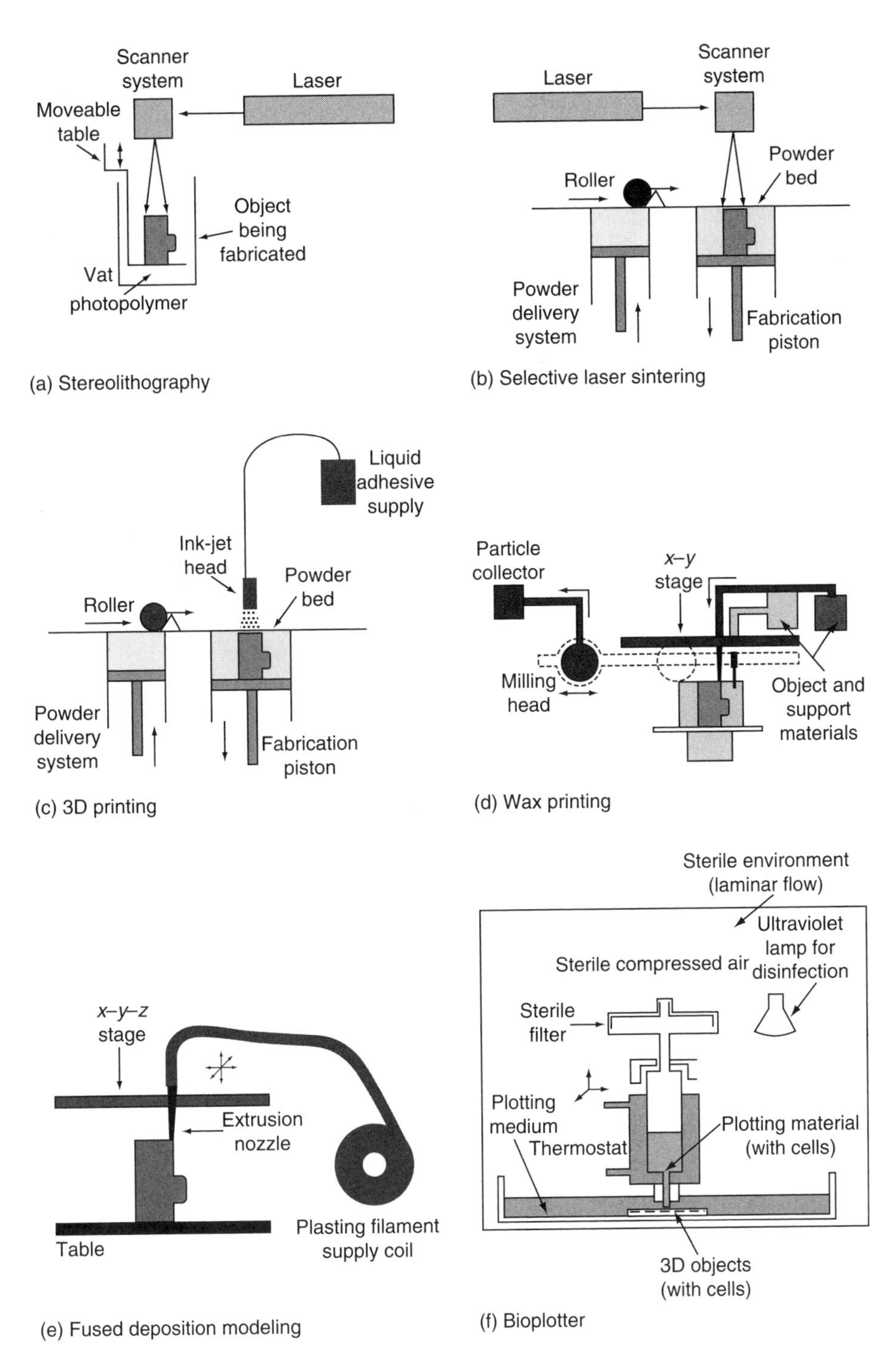

FIGURE 1.2 SFF systems categorized by the processing technique. (a,b) Laser-based processing systems include (a) the stereolithography system, which photopolymerizes a liquid and (b) the selective laser sintering (SLS) systems, which sinter powdered material. In each system, material is swept over a built platform that is lowered for each layer. (c,d) Printing-based systems, including (c) 3D printing and (d) a wax printing machine. 3D printing of a chemical binder onto a powder bed. The wax-based system prints two types of wax material in sequence. (e,f) Nozzle-based systems. (e) The fused deposition modeler prints a thin filament of material that is heated through a nozzle. (f) The Bioplotter prints material that is processed either thermally or chemically. (From Hollister, S.J., *Nat. Mater.*, 4, 7, 518, 2005. With permission.)

have been used as binders (Giordano et al. 1996); however, recent examples stress the use of biocompatible materials. In one example, hydroxyapatite powder was used to prepare bone repair scaffolds using a binder composed of 25% v/v polyacrylic acid in a water–glycerol mixture (Dutta Roy et al. 2003). Aqueous citric acid solution has also been used as a binder in the preparation of calcium phosphate–based ceramics (Khalyfa et al. 2007).

1.3.2.2.3 Selective Laser Sintering

Resembling 3D printing, the SLS process also begins by applying a thin layer of powder to a surface. A laser beam sinters the powder particles together in the desired pattern. Upon completion of the layer patterning, a new layer of powder is deposited and the process repeated. This technique has been used to prepare scaffolds from the biodegradable polymers polyetheretherketone, poly (vinyl alcohol), polycaprolactone (Williams et al. 2005), and poly(L-lactic acid) (Tan et al. 2005). Composites of some of these polymers and hydroxyapatite have also been prepared using SLS (Chua et al. 2004; Tan et al. 2005; Wiria et al. 2007).

1.3.2.2.4 Wax Printing

In the fabrication of 3D scaffolds using a wax printer, a negative mold is created by printing droplets of a build wax and a support wax on a surface, which harden after cooling. Once a layer is printed, the surface is milled flat and another layer is printed. This process continues until the structure is complete at which point the support wax is dissolved to yield a porous negative mold. The desired scaffold material is added to the mold as a casting solution and allowed to solidify, and the negative mold is dissolved or melted to release the scaffold. This technique has been used to prepare, for example, scaffolds for bone and cartilage replacement (Manjubala et al. 2005). Like most SFF processes, this technique was not originally designed for use in biological systems. The waxes and solvents used are often proprietary formulations and contain dyes, both of which can contaminate the scaffold with nonbiocompatible agents (Sachlos et al. 2003). Recent reports have made use of apparently biocompatible proprietary waxes (BioBuild and BioSupport) that can be orthogonally dissolved using ethanol and water (Sachlos et al. 2006); however, the identities of these materials have not been disclosed in the literature.

1.3.2.2.5 Stereolithography

Stereolithography relies on light-mediated chemical reactions to create a 3D object from a liquid polymer. In this process, a surface is lowered into a vat of photocurable polymer and the resultant layer of liquid polymer on the top of the surface is exposed to a laser to harden the polymer. The surface is then submerged slightly, which covers it with a new layer of liquid polymer that can be exposed to the laser. The surface can be raised or lowered as needed to create the 3D object. Biomaterials that have been in this application include poly(propylene fumarate) (Cooke et al. 2003; Lee et al. 2007), which contains photocrosslinkable double bonds, and acrylated poly(ethylene glycol) (Dhariwala et al. 2004; Arcaute et al. 2006).

1.3.2.3 Nanofibrous Scaffolds

1.3.2.3.1 Electrospinning

This technique produces nanofibers in a continuous fashion that are interconnected. The fiber diameter can range from 5 nm to more than 1 μm (Murugan and Ramakrishna 2006). Electrospinning differs from the current SFF technologies in that it produces a nanofibrous scaffold. Such a construct mimics the ECM by possessing high surface area, high aspect ratio, high porosity, small pore size, and low density (Murugan and Ramakrishna 2007). Due to the nature of the electrospinning process, randomly oriented fibers are produced (Matthews et al. 2002). Recent efforts have focused on electrospinning aligned fibers (Yang et al. 2005a). Both natural and synthetic materials have been electrospun into random and aligned meshes including collagen, gelatin, and chitosan (Murugan and Ramakrishna 2006).

1.3.2.3.2 Self-Assembling Scaffolds

Self-assembly relies on noncovalent interactions to achieve the goal of a spontaneously assembling 3D structure. Possessing this property, biopolymers such as peptides and nucleic acids are ideally suited for this role. Rationally designed peptides that spontaneously form 3D scaffolds in response to specific environmental triggers may have great potential in tissue engineering. Several elegant methods have been reported making use of peptides (Hartgerink et al. 2001, 2002; Beniash et al. 2005). Furthermore, these designer self-assembling peptide scaffolds have recently been demonstrated to repair nervous tissue, to stop bleeding in seconds, to repair infarctuated myocardia, as well as being useful medical devices for slow drug release (Zhang et al. 2005; Gelain et al. 2007). This concept has also been applied to DNA, where branched molecules were designed so that the arms of the DNA can hybridize with each other. In the presence of a DNA ligase, which serves to crosslink the DNA, the DNA molecules self-assembled into a hydrogel (Um et al. 2006).

1.3.2.4 Hybrid (Cell/Scaffold) Constructs

1.3.2.4.1 Conventional Cell-Laden Hydrogels

Hydrogels are swollen, typically crosslinked networks that are particularly useful for suspending cells in 3D. A variety of synthetic and natural polymers have been utilized for this application including poly(ethylene glycol) (PEG) and copolymers containing PEG (Tessmar and Gopferich 2007a), hyaluronic acid (Baier Leach et al. 2003), chitosan (Leach et al. 2004), and alginate (Mosahebi et al. 2001). Photocrosslinkable systems have been used extensively to form the gels, and other methods have been developed including enzymatic (Um et al. 2006) and thermosensitive (Park et al. 2007) systems to avoid the use of potentially cytotoxic UV light and radicals. Hydrogels have been used extensively to prevent adhesions due to their relative lack of cell adhesiveness (Sawada et al. 2001; Yeo et al. 2006). Consequently, cell adhesion proteins have been incorporated into hydrogels to promote cell adhesion (Hern and Hubbell 1998; Rowley et al. 1999; Shu et al. 2004). Degradation of hydrogels generally occurs by hydrolysis; however, enzymatically degrading hydrogels have also been reported (He and Jabbari 2007). The mechanical properties of hydrogels are generally weak so there have been efforts to create strong hydrogels (Kaneko et al. 2005). In order to form a 3D structure, conventional cell-laden hydrogels utilize a mold into which the cell-laden hydrogel is cast.

1.3.2.4.2 3D Patterning of Cell-Laden Hydrogels

In order to achieve more control over the 3D placement of cells within hydrogels and realize patient-specific geometries, a number of SFF technologies have been adapted for use with cell-laden hydrogels. Laser-guided direct writing uses a weakly focused laser beam to trap cells and then deposit them on a surface (Odde and Renn 2000; Nahmias et al. 2005). This technique allows single-cell resolution patterning and has been used to directly write endothelial cells that self-assemble into vascular structures (Nahmias et al. 2005). Using a modified ink-jet printer and thermosensitive gels, multiple layers of different cell types can be printed to create a 3D organ (organ printing) (Boland et al. 2003). This technology attempts to mimic the architecture of organs, which consist of complex structures containing many cell types positioned in precise locations. Bioplotter, a commercially available instrument, utilizes a needle to dispense a material in a layer-by-layer fashion into a plotting medium, which causes the material to solidify. A polymer–cell mixture can be dispensed using this technique leading to the formation of a cell-laden hydrogel (Landers and Mülhaupt 2000). Microfluidics have also been used to create 3D structures using a layer-by-layer approach (Tan and Desai 2004). In this method, pressure-driven microfluidics are used to transport cell–polymer solutions, which are deposited as layers within the microchannels. Photolithography has been utilized to create patterned hydrogel structures that encapsulate cells (Liu and Bhatia 2002). Cell-laden hydrogels have also been created from photopolymerizable polymer solutions where the cells are first localized using dielectrophoretic forces and then locked into place by light-mediated hydrogel formation (Albrecht et al. 2006). Recent efforts have also focused on adapting traditional SFF technologies for use with hydrogels, for example by using stereolithography to create complex PEG hydrogels (Arcaute et al. 2006).

1.4 DELIVERY OF TISSUE-INDUCING FACTORS

1.4.1 POTENTIAL OF CONTROLLED RELEASE SYSTEM TO ENHANCE TISSUE FORMATION

Since the inception of molecular biology, biologists have steadily worked on identifying and isolating molecular agents responsible for tissue formation and repair. Mechanisms of development and wound healing are continually being elucidated and their molecular bases constantly being explored for therapeutic exploitation. Cellular therapies important for tissue engineering are also highly dependent on cell-signaling factors, as the culture of cells appropriately differentiated or undifferentiated often requires the addition of isolated molecular agents that promote maintenance of specific cell phenotypes. For example, progenitor cells isolated from the retina can be maintained in an undifferentiated state for long periods of time in the presence of recombinant epidermal growth factor (EGF). These cells can then be cultured with the EGF replaced by nerve growth factor (NGF), brain-derived neurotrophic factor (BDNF), and basic fibroblast growth factor (bFGF) to induce differentiation into cells that express neuronal and glial markers (Tomita et al. 2006). In tissue engineering, the goal is to supply cells with the factors necessary to induce proliferation and/or differentiation so the cells then may secrete the appropriate extracellular components for tissue formation. The supplication of tissue-inducing factors in growth medium may be sufficient for growth of tissues in vitro, but administration of such biomolecules to induce tissue formation in vivo is generally not sufficient, as the molecules diffuse rapidly away from the desired location and are quickly degraded. For this reason, controlled drug-delivery systems are usually necessary.

As tissue engineering approaches were taking shape in the mid-1970s, advancements were being made in a seemingly disparate field that would eventually be considered one of the most important enabling technologies in tissue engineering. In 1976, Langer and Folkman, seeking experimental methods to assay angiogenesis factors in vivo, demonstrated sustained release of an enzyme from synthetic polymers (Langer and Folkman 1976). This advance helped pave the way for the development of versatile delivery systems capable of delivering biomolecules in a controlled fashion. Polymeric delivery systems have been reported in the form of tablets, wafers, fibers, extruded implants, films, microparticles, and many others (Wise 2000). Indeed, polymeric controlled release technology has progressed such that for nearly every macromolecule undergoing therapeutic development, polymeric release has probably at least been considered (Schwendeman 2002). Researchers have frequently experimented with drug delivery systems fabricated from biodegradable polymers which have been approved by the FDA for other uses. Coincidentally, early tissue engineering research groups began to use the same materials for scaffold fabrication, and it was not long before the scaffold began to be utilized as a delivery device for both cells and tissue-inducing biomolecules (Sokolsky-Papkov et al. 2007).

1.4.2 TYPES OF TISSUE-INDUCING FACTORS

A large range of biomolecules have been investigated for the controlled induction of tissue formation, but the majority of important factors can be divided into three classes: small molecules, proteins and polypeptides, and oligonucleotides.

1. Small molecules (arbitrarily lumped into one category here for convenience) are important components in numerous cell signaling cascades, both in intercellular communication, as in the case of corticosteroids and other hormones, and in intracellular signaling. They typically trigger intracellular signaling cascades by binding to specific protein receptors, which leads to gene transcription.

2. Polypeptides, most often as whole proteins, may act on cells as mitogens, morphogens, growth factors, survival factors, and cytokines (these terms are technically neither synonymous nor mutually exclusive; however, for the purposes of this chapter they will not distinguished). These tissue-inducing proteins may be soluble or bound to the ECM, and typically act upon cells through receptor–ligand binding.
3. Oligonucleotides, either as DNA or RNA, can either bind to DNA to affect gene transcription, RNA to affect gene translation, or when delivered as whole genes, become incorporated directly into the cell's genome.

The challenges and strategies to deliver these classes of compounds vary due to the differences in their physical and chemical properties. Here, we briefly survey the important considerations of delivering each and provide some prominent examples. However, because the vast majority of tissue-inducing factors investigated for tissue engineering have been macromolecules, the greater part of this section will be devoted to discussing proteins and gene delivery.

1.4.3 Small Molecule Delivery for Tissue Engineering

Due to their size, small molecules tend to diffuse rapidly, so controlled delivery depends largely on strategies to slow or prevent diffusion. These strategies vary widely depending on the structure of the molecule to be delivered and the target environment. Some methods to retard diffusion include making use of ionic interactions to form insoluble complexes or matching hydrophilic and lipophilic properties of the drug and delivery device to retard release. More advanced techniques involve chemically modifying the molecule or attaching it to the delivery device. For example, Nuttelman and colleagues recently synthesized PEG hydrogels containing dexamethasone covalently linked to the hydrogel backbone with degradable lactide units. Dexamethasone is a corticosteroid which reliably promotes osteogenic differentiation of human mesenchymal stem cells (hMSCs). The authors showed that dexamethasone was released slowly from the hydrogel and induced hMSCs to express osteocytic phenotypes (Nuttelman et al. 2006).

Small molecules serve as important components of intercellular communications, as in the case of hormones, and intracellular signaling, particularly as second messengers. It is often in this second messenger capacity that small molecules have been utilized for tissue induction, as evidenced by their codelivery with other factors. For example, cyclic adenosine monophosphate (cAMP) has been found to act synergistically when coadministered with Schwann cell implants in the injured spinal cord (Pearse et al. 2004). Another research group delivered cAMP along with a neuronal growth factor via microparticles injected into the eye and found the combination to be effective for promoting optic nerve regeneration (Yin et al. 2006). As delivery of multiple factors with distinct release profiles becomes more common, it is likely that the role of small molecules in tissue engineering will be augmented.

1.4.4 Protein Delivery for Tissue Engineering

1.4.4.1 Challenges for Controlled Protein Delivery

The controlled delivery of proteins and polypeptides has been investigated extensively for a wide range of applications and has been met with some success; however, significant challenges remain. Because proteins are highly complex, ordered molecules whose functions depend on chemical and structural integrity, the greatest difficulties in devising controlled protein release systems arise from instability of the protein, formulation, storage, and release (Fu et al. 2000). Much work has been undertaken to elucidate mechanisms of protein degradation and inactivation and therefore to devise methods to mitigate these processes. The major mechanisms of protein inactivation in polymeric delivery systems include aggregation due to dehydration and rehydration, protein unfolding or aggregation along hydrophobic surfaces or at aqueous–organic interfaces, and acidification of the microclimate within the delivery system (Schwendeman 2002). Reported methods of overcoming these challenges include zinc complexation (Johnson et al. 1997) or addition of lyoprotectants (Prestrelski et al. 1993) to inhibit moisture-related aggregation, choosing protein-friendly processing techniques to prevent aggregation at polymer surfaces and interfaces (Herbert et al. 1998; Burke 2000), and addition of antacids (Zhu et al. 2000) or pore-forming excipients such as PEG (Jiang and Schwendeman 2001) to prevent microclimate acidification. While these advances have been instructive, different proteins are susceptible to different forms of instability; thus, it is necessary to optimize each delivery system for its specific application.

1.4.4.2 Strategies for Protein Delivery

Controlled protein delivery for tissue engineering can be achieved using multiple delivery vehicles, including transplanted cells that are genetically modified (Chang et al. 1999; Tresco et al. 2000), polymer microparticles (Edelman et al. 1991; Krewson et al. 1996; Oldham et al. 2000; Lu, Yaszemski, and Mikos 2001), and scaffolds. The main advantage of the scaffold-free approach is that requirements of the delivery system may be met independently from those of the scaffold. However, researchers have increasingly turned toward utilizing the scaffolds themselves as delivery vehicles. In order to accomplish this, proteins have been adsorbed to the surface of the scaffold, encapsulated in the bulk of the scaffold, or covalently attached to the scaffold (Tessmar and Gopferich 2007b). Maintaining stability of the protein in these cases is far from trivial and obviously essential. Protein release kinetics will naturally vary among these techniques, and utilizing combinations of release techniques has proved to be advantageous. For example, one research group demonstrated release of two growth factors with distinct kinetics by incorporating gelatin microparticles within a hydrogel scaffold (Holland et al. 2005). Another group demonstrated dual release kinetics by using a sequential emulsion technique to form protein-containing coatings on a preformed scaffold (Sohier et al. 2006).

1.4.4.3 Controlled Release of Growth Factors to Enhance Tissue Formation

Delivery of an enormous variety of proteins has been investigated for tissue engineering applications, due in part to the extreme importance of proteins in cellular signaling. The groups of tissue-inducing proteins often collectively called growth factors have been widely studied (Tessmar and Gopferich 2007b). Growth-factor strategies for tissue formation include promotion of cell proliferation, differentiation into the desired tissue-forming cell types, migration into the desired locations, cell growth along with secretion of matrix for tissue formation, and generation of a blood supply (Boontheekul and Mooney 2003; Tabata 2003; Tessmar and Gopferich 2007b). Growth factors often work in concert with one another, and each may act upon numerous tissue types and produce varying effects on different cell types. Moreover, they act in diverse manners, for example by binding cell surface receptors before internalization or binding the ECM before cell interaction, and cells often respond to soluble growth factors according to a concentration gradient (Boontheekul and Mooney 2003). For these reasons, the engineering of a delivery system used to present these biomolecules to the tissue-forming cells is of utmost importance.

Scaffolds and delivery systems have proven effective for generating relatively simple neotissue constructs in vitro and in vivo. However, developing and healing tissues often respond to transient or gradient concentrations of signaling molecules, so the formation of spatially complex tissues will likely require presentation of biomolecules in a spatiotemporally controlled manner (Saltzman and Olbricht 2002). While difficult, this challenge has spurred impressive innovations. New methods of scaffold seeding such as layer-by-layer film deposition can help lead to complex temporal files within tissues, by incorporating growth factors, proteins, and other important cellular components within specific layers (Wood et al. 2005). This can help control the speed and time spans of proliferation and differentiation during the culture process. Scaffolds endowed with a spatial gradient of NGF have shown promise in directing axonal outgrowth (Moore et al. 2006), and those with a gradient of bFGF promoted the directed migration of vascular smooth muscle cells (DeLong et al. 2005). One research group developed a method of sintering protein-containing microspheres that allowed multiple proteins to be released with distinct rates from distinct zones within a scaffold (Suciati et al. 2006). Spatiotemporal control has also shown effectiveness in vivo. For example, Chen and colleagues demonstrated release of two angiogenic growth factors with distinct release kinetics from within specific regions of a scaffold. When implanted in an ischemic hindlimb, the region of the scaffold that released the two factors sequentially promoted development of a mature vascular network that was superior to that found in the region delivering only one growth factor (Chen et al. 2007). Furthermore, for tissues that are subject to mechanical stimuli, such as bone, muscle, and blood vessels, Lee et al. demonstrated a controlled growth factor release system that can respond to repeated mechanical stimuli (Lee et al. 2000). VEGF encapsulated in alginate hydrogels was released in response to compression both in vitro and in vivo, and mechanical stimulation was shown to upregulate the blood vessel formation in vivo.

1.4.5 Nucleic Acid Delivery for Tissue Engineering

1.4.5.1 Techniques for Gene Delivery

Advances in gene delivery have provided an alternative venue to recapitulate the natural tissue development process for tissue engineering purposes. For various tissue engineering applications, a target gene can be transferred to specific cell types, such as stem cells, to promote the desired cell differentiation and tissue formation. Cells can either be transfected ex vivo and then seeded onto 3D scaffolds for in vivo implantation, or they can be transfected in vivo directly.

The techniques for delivering genetic materials into mammalian cells can be broadly divided into two categories: viral-based and synthetic nonviral methods. Both approaches have their own advantages and disadvantages, and no single vector is suitable for all gene delivery applications. The viral approach employs a key property of viruses, which deliver their genome into target cells. Many different types of viruses, such as retrovirus, adenovirus, and lentivirus, can be transformed into gene delivery vehicles by replacing part of the viral genome with a target gene (Vile et al. 1996; During 1997). As the viral approach essentially utilizes the naturally evolved mechanism for viral self-replication, it is typically very efficient and thus has been the major approach undertaken for most applications. In fact, 69% of the ongoing clinical trials employ a viral-based approach (Gene Therapy Clinical Trials online, http://www.wiley.co.uk/genetherapy/clinical/, Accessed 2005). Despite the high transfection efficiency, the viral-based approach is also associated with several major limitations. As viruses are inherently immunogenic and potentially pathogenic, safety concerns have always been a major issue for the clinical applications of viral-based gene delivery. Furthermore, viral vectors do not facilitate the design of target cell specificity and are associated with relatively high manufacturing costs.

Nonviral synthetic vectors, mostly cationic polymers and lipids, provide another attractive vehicle for gene delivery. In general, synthetic vectors are cationic materials that can electrostatically bind to DNA or RNA to form condensed nanoparticles (polyplexes or lipoplexes). The biomaterials that have been explored include cationic polymers, cationic lipids, liposomes, chitosans, dendrimers, and inorganic nanoparticles (Merdan et al. 2002; Partridge and Oreffo 2004; Wagner et al. 2004). These synthetic vectors overcome the problems associated with a viral-based approach and are nonimmunogenic. They also enable greater flexibility in structure design and integrating a targeting moiety, as well as relatively easy synthesis and lower manufacturing costs. However, nonviral-based vectors have suffered from low transfection efficiency and occasional toxicity, and most synthetic vectors are unstable in the presence of serum, thus severely hindering their applications in vivo.

1.4.5.2 Major Barriers in Gene Delivery and Conventional Solutions

To enhance the delivery of target genes into the cell nucleus, it is very important to understand the major barriers that gene vectors need to overcome. Before reaching the target cell nucleus, polyplexes must first attach to cell surface, be internalized through endocytosis, escape from the resulting endosome/lysosome, navigate through the cytoplasm toward cell nucleus, and finally cross the nuclear membrane (Pack et al. 2005). Furthermore, polyplexes must be unpackaged at a certain time point so that the DNA can be released.

Overcoming the extracellular barriers requires efficient condensing of plasmid DNA, stability of nanoparticles in the blood stream and surrounding tissue, and specific targeting to the cells of interest. Polyplexes form spontaneously upon mixing of cationic polymers with DNA and condense into nanoparticles with a size ranging from thirty to several hundred nanometers. Polyplexes protect the naked DNA from being degraded by DNase. The stability of polyplexes in serum depends on the polymer chemistry and the DNA/polymer charge ratio. In general, positively charged polyplexes show better stability under physiological salt conditions in comparison to neutral polyplexes. However, in the presence of serum, negatively charged proteins such as albumin can adsorb onto the nanoparticles and cause aggregation, which leads to clearance of nanoparticles by phagocytic cells (Dash et al. 1999).

Once attached to the cell surface, polyplexes are internalized, either by cell-surface receptor-mediated endocytosis or by adsorptive pinocytosis (Mislick and Baldeschwieler 1996). Polyplexes will then become localized in endosomes, which are vesicles that rapidly acidify to pH 5–6 due to the action of an ATPase proton-pump enzyme in the vesicle membrane. Polyplexes can subsequently be transported to lysosomes, which are organelles with an internal pH of $\sim$4.5 and an abundance of degradative enzymes. Significant amounts of DNA are believed to be degraded during the endosome/lysosome phase, and only those that escape into the cytoplasm can reach the cell nucleus. One way to overcome this barrier is by using "proton-sponge" polymers, such as polyethylenimine (PEI) and polyamidoamine (PAMAM) dendrimers (Haensler and Szoka 1993; Boussif et al. 1995). These polymers contain many secondary and tertiary amines, and undergo large changes in protonation during endocytic trafficking. This process is accompanied by an increased influx of counterions, increased osmotic pressure, and vesicle rupture, which releases the polyplexes into the cytoplasm (Behr 1997). In an effort to mimic the endosomal escape mechanisms utilized by viruses, membrane-active peptides, such as the HIV TAT sequence and influenza virus hemagglutinin subunit HA-2, have also been incorporated into polycationic polymers (Plank et al. 1998; Beerens et al. 2003). In addition, nuclear localization sequence can also be conjugated to the polymer vector to enhance DNA targeting to the cell nucleus (Cartier and Reszka 2002; Chan and Jans 2002).

1.4.5.3 High-Throughput Approach to Identify Novel Biodegradable Materials for Gene Delivery

To improve the biocompatibility and DNA release, recent research efforts in polymer-based gene delivery have incorporated biodegradable components, such as hydrolyzable ester bonds, into the structural design. Several biodegradable polymeric gene vectors have been synthesized including poly(amino-ester) (Lim et al. 2002), poly-amino acid (Guping et al. 2005), and poly(β-amino esters) (Lynn and Langer 2000; Lynn et al. 2001; Akinc et al. 2003a,b; Anderson et al. 2004; Kim et al. 2005b). Among these, poly(β-amino esters) are particularly attractive due to their facile synthesis, high transfection efficiency, and low toxicity. In contrast to the conventional approach for synthesizing polymers, which often involves multistep purifications and protection/deprotection steps, these polymers can easily be synthesized by the conjugate addition of primary amines or bis-secondary amines to diacrylate compounds (Lynn and Langer 2000). Furthermore, polymers used to be synthesized and screened on an individual basis, a process that is slow, labor intensive, and inefficient. High throughput synthesis and screening of a large polymer library using combinatorial methods enables faster discovery of potential polymer vectors, better understanding of the structure/property relationship, and rational design of novel polymers for gene delivery. Recently, Anderson et al. reported a semiautomated, solution-phase parallel synthesis and screening of a large library of 2350 structurally diverse, degradable poly(β-amino esters) using commercially available monomers (Anderson et al. 2003). High throughput screening discovered 47 polymers that demonstrate better transfection efficiency than PEI, the best commercially available polymer transfection reagent (Anderson et al. 2003). Further structure/property analyses demonstrated structural similarity in the top-performing polymers, which are all formed from amino alcohols. Such structural convergence offers valuable insight on rational design of polymer vectors for gene delivery (Anderson et al. 2005). The polymer with the highest transfection efficiency, C32, is an aminopentanol-terminated polymer with a molecular weight around 18 kDa relative to polystyrene standards. When injected intratumorally in vivo in mice, the C32 polymer demonstrates high biocompatibility and significantly reduces tumor size, a property that is attributable to cell apoptosis (Anderson et al. 2004; Peng et al. 2007). The top-performing poly(β-amino esters) also showed great efficacy and low cytotoxicity in transfecting primary human vascular endothelial cells (HUVEC) in the presence of serum, which have been a great challenge (Green et al. 2006). These results demonstrated great potential of using these polymers for vascular tissue engineering applications.

Amine-terminated poly(β-amino esters) have been shown to be generally more efficient in transfection efficiency. To examine further the effect of the type of amine group at the end chain on gene delivery, a generalized method has been presented to modify poly(β-amino esters) without the need for purification (Zugates et al. 2007). This system enables the rapid synthesis and screening of many structural variations at the polymer chain terminus. End modification of C32 significantly enhances its in vitro transfection efficiency. Most notably, the end-modification strategy has led to the discovery of many effective polymers that work very well in the presence of serum, which overcomes a great obstacle in using nonviral vectors for gene delivery. In vivo, intraperitoneal (IP) gene delivery using end-modified C32 polymers leads to expression levels over one order of magnitude higher than the levels attained by using unmodified C32.

1.4.5.4 Sustained DNA Release from Polymeric Scaffolds for Tissue Engineering

The delivery of DNA plasmids using polymeric scaffolds allows for localized delivery over an extended time period. For this approach, DNA plasmids, either naked or condensed, can be encapsulated into polymeric scaffolds and used with or without cells (Figure 1.3). Polymeric scaffolds have been extensively used for controlled drug release purposes and knowledge learnt from those applications can also be applied to DNA delivery. For example, local delivery of growth factors can promote desired cell differentiation and tissue formation, but is usually associated with problems such as burst release profile and loss of protein activity after encapsulation. In contrast, controlled release of DNA plasmids encoding those growth factors using polymeric scaffolds can overcome the above limitations. Both biocompatible synthetic and natural polymers have been employed for DNA delivery purposes such as poly(lactide-co-glycolide) (PLGA), collagen and hyaluronan (Cohen et al. 2000; Walter et al. 2001; Eliaz and Szoka 2002; Huang et al. 2003, 2005a; Segura et al. 2005). Localized gene delivery using scaffold has been shown to improve bone regeneration (Huang et al. 2005b), angiogenesis (Shea et al. 1999), as well as skin and nerve regeneration (Tyrone et al. 2000; Berry et al. 2001).

1.4.5.5 Targeted Gene Delivery for In Vivo Applications

For in vivo gene delivery, specificity is critical and one approach to achieve specificity is to attach targeting ligands to the surface of the nanoparticles so that only the targeted cell type will be transfected. One major challenge concerns the change in biophysical properties of nanoparticles after coating (Suh et al. 2002; Kursa et al. 2003). Recently, a general method of coating polymer/DNA nanoparticles was developed, and peptide coated nanoparticles were found to have

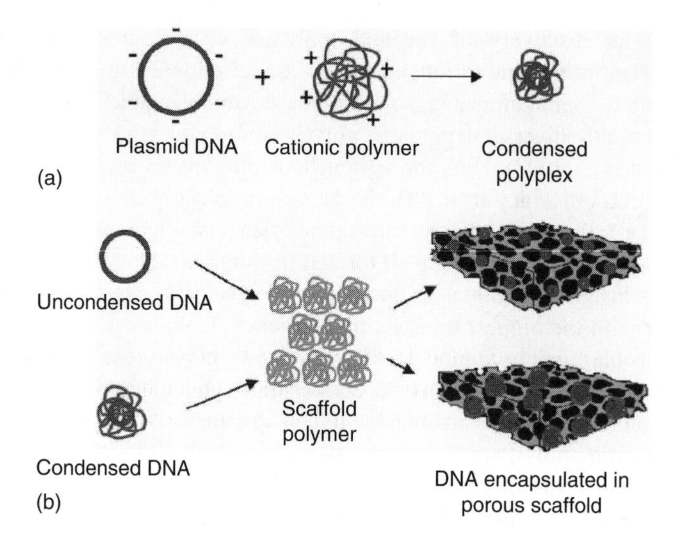

FIGURE 1.3 Schematic of DNA condensation and encapsulation into polymeric depot systems. (a) DNA complexation with cationic polymers leads to the formation of nanometer sized polyplexes. (b) Condensed or uncondensed DNA can be encapsulated into polymeric scaffolds for sustained delivery. (From Sorrie, H. et al., *Adv. Drug Deliv. Rev.*, 58, 500, 2006. With permission.)

favorable biophysical characteristics including small particle size, near-neutral zeta potential, and stability in serum (Green et al. 2007). At appropriate formulation conditions including near-neutral charge ratio, the coated nanoparticles enable effective ligand-specific gene delivery to human primary endothelial cells in serum-containing media. As this nanoparticulate drug delivery system has high efficacy, ligand-based specificity, biodegradability, and low cytotoxicity, it potentially may be useful in several clinical applications.

1.4.5.6 Antisense Oligonucleotides and siRNA Delivery

In contrast to DNA delivery, where certain genes are turned on, oligonucleotides containing a sequence complementary to certain gene or mRNA can also be delivered to initiate their degradation and knockdown. Antisense oligonucleotides have been shown to silence the expression of specific genes to achieve a desired cellular function (Tsuboi et al. 2007; Wilton et al. 2007). Recent discovery of the RNA interference (RNAi) pathway has broadened the area of gene delivery and opened up new venues to regulate cell phenotype. The RNAi pathway was first discovered and described in *Caenorhabditis elegans* in 1998 (Fire et al. 1998), and the phenomenon was then found to be present in mammalian cells as well (Elbashir et al. 2001). RNAi is the process of sequence-specific, posttranscriptional gene silencing mediated by small interfering RNA (siRNA), a class of double-stranded RNA molecules that are typically 20–25 nucleotides long. Several groups have demonstrated that RNAi can be elicited in mammalian cells using exogenously derived siRNA (Caplen et al. 2001; Elbashir et al. 2001). Due to its high efficiency in silencing gene expressions and ease of use, siRNA has rapidly drawn significant attention in functional genomics, pathway analysis, and drug target validation experiments.

The safe and efficient delivery of nucleic acids remains the major challenge for nucleic acid-based therapeutics. Several methods can be employed to deliver siRNA to mammalian cells including electroporation, reverse transfection, and chemical transfection. Nonviral delivery using lipids or cationic polymers are safe and promising, but the current collection of available delivery materials is still limited and the transfection efficiency is not yet ideal. In the future, high throughput approaches to the synthesis and screening of large libraries of potential nonviral delivery molecules hopefully will lead to identification of novel materials that can have broad applications for delivery of siRNA or antisense oligonucleotide therapeutics.

1.5 BIOREACTORS IN TISSUE ENGINEERING

1.5.1 Requirements for Bioreactors in Tissue Engineering

With the rapid advancement in the field of tissue engineering, bioreactors have gained increasing attention as a powerful tool to provide additional exogenous stimuli for the engineered tissue construct to achieve long-term success. As a vessel in which various parameters can be precisely controlled, bioreactors can be designed to provide the desired conditions for the cells to regenerate functional tissue. Examples of such variables include mechanical signal, temperature, media flow rate, oxygen and carbon dioxide concentrations, and other tissue-specific stimuli. To promote tissue regeneration in vitro, several requirements need to be satisfied when designing bioreactors. These are best described as (1) a need to simulate the in vivo environment required for cell proliferation and differentiation, (2) the importance of a uniform cellular distribution in 3D scaffolds, (3) necessary maintenance of adequate nutrient concentrations, (4) appropriate mass transfer of nutrients to developing tissues, and (5) exposure to physical stimuli simulating in vivo conditions in tissues (Freed and Vunjak-Novakovic 2000; Ellis, Jarman-Smith, and Chaudhuri 2005).

To evaluate the performance of a designed bioreactor, the engineered tissues cultivated within the bioreactor should be evaluated both structurally and functionally, and specific assessment depends on the target tissue type. In general, all tissues must have an adequate distribution of cells, ECM, and certain characteristic components of the tissue. For example, a specific structural requirement for engineered cartilage is the amount of glucosaminoglycan (GAG) present (Freed and Vunjak-Novakovic 2000). To determine the functional properties of specific tissues, such as cartilage and cardiac tissue, mechanical and electrophysiological properties, respectively, should also be examined (Bursac et al. 1999; Vunjak-Novakovic et al. 1999). Aside from the general requirements for all bioreactors, specific structural and functional requirements of target tissues must be considered when designing and deciding which bioreactor to use.

1.5.2 Bioreactors for Dynamic Cell Seeding

Bioreactors can be used for the engineered construct in various aspects including enhancing cell seeding, increasing construct size and cellularity, and promoting ECM deposition. Generally, the initial step in engineering a tissue includes seeding the appropriate cells onto scaffolds. These scaffolds are usually biodegradable 3D polymeric constructs and are typically comprised of the synthetic polymers: polyglycolide, polylactide, and polylactide coglycolide (Griffith and Naughton 2002). The conventional approach of seeding cells onto scaffolds is done in a static manner, often resulting in low seeding efficiency and a heterogeneous distribution of the seeded cells. However, uniform tissue growth requires a high yield process in cell seeding and a great degree of uniformity in cell attachment. It has been shown that uniform cell seeding at high densities leads to optimized 3D tissue formation upon cultivation (Martin et al. 2004). In contrast to static seeding, dynamic seeding would induce flow in the vessel, either by convective mixing using spinner flask bioreactors or convective flow using perfused bioreactors. This method causes mass transfer to occur through convection to the scaffold surface and then primarily by diffusion through the scaffold. Due to the incorporation of convective mass transfer, dynamic seeding has been shown to yield higher seeding efficiencies (Martin et al. 2004). Despite these higher efficiencies, convective mass transfer may not lead to highly uniform distributions of cells. To ensure uniformity, flow conditions or rotational speeds, amongst several other parameters, must be optimized based on the desired tissue characteristics and scaffolds type, etc. It has been shown, for example, that direct perfusion in seeding systems increases the level of uniformity of seeding and subsequent tissues, compared to static and stirred-flask vessels (Wendt et al. 2003). In this study, the direct perfusion took place within the bioreactor used for subsequent cultivation of the tissue as well. This approach eliminates all difficulties associated with transferring seeded scaffolds to the bioreactor for cultivation. Additionally, for cells that are shear sensitive, the time spent in suspension during seeding must be minimized (Freed and Vunjak-Novakovic 2000). The exact mode of cell seeding chosen is heavily dependent on the cell and tissue type. Studies have shown that mixed flasks work well for cartilage, where the kinetic rate and cell deformation rates are minimized (Vunjak-Novakovic et al. 1998). In the case of cardiac tissues, rotating bioreactor vessels have been accompanied with high metabolic activity and thus are the seeding and culture method of choice (Carrier et al. 1999).

1.5.3 Bioreactors to Improve Mass Transfer

One of the greatest challenges in engineering tissue constructs is enabling adequate mass transfer of nutrients to the seeded scaffold. It has been shown that nutrient supplying vasculature is usually within 100–200 μm of a living tissue in vivo (Yang et al. 2001) whereas engineered tissues must be at least on the scale of several millimeters in size to be useful (Martin et al. 2004). Ensuring oxygen and nutrient transport over this length scale is very important in creating healthy multilayer tissues. To address this problem, many different bioreactors have been developed including static/mixed spinner flasks, slow turning lateral vessels (STLV), high aspect ratio rotating vessels (HARV), rotating wall perfused vessels (RWPV), perfused columns, and perfused chambers (for a review, see Freed and Vunjak-Novakovic (2000)). In perfusion vessels, cells are retained within a chamber rather than continually removed and have a continuous supply of nutrients leading to high cell densities within the tissue. All stirred vessels also lead to high cell densities but care must be taken to ensure that mixing is not too vigorous in the case of shear-sensitive cells. Rotating vessels, such as the STLV and RWPV, were initially developed by the National Aeronautics and Space Administration (NASA) to use in microgravity experiments, yielding free-floating scaffolds and highly laminar flow conditions. With highly laminar flows, mass transfer limitations to the scaffold surface are minimized and efficient nutrient transfer can occur (Ellis et al. 2005). The effects of using STLVs and RWPVs have been examined in cartilage and skin cultures, both of which yielded tissues of better overall properties than when other vessels were used (Ellis et al. 2005).

When designing and deciding which bioreactors to implement, it is important to consider the necessary length scales and balances between convective and diffusive mass transfer (Peclet number, Pe), especially in the case of cell seeding. During the cultivation process, it is critical to consider the balance between reaction rates and diffusional mass transfer (Damköhler number, Da), the dominant mode of mass transfer within the scaffold. The relevant reaction rates are those of nutrient consumption. Considering the tissue requirements, experimenting with different Pe and Da can help determine the optimal values of bioreactor parameters, such as residence times and nutrient feed compositions.

1.5.4 Bioreactors to Provide Mechanical Stimuli for Enhanced Tissue Formation

Bioreactors also allow studies of mechanical stimuli on the cells and 3D tissue structures. Mechanical stimuli, such as shear stress due to flow characteristics, have been shown to have a great effect on the development of tissues (Ellis et al. 2005). For example, cardiac muscle in vivo encounters strong pulsatile flows, whereas bone constantly encounters mechanical stress and compression. Exposure to various mechanical stimuli during seeding and cultivation in bioreactors, in attempts to simulate in vivo conditions, has led to significant enhancement in the functions of engineered tissues. For example, exposure of cardiac cells to cyclical mechanical stretching results in a marked improvement in cell proliferation and distribution as well as ECM organization, which ultimately leads to a greater increase in the strength of the tissue (Akhyari et al. 2002). In the case of chondrocytes, both dynamic compression and shearing improve ECM production and the tissue mechanical properties (Waldman et al. 2003). Interestingly, it was also noted that imparting shear stresses on the developing cartilage tissue leads to a greater improvement on the mechanical properties as compared to the compression of subjected tissues (Waldman et al. 2003). This result supports the notion that the exact effects of different mechanical stimuli on these tissues vary greatly with the type of stimuli applied and the tissue type, and it also emphasizes the need to study in seclusion the effects of individual parameters on the development of tissues.

1.5.5 Future Directions for Using Bioreactors in Tissue Engineering

Bioreactors are powerful tools to provide a more favorable environment for engineering tissue constructs in vitro. Their applications for cultivating engineered tissues for use in biomedical applications represent current and future directions that have gained large appeal. The first product which demonstrated large-scale use of bioreactors was Dermagraft, a skin graft developed by Advanced Tissue Sciences (Martin et al. 2004). In addition to its great potential to create tissue grafts for clinical applications, bioreactors have and will continue to be very useful tools for studying tissue growth in general. Unlike conditions in vivo, bioreactors enable the control and examination of the effects of certain factors on tissue development individually or in various combinations. The knowledge gained from such mechanistic studies will in turn provide guidance for research in tissue manufacture for clinical uses. Another interesting area where bioreactors can be used as a valuable tool is for studying the effect of different parameters on tissue development during pathological processes and various diseases (Griffith and Naughton 2002). Novel ideas such as these along with development of new on-sight control systems and computational fluid dynamic studies will lead to development of more advanced bioreactors and engineered tissues in the future.

1.6 CONCLUSIONS

The field of tissue engineering has been growing rapidly for the past two decades, driven by the enormous demand and realistic potential of this new discipline. Much progress has been made including the isolation and utilization of adult and ES cells, development of biodegradable scaffolds, delivery of various tissue-inducing factors, and applications of bioreactors to promote tissue formation. Although significant advances have been accomplished, most regenerative therapies are still in the developmental phase. Understanding the fundamental biology associated with normal tissue development is critical for the development of more powerful approaches to achieve controlled cell differentiation and tissue formation. More quantitative approach such as system biology and computational modeling may also shed light on deciphering the complex signaling network. Advances in microfabrication technology might also help design artificial scaffolds and enable mechanistic studies of spatial cues and gradients, etc. Generating vascularized tissues is an essential prerequisite for most tissue types to be clinically useful. In summary, further progress in the field will rely on the advancement and close interactions among multiple disciplines, such as developmental biology, nanotechnology, material sciences, immunology, and computational biology.

REFERENCES

Aggarwal, S. and M.F. Pittenger. 2005. Human mesenchymal stem cells modulate allogeneic immune cell responses. *Blood* 105(4):1815–1822.

Akhyari, P., P.W.M. Fedak, R.D. Weisel, T.Y.J. Lee, S. Verma, D.A.G. Mickle, and R.K. Li. 2002. Mechanical stretch regimen enhances the formation of bioengineered autologous cardiac muscle grafts. *Circulation* 106:I137–I142.

Akinc, A., D.G. Anderson, D.M. Lynn, and R. Langer. 2003a. Synthesis of poly(beta-amino ester)s optimized for highly effective gene delivery. *Bioconjug Chem* 14(5):979–988.

Akinc, A., D.M. Lynn, D.G. Anderson, and R. Langer. 2003b. Parallel synthesis and biophysical characterization of a degradable polymer library for gene delivery. *J Am Chem Soc* 125(18):5316–5323.

Albrecht, D.R., G.H. Underhill, T.B. Wassermann, R.L. Sah, and S.N. Bhatia. 2006. Probing the role of multicellular organization in three-dimensional microenvironments. *Nat Methods* 3(5):369–375.

Anderson, D.G., A. Akinc, N. Hossain, and R. Langer. 2005. Structure/property studies of polymeric gene delivery using a library of poly(beta-amino esters). *Mol Ther* 11(3):426–434.

Anderson, D.G., D.M. Lynn, and R. Langer. 2003. Semi-automated synthesis and screening of a large library of degradable cationic polymers for gene delivery. *Angew Chem Int Ed Engl* 42(27):3153–3158.

Anderson, D.G., W. Peng, A. Akinc, N. Hossain, A. Kohn, R. Padera, R. Langer, and J.A. Sawicki. 2004. A polymer library approach to suicide gene therapy for cancer. *Proc Natl Acad Sci U S A* 101(45): 16028–16033.

Arcaute, K., B.K. Mann, and R.B. Wicker. 2006. Stereolithography of three-dimensional bioactive poly (ethylene glycol) constructs with encapsulated cells. *Ann Biomed Eng* 34(9):1429–1441.

Aubin, J.E. 1998. Bone stem cells. *J Cell Biochem Suppl* 30–31:73–82.

Baier Leach, J., K.A. Bivens, C.W. Patrick, Jr., and C.E. Schmidt. 2003. Photocrosslinked hyaluronic acid hydrogels: Natural, biodegradable tissue engineering scaffolds. *Biotechnol Bioeng* 82(5):578–589.

Baksh, D., J.E. Davies, and P.W. Zandstra. 2003. Adult human bone marrow-derived mesenchymal progenitor cells are capable of adhesion-independent survival and expansion. *Exp Hematol* 31(8):723–732.

Baksh, D., R. Yao, and R.S. Tuan. 2007. Comparison of proliferative and multilineage differentiation potential of human mesenchymal stem cells derived from umbilical cord and bone marrow. *Stem Cells* 25(6):1384–1392.

Beerens, A.M., A.F. Al Hadithy, M.G. Rots, and H.J. Haisma. 2003. Protein transduction domains and their utility in gene therapy. *Curr Gene Ther* 3(5):486–494.

Behr, J.P. 1997. The proton sponge: A trick to enter cells the viruses did not exploit. *Chimia* 51(1–2):34–36.

Beniash, E., J.D. Hartgerink, H. Storrie, J.C. Stendahl, and S.I. Stupp. 2005. Self-assembling peptide amphiphile nanofiber matrices for cell entrapment. *Acta Biomater* 1(4):387–397.

Berry, M., A.M. Gonzalez, W. Clarke, L. Greenlees, L. Barrett, W. Tsang, L. Seymour, J. Bonadio, A. Logan, and A. Baird. 2001. Sustained effects of gene-activated matrices after CNS injury. *Mol Cell Neurosci* 17(4):706–716.

Bissell, M.J., H.G. Hall, and G. Parry. 1982. How does the extracellular matrix direct gene expression? *J Theor Biol* 99(1):31–68.

Boland, T., V. Mironov, A. Gutowska, E.A. Roth, and R.R. Markwald. 2003. Cell and organ printing 2: Fusion of cell aggregates in three-dimensional gels. *Anat Rec A Discov Mol Cell Evol Biol* 272(2): 497–502.

Boontheekul, T. and D.J. Mooney. 2003. Protein-based signaling systems in tissue engineering. *Curr Opin Biotechnol* 14(5):559–565.

Boussif, O., F. Lezoualc'h, M.A. Zanta, M.D. Mergny, D. Scherman, B. Demeneix, and J.P. Behr. 1995. A versatile vector for gene and oligonucleotide transfer into cells in culture and in vivo: Polyethylenimine. *Proc Natl Acad Sci U S A* 92(16):7297–7301.

Brittberg, M., A. Lindahl, A. Nilsson, C. Ohlsson, O. Isaksson, and L. Peterson. 1994. Treatment of deep cartilage defects in the knee with autologous chondrocyte transplantation. *N Engl J Med* 331(14): 889–895.

Bruder, S.P., D.J. Fink, and A.I. Caplan. 1994. Mesenchymal stem cells in bone development, bone repair, and skeletal regeneration therapy. *J Cell Biochem* 56(3):283–294.

Burke, P.A. 2000. Controlled release protein therapeutic effects of process and formulation on stability. In *Handbook of Pharmaceutical Release Technology*, edited by D.L. Wise. New York: Marcel Dekker.

Bursac, N., M. Papadaki, R.J. Cohen, F.J. Schoen, S.R. Eisenberg, R. Carrier, G. Vunjak-Novakovik, and L.E. Freed. 1999. Cardiac muscle tissue engineering: Toward an in vitro model for electrophysiological studies. *Am J Physiol Heart Circ Physiol* 277(2):433–444.

Caplan, A.I. and S.P. Bruder. 2001. Mesenchymal stem cells: Building blocks for molecular medicine in the 21st century. *Trends Mol Med* 7(6):259–264.

Caplen, N.J., S. Parrish, F. Imani, A. Fire, and R.A. Morgan. 2001. Specific inhibition of gene expression by small double-stranded RNAs in invertebrate and vertebrate systems. *Proc Natl Acad Sci U S A* 98(17): 9742–9747.

Carrier, R., M. Papadaki, M. Rupnick, F.J. Schoen, N. Bursac, R. Langer, L.E. Freed, and G. Vunjak-Novakovic. 1999. Cardiac tissue engineering: Cell seeding, cultivation parameters and tissue construct characterization. *Biotechnol Bioeng* 64(5):580–589.

Cartier, R. and R. Reszka. 2002. Utilization of synthetic peptides containing nuclear localization signals for nonviral gene transfer systems. *Gene Ther* 9(3):157–167.

Cetrulo, C.L. Jr. 2006. Cord-blood mesenchymal stem cells and tissue engineering. *Stem Cell Rev* 2(2): 163–168.

Chan, C.K. and D.A. Jans. 2002. Using nuclear targeting signals to enhance non-viral gene transfer. *Immunol Cell Biol* 80(2):119–130.

Chang, P.L., J.M. Van Raamsdonk, G. Hortelano, S.C. Barsoum, N.C. MacDonald, and T.L. Stockley. 1999. The in vivo delivery of heterologous proteins by microencapsulated recombinant cells. *Trends Biotechnol* 17(2):78–83.

Chen, R.R., E.A. Silva, W.W. Yuen, and D.J. Mooney. 2007. Spatio-temporal VEGF and PDGF delivery patterns blood vessel formation and maturation. *Pharm Res* 24(2):258–264.

Chua, C.K., K.F. Leong, K.H. Tan, F.E. Wiria, and C.M. Cheah. 2004. Development of tissue scaffolds using selective laser sintering of polyvinyl alcohol/hydroxyapatite biocomposite for craniofacial and joint defects. *J Mater Sci Mater Med* 15(10):1113–1121.

Cohen, H., R.J. Levy, J. Gao, I. Fishbein, V. Kousaev, S. Sosnowski, S. Slomkowski, and G. Golomb. 2000. Sustained delivery and expression of DNA encapsulated in polymeric nanoparticles. *Gene Ther* 7(22):1896–1905.

Cooke, M.N., J.P. Fisher, D. Dean, C. Rimnac, and A.G. Mikos. 2003. Use of stereolithography to manufacture critical-sized 3D biodegradable scaffolds for bone ingrowth. *J Biomed Mater Res B Appl Biomater* 64B(2):65–69.

Cowan, C.M., Y.Y. Shi, O.O. Aalami, Y.F. Chou, C. Mari, R. Thomas, N. Quarto, C.H. Contag, B. Wu, and M.T. Longaker. 2004. Adipose-derived adult stromal cells heal critical-size mouse calvarial defects. *Nat Biotechnol* 22(5):560–567.

Crump, S.S. 1992. Apparatus and method for creating three-dimensional objects. Stratasys Inc., assignee. U.S. Patent 5,121,329.

Dagalakis, N., J. Flink, P. Stasikelis, J.F. Burke, and I.V. Yannas. 1980. Design of an artificial skin. Part III. Control of pore structure. *J Biomed Mater Res* 14(4):511–528.

Dash, P.R., M.L. Read, L.B. Barrett, M.A. Wolfert, and L.W. Seymour. 1999. Factors affecting blood clearance and in vivo distribution of polyelectrolyte complexes for gene delivery. *Gene Ther* 6(4):643–650.

DeLong, S.A., J.J. Moon, and J.L. West. 2005. Covalently immobilized gradients of bFGF on hydrogel scaffolds for directed cell migration. *Biomaterials* 26(16):3227–3234.

Dhariwala, B., E. Hunt, and T. Boland. 2004. Rapid prototyping of tissue-engineering constructs, using photopolymerizable hydrogels and stereolithography. *Tissue Eng* 10(9–10):1316–1322.

During, M.J. 1997. Adeno-associated virus as a gene delivery system. *Adv Drug Deliv Rev* 27(1):83–94.

Dutta Roy, T., J.L. Simon, J.L. Ricci, E.D. Rekow, V.P. Thompson, and J.R. Parsons. 2003. Performance of hydroxyapatite bone repair scaffolds created via three-dimensional fabrication techniques. *J Biomed Mater Res A* 67(4):1228–1237.

Edelman, E.R., E. Mathiowitz, R. Langer, and M. Klagsbrun. 1991. Controlled and modulated release of basic fibroblast growth factor. *Biomaterials* 12(7):619–626.

Elbashir, S.M., J. Harborth, W. Lendeckel, A. Yalcin, K. Weber, and T. Tuschl. 2001. Duplexes of 21-nucleotide RNAs mediate RNA interference in cultured mammalian cells. *Nature* 411(6836):494–498.

Eliaz, R.E. and F.C. Szoka, Jr. 2002. Robust and prolonged gene expression from injectable polymeric implants. *Gene Ther* 9(18):1230–1237.

Elisseeff, J., A. Ferran, S. Hwang, S. Varghese, and Z. Zhang. 2006. The role of biomaterials in stem cell differentiation: Applications in the musculoskeletal system. *Stem Cells Dev* 15(3):295–303.

Ellis, M., M. Jarman-Smith, and J.B. Chaudhuri. 2005. Bioreactor systems for tissue engineering: A four-dimensional challenge. In *Bioreactors for Tissue Engineering*, edited by J. Chaudhuri and M. Al-Rubeai. Dordrecht: Springer.

Fire, A., S. Xu, M.K. Montgomery, S.A. Kostas, S.E. Driver, and C.C. Mello. 1998. Potent and specific genetic interference by double-stranded RNA in *Caenorhabditis elegans*. *Nature* 391(6669):806–811.

Folkman, J. 1971. Tumor angiogenesis: Therapeutic implications. *N Engl J Med* 285(21):1182–1186.

Freed, L.E. and G. Vunjak-Novakovic. 2000. Tissue engineering bioreactors. In *Principles of Tissue Engineering*, edited by R.P. Lanza, R. Langer, and J. Vacanti. San Diego: Academic Press.

Fu, K., A.M. Klibanov, and R. Langer. 2000. Protein stability in controlled-release systems. *Nat Biotechnol* 18(1):24–25.

Gelain, F., A. Horii, and S. Zhang. 2007. Designer self-assembling peptide scaffolds for 3-d tissue cell cultures and regenerative medicine. *Macromol Biosci* 7(5):544–551.

Gene Therapy Clinical Trials online, http://www.wiley.co.uk/genetherapy/clinical/. Accessed 2005.

Giordano, R.A., B.M. Wu, S.W. Borland, L.G. Cima, E.M. Sachs, and M.J. Cima. 1996. Mechanical properties of dense polylactic acid structures fabricated by three dimensional printing. *J Biomater Sci Polym Ed* 8(1):63–75.

Green, J.J., E. Chiu, E.S. Leshchiner, J. Shi, R. Langer, and D.G. Anderson. 2007. Electrostatic ligand coatings of nanoparticles enable ligand-specific gene delivery to human primary cells. *Nano Lett* 7(4): 874–879.

Green, J.J., J. Shi, E. Chiu, E.S. Leshchiner, R. Langer, and D.G. Anderson. 2006. Biodegradable polymeric vectors for gene delivery to human endothelial cells. *Bioconjug Chem* 17(5):1162–1169.

Griffith, L.G. and G. Naughton. 2002. Tissue engineering-current challenges and expanding opportunities. *Science* 295:1009–1016.

Guping, Y., T. Guping, and Yanjie. 2005. A new biodegradable poly-amino acid: Alpha,beta-poly[(N-hydroxypropyl/aminoethyl)-DL-aspartamide-co-L-lysine], a potential nonviral vector for gene delivery. *Drug Deliv* 12(2):89–96.

Haensler, J. and F.C Szoka. 1993. Polyamidoamine cascade polymers mediate efficient transfection of cells in culture. *Bioconjug Chem* 4(5):372–379.

Hartgerink, J.D., E. Beniash, and S.I. Stupp. 2001. Self-assembly and mineralization of peptide-amphiphile nanofibers. *Science* 294(5547):1684–1688.

Hartgerink, J.D., E. Beniash, and S.I. Stupp. 2002. Peptide-amphiphile nanofibers: A versatile scaffold for the preparation of self-assembling materials. *Proc Natl Acad Sci U S A* 99(8):5133–5138.

He, X. and E. Jabbari. 2007. Material properties and cytocompatibility of injectable MMP degradable poly (lactide ethylene oxide fumarate) hydrogel as a carrier for marrow stromal cells. *Biomacromolecules* 8(3):780–792.

Healy, K.E., A. Rezania, and R.A. Stile. 1999. Designing biomaterials to direct biological responses. *Ann N Y Acad Sci* 875:24–35.

Helmlinger, G., F. Yuan, M. Dellian, and R.K. Jain. 1997. Interstitial pH and pO_2 gradients in solid tumors in vivo: High-resolution measurements reveal a lack of correlation. *Nat Med* 3(2):177–182.

Herbert, P., K. Murphy, O. Johnson, N. Dong, W. Jaworowicz, M.A. Tracy, J.L. Cleland, and S.D. Putney. 1998. A large-scale process to produce microencapsulated proteins. *Pharm Res* 15(2):357–361.

Hern, D.L. and J.A. Hubbell. 1998. Incorporation of adhesion peptides into nonadhesive hydrogels useful for tissue resurfacing. *J Biomed Mater Res* 39(2):266–276.

Hill, E., T. Boontheekul, and D.J. Mooney. 2006. Regulating activation of transplanted cells controls tissue regeneration. *Proc Natl Acad Sci U S A* 103(8):2494–2499.

Holland, T.A., Y. Tabata, and A.G. Mikos. 2005. Dual growth factor delivery from degradable oligo (poly(ethylene glycol) fumarate) hydrogel scaffolds for cartilage tissue engineering. *J Control Release* 101(1–3):111–125.

Hollister, S.J. 2005. Porous scaffold design for tissue engineering. *Nat Mater* 4(7):518–524.

Hollister, S.J., R.A. Levy, T.M. Chu, J.W. Halloran, and S.E. Feinberg. 2000. An image-based approach for designing and manufacturing craniofacial scaffolds. *Int J Oral Maxillofac Surg* 29(1):67–71.

Hoque, M.E., D.W. Hutmacher, W. Feng, S. Li, M.H. Huang, M. Vert, and Y.S. Wong. 2005. Fabrication using a rapid prototyping system and in vitro characterization of PEG–PCL–PLA scaffolds for tissue engineering. *J Biomater Sci Polym Ed* 16(12):1595–1610.

Huang, Y.C., M. Connell, Y. Park, D.J. Mooney, and K.G. Rice. 2003. Fabrication and in vitro testing of polymeric delivery system for condensed DNA. *J Biomed Mater Res A* 67(4):1384–1392.

Huang, Y.C., K. Riddle, K.G. Rice, and D.J. Mooney. 2005a. Long-term in vivo gene expression via delivery of PEI-DNA condensates from porous polymer scaffolds. *Hum Gene Ther* 16(5):609–617.

Huang, Y.C., C. Simmons, D. Kaigler, K.G. Rice, and D.J. Mooney. 2005b. Bone regeneration in a rat cranial defect with delivery of PEI-condensed plasmid DNA encoding for bone morphogenetic protein-4 (BMP-4). *Gene Ther* 12(5):418–426.

Hubbell, J.A. 1995. Biomaterials in tissue engineering. *Biotechnology (N Y)* 13(6):565–576.

Hubbell, J.A. 1999. Bioactive biomaterials. *Curr Opin Biotechnol* 10(2):123–129.

Hulbert, S.F., F.A. Young, R.S. Mathews, J.J. Klawitter, C.D. Talbert, and F.H. Stelling. 1970. Potential of ceramic materials as permanently implantable skeletal prostheses. *J Biomed Mater Res* 4(3):433–456.

Hutmacher, D.W., T. Schantz, I. Zein, K.W. Ng, S.H. Teoh, and K.C. Tan. 2001. Mechanical properties and cell cultural response of polycaprolactone scaffolds designed and fabricated via fused deposition modeling. *J Biomed Mater Res* 55(2):203–216.

Hwang, N.S., M.S. Kim, S. Sampattavanich, J.H. Baek, Z. Zhang, and J. Elisseeff. 2006. Effects of three-dimensional culture and growth factors on the chondrogenic differentiation of murine embryonic stem cells. *Stem Cells* 24(2):284–291.

Hwang, W.S., Y.J. Ryu, J.H. Park, E.S. Park, E.G. Lee, J.M. Koo, H.Y. Jeon, B.C. Lee, S.K. Kang, S.J. Kim, C. Ahn, J.H. Hwang, K.Y. Park, J.B. Cibelli, and S.Y. Moon. 2004. Evidence of a pluripotent human embryonic stem cell line derived from a cloned blastocyst. *Science* 303(5664):1669–1674.

Ingber, D.E., V.C. Mow, D. Butler, L. Niklason, J. Huard, J. Mao, I. Yannas, D. Kaplan, and G. Vunjak-Novakovic. 2006. Tissue engineering and developmental biology: Going biomimetic. *Tissue Eng* 12(12): 3265–3283.

Itala, A.I., H.O. Ylanen, C. Ekholm, K.H. Karlsson, and H.T. Aro. 2001. Pore diameter of more than 100 microm is not requisite for bone ingrowth in rabbits. *J Biomed Mater Res* 58(6):679–683.

Izadpanah, R., C. Trygg, B. Patel, C. Kriedt, J. Dufour, J.M. Gimble, and B.A. Bunnell. 2006. Biologic properties of mesenchymal stem cells derived from bone marrow and adipose tissue. *J Cell Biochem* 99(5):1285–1297.

Jiang, W. and S.P. Schwendeman. 2001. Stabilization and controlled release of bovine serum albumin encapsulated in poly(D,L-lactide) and poly(ethylene glycol) microsphere blends. *Pharm Res* 18(6): 878–885.

Johnson, M.E., D. Blankschtein, and R. Langer. 1997. Evaluation of solute permeation through the stratum corneum: Lateral bilayer diffusion as the primary transport mechanism. *J Pharm Sci* 86(10): 1162–1172.

Kaneko, T., S. Tanaka, A. Ogura, and M. Akashi. 2005. Tough, thin hydrogel membranes with giant crystalline domains composed of precisely synthesized macromolecules. *Macromolecules* 38(11):4861–4867.

Karageorgiou, V. and D. Kaplan. 2005. Porosity of 3D biomaterial scaffolds and osteogenesis. *Biomaterials* 26(27):5474–5491.

Karp, J.M., F. Sarraf, M.S. Shoichet, and J.E. Davies. 2004. Fibrin-filled scaffolds for bone-tissue engineering: An in vivo study. *J Biomed Mater Res A* 71(1):162–171.

Khalyfa, A., S. Vogt, J. Weisser, G. Grimm, A. Rechtenbach, W. Meyer, and M. Schnabelrauch. 2007. Development of a new calcium phosphate powder-binder system for the 3D printing of patient specific implants. *J Mater Sci Mater Med* 18(5):909–916.

Kim, M.S., N.S. Hwang, J. Lee, T.K. Kim, K. Leong, M.J. Shamblott, J. Gearhart, and J. Elisseeff. 2005a. Musculoskeletal differentiation of cells derived from human embryonic germ cells. *Stem Cells* 23(1): 113–123.

Kim, T.I., H.J. Seo, J.S. Choi, J.K. Yoon, J.U. Baek, K. Kim, and J.S. Park. 2005b. Synthesis of biodegradable cross-linked poly(beta-amino ester) for gene delivery and its modification, inducing enhanced transfection efficiency and stepwise degradation. *Bioconjug Chem* 16(5):1140–1148.

Krewson, C.E., R. Dause, M. Mak, and W.M. Saltzman. 1996. Stabilization of nerve growth factor in controlled release polymers and in tissue. *J Biomater Sci Polym Ed* 8(2):103–117.

Kruyt, M.C., J.D. de Bruijn, C.E. Wilson, F.C. Oner, C.A. van Blitterswijk, A.J. Verbout, and W.J. Dhert. 2003. Viable osteogenic cells are obligatory for tissue-engineered ectopic bone formation in goats. *Tissue Eng* 9(2):327–336.

Kursa, M., G.F. Walker, V. Roessler, M. Ogris, W. Roedl, R. Kircheis, and E. Wagner. 2003. Novel shielded transferrin-polyethylene glycol-polyethylenimine/DNA complexes for systemic tumor-targeted gene transfer. *Bioconjug Chem* 14(1):222–231.

Landers, R. and R. Mülhaupt. 2000. Desktop manufacturing of complex objects, prototypes and biomedical scaffolds by means of computer-assisted design combined with computer-guided 3D plotting of polymers and reactive oligomers. *Macromol Mater Eng* 282:17–21.

Langer, R. and J. Folkman. 1976. Polymers for the sustained release of proteins and other macromolecules. *Nature* 263(5580):797–800.

Langer, R. and D.A. Tirrell. 2004. Designing materials for biology and medicine. *Nature* 428(6982):487–492.

Langer, R. and J.P. Vacanti. 1993. Tissue engineering. *Science* 260(5110):920–926.

Lanza, R.P., J.B. Cibelli, and M.D. West. 1999. Prospects for the use of nuclear transfer in human transplantation. *Nat Biotechnol* 17(12):1171–1174.

Lanza, R.P., J.B. Cibelli, M.D. West, E. Dorff, C. Tauer, and R.M. Green. 2001. The ethical reasons for stem cell research. *Science* 292(5520):1299.

Le Blanc, K., L. Tammik, B. Sundberg, S.E. Haynesworth, and O. Ringden. 2003. Mesenchymal stem cells inhibit and stimulate mixed lymphocyte cultures and mitogenic responses independently of the major histocompatibility complex. *Scand J Immunol* 57(1):11–20.

Leach, J.B., K.A. Bivens, C.N. Collins, and C.E. Schmidt. 2004. Development of photocrosslinkable hyaluronic acid-polyethylene glycol-peptide composite hydrogels for soft tissue engineering. *J Biomed Mater Res A* 70(1):74–82.

Lee, K.W., S. Wang, B.C. Fox, E.L. Ritman, M.J. Yaszemski, and L. Lu. 2007. Poly(propylene fumarate) bone tissue engineering scaffold fabrication using stereolithography: Effects of resin formulations and laser parameters. *Biomacromolecules* 8(4):1077–1084.

Lee, K.Y., M.C. Peters, K.W. Anderson, and D.J. Mooney. 2000. Controlled growth factor release from synthetic extracellular matrices. *Nature* 408(6815):998–1000.

Leong, K.F., C.M. Cheah, and C.K. Chua. 2003. Solid freeform fabrication of three-dimensional scaffolds for engineering replacement tissues and organs. *Biomaterials* 24(13):2363–2378.

Li, G., A.S. Virdi, D.E. Ashhurst, A.H. Simpson, and J.T. Triffitt. 2000. Tissues formed during distraction osteogenesis in the rabbit are determined by the distraction rate: Localization of the cells that express the mRNAs and the distribution of types I and II collagens. *Cell Biol Int* 24(1):25–33.

Lim, Y.B., S.M. Kim, H. Suh, and J.S. Park. 2002. Biodegradable, endosome disruptive, and cationic network-type polymer as a highly efficient and nontoxic gene delivery carrier. *Bioconjug Chem* 13(5):952–957.

Liu, V.A. and S.N Bhatia. 2002. Three-dimensional photopatterning of hydrogels containing living cells. *Biomed Microdev* 4(4):257–266.

Lo, H., M.S. Ponticiello, and K.W. Leong. 1995. Fabrication of controlled release biodegradable foams by phase separation. *Tissue Eng* 1(1):15–28.

Lu, L., M.J. Yaszemski, and A.G. Mikos. 2001. TGF-beta1 release from biodegradable polymer microparticles: Its effects on marrow stromal osteoblast function. *J Bone Joint Surg Am* 83-A Suppl 1(Pt 2):S82–S91.

Lutolf, M.P. and J.A. Hubbell. 2005. Synthetic biomaterials as instructive extracellular microenvironments for morphogenesis in tissue engineering. *Nat Biotechnol* 23(1):47–55.

Lynn, D.M., D.G. Anderson, D. Putnam, and R. Langer. 2001. Accelerated discovery of synthetic transfection vectors: Parallel synthesis and screening of a degradable polymer library. *J Am Chem Soc* 123(33): 8155–8156.

Lynn, D.M. and R. Langer. 2000. Degradable poly(beta-amino esters): Synthesis, characterization, and self-assembly with plasmid DNA. *J Am Chem Soc* 122(44):10761–10768.

Lysaght, M.J. and J. Reyes. 2001. The growth of tissue engineering. *Tissue Eng* 7(5):485–493.

Macpherson, H., P. Keir, S. Webb, K. Samuel, S. Boyle, W. Bickmore, L. Forrester, and J. Dorin. 2005. Bone marrow-derived SP cells can contribute to the respiratory tract of mice in vivo. *J Cell Sci* 118(Pt 11): 2441–2450.

Maitra, B., E. Szekely, K. Gjini, M.J. Laughlin, J. Dennis, S.E. Haynesworth, and O.N. Koc. 2004. Human mesenchymal stem cells support unrelated donor hematopoietic stem cells and suppress T-cell activation. *Bone Marrow Transplant* 33(6):597–604.

Manjubala, I., A. Woesz, C. Pilz, M. Rumpler, N. Fratzl-Zelman, P. Roschger, J. Stampfl, and P. Fratzl. 2005. Biomimetic mineral-organic composite scaffolds with controlled internal architecture. *J Mater Sci Mater Med* 16(12):1111–1119.

Mankovich, N.J., A.M. Cheeseman, and N.G. Stoker. 1990. The display of three-dimensional anatomy with stereolithographic models. *J Digit Imaging* 3(3):200–203.

Martin, I., D. Wendt, and M. Heberer. 2004. The role of bioreactors in tissue engineering. *Trends Biotechnol* 22(2):80–86.

Matthews, J.A., G.E. Wnek, D.G. Simpson, and G.L. Bowlin. 2002. Electrospinning of collagen nanofibers. *Biomacromolecules* 3(2):232–238.

McIntire, L.V. 2003. WTEC panel report on tissue engineering. *Tissue Eng* 9(1):3–7.

Merdan, T., J. Kopecek, and T. Kissel. 2002. Prospects for cationic polymers in gene and oligonucleotide therapy against cancer. *Adv Drug Deliv Rev* 54(5):715–758.

Mikos, A.G., S.W. Herring, P. Ochareon, J. Elisseeff, H.H. Lu, R. Kandel, F.J. Schoen, M. Toner, D. Mooney, A. Atala, M.E. Van Dyke, D. Kaplan, and G. Vunjak-Novakovic. 2006. Engineering complex tissues. *Tissue Eng* 12(12):3307–3339.

Mislick, K.A. and J.D. Baldeschwieler. 1996. Evidence for the role of proteoglycans in cation-mediated gene transfer. *Proc Natl Acad Sci U S A* 93(22):12349–12354.

Mooney, D.J., D.F. Baldwin, N.P. Suh, J.P. Vacanti, and R. Langer. 1996. Novel approach to fabricate porous sponges of poly(D,L-lactic-co-glycolic acid) without the use of organic solvents. *Biomaterials* 17(14): 1417–1422.

Mooney, D.J., G. Organ, J.P. Vacanti, and R. Langer. 1994. Design and fabrication of biodegradable polymer devices to engineer tubular tissues. *Cell Transplant* 3(2):203–210.

Moore, K., M. MacSween, and M. Shoichet. 2006. Immobilized concentration gradients of neurotrophic factors guide neurite outgrowth of primary neurons in macroporous scaffolds. *Tissue Eng* 12(2):267–278.

Mosahebi, A., M. Simon, M. Wiberg, and G. Terenghi. 2001. A novel use of alginate hydrogel as Schwann cell matrix. *Tissue Eng* 7(5):525–534.

Murugan, R. and S. Ramakrishna. 2006. Nano-featured scaffolds for tissue engineering: A review of spinning methodologies. *Tissue Eng* 12(3):435–447.

Murugan, R. and S. Ramakrishna. 2007. Design strategies of Ttissue engineering scaffolds with controlled fiber orientation. *Tissue Eng*.

Muschler, G.F., C. Nakamoto, and L.G. Griffith. 2004. Engineering principles of clinical cell-based tissue engineering. *J Bone Joint Surg Am* 86-A(7):1541–1558.

Nahmias, Y., R.E. Schwartz, C.M. Verfaillie, and D.J. Odde. 2005. Laser-guided direct writing for three-dimensional tissue engineering. *Biotechnol Bioeng* 92(2):129–136.

Nair, L.S. and C.T. Laurencin. 2006. Polymers as biomaterials for tissue engineering and controlled drug delivery. *Adv Biochem Eng Biotechnol* 102:47–90.

Nelson, C.M. and M.J. Bissell. 2006. Of extracellular matrix, scaffolds, and signaling: Tissue architecture regulates development, homeostasis, and cancer. *Annu Rev Cell Dev Biol* 22:287–309.

Nickerson, J. 2001. Experimental observations of a nuclear matrix. *J Cell Sci* 114(Pt 3):463–474.

Nuttelman, C.R., M.C. Tripodi, and K.S. Anseth. 2006. Dexamethasone-functionalized gels induce osteogenic differentiation of encapsulated hMSCs. *J Biomed Mater Res A* 76(1):183–195.

Odde, D.J. and M.J. Renn. 2000. Laser-guided direct writing of living cells. *Biotechnol Bioeng* 67(3): 312–318.

Oldham, J.B., L. Lu, X. Zhu, B.D. Porter, T.E. Hefferan, D.R. Larson, B.L. Currier, A.G. Mikos, and M.J. Yaszemski. 2000. Biological activity of rhBMP-2 released from PLGA microspheres. *J Biomech Eng* 122(3):289–292.

Osawa, M., K. Hanada, H. Hamada, and H. Nakauchi. 1996. Long-term lymphohematopoietic reconstitution by a single CD34-low/negative hematopoietic stem cell. *Science* 273(5272):242–245.

Pack, D.W., A.S. Hoffman, S. Pun, and P.S. Stayton. 2005. Design and development of polymers for gene delivery. *Nat Rev Drug Discov* 4(7):581–593.

Park, K., H.H. Jung, J.S. Son, J.-W. Rhie, K.D. Park, K.-D. Ahn, and D.K. Han. 2007. Thermosensitive and cell-adhesive pluronic hydrogels for human adipose-derived stem cells. *Key Engineering Materials* 342–343 (Advanced Biomaterials VII):301–304.

Partridge, K.A. and R.O. Oreffo. 2004. Gene delivery in bone tissue engineering: Progress and prospects using viral and nonviral strategies. *Tissue Eng* 10(1–2):295–307.

Pearse, D.D., F.C. Pereira, A.E. Marcillo, M.L. Bates, Y.A. Berrocal, M.T. Filbin, and M.B. Bunge. 2004. cAMP and Schwann cells promote axonal growth and functional recovery after spinal cord injury. *Nat Med* 10(6):610–616.

Peng, W., D.G. Anderson, Y. Bao, R.F. Padera, Jr., R. Langer, and J.A. Sawicki. 2007. Nanoparticulate delivery of suicide DNA to murine prostate and prostate tumors. *Prostate* 67(8):855–862.

Peppas, N.A. and R. Langer. 1994. New challenges in biomaterials. *Science* 263(5154):1715–1720.

Peterson, L., T. Minas, M. Brittberg, A. Nilsson, E. Sjogren-Jansson, and A. Lindahl. 2000. Two- to 9-year outcome after autologous chondrocyte transplantation of the knee. *Clin Orthop Relat Res* (374):212–234.

Petite, H., V. Viateau, W. Bensaid, A. Meunier, C. de Pollak, M. Bourguignon, K. Oudina, L. Sedel, and G. Guillemin. 2000. Tissue-engineered bone regeneration. *Nat Biotechnol* 18(9):959–963.

Pittenger, M.F., A.M. Mackay, S.C. Beck, R.K. Jaiswal, R. Douglas, J.D. Mosca, M.A. Moorman, D.W. Simonetti, S. Craig, and D.R. Marshak. 1999. Multilineage potential of adult human mesenchymal stem cells. *Science* 284(5411):143–147.

Plank, C., W. Zauner, and E. Wagner. 1998. Application of membrane-active peptides for drug and gene delivery across cellular membranes. *Adv Drug Deliv Rev* 34(1):21–35.

Prestrelski, S.J., N. Tedeschi, T. Arakawa, and J.F. Carpenter. 1993. Dehydration-induced conformational transitions in proteins and their inhibition by stabilizers. *Biophys J* 65(2):661–671.

Rowley, J.A., G. Madlambayan, and D.J. Mooney. 1999. Alginate hydrogels as synthetic extracellular matrix materials. *Biomaterials* 20(1):45–53.

Sachlos, E., D. Gotora, and J.T. Czernuszka. 2006. Collagen scaffolds reinforced with biomimetic composite nano-sized carbonate-substituted hydroxyapatite crystals and shaped by rapid prototyping to contain internal microchannels. *Tissue Eng* 12(9):2479–2487.

Sachlos, E., N. Reis, C. Ainsley, B. Derby, and J.T. Czernuszka. 2003. Novel collagen scaffolds with predefined internal morphology made by solid freeform fabrication. *Biomaterials* 24(8):1487–1497.

Sachs, E.M., S.H. John, J.C. Michael, and A.W. Paul, 1993. Three-dimensional printing techniques. Massachusetts Institute of Technology, assignee. U.S. Patent 5,204,055.

Saltzman, W.M. and W.L. Olbricht. 2002. Building drug delivery into tissue engineering. *Nat Rev Drug Discov* 1(3):177–186.

Sanders, J.E., S.G. Malcolm, S.D. Bale, Y.N. Wang, and S. Lamont. 2002. Prevascularization of a biomaterial using a chorioallontoic membrane. *Microvasc Res* 64(1):174–178.

Sawada, T., K. Tsukada, K. Hasegawa, Y. Ohashi, Y. Udagawa, and V. Gomel. 2001. Cross-linked hyaluronate hydrogel prevents adhesion formation and reformation in mouse uterine horn model. *Hum Reprod* 16(2):353–356.

Schnabel, M., S. Marlovits, G. Eckhoff, I. Fichtel, L. Gotzen, V. Vecsei, and J. Schlegel. 2002. Dedifferentiation-associated changes in morphology and gene expression in primary human articular chondrocytes in cell culture. *Osteoarthritis Cartilage* 10(1):62–70.

Schwendeman, S.P. 2002. Recent advances in the stabilization of proteins encapsulated in injectable PLGA delivery systems. *Crit Rev Ther Drug Carrier Syst* 19(1):73–98.

Segura, T., P.H. Chung, and L.D. Shea. 2005. DNA delivery from hyaluronic acid-collagen hydrogels via a substrate-mediated approach. *Biomaterials* 26(13):1575–1584.

Shamblott, M.J., J. Axelman, S. Wang, E.M. Bugg, J.W. Littlefield, P.J. Donovan, P.D. Blumenthal, G.R. Huggins, and J.D. Gearhart. 1998. Derivation of pluripotent stem cells from cultured human primordial germ cells. *Proc Natl Acad Sci U S A* 95(23):13726–13731.

Shea, L.D., E. Smiley, J. Bonadio, and D.J. Mooney. 1999. DNA delivery from polymer matrices for tissue engineering. *Nat Biotechnol* 17(6):551–554.

Shu, X.Z., K. Ghosh, Y. Liu, F.S. Palumbo, Y. Luo, R.A. Clark, and G.D. Prestwich. 2004. Attachment and spreading of fibroblasts on an RGD peptide-modified injectable hyaluronan hydrogel. *J Biomed Mater Res A* 68(2):365–375.

Sohier, J., T.J. Vlugt, N. Cabrol, C. Van Blitterswijk, K. de Groot, and J.M. Bezemer. 2006. Dual release of proteins from porous polymeric scaffolds. *J Control Release* 111(1–2):95–106.

Sokolsky-Papkov, M., K. Agashi, A. Olaye, K. Shakesheff, and A.J. Domb. 2007. Polymer carriers for drug delivery in tissue engineering. *Adv Drug Deliv Rev* 59(4–5):187–206.

Sottile, V., A. Thomson, and J. McWhir. 2003. In vitro osteogenic differentiation of human ES cells. *Cloning Stem Cells* 5(2):149–155.

Street, J., D. Winter, J.H. Wang, A. Wakai, A. McGuinness, and H.P. Redmond. 2000. Is human fracture hematoma inherently angiogenic? *Clin Orthop Relat Res* (378):224–237.

Suciati, T., D. Howard, J. Barry, N.M. Everitt, K.M. Shakesheff, and F.R. Rose. 2006. Zonal release of proteins within tissue engineering scaffolds. *J Mater Sci Mater Med* 17(11):1049–1056.

Suh, W., S.O. Han, L. Yu, and S.W. Kim. 2002. An angiogenic, endothelial-cell-targeted polymeric gene carrier. *Mol Ther* 6(5):664–672.

Sun, J.J., C.J. Bae, Y.H. Koh, H.E. Kim, and H.W. Kim. 2007. Fabrication of hydroxyapatite-poly(epsilon-caprolactone) scaffolds by a combination of the extrusion and bi-axial lamination processes. *J Mater Sci Mater Med* 18(6):1017–1023.

Tabata, Y. 2003. Tissue regeneration based on growth factor release. *Tissue Eng* 9(Suppl 1):S5–S15.

Tan, K.H., C.K. Chua, K.F. Leong, C.M. Cheah, W.S. Gui, W.S. Tan, and F.E. Wiria. 2005. Selective laser sintering of biocompatible polymers for applications in tissue engineering. *Biomed Mater Eng* 15(1–2): 113–124.

Tan, W. and T.A. Desai. 2004. Layer-by-layer microfluidics for biomimetic three-dimensional structures. *Biomaterials* 25(7–8):1355–1364.

Tessmar, J.K. and A.M. Gopferich. 2007a. Customized PEG-derived copolymers for tissue-engineering applications. *Macromol Biosci* 7(1):23–39.

Tessmar, J.K. and A.M. Gopferich 2007b. Matrices and scaffolds for protein delivery in tissue engineering. *Adv Drug Deliv Rev* 59(4–5):274–291.

Thomson, J.A., J. Itskovitz-Eldor, S.S. Shapiro, M.A. Waknitz, J.J. Swiergiel, V.S. Marshall, and J.M. Jones. 1998. Embryonic stem cell lines derived from human blastocysts. *Science* 282(5391):1145–1147.

Tomita, M., T. Mori, K. Maruyama, T. Zahir, M. Ward, A. Umezawa, and M.J. Young. 2006. A comparison of neural differentiation and retinal transplantation with bone marrow-derived cells and retinal progenitor cells. *Stem Cells* 24(10):2270–2278.

Tresco, P.A., R. Biran, and M.D. Noble. 2000. Cellular transplants as sources for therapeutic agents. *Adv Drug Deliv Rev* 42(1–2):3–27.

Tsuboi, R., M. Yamazaki, Y. Matsuda, K. Uchida, R. Ueki, and H. Ogawa. 2007. Antisense oligonucleotide targeting fibroblast growth factor receptor (FGFR)-1 stimulates cellular activity of hair follicles in an in vitro organ culture system. *Int J Dermatol* 46(3):259–263.

Tyrone, J.W., J.E. Mogford, L.A. Chandler, C. Ma, Y. Xia, G.F. Pierce, and T.A. Mustoe. 2000. Collagen-embedded platelet-derived growth factor DNA plasmid promotes wound healing in a dermal ulcer model. *J Surg Res* 93(2):230–236.

Um, S.H., J.B. Lee, N. Park, S.Y. Kwon, C.C. Umbach, and D. Luo. 2006. Enzyme-catalysed assembly of DNA hydrogel. *Nat Mater* 5(10):797–801.

Vacanti, J.P., M.A. Morse, W.M. Saltzman, A.J. Domb, A. Perez-Atayde, and R. Langer. 1988. Selective cell transplantation using bioabsorbable artificial polymers as matrices. *J Pediatr Surg* 23(1) (Pt 2):3–9.

Velema, J. and D. Kaplan. 2006. Biopolymer-based biomaterials as scaffolds for tissue engineering. *Adv Biochem Eng Biotechnol* 102:187–238.

Vile, R.G., A. Tuszynski, and S. Castleden. 1996. Retroviral vectors. From laboratory tools to molecular medicine. *Mol Biotechnol* 5(2):139–158.

Viola, J., B. Lal, and O. Grad. 2003. *The Emergence of Tissue Engineering as a Research Field.* Arlington, VA: National Science Foundation.

Vunjak-Novakovic, G. and D.L. Kaplan. 2006. Tissue engineering: The next generation. *Tissue Eng* 12(12): 3261–3263.

Vunjak-Novakovic, G., I. Martin, B. Obradovic, S. Treppo, A.J. Grodzinsky, R. Langer, and L.E. Freed. 1999. Bioreactor cultivation conditions modulate the composition and mechanical properties of tissue engineered cartilage. *J Orthop Res* 17(1):130–138.

Vunjak-Novakovic, G., B. Obradovic, P. Bursac, I. Martin, R. Langer, and L.E. Freed. 1998. Dynamic seeding of polymer scaffolds for cartilage tissue engineering. *Biotechnol Prog* 14(2):193–202.

Wagner, E., R. Kircheis, and G.F. Walker. 2004. Targeted nucleic acid delivery into tumors: New avenues for cancer therapy. *Biomed Pharmacother* 58(3):152–161.

Wald, H.L., G. Sarakinos, M.D. Lyman, A.G. Mikos, J.P. Vacanti, and R. Langer. 1993. Cell seeding in porous transplantation devices. *Biomaterials* 14(4):270–278.

Waldman, S.D., C.G. Spiteri, M.D. Grynpas, R.M. Pilliar, J. Hong, and R.A. Kandel. 2003. Effect of biomechanical conditioning on cartilaginous tissue formation in vitro. *J Bone Joint Surg Am* 85:101–105.

Walter, E., D. Dreher, M. Kok, L. Thiele, S.G. Kiama, P. Gehr, and H.P. Merkle. 2001. Hydrophilic poly(DL-lactide-co-glycolide) microspheres for the delivery of DNA to human-derived macrophages and dendritic cells. *J Control Release* 76(1–2):149–168.

Wendt, D., A. Marsano, M. Jakob, M. Heberer, and I. Martin. 2003. Oscillating perfusion of cell suspensions through three-dimensional scaffolds enhances cell seeding efficiency and uniformity. *Biotechnol Bioeng* 84(2):205–214.

Wettergreen, M.A., B.S. Bucklen, W. Sun, and M.A. Liebschner. 2005. Computer-aided tissue engineering of a human vertebral body. *Ann Biomed Eng* 33(10):1333–1343.

Williams, J.M., A. Adewunmi, R.M. Schek, C.L. Flanagan, P.H. Krebsbach, S.E. Feinberg, S.J. Hollister, and S. Das. 2005. Bone tissue engineering using polycaprolactone scaffolds fabricated via selective laser sintering. *Biomaterials* 26(23):4817–4827.

Wilson, C.E., W.J. Dhert, C.A. Van Blitterswijk, A.J. Verbout, and J.D. De Bruijn. 2002. Evaluating 3D bone tissue engineered constructs with different seeding densities using the alamarBlue assay and the effect on in vivo bone formation. *J Mater Sci Mater Med* 13(12):1265–1269.

Wilton, S.D., A.M. Fall, P.L. Harding, G. McClorey, C. Coleman, and S. Fletcher. 2007. Antisense oligonucleotide-induced exon skipping across the human dystrophin gene transcript. *Mol Ther* 15(7): 1288–1296.

Wiria, F.E., K.F. Leong, C.K. Chua, and Y. Liu. 2007. Poly-epsilon-caprolactone/hydroxyapatite for tissue engineering scaffold fabrication via selective laser sintering. *Acta Biomater* 3(1):1–12.

Wise, D.L., ed. 2000. *Handbook of Pharmaceutical Controlled Release Technology*. New York: Marcel Dekker.

Wood, K.C., S.R. Little, R. Langer, and P.T. Hammond. 2005. A family of hierarchically self-assembling linear-dendritic hybrid polymers for highly efficient targeted gene delivery. *Angew Chem Int Ed Engl* 44(41):6704–6708.

Wray, J.B. 1970. The biochemical characteristics of the fracture hematoma in man. *Surg Gynecol Obstet* 130(5):847–852.

Yang, F., R. Murugan, S. Wang, and S. Ramakrishna. 2005a. Electrospinning of nano/micro scale poly(L-lactic acid) aligned fibers and their potential in neural tissue engineering. *Biomaterials* 26(15):2603–2610.

Yang, F., C.G. Williams, D.A. Wang, H. Lee, P.N. Manson, and J. Elisseeff. 2005b. The effect of incorporating RGD adhesive peptide in polyethylene glycol diacrylate hydrogel on osteogenesis of bone marrow stromal cells. *Biomaterials* 26(30):5991–5998.

Yang, S., K.F. Leong, Z. Du, and C.K. Chua. 2001. The design of scaffolds for use in tissue engineering. Part I. Traditional factors. *Tissue Eng* 7(6):679–689.

Yeo, Y., C.B. Highley, E. Bellas, T. Ito, R. Marini, R. Langer, and D.S. Kohane. 2006. In situ cross-linkable hyaluronic acid hydrogels prevent post-operative abdominal adhesions in a rabbit model. *Biomaterials* 27(27):4698–4705.

Yin, Y., M.T. Henzl, B. Lorber, T. Nakazawa, T.T. Thomas, F. Jiang, R. Langer, and L.I. Benowitz. 2006. Oncomodulin is a macrophage-derived signal for axon regeneration in retinal ganglion cells. *Nat Neurosci* 9(6):843–852.

Zhang, S., F. Gelain, and X. Zhao. 2005. Designer self-assembling peptide nanofiber scaffolds for 3D tissue cell cultures. *Semin Cancer Biol* 15(5):413–420.

Zhu, G., S.R. Mallery, and S.P. Schwendeman. 2000. Stabilization of proteins encapsulated in injectable poly (lactide-co-glycolide). *Nat Biotechnol* 18(1):52–57.

Zugates, G.T., W. Peng, A. Zumbuehl, S. Jhunjhunwala, Y.H. Huang, R. Langer, J.A. Sawicki, and D.G. Anderson. 2007. Rapid optimization of gene delivery by parallel end-modification of poly(beta-amino ester)s. *Mol Ther* 15(7):1306–1312.

Zuk, P.A., M. Zhu, H. Mizuno, J. Huang, J.W. Futrell, A.J. Katz, P. Benhaim, H.P. Lorenz, and M.H. Hedrick. 2001. Multilineage cells from human adipose tissue: Implications for cell-based therapies. *Tissue Eng* 7(2):211–228.

2 Top–Down Analysis of a Dynamic Environment: Extracellular Matrix Structure and Function

Mandana Veiseh, Eva A. Turley, and Mina J. Bissell

CONTENTS

2.1 INTRODUCTION

The extracellular matrix (ECM), and its specialized form, the basement membrane (BM), are dynamic and indispensable components of all organs, providing structural scaffolding and contextual information [1,2]. ECM, as a component of tissues, is a three-dimensional (3D), heterogeneous chemical composite made of multiple structural and functional units that resembles a multivalent ligand. The concept that ECM functions beyond providing a static or structural environment was originally demonstrated in developmental biology studies [3] and the existence of cross-talk between the ECM and chromatin, termed "dynamic reciprocity," was initially proposed in a theoretical model in 1982 [4]. These findings are now being translated in tissue engineering: "An interdisciplinary field that applies the principles of engineering and life sciences toward the development of biological substitutes for the repair or regeneration of tissue or organ functions" [5,6].

Detailed understanding of structure and function of normal tissues, as well as the ability to precisely imitate functions in situ is central to development of artificial organs and tissue engineering strategies.

This is not feasible without taking into account the roles that ECM play in its interactions with cell and cell-biomaterials. Once cells are placed in contact with an appropriate biomaterial [7], they continue to multiply, proliferate, differentiate, secrete factors including ECM molecules, and remodel their environment. To arrive at a molecular understanding of this process, cell and molecular biologists for decades have relied on culturing cells on two-dimensional (2D) substrata. However, ECM–cell functions are significantly different on 2D versus 3D or even 3D versus in vivo systems [8–12]. A number of recent reviews from our laboratory have described the importance of the microenvironment, ECM, and 3D architecture in determining organ specificity in more detail than will be considered here [13–15].

Our current understanding of the importance of microenvironment-mediated signaling in determining tissue specificity and architecture, in parallel with advances in nanoscale technology have brought about the possibility of devising tissue-engineered products with properties more similar to those of native tissues. Whereas the context of microenvironment and tissue architecture has been previously considered in mammary glands, here we bring attention to the "nanoenvironment" by considering the main functional nanostructures that constitute the ECM. In reality, many biological molecules (such as proteins and focal adhesions) are nanoscale structures; therefore cells in our bodies are programmed to interact with nanophase materials. Using a top–down approach, which in nanotechnology is defined as methods and tools to build small components starting from larger regular material [16], we dissect tissue microenvironment (larger material) into its smaller components (ECM components) and analyze some of their regulatory roles in tissue architecture and function in this chapter. Specifically, we describe the key principles governing the structure and function of normal organs and tissues using mammary gland morphogenesis as an example (although the fundamentals extend beyond this organ), as well as those of ECM interaction with cells within tissues or solid biomaterials. We conclude that although we know the alphabet for and some of the language of the genes involved in these processes, we do not yet understand the alphabet and the logic of organ architecture at physiological nanoscale, which is necessary to precisely imitate organ function in tissue engineering.

2.2 STRUCTURE AND FUNCTION OF NORMAL ORGANS AND TISSUES

Organs are composed of various cell types and ECM organized in spatially complex architectural patterns. Each organ has a characteristic form (i.e., unique structural pattern) tailored to coordinate its physiological functions. For instance, skeletal muscle is composed of myoblasts fused into large syncytial myofibers, which are then arranged in dense bundles, an organization that permits coordinated transmission of mechanical forces over long distances, a task not feasible for individual myofibroblasts. Unique relationships between form and function are manifested in tissues with absorptive or secretory functions such as lung, kidney, intestine, and mammary gland, where the epithelium is organized in architectural units designed to increase the surface area for either absorption or secretion [17]. Thus the structure of all organs is made up of specific cellular and extracellular elements and the architecture is dictated by functional necessity. The cell and ECM components of organs retain a memory of the developmental processes that gave rise to them, and these remain stable as long as physiological cell–cell and cell–ECM interactions along with appropriate humoral signals are not interrupted. Loss or misregulation of these interactions can lead to pathological conditions that result in loss of architectural integrity and consequently a disruption of tissue function.

Although most organs reach their development steady-state at the end of embryogenesis, some also undergo major developmental changes in adulthood. The mammary gland and intestine are two examples, and both provide excellent models of tissue morphogenesis and homeostasis in 3D environments including the designer scaffolds used in tissue engineering. In this chapter, we focus on the mammary gland (Figure 2.1) [18]. Regardless of its developmental state, the mammary epithelium is a polarized bilayered structure composed of an inner continuous assembly of luminal epithelial cells surrounded by an outer layer of myoepithelial cells. This bilayer is separated from the

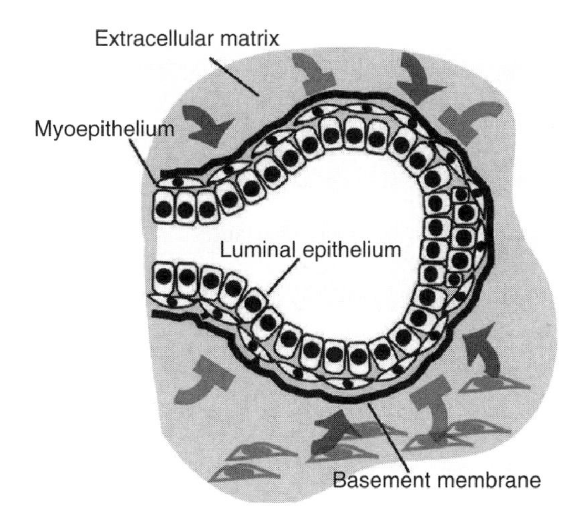

FIGURE 2.1 Normal mammary gland architecture in the "virgin" state. The unit of mammary structure is the acinus where a continuous layer of luminal epithelial cells is surrounded by a layer of myoepithelial cells, and this bilayered structure is enclosed by the BM. (Modified from Bissell, M.J. and LaBarge, M.A., *Cancer Cell*, 7, 17, 2005. With permission.)

stroma by a laminin-rich basement membrane (lrBM). The stroma is a connective milieu composed of fibroblasts and adipocytes, nerves, blood vessels, lymphatics, and immune cells (for details, see [19–21]). During puberty, the mammary gland epithelium branches into numerous ducts with terminal structures known as acini [terminal ductal lobular units (TDLU) in humans].

This structural pattern is maintained in the postpubertal virgin gland but is modified during pregnancy, lactation, and involution. The lactating mammary gland reaches its fully developed and fully functional state only after parturition. Milk secretion via the nipple is induced by suckling and results from synchronized contraction of the myoepithelial cells (themselves induced by oxytocin from the pituitary) that surround the milk-producing cells (luminal epithelial cells) and that squeeze the milk through the mammary tree. The mammary gland remodels back to a resting virgin-like state during involution: many epithelial cells die via apoptosis, the alveoli collapse and lose their ability to produce milk. Thus coordinated action of hormones, metalloproteinases, and apoptotic molecules and other gene products [22–24] results in manifestation of organ-specific functions.

2.3 EXTRACELLULAR MATRIX

Virtually present in all tissues and particularly dominant in connective tissues, ECM falls in two main categories: interstitial/stromal ECM and BM [25]. The interstitial ECM exists as an intercellular substance between and around cells of the connective tissue, whereas BM is present at the basolateral surface of many epithelial cell types in a number of tissues [2]. BM resembles a tough, thin, sheet-like substratum [26] to which cells adhere, and provides both structural support and contextual information. In particular it is an important regulator of cell polarity, differentiation, growth, apoptosis, and gene expression [27]. Mesenchymal cell types such as fibroblasts and smooth muscle cells actively produce interstitial ECM while the BM in glandular epithelial organs is produced via cooperation of different cell types [28].

2.3.1 Composition of ECM

Chemically, ECM is a hydrated network of molecules composed of proteins and glycoproteins, proteoglycans, and soluble and insoluble signaling molecules. BM is composed mainly of laminins

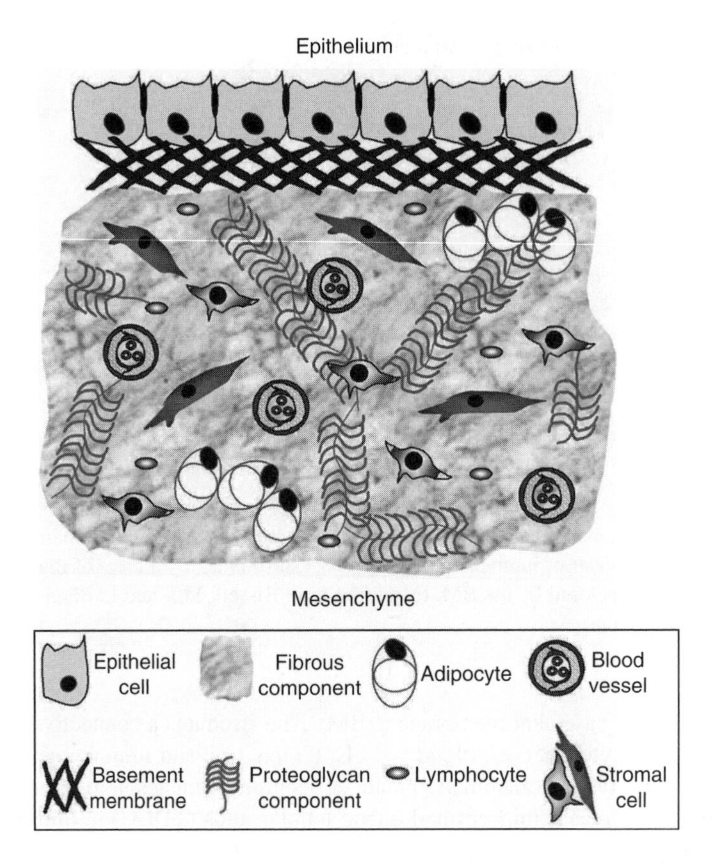

FIGURE 2.2 Schematic view of epithelium and mesenchyme including their structural elements (not drawn to actual scales).

(Lns), type IV collagen, entactin/nidogen, and proteoglycans such as heparin sulfate [2]. Schematically, ECM resembles a fiber-reinforced 3D-composite in which proteins such as collagen and elastin arrange into fibrous structures while proteoglycans, glycoproteins, as well as soluble and insoluble signaling molecules form the interfibrous milieu (Figure 2.2).

2.3.2 Structural Elements of ECM

2.3.2.1 Nonfibrous Components

Protein bound carbohydrates and some types of collagen (e.g., type IV, VI, VIII, XVIII, and XV), which are present in interstitial tissue and BMs are nonfibrous components of ECM. The carbohydrates are composed of long-chained and negatively charged polysaccharides known as glycosaminoglycans (GAGs). GAGs are highly charged (and often sulfated) polysaccharide chains made of repeating unbranched disaccharide units (one of which is an amino sugar). These are classified into four main categories:

1. Hyaluronic acid [hyaluronan (HA), the only nonsulfated GAG that is also usually not covalently attached to protein]
2. Chondroitin sulfate and dermatan sulfate
3. Heparan sulfate and heparin
4. Keratin sulfate (which is composed of repeating disaccharides of *n*-acetylglucosamine and galactose instead of a uronic acid, and can be sulphonated and be linked to proteins)

With the possible exception of HA, GAGs are covalently attached to a protein backbone to form proteoglycans with a "bottle brush" structure. Proteoglycans are diverse due to their different core proteins and different glycosaminoglycans: they are classified by the predominant glycosaminoglycan modifying the protein core. For example, syndecans are examples of heparan sulfate proteoglycans. In addition to their ECM structural roles, most proteoglycans also modulate cell growth and differentiation by regulating cell signaling. Many but not all of these are integral membrane proteins. For instance, the core protein of syndecans has membrane spanning sequence and a short cytosolic tail. The large external domain is modified by a small number of heparin sulfate chains. Syndecans bind to ECM collagen, fibronectin, and thrombospondin and like many other proteoglycans modulate the activity of growth factors [29].

Hyaluronan is a nonsulfated, linear GAG composed of repeating disaccharide units of β-D-glucuronic acid and *N*-acetyl-D-glucosamine. Physiologically it is most often a mixture of molecular weights ranging from oligosaccharides to high molecular forms (e.g., >1000 kDa) that result from a balance between synthesis by one or more of three distinct HA synthases (HAS) [30–32]. Unlike other GAG synthases, HAS1–3 are located at the plasma membrane. HA is a highly expressed component of ECM, particularly interstitial ECM. Interestingly, it is perhaps the most extensively used GAG in tissue engineering [33] but is the least understood GAG in terms of the mechanisms by which it affects tissue function. This is likely due to an initial research focus on its remarkable viscoelastic and water binding/hydrating properties. For example, it forms a viscous hydrated gel, which gives connective tissue turgor pressure (i.e., pressure of the cell contents against the cell wall) and an ability to resist compression at the same time as providing resilience. The viscous properties of HA also contribute to its unique lubricating functions which are particularly important in joint cartilages. Within tissues HA forms a pericellular coat particularly around migrating cells but also sedentary cells such as chondrocytes [34]. It is essential to embryonic development since genetic deletion of HAS2 results in embryonic lethality, a defect that is due at least in part to a defect in migration and epithelial-to-mesenchymal transformation (EMT) during heart development. It also plays multifunctional roles in a number of physiological and pathological events particularly chronic inflammation, vascular disease, and tumor progression [34–38]. HA–cell interactions are mediated by cell surface receptors, which include CD44, RHAMM/HMMR [39], LYVE-1, a lymphatic HA receptor, layilin, a C-type lectin [40], and Toll-like receptor 2,4 which bind to HA oligosaccharides [41]. HA-mediated cell signaling is best characterized for CD44 and RHAMM/HMMR [34,37,42].

2.3.2.2 Fibrous Components

Fibrous components of the ECM perform scaffolding functions, provide stiffness control and tension resistance, but also bind and sequester adhesion factors (e.g., fibronectin) and some growth factors (e.g., bone morphogenic protein-2) [43]. The primary structural fibrous proteins include collagen, fibrin, and elastin. These proteins are the basis of ECM in all multicellular organisms. Collagen, a family of at least 19 genetically and functionally distinct polypeptides [44], will be discussed in more detail here.

The superfamily of collagen proteins is characterized by a long stretch of triple helical domains that can only occur by having every third amino acid be a glycine and every fifth a proline (for reviews on collagen read [44,45]). Hydroxylation of about 50% of the proline residues provides stability to the helix conformation. Collagens are divided into a number of types based on differences in their polymeric structure [46]. For instance, collagen, type I, is a triple helical linear protein consisting of 2 $\alpha 1$ and one $\alpha 2$ chains 300 nm long; 0.5 nm wide; 67 nm periodicity [47], and collagen type IV is an important constituent of basement membranes. Types I, II and III, V, XI, XXIV, XXVII are all fibrous collagens [43]. Collagens are synthesized on membrane-bound polysomes as soluble procollagen precursor forms, which are proteolytically processed after secretion and self-assemble into fibrils. Cross linking occurs over time as collagens mature into insoluble fibrous proteins of the extracellular space. Unlike collagen, elastins can be secreted and mainly reside in pliable organs such as skin and arteries. ECM elastic fibers can be anchored to cell membranes by adaptor proteins such as Fibulin-2 [48].

2.3.2.3 Multidomain Proteins

Multidomain proteins, which include fibronectin, laminin, vitronectin, and entactin are responsible for the adhesion and motility of cells within the ECM and are used in artificial scaffolds for the same purpose. We briefly discuss two major multidomain ECM proteins: fibronectin (Fn) and laminin, which perform different extracellular functions.

Fibronectin is a highly expressed, high-molecular-weight glycoprotein ($\sim$440 kDa) possessing multiple binding sites for a variety of ECM components. It is synthesized as a soluble globular protein and unfolds upon secretion [49,50]. Fn is a dimer formed from two monomers linked at the C-terminal end by a disulfide bond, two to three nanometers thick and 60–70 nm long. Each monomer consists of a series of more than 25 tightly folded globular modules linearly connected by short interdomain chains of variable flexibility [50–52]. Each monomer consists of many domains (designated as Fn1, Fn2, and Fn3) [53] that exhibit distinct binding repertoires for other ECM ligands including collagens, fibrin, heparins, and chondroitin sulfate. Fn also contains binding regions for DNA and for a number of cell-surface receptors. The first identified cell receptor-binding domain of Fn was a 12 kDa fragment sequence, Arg–Gly–Asp (RGD), which exhibits binding preference for members of integrin receptors such as $\alpha_5\beta_1$ [54–57]. Fn is highly flexible and its shape can be markedly affected by deposition onto surfaces [51–58]. These properties and adhesion functions of fibronectin play key roles in blood clotting and control of adhesion and migration.

Laminins are designated as $\alpha_x\beta_y\gamma_z$ heterodimer glycoproteins, and are major adhesion and structural proteins of the basement membrane that are required for its assembly. To date, at least 15 laminins have been described that are derived from five α, three β, and three γ subunits. They are large ($\sim$900 kDa) glycoproteins composed of three distinct polypeptides in cross or T shaped configuration with three short arms (34 ± 4, $\sim 34 \pm 4$, 48 ± 4 nm) and one long arm (77 nm) [59,60]. The short arms are composed of parts of one chain, whereas the long arm is formed by parts of each of the three chains. All individual chains are held together by disulfide bonds. Like fibronectin, laminin is composed of a series of distinct domains with ECM and cell adhesion properties [61,62]: laminin binds to type IV collagen, cell surface integrins, dystroglycan receptors [63], syndecans, and sulfatides [62]. Specifically, LG1–3 and LG4 regions bind to $\beta_1$integrin and dystroglycan, respectively [64]. Laminin and its association with these cellular receptors are required for developmental processes, cellular organization, and cell differentiation [62–65]. For example, the interaction of Laminin 111 (i.e., $\alpha_1\beta_1\gamma_1$) with integrin and dystroglycan receptors leads to changes in cell shape and activation of signaling cascades, which regulate expression of milk proteins in the mammary gland [12,64,66,67] where it is produced essentially by myoepithelial cells [68]. It is this function of the myoepithelial cells that allow them to be an organizing principle for epithelial polarity and function [69].

2.4 FUNCTIONS OF EXTRACELLULAR MATRIX

ECM plays critical roles that affect all aspects of tissue and organism homeostasis. These include cell proliferation, differentiation, motility, and death. ECM affects these processes by providing structural support and cues or signals to cells. Both of these functions are essential for maintaining tissue architecture. It is now well accepted that ECM influences the form and fate of cells via its intricate structure linked to intracellular signaling cascades [4,67,70,71]. The binding of these receptors to ECM molecules induces a classic outside-in signal that activates signaling cascades controlling the organization of the cytoplasmic cytoskeleton and the nuclear skeleton to alter cellular functions such as gene expression [17]. Intriguingly, the status of the cell can also influence this signaling property of ECM; for example, modification of integrin receptors by intracellular events can change their affinity for ECM ligands, which in turn further modifies intracellular signaling. This dynamic interaction permits incredible fine-tuning of cellular activity in response to even small changes in ECM composition. ECM also provides the adhesion support for basement membranes in vivo and substrata in culture: this interaction also exerts control over cellular shape [12]. Thus in summary, although some ECM complexes are designed for a particular function such as strength (tendon), filtration (the BM in the kidney glomerulus), or adhesion (BM supporting most epithelia), they are more generally involved in:

1. Establishment of tissue micro/nanoenvironment
2. Mechanical and structural support
3. Determination of cell shape and cell polarity
4. Sequestration, storage, and presentation of soluble regulatory molecules
5. Outside-in biochemical signaling
6. Regulation of cell function including: proliferation, growth, survival, migration, and differentiation

ECM continuously remodels (even within a single tissue) throughout normal processes such as development and aging as well as during wound repair and disease progression. The remodeling process occurs as a result of exposure to various stimuli such as growth factors, cytokines, potent functional proteins, and hormones [25,72,73]. The extent of ECM remodeling, however, is quite different in normal processes compared to pathological states such as cancer. Whereas the former is regulated, the latter is unbalanced. For example, extensive fragmentation of ECM molecules occurs in tumor stroma because of matrix metalloprotineases (MMPs), other proteases and hyaluronidases. These fragments often possess biological activity not found in the intact ECM component that for example promotes migration and proliferation [34,38,74,75]. In contrast, processes of directed cell motility that occur during mammary gland morphogenesis involves highly localized and tightly regulated proteolytic events associated with only transient decreases in tight junctions between neighboring cells that permit the breaching of limited numbers of cells through the BM [74,76,77]. Nevertheless, aspects of ECM remodeling in wound repair are similar to that which occurs during tumor progression, and perhaps not surprisingly tumor cells grow more rapidly in a wound environment than in homeostatic tissue. One approach that has met with some success in decoding the mechanisms by which ECM remodeling exerts such versatile functions and the mechanisms in particular that are common to wound and tumor is to culture cells in biomimetic environments to recapitulate in vivo cues [67,78,79].

2.4.1 Cell–Extracellular Matrix Interactions

Cell–ECM interaction involves complex yet specific processes requiring initial recognition and resulting in adhesion, signaling, cytoskeletal reorganization, and gene expression. The ECM ligands, which cells generally interact with are immobilized and not in solution but historically the focus has been on the role of soluble ligands such as hormones and growth factors. For example, ECM binds and immobilizes a number of soluble growth factors and effectors such as transforming growth factor-β (TGFβ), vascular-endothelial-cell growth factor (VEGF), and hepatocyte growth factor (HGF) [80,81]. Since ECM also promotes cell adhesions, which represent checkpoints for regulation of cell shape and gene expression through their association with the cytoskeleton and the cellular signaling pathways, in effect it links growth factor and other effectors to signaling in a cell architecture-dependent manner. This coupling integrates soluble and insoluble factors to allow a reciprocal flow of mechanical and biochemical signals between the cell and its microenvironment (Figure 2.3) [17].

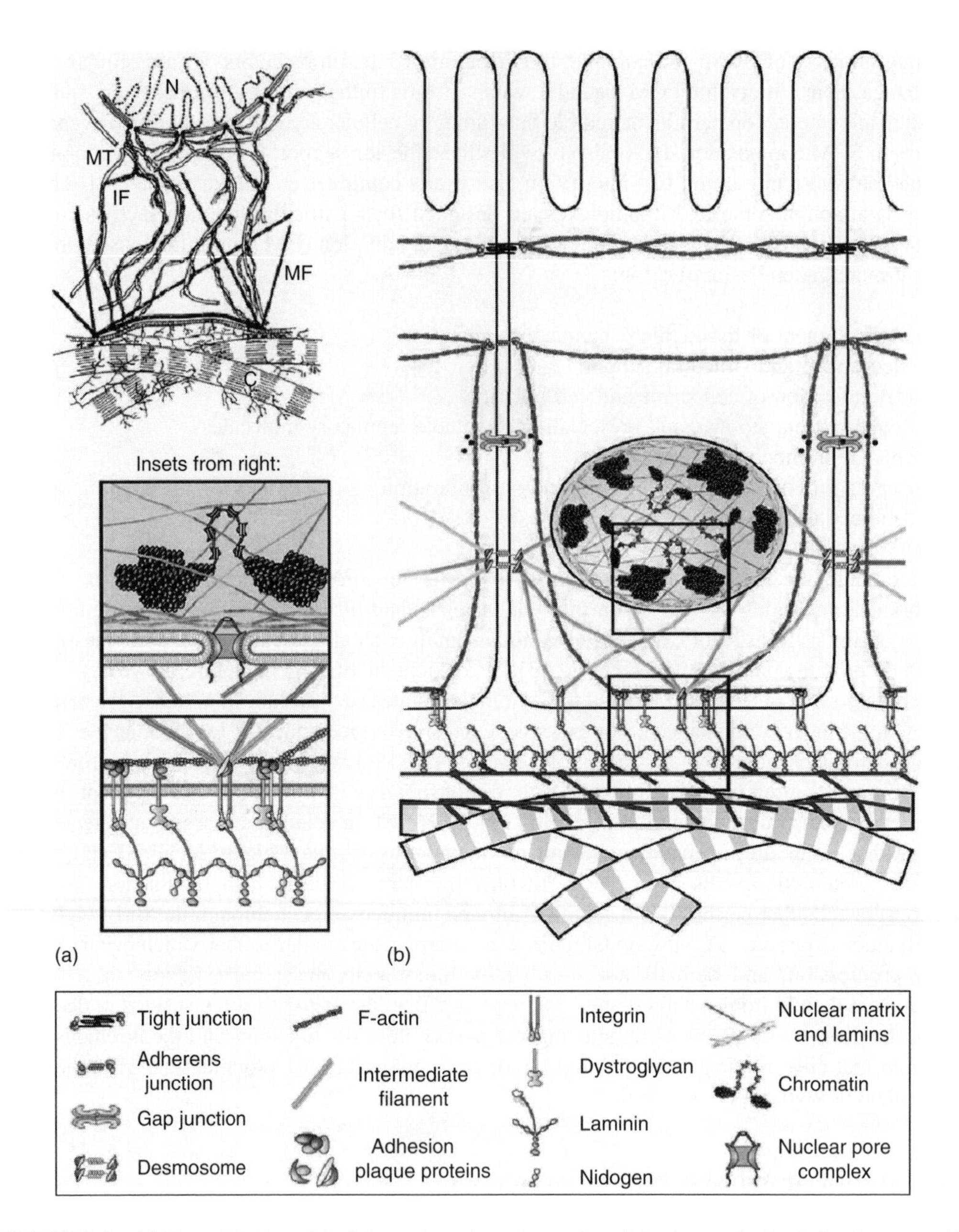

FIGURE 2.3 (a) The original model of dynamic reciprocity, or the minimum required unit for tissue-specific functions. N, nucleus; MT, microtubules; IF, intermediate filaments; MF, microfilaments; C, collagen. (b) A more complete view of dynamic reciprocity. (Modified from Nelson, C.M. and Bissell, M.J., *Ann. Rev. Cell Develop. Biol.*, 22, 287, 2006. With permission.)

Reciprocal flows have been demonstrated between epithelium and mesenchyme of several organs such as lung and kidney and the prostate, salivary, and mammary glands (reviewed in [82–86]). Such reciprocal information flow has also been demonstrated to occur between heterogeneous cell and tissue types, for example epithelium and mesenchyme of several organs, such as lung, kidney, prostate, and salivary and mammary glands (reviewed in [82–86]) and they appear to be required for establishing and sustaining the branching architecture of these organs; the

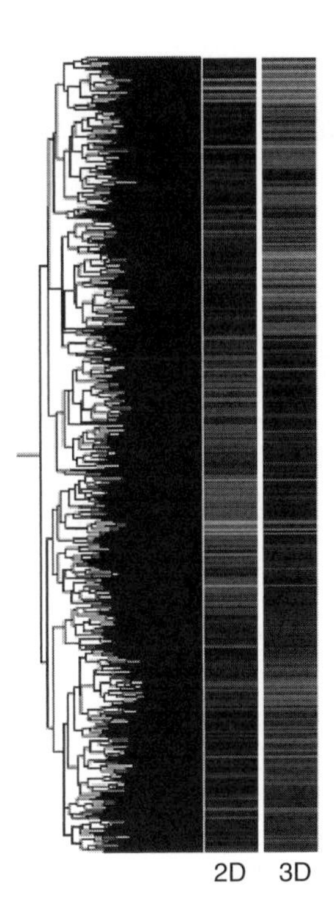

FIGURE 2.4 Gene expression analysis of nonmalignant human mammary epithelial cells (S1) cultured on plastic (2D) or lrECM (3D). 1873 Affymetrix probes detected transcripts that were differentially expressed (cutoff $\geq$ 1.5-fold). (From Kenny, P., PhD Thesis. With permission.)

mechanisms by which this happens are not completely understood but several molecular principles are emerging.

Cell–ECM interaction could either promote or inhibit differentiation process. For example, Fn and Ln appear to have opposing roles at least in the mammary gland. While, Ln has been required for acinus formation and normal function, Fn appeared to be instrumental in mammary malignancy [87]. Another functional influence of ECM on cells in particular is the altered patterns of gene expression through a series of feedback inhibitory mechanisms and selective processing [4]. It has been well established that ECM connects to nuclei or chromatin physically and biochemically [88,89] and that this results in a profound influence on transcription factor activation and chromatin remodeling. Figure 2.4 shows an example of such an ECM influence on gene expression profiles of cultured cells. As shown in the figure, cells grown on 3D lrECM express distinct patterns of genes compared to their 2D cultured counterparts grown on rigid plastic substrates. The data indicate that the microenvironmental context can have profound effects on the patterns of gene expression.

An additional effect of ECM is on cell morphology. Cells assume a distinct morphology when grown in 3D lrECM. Epithelial cells embedded within a malleable lrECM in the presence of lactogenic cues (such as hormones) form polarized 3D acinar structures that recapitulate in vivo alveolar phenotype (Figure 2.5) [90–93]; after 4 days, the central cells undergo apoptosis to create a central hollow lumina into which the milk proteins can be secreted [64,93–96]. However, this does not occur when same epithelial cells are grown on 2D plastic substrates; they spread as monolayer and cannot produce milk even in the presence of lactogenic hormones in the medium.

ECM plays crucial roles in growth factor and adhesion pathways. While, the growth factor and adhesion pathways talk to each other in 3D, i.e., that inhibition of one signaling pathway leads to down-modulation of the other [97,98], they act independently in 2D (Table 2.1).

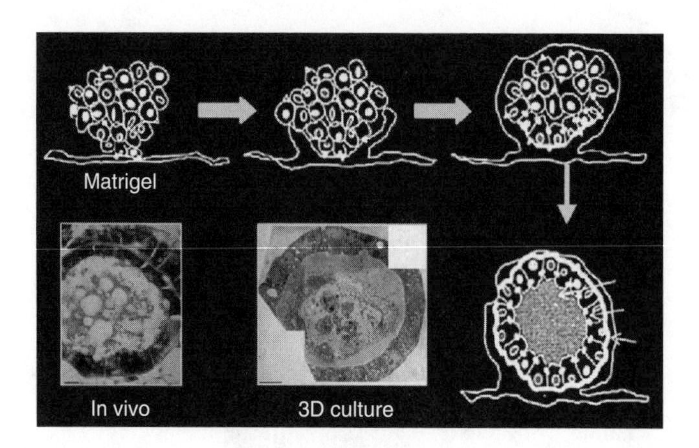

FIGURE 2.5 Formation of model mouse acini in cell culture on a laminin-rich ECM. Similarities can be seen between acini in vivo and in three-dimensional cell culture. (Modified from Streuli, C.H. and Bissell, M.J., *Regulatory Mechanisms in Breast Cancer*, Kluwer Academic Publishers, Boston, 1991; Barcellos-Hoff, M.H. and Bissell, M.J., *Autocrine and Paracrine Mechanisms in Reproductive Endocrinology*, Plenum Press, New York, 1989. With permission.)

When 3D T4 – 2 cells are treated with functional inhibitors of either β1 integrin or EGFR (T4 – 2 treated) to revert the cells, endogenous β1 and EGFR protein levels down-modulate coordinately. Finally, as a multivalent connecting environment, ECM can influence the clustering of growth factors and can modify the signals [99,100].

2.5 CELL-BIOMATERIALS INTERACTION: THE CENTRAL PLAYERS

2.5.1 ADHESION

Cell adhesion to ECM, other cells, and biomaterials [7] (e.g., hydrogel scaffolds [101–103]) is a complex and regulated process. The process of cell adhesion and spreading on/or within solid biomaterials cannot be studied without considering the rapid rate of protein adsorption to biomaterial surfaces [104–106]. In general cells seeded from a serum containing media more likely interact with an adsorbed protein layer than with the actual surface of the biomaterial. If the surface physicochemical properties of the scaffolds, the proteins and other deposited moieties that form the artificial ECM are appropriate, the cells adhere and function properly. If not, they remain

TABLE 2.1
Cross Modulation of β1 and EGFR Protein

	3D				2D		
	S1	T4-2	T4-2		T4-2	T4-2 "treated"	
Inhibitor added	–	–	β1-integrin inhibitor	EGFR inhibitor	–	β1-integrin inhibitor	EGFR inhibitor
β1 integrin (total levels)	+	+++	+	+	+++	+++	+++
EGFR (total levels)	+	++++	+	+	++++	++++	++++
EGFR (activated)	+	++++	+	+	++++	++++	+

Source: Modified from Bissell M.J., et al., *Cancer Res.*, 59, 1757s, 1999. With permission.

essentially spherical, exhibiting little or no adhesion to the surface. This behavior has been demonstrated by selective adhesion of specific ECM proteins onto regions of the biomaterial surfaces via several techniques [107–110]. For instance when macrophage cells were exposed to surfaces that were heterogeneously modified with adhesive ECM molecules against a nonfouling (biologically inert) poly(ethylene glycol) (PEG) background, they adhered selectively over ECM proteins and functioned properly for extended time [111].

As a cell comes into contact with a solid adhesive surface, adhesion occurs through specific adhesion plaques or sites known as focal adhesions [112–114]. The formation of these adhesion sites is strongly influenced by the types of proteins present on the surface that cells are attaching to. As the cell attaches, a specific sequence of events occur that depend upon the nature of cell surface receptor/ECM interaction. For example, integrin receptors (with a projected area of approximately 25 nm^2 for the nonclustered heterodimer) bind first to ECM proteins coated onto the surfaces [115]. Next, a group of intracellular cytoskeleton adaptor proteins (talin, vinculin, paxillin, α-actinin) with dimensions in the nanometer range accumulate within the forming adhesion plaque that connect integrins to actin cytoskeleton. This results in a functional linkage of the cellular cytoskeleton with the solid surfaces (Figure 2.6).

The cell cytoskeleton may link with the ECM through a single or multivalent ECM/integrin interaction. The latter is more stable and is typically found in adhesion plaques. The adhesive plaques at cell/cell, cell/ECM, and cell/solid biomaterial surfaces are all nanostructured. Although not all types of cell adhesions are found between cells and biomaterials, the following have been demonstrated at a cell/solid biomaterial interface [113,116]:

1. Focal adhesions: 10–15 nm spacing between lipid membrane and solid biomaterial
2. Close contacts (often surround focal adhesions): 30–50 nm spacing between cell surface and solid biomaterial
3. ECM contacts: 50–100 nm spacing between ventral cell wall and underlying solid biomaterial with sporadic connection of filamentous ECM components

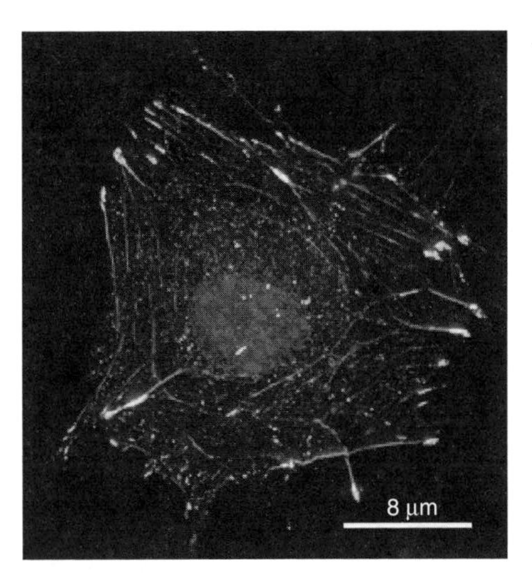

FIGURE 2.6 (See color insert following page 206.) Fluorescence image of human umbilical cord vein endothelial cells (HUVECs) adhered on a two-dimensional (2D) glass surface in the presence of ECM proteins and proteoglycans. Actin filaments (red) and integrins are connected through vinculin proteins (green). Blue: nucleus-DAPI, green: antivinculin–FITC, red: Alexa Fluor 594 phalloidin-F-actin. (From Veiseh, M. Unpublished observation. With permission.)

The nanometer sizes of the integrins, the focal adhesion associated proteins, and the resultant cluster structures involved in the adhesion process, together, indicate the necessity for incorporation and analysis of adhesive structures with nanoscale (or molecular) resolutions in tissue engineering designs.

2.5.2 Ultra-Signaling to Ultra-Patterns

Beyond a response to multivalent chemical interactions, cells also respond to specific spatial patterns and structural periodicity of ECM at both micron and nanoscale level. Both the spatial distances of repeating structures and their organization into higher orders depend upon the composition of the ECM [4,117]. Several lines of evidence indicate that cells sense and signal in response to repeating nanostructures and geometries. Vogel et al. [118] have defined "geometry sensing as the formation of signaling complexes by changes in the spacing of molecular recognition sites" and Yim et al. have shown that neuronal differentiation of human MSC was strongly influenced by nanotopography [119]. When Semino et al. weaved nanofibers of self-assembling peptides into a mesh with just the right structure for slow release of nutrients and essential biological molecules to embed liver stem cells, they noticed that the stem cells both self-renewed and differentiated into mature liver cells [120]. The molecular mechanisms behind these observations have not yet been identified but their identity will certainly illuminate concepts of how nanoenvironment and nanorange patterns provoke cell signaling.

2.6 CONCLUSIONS

Maintenance of tissue architecture and function requires an entrainment of cell survival, growth, and differentiation signaling cascades to cell–extracellular matrix cues. Cells respond profoundly to the hierarchical patterns and physicochemical properties of their surrounding environment. While much has been uncovered and understood about structure and functions of ECM in micron-scale, we believe a new road map for signaling needs to be defined within the actual physiological scale, i.e., within the nanometric scale.

2.7 FUTURE DIRECTION

The technological tools for visualization of biological structures have reached a level that can allow for fine tuning the analysis of the ECM composition, its nanostructures, and even possibly the nanoscale dynamic signaling. Insights from cell biology will aid in synthesis of native ECM and understanding the remodeling events. Fine characterization of biommimetic ECMs in nanoscale and their real time interactions (i.e., dynamic imaging of individual cells and ECM components within 3D environments) are imperative.

ACKNOWLEDGMENTS

We thank Dr. Ren Xu and Jamie L. Inman for critical reading and helpful suggestions on the manuscript; Dr. Richard Schwarz and Dr. Paraic A. Kenny for collagen section and Figure 2.4, respectively. These investigations were supported by grants and a distinguished fellowship award from the U.S. Department of Energy, Office of Biological and Environmental Research (DE-AC03 SF0098), by the Department of Defense, Breast Cancer Research Program (W81XWH0510338) to M.J. Bissell, by an innovator award from the Department of Defense Breast Cancer Research Program (DAMD170210438) to M.J. Bissell and by the National Cancer Institute (CA64736 to M.J. Bissell, CA57621 to Zenawerb and M.J. Bissell). E.A. Turley and M.J. Bissell are also the recipients of a Department of Defense Breast Cancer research program grant (BC044087). E.A. Turley is supported by the Breast Cancer Society of Canada.

REFERENCES

1. Badylak, S.E. 2002. The extracellular matrix as a scaffold for tissue reconstruction. *Seminars in Cell and Developmental Biology* 13(5): 377–383. < Go to ISI > ://000178731800008.
2. Kalluri, R. 2003. Basement Membranes: Structure, assembly and role in tumour angiogenesis. *Nature Reviews Cancer* 3(6): 422–433. < Go to ISI > ://000183424100015.
3. Gilbert, S.F. 2006. *Developmental Biology*. Eigth edition, Chapter 6. Sinauer Associates, Inc., USA.
4. Bissell, M.J., H.G. Hall, and G. Parry. 1982. How does the extracellular matrix direct gene expression? *Journal of Theortical Biology* 99: 31–68.
5. Langer, R. and J. Vacanti. 1993. Tissue engineering. *Science* 260: 920–926.
6. Murugan, R. and S. Ramakrishna. 2007. Design strategies of tissue engineering scaffolds with controlled fiber orientation. *Tissue Engineering* 13(8): 1845–1866. < Go to ISI >://000248742200006.
7. Ratner, B.D. and S.J. Bryant. 2004. Biomaterials: Where we have been and where we are going. *Annual Review of Biomedical Engineering* 6: 41–75. Abstract available at http://www.ncbi.nlm.nih.gov/entrez/query.fcgi?cmd = Retrieve&db = PubMed&dopt = Citation&list_uids = 15255762.
8. Bissell, D.M., L.E. Hammaker, and U.A. Meyer. 1973. Parenchymal cells from adult rat-liver in nonproliferating monolayer culture.1. Functional studies. *Journal of Cell Biology* 59(3): 722–734. < Go to ISI > ://A1973R469000016.
9. Bissell, M.J. 1981. The differentiated state of normal and malignant cells or how to define a "normal" cell in culture. *International Review of Cytology* 70: 27–100. Abstract available at http://www.ncbi.nlm.nih.gov/entrez/query.fcgi?cmd = Retrieve&db = PubMed&dopt = Citation&list_uids = 7228573.
10. Bissell, M.J., A. Rizki, and I.S. Mian. 2003. Tissue architecture: The ultimate regulator of breast epithelial function—commentary. *Current Opinion in Cell Biology* 15(6): 753–762. < Go to ISI > ://000187109200016.
11. Pampaloni, F., E.G. Reynaud, and E.H. Stelzer. 2007. The third dimension bridges the gap between cell culture and live tissue. *Nature Reviews. Molecular Cell Biology*. Abstract available at http://www.ncbi.nlm.nih.gov/entrez/query.fcgi?cmd = Retrieve&db = PubMed&dopt = Citation&list_uids = 17684528.
12. Roskelley, C.D., P.Y. Desprez, and M.J. Bissell. 1994. Extracellular matrix-dependent tissue-specific gene-expression in mammary epithelial-cells requires both physical and biochemical signal-transduction. *Proceedings of the National Academy of Sciences of the United States of America* 91(26): 12378–12382. < Go to ISI > ://A1994PY29400006.
13. Bissell, M.J., D.C. Radisky, A. Rizki, V.M. Weaver, and O.W. Petersen. 2002. The organizing principle: Microenvironmental influences in the normal and malignant breast. *Differentiation* 70(9–10): 537–546. Abstract available at http://www.ncbi.nlm.nih.gov/entrez/query.fcgi?cmd = Retrieve&db = PubMed&dopt = Citation&list_uids = 12492495.
14. Nelson, C.M. and M.J. Bissell. 2006. Of extracellular matrix, scaffolds, and signaling: Tissue architecture regulates development, homeostasis, and cancer. *Annual Review of Cell and Developmental Biology* 22: 287–309. < Go to ISI > ://000242325100013.
15. Schmeichel, K.L. and M.J. Bissell. 2003. Modeling tissue-specific signaling and organ function in three dimensions. *Journal of Cell Science* 116(12): 2377–2388. Abstract available at http://www.ncbi.nlm.nih.gov/entrez/query.fcgi?cmd = Retrieve&db = PubMed&dopt = Citation&list_uids = 12766184.
16. Gordon, A.T., G.E. Lutz, M.L. Boninger, and R.A. Cooper. 2007. Introduction to nanotechnology: Potential applications in physical medicine and rehabilitation. *American Journal of Physical Medicine & Rehabilitation* 86(3): 225–241. Abstract available at http://www.ncbi.nlm.nih.gov/entrez/query.fcgi?cmd = Retrieve&db = PubMed&dopt = Citation&list_uids = 17314708.
17. Hagios, C., A. Lochter, and M.J. Bissell. 1998. Tissue architecture: The ultimate regulator of epithelial function? *Philosophical Transactions of the Royal Society of London B* 353: 857–870.
18. Bissell, M.J. and M.A. LaBarge. 2005. Context, tissue plasticity, and cancer: Are tumor stem cells also regulated by the microenvironment? *Cancer Cell* 7(1): 17–23. < Go to ISI > ://000226533000004.
19. Drife, J.O. 1986. Breast development in puberty. *Annals of the New York Academy of Sciences* 464: 58–65. < Go to ISI > ://A1986E223100006.
20. Ronnov-Jessen, L., O.W. Petersen, and M.J. Bissell. 1996. Cellular changes involved in conversion of normal to malignant breast: Importance of the stromal reaction. *Physiological Review* 76(1): 69–125. Abstract available at http://www.ncbi.nlm.nih.gov/entrez/query.fcgi?cmd = Retrieve&db = PubMed&dopt = Citation&list_uids = 8592733.

21. Sakakura, T., Y. Nishizuka, and C.J. Dawe. 1976. Mesenchyme-dependent morphogenesis and epithelium-specific cytodifferentiation in mouse mammary-gland. *Science* 194(4272): 1439–1441. <Go to ISI>://A1976CP27400028.
22. Neville, M.C., T.B. McFadden, and I. Forsyth. 2002. Hormonal regulation of mammary differentiation and milk secretion. *Journal of Mammary Gland Biology and Neoplasia* 7(1): 49–66. <Go to ISI>://000175924200005.
23. Talhouk, R.S., M.J. Bissell, and Z. Werb. 1992. Coordinated expression of extracellular matrix-degrading proteinases and their inhibitors regulates mammary epithelial function during involution. *Journal of Cell Biology* 118(5): 1271–1282. <Go to ISI>://A1992JL55000023.
24. Talhouk, R.S., J.R. Chin, E.N. Unemori, Z. Werb, and M.J. Bissell. 1991. Proteinases of the mammary-gland-developmental regulation in vivo and vectorial secretion in culture. *Development* 112(2): 439–449. <Go to ISI>://A1991FR42700008.
25. Guo, W.J. and F.G. Giancotti. 2004. Integrin signalling during tumour progression. *Nature Reviews Molecular Cell Biology* 5(10): 816–826. <Go to ISI>://000224558200015.
26. Schoen, F.J. and R.N. Mitchell. 2004. Tissues, the extracellular matrix, and cell-biomaterial interactions. In *Biomaterials Sciences: An Introduction to Materials in Medicine*, B.D. Ratner, A.S. Hoffman, F.J. Schoen and J.E. Lemons (Eds.), pp. 260–272: Elsevier Academic Publishers.
27. Adams, J.C. and F.M. Watt. 1993. Regulation of development and differentiation by the extracellular-matrix. *Development* 117(4): 1183–1198. <Go to ISI>://A1993LD08900001.
28. Kedinger, M., P.M. Simon, J.F. Grenier, and K. Haffen. 1981. Role of epithelial–mesenchymal interactions in the ontogenesis of intestinal brush-border enzymes. *Developmental Biology* 86(2): 339–347. Abstract available at http://www.ncbi.nlm.nih.gov/entrez/query.fcgi?cmd=Retrieve&db=PubMed&dopt=Citation&list_uids=6793427.
29. Ratner, B.D., A.S. Hoffman, F.J. Schoen, and J.E. Lemons (Eds.). 2004. *Biomaterials Sciences: An Introduction to Materials in Medicine*, Second edition, Chapter 3. Elsevier Academic Publishers, USA.
30. McDonald, J.A. and T.D. Camenisch. 2002. Hyaluronan: Genetic insights into the complex biology of a simple polysaccharide. *Glycoconjugate Journal* 19(4–5): 331–339. Abstract available at http://www.ncbi.nlm.nih.gov/entrez/query.fcgi?cmd=Retrieve&db=PubMed&dopt=Citation&list_uids=12975613.
31. Stern, R. and M.J. Jedrzejas. 2006. Hyaluronidases: Their genomics, structures, and mechanisms of action. *Chemical Reviews* 106(3): 818–839. Abstract available at http://www.ncbi.nlm.nih.gov/entrez/query.fcgi?cmd=Retrieve&db=PubMed&dopt=Citation&list_uids=16522010.
32. Weigel, P.H. 2002. Functional characteristics and catalytic mechanisms of the bacterial hyaluronan synthases. *IUBMB Life* 54(4): 201–211. Abstract available at http://www.ncbi.nlm.nih.gov/entrez/query.fcgi?cmd=Retrieve&db=PubMed&dopt=Citation&list_uids=12512859.
33. Allison, D.D. and K.J. Grande-Allen. 2006. Review. Hyaluronan: A powerful tissue engineering tool. *Tissue Engineering* 12(8): 2131–2140. Abstract available at http://www.ncbi.nlm.nih.gov/entrez/query.fcgi?cmd=Retrieve&db=PubMed&dopt=Citation&list_uids=16968154.
34. Toole, B.P. 2004. Hyaluronan: From extracellular glue to pericellular Cue. *Nature Reviews Cancer* 4(7): 528–539. <Go to ISI>://000222435500013.
35. Adamia, S., C.A. Maxwell, and L.M. Pilarski. 2005. Hyaluronan and hyaluronan synthases: Potential therapeutic targets in cancer. *Current Drug Targets. Cardiovascular & Haematological Disorders* 5(1): 3–14. Abstract available at http://www.ncbi.nlm.nih.gov/entrez/query.fcgi?cmd=Retrieve&db=PubMed&dopt=Citation&list_uids=15720220.
36. Bartolazzi, A., R. Peach, A. Aruffo, and I. Stamenkovic. 1994. Interaction between Cd44 and hyaluronate is directly implicated in the regulation of tumor development. *The Journal of Experimental Medicine* 180(1): 53–66. Abstract available at http://www.ncbi.nlm.nih.gov/entrez/query.fcgi?cmd=Retrieve&db=PubMed&dopt=Citation&list_uids=7516417.
37. Evanko, S.P., M.I. Tammi, R.H. Tammi, and T.N. Wight. 2007. Hyaluronan-dependent pericellular matrix. *Advanced Drug Delivery Reviews* 59(13): 1351–1365. Abstract available at http://www.ncbi.nlm.nih.gov/entrez/query.fcgi?cmd=Retrieve&db=PubMed&dopt=Citation&list_uids=17804111.
38. Tammi, M.I., A.J. Day, and E.A. Turley. 2002. Hyaluronan and homeostasis: A balancing act. *The Journal of Biological Chemistry* 277(7): 4581–4584. Abstract available at http://www.ncbi.nlm.nih.gov/entrez/query.fcgi?cmd=Retrieve&db=PubMed&dopt=Citation&list_uids=11717316.

39. Hamilton, S.R., S.F. Fard, F.F. Paiwand, et al. 2007. The hyaluronan receptors Cd44 and Rhamm (Cd168) form complexes with Erk1,2 that sustain high basal motility in breast cancer cells. *The Journal of Biological Chemistry* 282(22): 16667–16680. Abstract available at http://www.ncbi.nlm.nih.gov/entrez/query.fcgi?cmd = Retrieve&db = PubMed&dopt = Citation&list_uids = 17392272.
40. Bono, P., K. Rubin, J.M. Higgins, and R.O. Hynes. 2001. Layilin, a novel integral membrane protein, is a hyaluronan receptor. *Molecular Biology of Cell* 12(4): 891–900. Abstract available at http://www.ncbi.nlm.nih.gov/entrez/query.fcgi?cmd = Retrieve&db = PubMed&dopt = Citation&list_uids = 11294894.
41. Jiang, D., J. Liang, and P.W. Noble. 2006. Hyaluronan in tissue injury and repair. *Annual Review of Cell and Developmental Biology*. Abstract available at http://www.ncbi.nlm.nih.gov/entrez/query.fcgi?cmd = Retrieve&db = PubMed&dopt = Citation&list_uids = 17506690.
42. Turley, E.A., P.W. Noble, and L.Y. Bourguignon. 2002. Signaling properties of hyaluronan receptors. *The Journal of Biological Chemistry* 277(7): 4589–4592. Abstract available at http://www.ncbi.nlm.nih.gov/entrez/query.fcgi?cmd = Retrieve&db = PubMed&dopt = Citation&list_uids = 11717317.
43. Griffith, L.G. and M.A. Swartz. 2006. Capturing complex 3D tissue physiology in vitro. *Nature Reviews Molecular Cell Biology* 7(3): 211–224. < Go to ISI > ://000235590500016.
44. Prockop, D.J. 1995. Collagensm: Molecular biology, diseases, and potentials for therapy. *Arm. Rev Biochem* 64: 403–434.
45. Wess, T.J. 2005. Collagen fibril form and function. *Advances in Protein Chemistry* 70: 341–374. < Go to ISI > ://000229968800010.
46. Hulmes, D.J. 1992. The collagen superfamily—diverse structures and assemblies. *Essays in Biochemistry* 27: 49–67. Abstract available at http://www.ncbi.nlm.nih.gov/entrez/query.fcgi?cmd = Retrieve&db = PubMed&dopt = Citation&list_uids = 1425603.
47. Gutsmann, T., G.E. Fantner, J.H. Kindt, et al. 2004. Force spectroscopy of collagen fibers to investigate their mechanical properties and structural organization. *Biophysical Journal* 86(5): 3186–3193. Abstract available at http://www.ncbi.nlm.nih.gov/entrez/query.fcgi?cmd = Retrieve&db = PubMed&dopt = Citation&list_uids = 15111431.
48. Sasaki, T., K. Mann, H. Wiedemann, et al. 1997. Dimer model for the microfibrillar protein fibulin-2 and identification of the connecting disulfide bridge. *The EMBO Journal* 16(11): 3035–3043. < Go to ISI > ://A1997XE74800007.
49. Erickson, H.P. and N.A. Carrell. 1983. Fibronectin in extended and compact conformations—electron-microscopy and sedimentation analysis. *Journal of Biological Chemistry* 258(23): 4539–4544. < Go to ISI > ://A1983RT93700079.
50. Johnson, K.J., H. Sage, G. Briscoe, and H.P. Erickson. 1999. The compact conformation of fibronectin is determined by intramolecular ionic interactions. *Journal of Biological Chemistry* 274(22): 15473–15479. < Go to ISI > ://000080560100027.
51. Hynes, R.O. (Ed.). 1989. *Fibronectins*. Edited by A. Rich. New York: Springer-Verlag.
52. Williams, E.C., P.A. Janmey, J.D. Ferry, and D.F. Mosher. 1982. Conformational states of fibronectin. *The Journal of Biological Chemistry* 257(24): 14973–14978.
53. Potts, J.R. and I.D. Campbell. 1996. Structure and function of fibronectin modules. *Matrix Biology: Journal of the International Society for Matrix Biology* 15(5): 313–320; discussion 21. Abstract available at http://www.ncbi.nlm.nih.gov/entrez/query.fcgi?cmd = Retrieve&db = PubMed&dopt = Citation&list_uids = 8981327.
54. Bokel, C. and N.H. Brown. 2002. Integrins in development: Moving on, responding to, and sticking to the extracellular matrix. *Developmental Cell* 3(3): 311–321. < Go to ISI > ://000178132800006.
55. Hynes, R.O. 1992. Integrins: Versatility, modulation, and signaling in cell adhesion. *Cell* 69: 11–25.
56. Pasqualini, R., E. Koivunen, and E. Ruoslahti. 1996. Peptides in cell adhesion: Powerful tools for the study of integrin-ligand interactions. *Brazilian Journal of Medical and Biological Research* 29(9): 1151–1158. Abstract available at http://www.ncbi.nlm.nih.gov/entrez/query.fcgi?cmd = Retrieve&db = PubMed&dopt = Citation&list_uids = 9181058.
57. Ruoslahti, E. and M.D. Pierschbacher. 1986. Arg-Gly-Asp: A versatile cell recognition signal. *Cell* 44(4): 517–518. Abstract available at http://www.ncbi.nlm.nih.gov/entrez/query.fcgi?cmd = Retrieve&db = PubMed&dopt = Citation&list_uids = 2418980.
58. Jonsson, U., Ivarsson, B., Lundstrom, I., Berghem, L. 1982. Adsorption behavior of fibronectin on well-characterized silica surfaces. *Journal of Colloid and Interface Science* 90(1): 148–163.

59. Bruch, M., R. Landwehr, and J. Engel. 1989. Dissection of laminin by cathepsin G into its long-arm and short-arm structures and localization of regions involved in calcium dependent stabilization and self-association. *European Journal of Biochemistry* 185(2): 271–279. Abstract available at http://www.ncbi.nlm.nih.gov/entrez/query.fcgi?cmd = Retrieve&db = PubMed&dopt = Citation&list_uids = 2511014.
60. Engel, J., E. Odermatt, A. Engel, et al. 1981. Shapes, domain organizations and flexibility of laminin and fibronectin, two multifunctional proteins of the extracellular matrix. *Journal of Molecular Biology* 150(1): 97–120. Abstract available at http://www.ncbi.nlm.nih.gov/entrez/query.fcgi?cmd = Retrieve&db = PubMed&dopt = Citation&list_uids = 6795355.
61. Aumailley, M., L. Bruckner-Tuderman, W.G. Carter, et al. 2005. A simplified laminin nomenclature. *Matrix Biology* 24(5): 326–332. < Go to ISI > ://000231205300002.
62. Li, S.H., D. Edgar, R. Fassler, W. Wadsworth, and P.D. Yurchenco. 2003. The role of laminin in embryonic cell polarization and tissue organization. *Developmental Cell* 4(5): 613–624. < Go to ISI > ://000185787200005.
63. Weir, M.L., M.L. Oppizzi, M.D. Henry, et al. 2006. Dystroglycan loss disrupts polarity and beta-casein induction in mammary epithelial cells by perturbing laminin anchoring. *Journal of Cell Science* 119(19): 4047–4058. < Go to ISI > ://000240630300015.
64. Streuli, C.H., C. Schmidhauser, N. Bailey, et al. 1995b. Laminin mediates tissue-specific gene-expression in mammary epithelia. *Journal of Cell Biology* 129(3): 591–603. < Go to ISI > ://A1995QW14200004.
65. Beck, K., I. Hunter, and J. Engel. 1990. Structure and function of laminin—anatomy of a multidomain glycoprotein. *Faseb Journal* 4(2): 148–160. < Go to ISI > ://A1990CM41800003.
66. Muschler, J., A. Lochter, C.D. Roskelley, P. Yurchenco, and M.J. Bissell. 1999. Division of labor among the alpha 6 beta 4 integrin, beta 1 integrins, and an e3 laminin receptor to signal morphogenesis and beta-casein expression in mammary epithelial cells. *Molecular Biology of the Cell* 10(9): 2817–2828. < Go to ISI > ://000082598500003.
67. Streuli, C.H., N. Bailey, and M.J. Bissell. 1991. Control of mammary epithelial differentiation—basement-membrane induces tissue-specific gene-expression in the absence of cell–cell interaction and morphological polarity. *Journal of Cell Biology* 115(5): 1383–1395. < Go to ISI > ://A1991GT48800018.
68. Gudjonsson, T., L. Ronnov-Jessen, R. Villadsen, et al. 2002. Normal and tumor-derived myoepithelial cells differ in their ability to interact with luminal breast epithelial cells for polarity and basement membrane deposition. *Journal of Cell Science* 115(1): 39–50. Abstract available at http://www.ncbi.nlm.nih.gov/entrez/query.fcgi?cmd = Retrieve&db = PubMed&dopt = Citation&list_uids = 11801722.
69. Gudjonsson, T., M.C. Adriance, M.D. Sternlicht, O.W. Petersen, and M.J. Bissell. 2005. Myoepithelial cells: Their origin and function in breast morphogenesis and neoplasia. *Journal of Mammary Gland Biology and Neoplasia* 10(3): 261–272. < Go to ISI > ://000238224600007.
70. Muschler, J., D. Levy, R. Boudreau, et al. 2002. A role for dystroglycan in epithelial polarization: Loss of function in breast tumor cells. *Cancer Research* 62(23): 7102–7109. Abstract available at http://www.ncbi.nlm.nih.gov/entrez/query.fcgi?cmd = Retrieve&db = PubMed&dopt = Citation&list_uids = 12460932.
71. Saunders, S., M. Jalkanen, S. O'Farrell, and M. Bernfield. 1989. Molecular cloning of syndecan, an integral membrane proteoglycan. *The Journal of Cell Biology* 108(4): 1547–1556. Abstract available at http://www.ncbi.nlm.nih.gov/entrez/query.fcgi?cmd = Retrieve&db = PubMed&dopt = Citation&list_uids = 2494194.
72. Labat-Robert, J. 2003. Age-dependent remodeling of connective tissue: Role of fibronectin and laminin. *Pathologie-Biologies* (*Paris*) 51(10): 563–568. Abstract available at http://www.ncbi.nlm.nih.gov/entrez/query.fcgi?cmd = Retrieve&db = PubMed&dopt = Citation&list_uids = 14622946.
73. Mott, J.D. and Z. Werb. 2004. Regulation of matrix biology by matrix metalloproteinases. *Current Opinion in Cell Biology* 16(5): 558–564. Abstract available at http://www.ncbi.nlm.nih.gov/entrez/query.fcgi?cmd = Retrieve&db = PubMed&dopt = Citation&list_uids = 15363807.
74. Lochter, A., S. Galosy, J. Muschler, et al. 1997. Matrix metalloproteinase stromelysin-1 triggers a cascade of molecular alterations that leads to stable epithelial-to-mesenchymal conversion and a pre-malignant phenotype in mammary epithelial cells. *Journal of Cell Biology* 139(7): 1861–1872. < Go to ISI > ://000071266100023.
75. Lochter, A., M. Navre, Z. Werb, and M.J. Bissell. 1999. Alpha1 and alpha2 integrins mediate invasive activity of mouse mammary carcinoma cells through regulation of stromelysin-1 expression. *Molecular Biology of the Cell* 10(2): 271–282. Abstract available at http://www.ncbi.nlm.nih.gov/entrez/query.fcgi?cmd = Retrieve&db = PubMed&dopt = Citation&list_uids = 9950676.

76. Fata, J.E., Z. Werb, and M.J. Bissell. 2003. Regulation of mammary gland branching morphogenesis by the extracellular matrix and its remodeling enzymes. *Breast Cancer Research* 6(1): 1–11. < Go to ISI > ://000187112000006.
77. Simian, M., Y. Hirai, M. Navre, et al. 2001. The interplay of matrix metalloproteinases, morphogens and growth factors is necessary for branching of mammary epithelial cells. *Development* 128(16): 3117–3131. < Go to ISI > ://000170825900007.
78. Boudreau, N., C.J. Sympson, Z. Werb, and M.J. Bissell. 1995. Suppression of ice and apoptosis in mammary epithelial-cells by extracellular-matrix. *Science* 267(5199): 891–893. < Go to ISI > :// A1995QG20700060.
79. Hahn, U., A. Stallmach, E.G. Hahn, and E.O. Riecken. 1990. Basement membrane components are potent promoters of rat intestinal epithelial cell differentiation in vitro. *Gastroenterology* 98(2): 322–335. Abstract available at http://www.ncbi.nlm.nih.gov/entrez/query.fcgi?cmd = Retrieve&db = PubMed&dopt = Citation&list_uids = 2295387.
80. Paralkar, V.M., S. Vukicevic, and A.H. Reddi. 1991. Transforming growth factor beta type 1 binds to collagen IV of basement membrane matrix: Implications for development. *Developmental Biology* 143(2): 303–308. Abstract available at http://www.ncbi.nlm.nih.gov/entrez/query.fcgi?cmd = Retrieve& db = PubMed&dopt = Citation&list_uids = 1991553.
81. Ruhrberg, C., H. Gerhardt, M. Golding, et al. 2002. Spatially restricted patterning cues provided by heparin-binding vegf-a control blood vessel branching morphogenesis. *Genes & Development* 16(20): 2684–2698. Abstract available at http://www.ncbi.nlm.nih.gov/entrez/query.fcgi?cmd = Retrieve&db = PubMed&dopt = Citation&list_uids = 12381667.
82. Cardoso, W.V. 2001. Molecular regulation of lung development. *Annual Review of Physiology* 63: 471–494. < Go to ISI > ://000168071900020.
83. Hieda, Y. and Y. Nakanishi. 1997. Epithelial morphogenesis in mouse embryonic submandibular gland: Its relationships to the tissue organization of epithelium and mesenchyme. *Development Growth & Differentiation* 39(1): 1–8. < Go to ISI > ://A1997WX47100001.
84. Marker, P.C., A.A. Donjacour, R. Dahiya, and G.R. Cunha. 2003. Hormonal, cellular, and molecular control of prostatic development. *Developmental Biology* 253(2): 165–174. < Go to ISI > :// 000180688400001.
85. Parmar, H. and G.R. Cunha. 2004. Epithelial–stromal interactions in the mouse and human mammary gland in vivo. *Endocrine-Related Cancer* 11(3): 437–458. < Go to ISI > ://000224343000004.
86. Yu, J., A.P. McMahon, and M.T. Valerius. 2004. Recent genetic studies of mouse kidney development. *Current Opinion in Genetics & Development* 14(5): 550–557. < Go to ISI > ://000224337700014.
87. Sandal, T., K. Valyi-Nagy, V.A. Spencer, et al. 2007. Epigenetic reversion of breast carcinoma phenotype is accompanied by changes in DNA sequestration as measured by Alui restriction enzyme. *American Journal of Pathology* 170(5): 1739–1749. < Go to ISI > ://000246050400029.
88. Maniotis, A.J., C.S. Chen, and D.E. Ingber. 1997. Demonstration of mechanical connections between integrins, cytoskeletal filaments, and nucleoplasm that stabilize nuclear structure. *Proceedings of the National Academy of Sciences of the United States of America* 94(3): 849–854. Abstract available at http://www.ncbi.nlm.nih.gov/entrez/query.fcgi?cmd = Retrieve&db = PubMed&dopt = Citation& list_uids = 9023345.
89. Xu, R., V.A. Spencer, and M.J. Bissell. 2007. Extracellular matrix-regulated gene expression requires cooperation of Swi/Snf and transcription factors. *Journal of Biological Chemistry* 282(20): 14992–14999.
90. Aggeler, J., J. Ward, L.M. Blackie, et al. 1991. Cytodifferentiation of mouse mammary epithelial cells cultured on a reconstituted basement membrane reveals striking similarities to development in vivo. *Journal of Cell Science* 99(Pt 2): 407–417. Abstract available at http://www.ncbi.nlm.nih.gov/entrez/query.fcgi?cmd = Retrieve&db = PubMed&dopt = Citation&list_uids = 1885677.
91. Barcellos-Hoff, M.H., J. Aggeler, T.G. Ram, and M.J. Bissell. 1989. Functional differentiation and alveolar morphogenesis of primary mammary cultures on reconstituted basement membrane. *Development* 105(2): 223–235. Abstract available at http://www.ncbi.nlm.nih.gov/entrez/query.fcgi?cmd = Retrieve&db = PubMed&dopt = Citation&list_uids = 2806122.
92. Emerman, J.T. and D.R. Pitelka. 1977. Maintenance and induction of morphological-differentiation in dissociated mammary epithelium on floating collagen membranes. *In vitro* 13(5): 316–328. < Go to ISI > ://A1977DL36900008.

93. Lee, E.Y.H.P., W.H. Lee, C.S. Kaetzel, G. Parry, and M.J. Bissell. 1985. Interaction of mouse mammary epithelial-cells with collagen substrata-regulation of casein gene-expression and secretion. *Proceedings of the National Academy of Sciences of the United States of America* 82(5): 1419–1423. < Go to ISI > ://A1985ADL3900028.
94. Lee, E.Y.H., G. Parry, and M.J. Bissell. 1984. Modulation of secreted proteins of mouse mammary epithelial-cells by the collagenous substrata. *Journal of Cell Biology* 98(1): 146–155. < Go to ISI > ://A1984RX39500017.
95. Streuli, C.H. and M.J Bissell. 1991. Mammary epithelial cells, extracellular matrix, and gene expression. In *Regulatory Mechanisms in Breast Cancer*, edited by M. Lippman and R. Dickson, pp. 365–381. Boston: Kluwer Academic Publishers.
96. Barcellos-Hoff, M.H. and Bissell M.J. 1989. A role for the extracellular matrix in autocrine and paracrine regulation of tissue-specific functions. In *Autocrine and Paracrine Mechanisms in Reproductive Endocrinology*, edited by L. Krey, B.J. Gulyas, and J.A McCrakaren, pp. 137–155. New York: Plenum Press.
97. Bissell M.J, V.M. Weaver, S.A. Lelievre, et al. 1999. Tissue structure, nuclear organization, and gene expression in normal and malignant breast. *Cancer Research* 59(7): 1757s–1763s. < Go to ISI > ://000079582000026.
98. Wang, F., V.M. Weaver, O.W. Petersen, et al. 1998. Reciprocal interactions between beta 1-integrin and epidermal growth factor receptor in three-dimensional basement membrane breast cultures: A different perspective in epithelial biology. *Proceedings of the National Academy of Sciences of the United States of America* 95(25): 14821–14826. < Go to ISI > ://000077436700043.
99. Schwartz, M.A. and V. Baron. 1999. Interactions between mitogenic stimuli, or, a thousand and one connections. *Current Opinion in Cell Biology* 11(2): 197–202. Abstract available at http://www.ncbi.nlm.nih.gov/entrez/query.fcgi?cmd = Retrieve&db = PubMed&dopt = Citation&list_uids = 10209147.
100. Trusolino, L., A. Bertotti, and P.M. Comoglio. 2001. A signaling adapter function for alpha6beta4 integrin in the control of hgf-dependent invasive growth. *Cell* 107(5): 643–654. Abstract available at http://www.ncbi.nlm.nih.gov/entrez/query.fcgi?cmd = Retrieve&db = PubMed&dopt = Citation&list_uids = 11733063.
101. Hoffman, A.S. 2001. Hydrogels for biomedical applications. *Annals of the New York Academy of Sciences* 944: 62–73. Abstract available at http://www.ncbi.nlm.nih.gov/entrez/query.fcgi?cmd = Retrieve&db = PubMed&dopt = Citation&list_uids = 11797696.
102. Pratt, A.B., F.E. Weber, H.G. Schmoekel, R. Muller, and J.A. Hubbell. 2004. Synthetic extracellular matrices for in situ tissue engineering. *Biotechnology and Bioengineering* 86(1): 27–36. Abstract available at http://www.ncbi.nlm.nih.gov/entrez/query.fcgi?cmd = Retrieve&db = PubMed&dopt = Citation&list_uids = 15007838.
103. Wichterle, O. and D. Lim. 1960. Hydrophilic gels for biological use. *Nature* 185: 117–118.
104. Horbett, T.A. 1982. Protein adsorption on biomaterials. In *Biomaterials: Interfacial Phenomena and Applications*, edited by S.L. Cooper, N.A. Peppas, A.S. Hoffman, and B.D. Ratner, pp. 233–243. Washington DC: American Chemical Society.
105. Horbett, T.A. 1993. Principles underlying the role of adsorbed plasma proteins in blood interactions with foreign materials. *Cardiovascular Pathology* 2(3): 137s–148s.
106. Horbett, T.A. 1994. The role of adsorbed proteins in animal cell. *Colloids and Surfaces B: Biointerfaces* 2: 225–240.
107. Chen, C.S., M. Mrksich, S. Huang, G.M. Whitesides, and D.E. Ingber. 1997. Geometric control of cell life and death. *Science* 276: 1425–1428.
108. Folch, A. and M. Toner. 2000. Microengineering of cellular interactions. *Annual Review of Biomedical Engineering* 2: 227–256.
109. Veiseh, M., M.H. Zareie, and M. Zhang. 2002. Highly selective protein patterning on gold–silicon substrates for biosensor applications. *Langmuir* 18(17): 6671–6678. < Go to ISI > ://000177487600033.
110. Xia, Y. and G.M. Whitesides. 1998. Soft lithography. *Angewandte Chemie* 110(5): 568–594.
111. Veiseh, M. and M. Zhang. 2006. Effect of silicon oxidation on long-term cell selectivity of cell-patterned Au/Sio2 platforms. *Journal of American Chemical Society* 128(4): 1197–1203.
112. Bray, D. 1992. *Cell Movements*. New York: Garland Publishing, Inc.
113. Hjortso, M.A. and Joseph W.R. (Eds.). 1994. *Cell Adhesion: Fundamentals and Biotechnological Applications*: CRC.

114. Zimerman, B., M. Arnold, J. Ulmer, et al. 2004. Formation of focal adhesion-stress fibre complexes coordinated by adhesive and non-adhesive surface domains. *IEE Proceeding. Nanobiotechnology* 151 (2): 62–66. Abstract available at http://www.ncbi.nlm.nih.gov/entrez/query.fcgi?cmd = Retrieve&db = PubMed&dopt = Citation&list_uids = 16475844.
115. Ballestrem, C., B. Hinz, B.A. Imhof, and B. Wehrle-Haller. 2001. Marching at the front and dragging behind: Differential Alphavbeta3-integrin turnover regulates focal adhesion behavior. *The Journal of Cell Biology* 155(7): 1319–1332. Abstract available at http://www.ncbi.nlm.nih.gov/entrez/query.fcgi?cmd = Retrieve&db = PubMed&dopt = Citation&list_uids = 11756480.
116. Ratner, B.D., A.S. Hoffman, F.J. Schoen, and J.E. Lemons. 1996. *Biomaterial Science: An Interdisciplinary Endeavor*. San Diego: Academic Press.
117. Gordon, J.R. and M.R. Bernfield. 1980. The basal lamina of the postnatal mammary epithelium contains glycosaminoglycans in a precise ultrastructural organization. *Developmental Biology* 74(1): 118–135. Abstract available at http://www.ncbi.nlm.nih.gov/entrez/query.fcgi?cmd = Retrieve&db = PubMed&dopt = Citation&list_uids = 7350005.
118. Vogel, V. and M. Sheetz. 2006. Local force and geometry sensing regulate cell functions. *Nature Reviews. Molecular Cell Biology* 7(4): 265–275. Abstract available at http://www.ncbi.nlm.nih.gov/entrez/query.fcgi?cmd = Retrieve&db = PubMed&dopt = Citation&list_uids = 16607289.
119. Yim, E.K.F., S.W. Pang, and K.W. Leong. 2007. Synthetic nanostructures inducing differentiation of human mesenchymal stem cells into neuronal lineage. *Experimental Cell Research* 313(9): 1820–1829. < Go to ISI > ://000246348400007.
120. Semino, C.E., J.R. Merok, G.G. Crane, G. Panagiotakos, and S. Zhang. 2003. Functional differentiation of hepatocyte-like spheroid structures from putative liver progenitor cells in three-dimensional peptide scaffolds. *Differentiation* 71(4–5): 262–270. Abstract available at http://www.ncbi.nlm.nih.gov/entrez/query.fcgi?cmd = Retrieve&db = PubMed&dopt = Citation&list_uids = 12823227.

3 Functions and Requirements of Synthetic Scaffolds in Tissue Engineering

Tejas Shyam Karande and C. Mauli Agrawal

CONTENTS

3.1 INTRODUCTION

Tissue engineering is an interdisciplinary field that has emerged to address the needs created by a number of interrelated problems including shortage of donor organs, donor site morbidity, and failure of mechanical devices [1]. These imperfect solutions continue to give great impetus to the relatively new field that applies the principles of engineering and the life sciences toward developing biological substitutes for the restoration, maintenance, or improvement of tissue function [1]. Most tissue engineering techniques utilize a three-dimensional porous scaffold seeded with cells. These scaffolds play a vital role in the development of new tissue.

The goal of this review is to discuss the functions and requirements of scaffolds in tissue engineering as well as their modification. Different types of synthetic scaffolds are discussed in detail along with their methods of fabrication.

3.2 FUNCTIONS AND REQUIREMENTS OF SCAFFOLDS

Scaffolds serve numerous functions critical for the success of tissue regeneration, which include [2]:

1. Serving as space-holders to prevent encroachment of tissues from the immediate vicinity into the affected site.
2. Providing a temporary support structure for the tissue that they are intended to replace.
3. Creating a substrate for cells to attach, grow, proliferate, migrate, and differentiate on.
4. Serving as a delivery vehicle for cells, facilitating their retention and distribution in the region where new tissue growth is desired.
5. Providing space for vascularization, neotissue formation, and remodeling to occur.
6. Enabling the efficient transport of nutrients, growth factors, blood vessels, and removal of waste material.

In order for scaffolds to perform the above functions, they need to meet some basic requirements, which necessitate them to [3–5]

1. Be biocompatible, i.e., not produce an unfavorable physiological response.
2. Be biodegradable, i.e., get broken down eventually and eliminated from the body via naturally occurring processes.
3. Degrade at a rate proportional to the regrowth of new tissue.
4. Have mechanical properties that are consistent with the tissue they are replacing.
5. Have the desired surface properties to enable cell attachment, growth, proliferation, and differentiation as well as extracellular matrix formation.
6. Have the optimum architectural properties in terms of pore size, porosity, pore interconnectivity, and permeability and to allow for efficient delivery of nutrients, growth factors, blood vessels, and removal of waste.
7. Be easily processed into three-dimensional complex shapes in a well-controlled and reproducible manner.

These requirements are discussed in greater detail below.

3.2.1 BIOCOMPATIBILITY

This is the primary requirement for any type of scaffold. The scaffold is required to elicit a beneficial response from the cells with which it is seeded and an appropriate immune response from the host tissue on implantation, meaning that the interactions that take place between the scaffold, cells, and host tissue should be favorable without any potential for harm due to induced cytotoxicity, generation of an adverse immune response, or activation of the blood clotting or complement pathways [6]. A number of factors contribute to the kind of tissue response generated by the biomaterial including the shape and size of the implant; its chemical reactivity; the mechanism, rate, and byproducts of degradation; site of implantation; and the host species [6]. Taking it a step further, one can expand the meaning of the term biocompatibility to include biofunctionality, which indicates the ability of the material to support and promote cell–material interactions according to the local tissue-specific application [7].

3.2.2 Biodegradation

Since synthetic scaffolds serve as temporary structures that are replaced by native tissue subsequently, they need to be gradually removed from the implant site by a process commonly referred to as biodegradation. The terms biodegradable, bioresorbable, bioerodable, and bioabsorbable are often used incorrectly and/or interchangeably in tissue engineering literature [8]. Biodegradable indicates breakdown due to macromolecular degradation caused by biological elements resulting in fragments or other degradation by-products that are not necessarily eliminated from the body; bioresorbable implies complete elimination of foreign material and bulk degradation by-products via resorption within the body, i.e., by natural pathways like filtration or metabolization; bioerodable signifies surface degradation whereas bioabsorbable means dissolution in body fluids without any polymer chain cleavage or decrease in molecular mass [9].

Generally, polymers of the poly(α-hydroxy acid) group undergo bulk degradation. Thus, their molecular weight commences to decrease immediately upon contact with aqueous media but their mass reduces much more slowly owing to the time required by molecular chains to decrease to a size appropriate for them to freely diffuse out of the polymer matrix. This phenomenon results in an initially delayed but then rapid disintegration of the implant accompanied with a simultaneous increase in the release of acidic degradation by-product. *In vivo*, not only can this result in inflammatory reactions but the sudden drop in pH can further compromise the biocompatibility of the implant unless there is sufficient buffering provided by the surrounding body fluids and vasculature [8,10]. Filler materials influence the degradation mechanism by preventing autocatalytic effect of the acidic end groups that occurs as a result of polymer chain hydrolysis [5]. Thus, to control acceleration of acidic degradation, researchers in the musculoskeletal tissue engineering arena have incorporated filler materials like tricalcium phosphate (TCP) [11] or Bioglass [12] or basic salts [10] into the polymer matrix to produce a composite material with the idea that the resorption products of these additives will buffer the acidic resorption by-products of the original polymer matrix thereby restoring biocompatibility and preventing inflammation [8].

3.2.3 Matching Rates of Degradation of Scaffold and Regrowth of New Tissue

This criterion is extremely important but difficult to achieve. It is essential for the scaffold to gradually transfer the function of load bearing and support to the newly growing tissue, especially in musculoskeletal applications where the scaffolds are generally subjected to higher loads compared to other areas. Ideally, the rate of bioresorption or bioerosion or biodegradation of the scaffold should match the rate of regrowth of new tissue at the site of scaffold placement to provide an almost seamless transition of load from the disintegrating scaffold to the strengthening, developing tissue without compromising the integrity of the implant. There are drawbacks associated with either a very fast rate of scaffold degradation or a very slow rate, relative to the rate of regrowth of new tissue. If the rate of scaffold disintegration is high, the newly forming tissue will be suddenly exposed to forces greater than what it can tolerate as it will not have had enough time to get conditioned to bear the new forces and can thus be adversely affected. On the other hand, if the rate of scaffold degradation is extremely slow, it can result in stress shielding of the growing tissue thereby protecting it from the forces that are meant to strengthen it during its development, thereby making it more susceptible to injury later on. Hutmacher [8] has outlined two strategies for selection of the polymeric scaffold material in musculoskeletal tissue engineering applications depending on the time up to which the scaffold needs to assume the role of load bearing. In the case where the scaffold material is required to play the major supporting role till the time the construct is completely remodeled by the host tissue, it needs to be designed to retain its strength till the time the developing tissue can begin to assume its structural role. In case of bone, the scaffold is required to retain its mechanical properties for at least 6 months, i.e., 24 weeks (3 weeks for cell seeding, 3 weeks for growth of premature tissue in a dynamic environment, 9 weeks for growth of mature tissue in a bioreactor, and 9 weeks *in situ*) after which it will gradually start losing its mechanical properties and should be metabolized by the body without a foreign body reaction after 12–18 months. In the second strategy, the scaffold plays the primary role of mechanically supporting cell proliferation and differentiation only till the time that the premature tissue is placed in a bioreactor, after which the function is taken over by the extracellular matrix (ECM) secreted by the cells while the scaffold degrades. Thus, careful selection of various parameters related to a scaffold and its composition, depending on the size of defect and anatomical location, are crucial. These include hydrophobicity, crystallinity, mechanical strength, molecular weight, kind of breakdown, etc.

The use of composite materials is now on the rise to tailor degradation rates and resorption kinetics [8]. Shikinami and Okuno [13] used a composite of uncalcined and unsintered hydroxyapatite (HA) with poly-L-lactide (PLL) to not only gain better control over resorption but also to enhance mechanical strength. Roether et al. [14] fabricated poly(DL-lactic acid) (PDLLA) foams coated with and impregnated by Bioglass as scaffolds for bone tissue engineering. The Bioglass coating on the pore walls affected the rate and extent of polymer degradation by acting as a protective barrier against hydrolysis [14,15]. The rapid exchange of protons in water for alkali in glass provides a pH buffering effect at the polymer surface, thereby slowing down degradation as a result of small pH changes during dissolution of bioactive glass [16]. However, Ang et al. [17] found that when HA was incorporated as a filler in a polycaprolactone (PCL) matrix, the matrices with higher concentrations of HA degraded much faster than those with a lower concentration although their mechanical properties and bioactivity improved initially. This could be attributed to the random hydrolytic chain cleavages in the amorphous regions of the PCL scaffold or the increase in hydrophilicity imparted by the addition of HA to the PCL scaffolds that were placed in a highly basic medium of 5M NaOH to actually accelerate the slow degradation rate of PCL [17]. Even in controlled settings, the chemical and mechanical degradation of a polymer can vary significantly between species, individuals, and anatomic locations, thus making it extremely difficult to define an ideal degradation rate [2]. Most design strategies favor extending degradation time over months to minimize the risk of early construct failure rather than the risk of delayed resorption [2].

3.2.4 Mechanical Properties

A scaffold seeded with cells and growth factors is commonly referred to as a construct. The mechanical properties (strength, modulus, toughness, and ductility) of the construct should match those of the host tissue as closely as possible at the time of implantation so that tissue healing is not compromised by mechanical failure of the scaffold before new tissue generation occurs [2,8]. Freeman et al. [18] combined the techniques of polymer fiber braiding and twisting to fabricate a poly(L-lactic acid) (PLLA) braid-twist scaffold for anterior cruciate ligament reconstruction. This addition of fiber twisting to the braided scaffold resulted in significantly better mechanical properties (ultimate tensile strength, ultimate strain, and greater toe region) over scaffolds that were braided. Webster and Ahn [19] advocate the use of nanostructured HA as the foundation for bone tissue engineering since in addition to being mechanically robust, the nanophase substrates enhance adhesion and other osteoblast functions as well as provide chemical and structural stability. Horch et al. [20] incorporated functionalized alumoxane nanoparticles into poly(propylene fumarate)/poly(propylene fumarate)-diacrylate (PPF/PPF-DA) to attain a composite with up to a threefold increase in flexural modulus compared to the polymer resin.

They attributed the significant improvement in flexural properties to the uniform dispersion of nanoparticles within the polymer as well as greater covalent interactions between the functionalized surface of the filler and polymer chains. Thus, researchers are altering existing methods to optimize the desired properties by modifying some aspect of scaffold fabrication, be it combining techniques or using combinations of polymers or reducing the size of the components.

Controlling the mechanical properties of the construct over time is extremely challenging. Scaffolds made of metal and ceramics do not degrade and would provide ideal mechanical characteristics in specific circumstances, but in general would compromise tissue repair and function due to stress shielding, possible fracture at tissue–implant interface, and diminished space for new tissue growth due to presence of a permanent implant [2]. Generally, polymers lack bioactive function, i.e., the ability to produce a strong interfacial bond with the growing tissue, for example, with bone tissue via the formation of a biologically active apatite layer [5,21]. This compounded with the flexible and weak nature of polymers limits their ability to meet the mechanical demands in surgery and the local environment thereafter, prompting the use of composites which comprise biodegradable polymers and bioactive ceramics [5,21]. Certain ceramic materials such as HA, TCP, and bioactive glass (Bioglass) form strong bonds with bone tissue through cellular activity in the presence of physiological fluids and are hence referred to as "bioactive" [14,22].

Another technique used to improve the load-bearing characteristics of the construct involves use of a bioreactor to mechanically precondition the implant and thus better prepare it to bear the loads it will be subjected to after implantation, in an efficient manner. A bioreactor can be defined as any device that tries to mimic and reproduce physiological conditions to maintain and encourage cell culture for tissue growth [23]. Bilodeau and Mantovani [23] have written an excellent review on the different characteristics of bioreactors designed to grow cartilage, bone, ligament, cardiac and vascular tissue, cardiac valve, and liver. They have discussed how mechanical stresses generated within bioreactors influence the quality of the ECM in case of bone, ligament, and cartilage and how other aspects like cell proliferation and differentiation are influenced more in case of the other tissues. Androjna et al. [24] investigated the effect of mechanically conditioning small intestine submucosa (SIS) scaffolds with and without tenocytes, *in vitro* within bioreactors, for enhancing tendon repair. They found the biomechanical properties (e.g., stiffness) of the cell-seeded scaffolds to increase as a result of cell-tensioning due to cyclic loading as compared to unseeded scaffolds and no- or static-load constructs (with or without cells). Reorganization of the matrix may have also contributed to this increased stiffness as the application of mechanical load may have reoriented the collagen architecture along the axis of the applied load [24]. Mahmoudifar and Doran [25] seeded chondrocytes on PGA scaffolds that were cultured in recirculation column bioreactors to produce cartilage constructs. The flow of media through the construct generates shear forces that provide mechanical stimuli to cells thereby improving the quality of cartilage produced [25,26]. Also, hydrostatic pressure produces compressive forces that are beneficial for cartilage formation [26]. Jeong et al. [27] successfully subjected smooth muscle cell-seeded PLCL [poly(lactide-co-caprolactone), 50:50] scaffolds to pulsatile strain and shear stress in a pulsatile perfusion bioreactor to stimulate vascular smooth muscle tissue development and retainment of their differentiated phenotype. Mechanical signals play a vital role in the engineering of constructs for cardiac tissue as well [28]. Akhyari et al. [29] subjected a cardiac cell-seeded gelatin matrix to a cyclical mechanical stretch regimen that not only improved cell proliferation and distribution but also increased the mechanical strength of the graft by an order of magnitude. Cell and tissue remodeling are important for achieving stable mechanical conditions at the implant site [8]. Thus, the necessity for the construct to maintain sufficient structural integrity during the *in vitro* and/or *in vivo* growth and remodeling phase [8].

3.2.5 Surface Properties

Most conventional polymers do not adequately meet the surface requirements of scaffolds thereby necessitating the modification of the surface of a biomaterial that already exhibits good bulk properties and biofunctionality [30]. Moroni et al. [31] developed a novel system to create scaffolds for cartilage repair with a biphasic polymer network made of poly[(ethylene oxide) terephthalate-co-poly(butylene) terephthalate] (PEOT/PBT) to obtain a shell–core fiber architecture, where the core provided the primary mechanical properties and organization to the scaffold while the shell worked as a coating to enhance the surface properties. The shell polymer contained a higher molecular weight of poly(ethylene glycol) (PEG) segments that were used in the copolymerization as well as a greater weight percentage of the PEOT domains relative to the core [31]. Liu et al. [32] fabricated surface-modified nanofibrous PLLA scaffolds using gelatin spheres as porogen. Gelatin molecules adhered to the scaffold surface during fabrication. This surface modification significantly improved initial osteoblast adhesion and proliferation as well as stimulated greater matrix secretion. Li et al. [33] explored a novel approach for the relatively uniform apatite coating of thick PLGA scaffolds even deep within the interior to enhance its osteoconductivity. They first coated apatite on the surface of paraffin spheres of the required size, which were then molded into a foam. PLGA/pyridine solution was made to penetrate the interspaces among the spheres. Cyclohexane was used to dissolve the spheres, resulting in highly porous PLGA scaffold with controlled pore size and excellent interconnectivity having a uniform apatite coating on the pore surface. Cai et al. [34] modified the surface of PDLLA scaffolds, prepared via thermally induced phase separation (TIPS), with baicalin using a physical entrapment method, to increase bone formation potential and biocompatibility that were histologically evaluated using a rabbit radialis defect model *in vivo*. Baicalin is a flavonoid compound and purified form of a Chinese herbal medicinal plant and possesses antioxidant as well as anti-inflammatory properties.

The local chemical environment controls the interactions between cells and scaffolds that occur at the surface. Generally, all implanted materials get immediately coated with proteins and lipids, which mediate the cellular response to these materials. Finally, it is the interaction between the scaffold surface and the biomolecules that adsorb on it, that dictates the net effect [2]. Koegler and Griffith [35] patterned the scaffold's surface chemistry and architecture to study cell response as it would facilitate the orderly development of new tissue. They evaluated how rat osteoblasts responded to PLGA scaffolds modified with poly(ethylene oxide) (PEO) and found that higher PEO concentrations decreased adhesion, proliferation, spreading, and migration but enhanced alkaline phosphatase activity.

Surface modification plays a very important role in tissue engineering techniques employing thin films or membranes. Tiaw et al. [36] subjected ultrathin PCL films to femtosecond and excimer laser ablation to produce drilled-through holes and blind holes, respectively, so as to enhance permeability for applications like epidermal tissue engineering. Laser treatment had made the membrane more hydrophilic thereby paving the way for further study in the area of membrane tissue engineering. Nakayama et al. [37] studied the effect of micropore density of scaffold films used in cardiovascular tissue engineering applications. They micropatterned four regions of a polyurethane film with different pore densities and used this to cover a stent that was implanted in arteries in a canine model as an *in vivo* model of transmural tissue ingrowth. Thrombus formation was maximum in nonporous regions and micropore regions of lowest density. They also found the thickness of the neointimal wall to decrease with a rise in micropore density.

3.2.6 Architectural Properties

The architectural properties of a scaffold mainly dictate the transport that occurs within it, which is primarily a function of diffusion. The transport issues comprise delivery of oxygen and other nutrients, removal of waste, transport of proteins and cell migration, which in turn are governed by scaffold porosity and permeability [38]. The size, geometry, orientation, interconnectivity, branching, and surface chemistry of pores and channels directly affect the extent and nature of nutrient diffusion and tissue ingrowth [39,40]. Generally, living tissue is observed in the outer regions of scaffolds whereas the interior fails to support viable tissue due to lack of adequate diffusion [41]. This may arise due to the fact that as cells within the pores of the scaffold begin to proliferate and secrete ECM, they simultaneously begin to block off the pores, thereby reducing the supply of nutrients to the interior. The formation of this surface layer of tissue with sparse matrix in the interior has been referred to as the "M&M effect," referring to the popular brand of candy having a hard crust and soft core [38].

3.2.6.1 Pore Size and Shape

A scaffold cannot be completely solid as cells need to grow within it and these need to be supplied with nutrients. Thus, the need for a scaffold to have holes or pores or channels seems obvious, but not so obvious is what their shape and dimensions should be. The pore size should at least be a few times the size of the cells that will be seeded on it to provide enough space for the entry and exit of nutrients and waste, respectively. Also, blood vessels and growth factors may need to enter the construct as well. There is no common pore or channel size range that is suitable for all types of tissue growth as cells making up different tissues have different dimensions. Sosnowski et al. [42] prepared PLL/PLGA scaffolds from microparticles with a bimodal pore size distribution. Macropores in the 50–400 μm range promoted osteoblast growth and proliferation within the scaffold whereas micropores in the range of 2 nm–5 μm in the scaffold walls allowed for diffusion of nutrients and metabolites as well as products of polyester hydrolysis. Draghi et al. [43] used three different porogens (gelatin microspheres, paraffin microspheres, and salt crystals) to fabricate scaffolds from commonly used biodegradable materials via the solvent casting/porogen leaching technique to see which allowed maximum control over scaffold morphology. Although all the porogens contributed to producing highly porous scaffolds, microsphere leaching produced well-defined spherical pores that resulted in better mechanical properties and lesser flow resistance.

Researchers have fabricated scaffolds with different pore sizes or even a range of pore sizes within the same scaffold to see their effect on cell growth and to mimic certain types of tissues. Oh et al. [44] fabricated cylindrical PCL scaffolds with gradually increasing pore sizes along the longitudinal axis using a novel centrifugation method to evaluate the effect of pore size on cell–scaffold interaction. The pore sizes within the scaffold gradually increased from 88 to 405 μm and the porosity from 80% to 94% due to the gradual increment of centrifugal force along the cylindrical axis. Chondrocytes, osteoblasts, and fibroblasts were evaluated for their interaction *in vitro* with this PCL scaffold and *in vivo* using calvarial defects in a rabbit model. The scaffold section having pore sizes in the 380–405 μm range showed better chondrocyte and osteoblast growth while the 186–200 μm range was better suited for fibroblast growth. Moreover, the scaffold section with a 290–310 μm range pore size seemed to be best suited for new bone formation. This shows the existence of pore ranges that are ideal for the growth of some cell types and that this range can change while the cells differentiate to form tissue.

Woodfield et al. [45] investigated the ability of anisotropic pore architectures to control the zonal organization of chondrocytes and ECM components in scaffolds made of poly(ethylene glycol)-terephthalate-poly(butylene terephthalate) (PEGT/PBT). They used a 3D fiber deposition technique to produce scaffolds with either uniformly spaced pores (fiber spacing of 1 mm and pore size of 680 μm diameter) or pore size gradients (fiber spacing of 0.5–2 mm and pore size range of 200–1650 μm diameter), but having a similar overall porosity of about 80%. They found the gradient to promote anisotropic cell distribution similar to that found in the upper, middle, and lower zones of immature bovine articular cartilage, irrespective of whether the method of cell seeding was static or dynamic. Additionally, they discovered a direct correlation between the zonal porosity and both DNA and glycosaminoglycan (GAG) content. Also, Harley et al. [46] produced cylindrical scaffolds with a radially aligned pore structure having a smaller mean pore size and lesser porosity toward the outside. Increasing the spinning time and/or velocity caused the formation of a large inner diameter hollow tube and a gradient of porosity along the radius due to increased sedimentation. Thus, an important underlying trend is the need for scaffolds to have an appropriate porosity.

3.2.6.2 Porosity

Porosity is the amount of void space within the scaffold structure. Several studies have reiterated the need for scaffolds to possess high porosity and high surface area-to-mass ratio for promoting uniform cell delivery and tissue ingrowth [47] as well as have an open pore network for optimal diffusion of nutrients and waste [48]. Another study indicated that a scaffold should ideally possess a porosity of 90% to allow for adequate diffusion during tissue culture and to provide adequate area for cell–polymer interactions [49]. However, Goldstein et al. [50] have suggested that polylactic–polyglycolic (PLG) acid scaffolds be prepared with a porosity not exceeding 80% for implantation into orthopaedic defects as it would otherwise compromise the scaffold integrity. Thus, in case of polymeric scaffolds there may be a conflict between optimizing porosity and maximizing mechanical properties. Moreover, Agrawal et al. [51] found that lower initial porosity and permeability result in a faster rate of degradation for PLG scaffolds and lower mechanical properties during the initial weeks. Wu and Ding [52] investigated the effects of porosity (80%–95%) and pore size (50–450 μm) on the degradation of 85/15 PLGA scaffolds, performed in phosphate-buffered saline (PBS) at 37°C up to 26 weeks. Scaffolds possessing a higher porosity or smaller pore size degraded more slowly than those with a lower porosity or larger pore size as the latter had thicker pore walls and smaller surface areas that prevented the diffusion of acidic degradation products resulting in greater acid-catalyzed hydrolysis.

Thus, in view of these contradictory factors, there is a need to optimize scaffolds for bone regeneration based on their specific mechanical requirements balanced with their desired useful life and diffusion characteristics. This could be achieved by optimizing porosity with respect to nutrient availability and using it with biomaterials that can provide adequate mechanical properties. Lin et al. [53] developed a general design optimization strategy for 3D internal scaffold architecture to have the required mechanical properties and porosity simultaneously, using the homogenization-based topology optimization algorithm for bone tissue engineering. Howk and Chu [54] showed that it was possible to increase the porosity and strength of a bone tissue engineering scaffold through simple iterations in architectural design using computer-aided design (CAD) software and finite element analysis. The goal of their optimization was to maintain the strength of a design constant while increasing its porosity. Xie et al. [55] selected mechano-active scaffolds that respond to applied compression stress without undergoing permanent deformation for engineering functional articular cartilage from a biomechanical point of view and then determined the best porosity. They used PLCL sponges (pore size: 300–500 μm, porosity: 71%–86%) as mechano-active scaffolds and determined that the lower their porosity, the nearer their mechanical properties came to those of native cartilage. Hence, the scaffold with a porosity of 71% was found to be the best suited for cartilage regeneration. Moroni et al. [56] varied pores in size and shape by altering fiber diameter, spacing, as well as orientation and layer thickness using the 3D fiber deposition method to study their influence on dynamic mechanical properties. They observed a reduction in elastic properties like dynamic stiffness and equilibrium modulus as well as an increase of viscous parameters like damping factor and creep unrecovered strain as porosity increased.

3.2.6.3 Pore Interconnectivity

It is not sufficient for a scaffold to be porous but the pores in the scaffold need to be interconnected for efficient delivery of nutrients to the interior and removal of waste to the exterior of the scaffold. Pore interconnectivity also has implications as far as transport of proteins, cell migration, and tissue ingrowth are concerned.

Griffon et al. [57] found chondrocyte proliferation and metabolic activity to improve with increasing interconnected pore size of chitosan sponges. Lee et al. [58] produced PPF scaffolds with controlled pore architecture to study the effects of pore size and interconnectivity on bone growth. They fabricated scaffolds with three pore sizes (300, 600, and 900 μm) and randomly closed 0%, 10%, 20%, or 30% of the pores. Porosity and permeability decreased as the number of closed pores increased, especially when the pore size was 300 μm, as a result of low porosity and pore occlusion. Suh et al. [59] compared the proliferation of chondrocytes on equally porous (95%) PLG scaffolds prepared by the solvent casting and particulate leaching (SCPL) technique using two different porogens: salt and gelatin. The scaffolds produced using gelatin exhibited better cell attachment and proliferation, and this was attributed to better pore interconnectivity at the same porosity. Hou et al. [60] suggested that extraction of salt particles in a salt leaching process implied that the resulting pores were interconnected. However, complete removal of the salt does not necessarily ensure a permeable structure as there might be dead-end spaces with only a single opening thereby not permitting end-to-end interconnectivity of the whole structure [38].

Traditional scaffold manufacturing techniques have been modified to increase pore interconnectivity. Murphy et al. [61] imparted improved pore interconnectivity to PLGA scaffolds by partially fusing the salt before creating the polymer matrix via either the solvent casting/salt leaching process or the gas foaming/salt leaching process. Gross and Rodriguez-Lorenzo [62] made spheroid salt particles in a flame and sintered them to produce an interconnected salt template, which was filled with a carbonated fluorapatite powder and polylactic polymer to form a composite scaffold. A larger pore size was possible with the use of large spherical salt particles and this technique could be used to successfully produce scaffolds with good interconnectivity and graded pore sizes. Hou et al. [63] fabricated highly porous (93%–98%) and interconnected scaffolds by freeze drying polymer solutions in the presence of a leachable template followed by leaching of the template itself. Sugar or salt particles were fused to form the well-connected template, the interstices of which were filled with a polymer solution in solvent, followed by freeze drying of the solvent and subsequent leaching of the template. This resulted in relatively large interconnected pores based on the template and smaller pores resulting from the freeze-drying process.

Darling and Sun [64] and Wang et al. [65] used microCT to quantify pore interconnectivity within their PCL scaffolds for bone tissue engineering that were manufactured by a type of solid freeform fabrication (SFF) technique called precision extrusion deposition. They achieved pore interconnectivity greater than 98% in their scaffolds. Moore et al. [66] also used microCT followed by a custom algorithm to nondestructively quantify pore interconnectivity. The program calculated accessible porosities over a range of minimum connection sizes. The accessible porosity varied with connection size as a function of porogen content. However, microCT is still not widely available and researchers have improvised, like Li et al. [67] who appreciated the difficulty in obtaining 3D information about pore interconnectivity through 2D images and devised a rather simple unique experiment to verify the same. They soaked porous HA in black pigment dispersion and centrifuged it. After removing the pigments, they sectioned, dried, and pictured the sample and found black colored pores to be accessible either directly or via adjacent pores.

3.2.6.4 Permeability

Permeability is a measure of the ease with which a fluid can flow through a structure. Generally, an increase in porosity leads to an increase in permeability, but for this to happen the pores need to be highly interconnected [38]. One of the authors (Agrawal) has previously shown that scaffolds can possess different permeabilities while maintaining similar porosity [51,68]. Thus, permeability should be treated as an independent scaffold design parameter. A high permeability can produce superior diffusion within the scaffold, which would facilitate the inflow of nutrients and the disposal of degradation products and metabolic waste [38]. Permeability is also affected by fluid–material interactions and thus influences the viscoelastic response of a scaffold. This, in turn, affects the fluid pumping movement of the scaffold which is important while designing scaffolds for articular cartilage repair, where mechanotransduction and cell apoptosis may be affected by hydrostatic pressure and flow-induced shear [69].

Scaffold porosity and permeability are clearly related to the physical and mechanical properties possessed by the scaffold. For example, better mechanical properties may be obtained for a scaffold if it is made more solid and less porous. Less obvious is the fact that porosity and permeability can also have a significant impact on the chemical behavior of the scaffold, especially its degradation characteristics [38]. For example, as stated earlier, it has been shown that low porosity and permeability PLG scaffolds degrade faster [51,70]. Also, such scaffolds exhibit a lower decrease in their mass, molecular weight, and mechanical properties under dynamic fluid flow conditions compared to static conditions [51]. This phenomenon has been attributed to the inhibition of autocatalytic degradation due to better diffusion or forced fluid flow.

Li et al. [71] proposed using the permeability/porosity ratio to describe the accessibility of inner voids in macroporous scaffolds as they found porosity and pore size to be inadequate descriptors. The above ratio is an indicator of the percolative efficiency per unit porous volume of a scaffold, where permeability can be termed as the conductance normalized by sample size and fluid viscosity. Good pore interconnectivity could lead to a positive correlation between porosity and permeability. Permeability could represent a combination of five important scaffold parameters: porosity, pore size and distribution, interconnectivity, fenestration size and distribution, and pore orientation.

Wang et al. [72] wanted to optimize scaffold morphology for connective tissue engineering to overcome the problem of disproportionately high tissue formation at surfaces of scaffolds grown in bioreactors relative to their interior. Thus, they determined geometric parameters for PEGT/PBT scaffolds using SEM, microCT, and flow permeability measurements and then seeded fibroblasts on these scaffolds under dynamic flow conditions for 2 weeks. Only scaffolds with an intermediate pore interconnectivity supported homogeneous tissue formation throughout the scaffold with complete filling of all pores. Hollister et al. [73] used an integrated image-based design along with SFF to create scaffolds with the desired elasticity and permeability from a variety of biomaterials including degradable polymer, titanium, and ceramics to fit any craniofacial defect. The scaffolds supported significant bone growth in minipig mandibles for a range of pore sizes from 300 to 1200 μm. Huang et al. [74] used scaffolds made of chitosan and PLGA with longitudinally oriented channels running through them to serve as guides for nerve generation. They found chitosan to be a better scaffold for nerve guidance compared to PLGA owing to its high permeability and characteristic porous structure.

In addition to traditionally used direct permeation experiments as conducted by Spain et al. [75] and Li et al. [71], researchers have begun to use magnetic resonance imaging (MRI) and microCT for measuring permeability as well. Neves et al. [76] used MRI to determine construct permeability to a low molecular weight MR contrast agent and correlate the findings with measurements of cell growth and energetics. They used perfusion bioreactors to seed mature sheep meniscal fibrochondrocytes on polyethylene terphthalate (PET) fabric to produce bioartificial meniscal cartilage constructs. Knackstedt et al. [77] used microCT with a resolution of 16.8 μm to measure a number of structural characteristics like pore volume-to-surface-area ratio, pore size distribution, permeability, tortuosity, diffusivity, and elastic modulus of coral bone graft samples.

3.2.7 Scaffold Fabrication

The successful generation of completely functional tissues should be addressed not only at the microscale to expose the cells to an environment conducive to their optimal functioning, but also at the macroscale for the tissue to possess suitable mechanical properties, facilitate nutrient transport, and promote coordination of multicellular processes [78].

3.2.7.1 Solvent Casting and Particulate Leaching

The most commonly used scaffold fabrication technique is solvent casting followed by particulate leaching, where the pore size of the resulting scaffold is controlled by the size of the porogen, and porosity is controlled by the porogen/polymer ratio. This method involves mixing a water-soluble porogen in a polymer solution followed by casting the mixture into a mold of the desired shape. The solvent is removed by evaporation or lyophilization and the porogen is leached out by immersion in deionized water. Widmer et al. [79] used solvent casting followed by extrusion, to form a tubular construct, and then leached the salt to generate PLGA and PLA scaffolds with a pore size of 5–30 μm and porosity in the 60%–90% range. Although salt is the most commonly used porogen, sugar as well as gelatin [32,59] and paraffin spheres [80] are also used and these are sometimes modified to enhance scaffold functionality [33]. In case paraffin spheres are used as the porogen, the solvent used is organic (like hexane) [80] and not water. This method is the most widely used owing to its simplicity. However, natural porogen dispersion allows little control over the internal scaffold architecture and pore interconnectivity. Also, the thickness of the scaffold that can be fabricated by this method is hindered by difficulty removing the porogen from deep within the scaffold interior [81]. This has led to the modification of the SCPL technique to produce greater pore interconnectivity in some cases [61,68,80,82], and to new techniques like rapid prototyping (RP), also known as solid freeform fabrication (SFF), in others [40,83–85]. Agrawal et al. [68] modified the technique by vibrating the mold while dissolving the salt, thereby preventing the particles from settling due to gravity, thereby enhancing permeability of the scaffold by creating better pore interconnectivity and more even distribution of pores.

3.2.7.2 Gas Foaming

This technique can be used to fabricate highly porous scaffolds in the absence of organic solvents. Carbon dioxide (CO_2) generally acts as the "porogen" in this method in its normal gaseous [86,87] or subcritical [88] or supercritical form [89,90]. Solid polymeric disks when exposed to high pressure CO_2 at room temperature get saturated with the gas. The solubility of the gas in the polymer is rapidly decreased by reducing the pressure to atmospheric levels, creating a thermodynamic instability for the dissolved CO_2 resulting in the nucleation and growth of gas bubbles in the interior of the polymer matrix. Mooney et al. [86] created PLGA scaffolds with a pore size of about 100 μm and porosity up to 93% using this method. However, this method resulted in a relatively nonporous skin layer due to rapid diffusion of the dissolved CO_2 from the surface and closed pore structure with limited pore interconnectivity. These drawbacks were improved upon by combining the above process with particulate leaching [87,91–93]. Harris et al. [87] compression molded PLGA and salt particles and then subjected them to gas foaming as described earlier. The polymer and salt particles fused to form a continuous matrix with the salt particles subsequently being leached out leaving behind a macroporous foam with good interconnectivity. Kim et al. [94] found that PLGA/HA scaffolds fabricated by gas foaming and particulate leaching enhanced bone regeneration compared to scaffolds fabricated by SCPL.

3.2.7.3 Emulsion Freeze Drying

This process involves creation of an emulsion by homogenization of a polymer solution and water mixture that is rapidly cooled to lock in the liquid state structure [81]. The solvent and water are then removed by freeze drying [81]. The disadvantage of this technique is that it yields scaffolds with a closed pore structure [95]. Whang et al. [82] investigated the effect of median pore size and protein loading on protein release kinetics from emulsion freeze-dried PLGA scaffolds. The profiles indicated an initial burst followed by a slower sustained release. The scaffold tortuosity and partition coefficient for protein adsorption significantly reduced protein diffusivity. The activity of the released protein was demonstrated by the successful delivery of rhBMP-2 from the scaffold to an ectopic site in a rat [96]. Moshfeghian et al. [97] evaluated the formation of chitosan/PLGA scaffold using controlled rate freezing and lyophilization. The microarchitecture of the scaffold was significantly influenced by the solvent and freezing temperature. Controlling the concentration of chitosan yielded scaffolds with a porosity exceeding 90%.

3.2.7.4 Thermally Induced Phase Separation

This process involves dissolution of the polymer in a solvent at a high temperature followed by a liquid–liquid or solid–liquid phase separation induced by lowering the solution temperature [81]. Subsequent use of sublimation causes removal of the solidified solvent-rich phase, resulting in a porous scaffold with good mechanical properties [81]. TIPS has been used to fabricate scaffolds covering a wide range of polymers and composites: from the regular PLLA, PDLLA, PDLLGA [98], and PLGA [99], to the more sophisticated poly(ester urethane) urea/collagen [100], amorphous calcium phosphate/PLLA [101], and PDLLA/Bioglass [102] to name a few. Rowland et al. [103] fabricated a PLGA/polyurethane (PU) composite scaffold using TIPS showing how this process can bring together two very different polymers, whose morphology can be manipulated by controlling the phase separation behavior of the initial homogeneous polymer solution. Helen et al. [104] found composite PDLLA/Bioglass foams prepared by TIPS to provide a suitable microenvironment for the culture and proliferation of bovine annulus fibrosus (BAF) cells as well as the production of sulphated glycosaminoglycans (sGAG), collagen type I and collagen type II, providing preliminary evidence of their suitability for treatment of intervertebral disks with damaged annulus fibrosus regions. Gong et al. [105] used TIPS to produce PLLA scaffolds, which were filled with chondrocytes entrapped in agar hydrogel, thereby resulting in an implant with suitable mechanical properties and macroscopic shape while possessing an interior that is analogous to native ECM. Mo et al. [106] used TIPS to produce a porous PCL solution coating on the outside of a PLGA fiber braided tube to produce a PCL/PLGA composite tubular scaffold for small diameter blood vessel tissue engineering. The porous PCL coating was used with the intention of providing a surface suitable for cell attachment, proliferation, and tissue regeneration. Cao et al. [107] compared the *in vitro* and *in vivo* degradation properties of PLGA scaffolds produced by TIPS and SCPL. TIPS produced far less changes in dimension, mass, internal architecture, and mechanical properties compared to SCPL over a 6 week period. Morphometric comparison indicated slightly better tissue ingrowth accompanied with greater loss of scaffold structure in SCPL scaffolds. Chun et al. [108] fabricated PLGA scaffolds using TIPS for the controlled delivery of plasmid DNA over a period of 21 days. The various parameters in TIPS fabrication directly affecting pore structure and pore interconnectivity, such as polymer concentration, solvent/nonsolvent ratio, quenching methods, as well as annealing time were also examined to determine their effects on sustained release of plasmid DNA.

3.2.7.5 Gravity and Microsphere Sintering

Qiu et al. [109] sintered HA-coated hollow ceramic microspheres that were developed in rotating-wall vessels, to create microcarriers for 3D bone tissue formation. Borden et al. [110] randomly packed PLGA microspheres to form a gel microsphere matrix, which had a high Young's modulus but a pore system less optimal for bone growth, and a sintered microsphere matrix, which had mechanical properties in the midrange of cancellous bone accompanied with a well-connected pore system. The sintered microsphere matrices were created by thermally fusing the PLGA microspheres into a 3D array without any HA. They went on to study the osteoconductivity and degradation profile of these scaffolds by evaluating how osteoblasts and fibroblasts interacted with these scaffolds and performing degradation studies [111,112]. The group went on to evaluate the matrices' efficacy by using it in a 15 mm ulnar defect in rabbits and found that it supported significant formation of bone at the implant–bone interface [113]. Jiang et al. [114] fabricated composite chitosan/PLGA scaffolds by sintering and found osteoblast-like cells to proliferate better on these as compared to PLGA scaffolds. The presence of chitosan on the microsphere surfaces upregulated gene expression of ALP, osteopontin, and bone sialoprotein as well as increased ALP activity. Kofron et al. [115] developed tubular PLGA/HA sintered microsphere matrices using solvent evaporation for bone regeneration. The tubular composites were made to more closely mimic the bone marrow cavity and were found to have mechanical properties similar to cylindrical composites of the same dimensions.

3.2.7.6 Rapid Prototyping/Solid Freeform Fabrication

RP techniques involve building 3D objects using layered manufacturing methods and offer several advantages over the traditional porogen leaching method, mainly independent control over the micro- and macroscale features enabling fabrication of complex structures customizable to the shape of the defect or injury [78]. Yang et al. [39] have reviewed the advantages and limitations of various RP techniques. Leong et al. [116] have tabulated the pros and cons of the conventional methods and discussed the capabilities and limitations of the important RP techniques. The process, in general, comprises the design of a scaffold model using CAD software, which is then expressed as a series of cross sections [117]. Corresponding to each cross section, the RP machine lays down a layer of material starting from the bottom and moving up a layer at a time to create the scaffold. Each new layer adheres to the one below it, thereby providing integrity to the finished product. Agrawal and Ray [3] and Yang et al. [118] have provided comprehensive reviews weighing the pros and cons of traditional scaffold materials and fabrication methods. The different types of techniques encompassed by SFF include fused deposition modeling (FDM), precision extrusion deposition (PED), selective laser sintering (SLS), stereolithography (STL), and 3D printing (3DP) [40].

FDM [4,85,119–121] utilizes a moving nozzle that extrudes a polymeric fiber in the horizontal plane and once a layer is completed, the plane is lowered and the procedure is repeated. PED is very similar to FDM, except that scaffold material in the form of granules or pellets is directly extruded and deposited in the form of fibers without the need of having to change these into precursor filaments as is the case with FDM [65].

Pressure-assisted microsyringe (PAM) [78] is like FDM but requires no heat and has greater resolution but cannot incorporate micropores using particulate leaching owing to the syringe dimensions. This method involves deposition of polymer solution in solvent through a syringe fitted with a 10–20 μm glass capillary needle. The solvent acts as the binding agent and the size of the polymer stream deposited can be altered by varying the syringe pressure, solution viscosity, tip diameter of the syringe, as well as speed of the motor [122].

SLS [123–126] involves building objects by sintering powder on a powder bed using a beam of infrared laser. The laser beam interacts with the powder to increase the local temperature to the glass transition temperature of the powder, causing the particles to fuse to each other as well as the layer underneath [39]. Laser power and scanning speed affect sintering significantly [127]. Also, control over the finished product can be achieved by varying the laser processing parameters as these, in turn, control the degree of particle fusion and porosity [127].

STL [128–130] uses an ultraviolet (UV) laser beam to selectively polymerize a liquid photocurable monomer, a layer at a time [117]. The CAD data guide the UV beam onto the liquid surface, which is then lowered to enable the liquid photopolymer to cover the surface. Arcaute et al. [130] encapsulated human dermal fibroblasts in bioactive PEG hydrogels that were photocrosslinked using STL.

3DP involves ink-jet printing of a binder onto a ceramic [131,132], polymer [117,133,134], or composite [135,136] powder surface, one layer at a time. The movement of the jet head, which dispenses the binder, is controlled by the CAD cross-sectional data. Adjacent powder particles join as the binder dissolves [117]. Indirect 3DP is sometimes used to overcome few of the pitfalls of 3DP. Lee et al. [137] used indirect 3DP, where the molds were printed first and the final material was cast into the mold cavity, to overcome some of the limitations of 3DP. These include higher pore sizes, due to the need to increase the thickness of each incremental layer to the porogen size range which can eventually compromise layer-to-layer connectivity resulting in lamination defects [137]. Also, shape complexity when powder material requires an organic solvent as the liquid binder, custom machines, proprietary control software, and extensive operator expertise make 3DP a helpful but sometimes difficult technique to employ [137].

Some researchers have combined two or more manufacturing techniques to optimize their scaffold designs. Taboas et al. [40] coupled SFF with conventional sponge scaffold fabrication techniques (phase separation, emulsion-solvent diffusion, and porogen leaching) to develop methods for casting scaffolds possessing designed and locally controlled as well as globally porous internal architectures. Dellinger et al. [138] used an SFF technique based on the robotic deposition of colloidal pastes to produce HA scaffolds of different architectures with porosities spanning multiple length scales. Macropores (100–600 μm) were obtained by spacing the HA rods appropriately whereas micropores (<30 μm) were produced by including polymer microsphere porogens in the HA paste and controlling the sintering of scaffolds. Moroni et al. [31] combined 3D fiber deposition and phase separation to create a shell–core fiber architecture by viscous encapsulation resulting in scaffolds with a biphasic polymer network.

Inspired by developmental biology, Varghese et al. [139] have combined RP procedures with cell encapsulation to print viable freeform structures using customized ink-jet printers, with the hope that this method might provide the required signals, rules, and framework for hierarchic self-assembly. They "printed" bovine aortic endothelial cells in culture media (which they termed "bioink") onto an alginate-coated frame that they used as a scaffold, to generate a 50 mm long tube with an outer diameter of 4 mm. Smith et al. [140] have co-extruded cells suspended in polymers using a direct-write 3D bioassembly tool to create viable, patterned tissue engineering constructs. Mironov et al. [141,142] have introduced the futuristic concept of organ printing, which is the computer-aided, jet-based, three-dimensional engineering of living human organs, to overcome the obstacles of generating vascularized organs. They propose using a cell printer capable of printing single cells, cell aggregates and gels on "printing paper" comprising sequentially arranged layers of a thermoreversible gel. Sun et al. [143] have given a broad overview of computer-aided tissue scaffold design, including biomimetic modeling as well as 3D cell and organ printing.

The main advantage of RP techniques is their ability to finely control the microstructure and macrostructure of scaffolds and thus produce complex topographies from a computer model; their main drawbacks are the low resolutions achievable by the current systems and the types of polymeric materials that can be used [39]. Sachlos and Czernuska [117] have not only discussed the conventional scaffold fabrication techniques and their drawbacks but have also described various SFF techniques and how they can overcome current scaffold design limitations. Tsang and Bhatia [78] have discussed the various fabrication techniques by dividing them based on their mode of assembly, i.e., fabrication with heat, binders, light, and molding whereas Hutmacher et al. [144] have described SFF techniques by dividing them based on their processing technology. Yeong et al. [145] have well articulated the various RP techniques and their emerging sub-branches as well as compared these methods and tabulated their strengths and weaknesses.

3.2.7.7 Hydrogels

Quite often acellular scaffolds are designed to provide the required mechanical properties but they end up being difficult to populate with cells uniformly, while constructs that are able to successfully achieve uniformly high cell distribution owing to their high porosity end up being mechanically weak [78]. Thus, researchers are increasingly trying to combine structural stability with high cell density while maintaining an *in vivo*-like environment to achieve the best of both worlds by fabricating hydrogels, which are crosslinked networks of hydrophilic polymers that are capable of absorbing large amounts of fluid. Hydrogels can be degradable or nondegradable and their water content influences the viability of encapsulated cells and, thus, the rate of tissue development [146]. They are increasing in popularity due to their high water content and mechanical properties that are similar to soft tissues like cartilage [147]. Solid scaffolds provide a substrate for cells to adhere to whereas liquid and gel scaffolds function to physically entrap cells [148]. Hydrogels can be formed *in situ* within a defect site and cells can be encapsulated during the hydrogel formation process. Their mechanical properties can be controlled by altering the comonomer composition, changing the crosslinking density, and modifying the polymerization conditions (reaction time, temperature, and amount and type of solvent) [149].

Photopolymerization is a commonly used technique for making hydrogels. Visible or UV light can react with certain light-sensitive compounds called photoinitiators to form crosslinked hydrogels *in vitro*, *in vivo*, or *in situ*. Thus, photopolymerization offers several advantages over traditional polymerization methods, namely, spatial as well as temporal control over polymerization, curing rates from less than a second to a few minutes at room or physiological temperatures, minimal heat production, as well as the ability to form complex shapes that adhere and conform to the defect site [147]. Although biological systems put constraints on the use of photopolymerization *in vivo*, owing to the limits of acceptable temperatures, pH, as well as toxicity of most monomers and organic solvents, these can generally be overcome by the implementation of mild polymerization conditions (low light intensity and organic solvent levels, short irradiation time, and physiological temperature) [147].

Although it may seem like hydrogel-based scaffold systems are at a disadvantage as far as mechanical properties of the skeletal system are concerned, they do provide an environment for accelerated tissue formation which in turn provides the desired mechanical stability [148]. Ferruti et al. [150] found amphoteric poly(amido-amine) (PAA)-based hydrogels containing carboxyl and amino groups in their repeating units, to have a good potential as scaffolds, based on their cytocompatibility with fibroblasts as well as noncytotoxic degradation products, but their mechanical properties needed improvement. They further modified the PAA hydrogels by introducing side guanidine groups to improve cell adhesion and proliferation and found the mechanical properties to improve when a second PAA carrying primary amino groups was used as a crosslinking agent [151]. Hydrogels are quite often modified with cell adhesion peptides to enhance cell attachment and spreading [152–154]. Sannino et al. [155] combined the photocrosslinking reaction with a foaming process to induce an interconnected porosity within PEG-based hydrogels that had been modified with peptide sequences for enhancing cell adhesion.

Several groups are working on making hydrogels with synthetic copolymers [156] or a combination of natural and synthetic polymers [157,158]. Martens et al. [156] photoencapsulated chondrocytes in a PEG-PVA copolymer network and found DNA, glycosaminoglycan (GAG), and total collagen content to increase with culture time, resulting in homogeneously distributed neocartilageneous tissue at the end of 6 weeks. Hiemstra et al. [159] prepared PEG-PLA hydrogels to draw upon the excellent antifouling properties and renal clearance below 30 kDa of PEG and biodegradability of PLA, as well as the biocompatibility of both, to engineer cartilage. Cascone et al. [160] prepared blends of nonbiodegradable poly(vinyl alcohol) (PVA) with different biological macromolecules like hyaluronic acid, dextran, and gelatin to improve the biocompatibility of PLA and thereby produce bioartificial hydrogels as potential tissue engineering scaffolds. A unique property about hydrogels that is being increasingly exploited by tissue engineers is the ability to make them bioresponsive and thus, "intelligent biomaterials" [161]. Wang et al. [161] synthesized a phosphoester-PEG (PhosPEG) hydrogel encapsulating marrow-derived mesenchymal stem cells (MSCs) for engineering bone. The rate of hydrolytic degradation of these phosphor-containing hydrogels increased in the presence of alkaline phosphatase (ALP), a bone-derived enzyme. The presence of phosphorus also increased mineralization and PhosPEG was also found to increase gene expression of bone-specific markers.

RP techniques have also begun to be used for fabricating hydrogels. Landers et al. [162] used 3D plotting, which is 3D dispensing in a liquid medium, to fabricate thermoreversible hydrogel scaffolds with a specific external shape and well-defined internal pore structure. They were also able to surface coat the scaffold to facilitate cell adhesion and growth. Dhariwala et al. [163] entrapped Chinese hamster ovary cells in a PEO hydrogel scaffold formed using STL. The cytotoxic effect of the initiator was minimized by using 50 μL of the initiator per milliliter of medium and the exposure time to the UV laser for the *in vitro* cytotoxicity experiments was longer than what would generally be used, giving the authors confidence in the low cytotoxic effect of the initiator. However, the elastic modulus was found to be comparable to values of soft tissue like breast tissue and not cartilage. Arcaute et al. [130] also used STL to fabricate PEG hydrogels encapsulating human dermal fibroblasts, with at least 87% found to be viable up to 24 h after fabrication.

Hoffman [164] has given an excellent overview of the important physiochemical parameters and properties of hydrogels relevant to their use as matrices in tissue engineering applications along with discussing their pros and cons. Drury and Mooney [165] have given a comprehensive review of hydrogels used as tissue engineering scaffolds. They have discussed the main synthetic and natural polymers used for making hydrogel scaffolds along with scaffold design variables and have identified three categories of scaffold applications: space filling agents, bioactive molecule delivery, and cell/tissue delivery. Brandl et al. [166] have described a rational approach for designing hydrogels for tissue engineering applications with an emphasis on physical properties and outlined their impact on cell function and tissue morphogenesis.

3.2.7.8 Electrospinning

Another scaffold fabrication technique receiving increasing importance is that of electrospinning of nanofibers for the production of scaffolds due to their resemblance scale-wise to native ECM [167]. In this process nanometer-scale diameter polymer fibers are produced using electrical forces [168]. When an applied electric field creates large enough forces at the surface of a polymer solution to overcome the surface tension, an electrically charged jet is ejected that solidifies into an electrically charged fiber, which can be manipulated into various shapes by electrical forces [168]. The commonly used polymers for this method of fabrication are the aliphatic polyesters [169]. For example, PLLA for neural stem cell adhesion and differentiation [170]; PLGA for viability, growth, and differentiation of human MSCs [171]; and PCL for human dermal fibroblast adhesion [172] as well as bone formation from MSCs [173]. Li et al. [167] studied the interaction of fibroblasts and bone-marrow derived MSCs on an electrospun 500–800 nm diameter PLG nanofibrous structure. Since pores in the structure were formed by randomly oriented fibers lying loosely upon one another, the cells while migrating through the pores could possibly push aside the surrounding unresisting, but mechanically strong, fibers thereby causing the pore to expand [167]. The authors hypothesized that this type of dynamic scaffold architecture allowed cells the freedom to adjust the pore diameter according to their liking and also let them pass through relatively small pores but cautioned that their theory needed further investigation. Li et al. [174] also evaluated electrospun 700 nm diameter PCL nanofibrous scaffolds for their ability to retain the functionality of chondrocytes and proposed their use as suitable scaffolds for cartilage tissue engineering. Yoshimoto et al. [175] too successfully cultured rat mesenchymal stem cells on electrospun 400 nm ($\pm$200 nm) diameter PCL scaffolds to show their potential as suitable scaffolds for bone tissue engineering. However, they found the fibers to have varying diameters along their lengths and irregular surfaces. Chen et al. [176] were able to achieve nanofibers up to a diameter of 117 nm but found them to contain beads, which in turn adversely affected cell adhesion and growth kinetics prompting them to conclude that the uniformity and diameter of the fibers played a crucial role in modulating cell attachment and proliferation. In spite of these minor drawbacks, nanofibers hold great promise as potential scaffolds owing to their high porosity and high surface area-to-volume ratio, which are favorable parameters for cell attachment, growth, and proliferation in addition to possessing favorable mechanical properties [167].

Deng et al. [177] investigated the morphology and biocompatibility of PLA-HA hybrid nanofibrous scaffolds prepared via electrospinning. They found the surface of the fibers to be coarse due to the formation of a new COO^- surface bond and saw improved MG-63 cell attachment and proliferation compared to pure PLA scaffolds. Meng et al. [178] found NIH 3T3 cells to adhere and grow more effectively on electrospun PHBV/collagen composite nanofibrous scaffolds relative to PHBV nanofibrous scaffolds. Li et al. [179] investigated the cytocompatibility of a co-electrospun PLGA, gelatin, and α-elastin composite scaffold (PGE) as a potential material for engineering of soft tissues like heart, lung, and blood vessels. Pan et al. [180] created a highly porous electrospun dextran/PLGA scaffold by physically blending the two polymers and characterized it for different cellular responses using dermal fibroblasts from the point of view of using this composite in enhancing the healing of chronic or trauma wounds. Townsend-Nicholson and Jayasinghe [181] employed a coaxial needle arrangement where a concentrated living cell suspension flowed through the inner needle and a medical-grade, highly viscous poly(dimethylsiloxane) (PDMS) with low electrical conductivity flowed through the outer needle, to form cell-containing composite microthreads. This bionanofabrication process did not seem to affect cell viability post-electrospinning, demonstrating the feasibility of using coaxial electrospinning to fabricate active biological scaffolds.

Researchers have studied different aspects of the electrospinning process to see how varying certain parameters associated with fiber production affect their properties. Thomas et al. [182] found that differences in collector rotation speeds affected tensile strength and modulus of aligned nanofibrous PCL meshes for bone scaffolds, due to increased fiber alignment and packing as well as decrease in interfiber pore size at higher uptake rates. Similarly, Li et al. [183] found that nanofiber organization was greatly influenced by the speed of the rotating target: greater the speed of rotation better the fiber alignment, which in turn had a profound effect on mechanical properties. Pham et al. [184] utilized a multilayering technique to construct a PCL scaffold comprising alternate layers of electrospun micro- and nanofibers to combine their advantages in a single structure. Microfibers offer the advantage of providing a greater pore size that facilitates cellular penetration and diffusion of nutrients within the structure while nanofibers provide a larger surface area for attachment and cell spreading [184]. Li et al. [185] characterized the physical and biological properties of six commonly used poly(α-hydroxy esters). Moroni et al. [186] studied the effect of different fiber diameters and their surface nanotopolgy on cell seeding, adhesion, and proliferation. They found smooth fibers with a diameter of 10 μm to support optimal cell seeding and adhesion within the range analyzed, while nanoporous surfaces were found to significantly enhance cell proliferation and spreading. Vaz et al. [187] used sequential multilayering electrospinning (ME) to produce a bilayered tubular scaffold comprising an outer stiff and aligned fibrous PLA layer and an inner pliable and randomly oriented fibrous PCL layer for engineering a blood vessel.

Researchers have also attempted to incorporate nanoparticles within the fibers or use nanocomposites to enhance the abilities of electrospun scaffolds. Wutticharoenmongkol et al. [188] fabricated novel electrospun scaffolds for bone tissue engineering using a PCL solution containing nanoparticles of calcium carbonate or HA and these were successfully evaluated *in vitro* for attachment, proliferation, and alkaline phosphatase activity using human osteoblasts. Lee et al. [189] combined a nanocomposite technique along with electrospinning to produce a scaffold with two pore sizes: nanosized pores for transport of nutrients and waste, and microsized pores for cell infiltration and blood vessel invasion. This was achieved by incorporating nanosized montmorillonite platelets into PLLA solution that was subsequently electrospun and subjected to cold compression molding followed by salt leaching/gas foaming to get the microsized pores. Thomas et al. [190] created a nanocomposite scaffold by electrostatic cospinning of nanofibrous PCL and nanohydroxyapatite (nanoHA) to better mimic the features of natural ECM.

Electrospun scaffolds have also found applications in the cardiovascular and skeletal muscle tissue engineering area. van Lieshout et al. [191] compared an electrospun valvular scaffold and a knitted valvular scaffold, both made from PCL, for their suitability in engineering of the aortic valve and concluded that the ideal scaffold would need to have the strength of the knitted structure combined with the cell-filtering ability of the spun structure. Zong et al. [192] examined the growth of cardiomyocytes on electrospun nanostructured PLGA membranes with different compositions to assess cell attachment, structure, and function on these potential heart tissue constructs. Riboldi et al. [193] evaluated the suitability of a commercially available electrospun degradable block polyesterurethane called DegraPol, as a scaffold for skeletal muscle tissue engineering by characterizing their morphological, degradative, and mechanical properties.

Electrospun scaffolds have also begun to be used for releasing drugs, growth factors, and DNA. Kim et al. [194] have successfully demonstrated the incorporation and sustained release of a hydrophilic antibiotic from electrospun PLGA-based nanofibrous scaffolds without the loss of structure and drug bioactivity. Luu et al. [195] have successfully demonstrated the incorporation and controlled release of plasmid DNA from an electrospun synthetic polymer/DNA composite demonstrating its potential use for therapeutic gene delivery. The synthetic polymer comprised PLGA random copolymer and PLA-PEG block copolymer.

Other publications have discussed electrospun scaffolds in detail. For example, Nair et al. [196] have reviewed recent advances in the development of synthetic biodegradable nanofibrous scaffolds fabricated via electrospinning. Boudriot et al. [197] have reviewed the spinning parameters relevant for making scaffolds as well as discussed scaffolds composed of nanofibers. Teo et al. [198] have discussed the ECM and how electrospinning techniques combined with surface modification and cross-linking of nanofibers can help one tailor the scaffold to meet the requirements of the tissue they wish to regenerate. Electrospinniing is not the only process by which nanofibrous scaffolds can be produced. Smith and Ma [199] have discussed how self-assembly, electrospinning, and phase separation can produce nanofibrous scaffolds spanning the entire range of sizes of ECM collagen.

3.3 MODIFICATION OF SCAFFOLDS

Scaffolds can also be modified to deliver biomolecules like proteins and growth factors as well as drugs [200]. Growth factors are polypeptides that either stimulate or inhibit cellular activities like proliferation, differentiation, migration, adhesion, and gene expression [201]. Growth factors can be incorporated directly into the scaffold during or after fabrication. Sheridan et al. [92] incorporated vascular endothelial growth factor (VEGF) into PLG scaffolds during the fabrication process and released it in a controlled manner. The released VEGF was found to retain over 90% of its bioactivity. Hu et al. [202] incorporated the osteoinductive growth factor bone morphogenetic protein (BMP) into composite scaffolds made of hydroxyapatite/collagen (HAC) and PLA, which were implanted in diaphyseal defects of dogs. Histological studies revealed that BMP not only promoted osteogenesis but also caused an accelerated degradation of the scaffold material. Williams et al. [124] seeded PCL scaffolds fabricated via SLS with BMP-7 transduced fibroblasts and implanted these constructs subcutaneously in mice to evaluate the biological properties. Histological evaluation and microCT analysis confirmed the generation of bone. Grondahl et al. [203] modified poly(3-hydroxybutyrate-co-3-hydroxyvalerate) (PHBV) with acrylic acid by graft copolymerization. PHBV is used in bone tissue engineering owing to its biocompatibility, favorable degradation characteristics, suitable mechanical properties, and support of osteoblast attachment. Acrylic acid was used to induce surface hydrophilicity, to eventually improve HA growth, and increase cell compatibility. Moreover, the carboxylic acid groups that were introduced on the PHBV surface by acrylic acid were linked to glucosamine, which is a model biomolecule, to show the ability of the material to be modified for tissue engineering applications. Similarly, Ma et al. [204] introduced a stable collagen layer on the PLLA scaffold surface, via grafting of polymethacrylic acid (PMAA), to incorporate basic fibroblast growth factor (bFGF) to improve biocompatibility, enhance cell growth, and more closely mimic the natural ECM. Ennet et al. [93] incorporated VEGF either directly into PLGA scaffolds or pre-encapsulated in PLGA microspheres that were used to fabricate the scaffolds using gas foaming. The pre-encapsulation led to VEGF being embedded more deeply within the scaffold, thereby resulting in a delayed release. *In vivo*, the released VEGF significantly enhanced local angiogenesis with negligible amounts being released in the systemic circulation. Park et al. [205] co-encapsulated bovine chondrocytes and gelatin microparticles loaded with transforming growth factor-β1 (TGF-β1) in a novel injectable hydrogel composite oligo(poly(ethylene glycol) fumarate) (OPF) for growing cartilage.

Protein and drug delivery using scaffolds for the purpose of enhancing cellular activity and treating local acute inflammation, respectively, are also being pursued. Lenza et al. [206] developed bioactive scaffolds that would allow the incorporation and delivery of proteins at controlled rates for promotion of cell function and growth of soft tissue. Yoon et al. [91] fabricated dexamethasone-containing porous PLGA scaffolds by a gas foaming/salt leaching method to create a biodegradable stent for reducing intimal hyperplasia in restenosis. Dexamethasone, which is a steroidal anti-inflammatory drug, was slowly released from the PLGA scaffold in a controlled manner for over a month without showing an initial burst release and was successful in drastically suppressing the proliferation of lymphocytes and smooth muscle cells *in vitro*.

A severe drawback to direct protein delivery is rapid degradation *in vivo* and limited stability even if encapsulated in a polymeric delivery vehicle [207]. A promising solution is the use of localized gene therapy to promote the creation of the required growth factor at the specific site of interest [207]. Thus, scaffolds can also be used as vehicles for gene delivery to promote localized transgene expression for inducing formation of functional tissue [208]. Jang et al. [208] employed substrate-mediated delivery that involves immobilization of DNA complexes on the polymer surface for eventual delivery to cells growing on the polymer. They studied the immobilization of polyethylenimine (PEI)/DNA complex and eventual cellular transfection on PLG scaffolds. With this technique they were able to uniformly distribute the DNA throughout the scaffold thereby transfecting more than 60% of the cells using low quantities of DNA at the surface.

3.4 TYPES OF SCAFFOLD MATERIALS

Scaffold materials are either fabricated from synthetic polymers or derived from naturally occurring ones. Synthetic polymers have the advantage of possessing highly controllable properties of strength, rate of degradation and microstructure, as well as batch-to-batch consistency [81,209]. Naturally derived materials possess the potential advantage of biological recognition that might enhance cell attachment and function [81]. However, they have certain disadvantages as well which include limited control of mechanical properties, biodegradability and batch-to-batch consistency, limited availability contributing to their being expensive, possible exhibition of immunogenicity, and possession of pathogenic impurities [81]. Velema and Kaplan [6] have written an excellent review discussing three important types from the two main classes of naturally derived polymers that are used for fabricating scaffolds. These are polysaccharides (alginate, chitosan, and hyaluronan) and fibrous proteins (collagen, silk fibroin, and elastin). We will only be discussing synthetic polymers here. Nair and Laurencin [210] have very elegantly discussed the primary synthetic and natural polymers used for tissue engineering and drug delivery while Gunatillake and Adhikari [209] have detailed the major classes of synthetic biodegradable polymers in tissue engineering and discussed them with regards to synthesis, properties, as well as their biocompatibility and biodegradability. Holland and Mikos [211] have discussed the different synthetic degradable polymers used for drug delivery in bone tissue engineering applications while Agrawal and Ray [3] have discussed the major biodegradable synthetic polymers for musculoskeletal tissue engineering applications and their methods of fabrication.

3.4.1 POLYESTERS

The most commonly used synthetic polymers are the aliphatic (α-hydroxy) polyesters. These include PGA, PLA, and their copolymer (PLGA), which have been widely used in a number of clinical applications, mainly as resorbable sutures as well as plates and fixtures for fracture fixation devices [209,210]. PGA is highly crystalline with a melting point exceeding 200°C and a glass transition temperature (T_g) around 35°C–40°C [209,210]. Owing to its high crystallinity, it possesses high tensile strength and modulus as well as low solubility in most organic solvents. PLA is generally used in the form of poly(L-lactic acid) (PLLA) and poly(DL-lactic acid) (PDLLA) to fabricate scaffolds. PLLA is semicrystalline with the degree of crystallinity depending on the molecular weight and processing parameters. It too possesses high tensile strength and modulus and has a melting point of 170°C and a T_g of 60°C–65°C. PDLLA, on the other hand, is amorphous with a T_g of 55°C–60°C. The chemical structures of PLA and PGA are similar except that PLA has a pendant methyl group making it more hydrophobic, and thereby more resistant to hydrolytic attack, than PGA. This produces differences in the degradation kinetics of the two and thus the degradation rate of their copolymer (PLGA) is controlled by the ratio in which these two are present [3]. PLA, PGA, and PLGA undergo bulk degradation by random hydrolysis of their ester linkages, whereby material is lost from the entire polymer volume simultaneously due to penetration of water into the bulk of the scaffold. PLA degrades to form lactic acid, which is normally present in the body, and enters the tricarboxylic acid (TCA) cycle to be excreted as water and carbon dioxide. PGA is broken down by hydrolysis and nonspecific esterases and carboxypeptidases. The glycolic acid monomer is either excreted via urine or enters the TCA cycle [3]. Some of the disadvantages of the aliphatic polyesters include inferior mechanical properties, release of acidic degradation products, and limited processability [209].

PCL is a semicrystalline polyester that melts at around 60°C and has a T_g of −60°C. It degrades at a much slower rate compared to the other aliphatic polyesters and has hence been widely used in the area of long-term controlled drug delivery. It is flexible, easy to process, as well as structurally and thermally stable, while simultaneously being less sensitive to environmental changes [85]. It possesses an exceptional ability to form blends with a variety of polymers thereby making it a widely used material for scaffold fabrication, especially pertaining to cartilage and bone tissue engineering applications [4]. PCL degrades hydrolytically via both bulk and surface erosion with 5-hydroxyhexanoic acids (caproic acids) being the degradation product [17].

3.4.2 Polyfumarates

Poly(propylene fumarate) (PPF) is the most widely studied in the category of copolyesters based on fumaric acid. It holds promise as a bone tissue engineering material [212] for filling skeletal defects as it has mechanical properties similar to that of trabecular bone and possesses the ability to cure *in situ* thereby providing skeletal defects of any shape or size to be filled with minimal intervention [213]. PPF-based polymers are available as injectable systems that employ chemical cross-linking thereby facilitating treatment of deep crevices in bone and defects of nonuniform shapes by being cross-linked *in situ* [214]. PPF possesses unsaturated sites in its backbone that are used in cross-linking reactions resulting in complex structures [209]. Since achieving PPF of high molecular weight is difficult owing to side reactions due to the presence of the backbone double bond, ceramics such as TCP, calcium carbonate, or calcium sulfate are incorporated to improve mechanical properties [209,214]. β-TCP not only increased mechanical strength but also acted as a buffer by minimizing pH change during degradation [214]. Peter et al. [215] concluded that injectable PPF/β-TCP pastes could be prepared with handling characteristics suitable for clinical orthopaedic applications and found the mechanical properties of the cured composites to be suitable for trabecular bone replacement. They also investigated the *in vivo* degradation and biocompatibility of PPF/β-TCP composites and found them to elicit a mild initial inflammatory response followed by thin fibrous encapsulation [216] and also found the composite to be osteoconductive *in vitro* [217]. PPF/β-TCP has also been cross-linked with other polymers, like poly(ethylene glycol)-dimethacrylate (PEG-DMA) [218], while some PPF composites have been reinforced with nanoparticles [20] to enhance their mechanical properties for orthopaedic tissue engineering applications. PPF composites have also been successfully employed as carriers of microspheres carrying model drugs [219] or microparticles encapsulating osteoblasts for bone tissue engineering applications [220]. The bioactivity of PPF was found to be augmented *in vivo* by the incorporation of nano-HA [221] while PPF coated with rhTGF-β1 was found to adequately induce bone formation in the cranium of rabbits [222]. Some novel fumarates, like poly(ε-caprolactone fumarate) [223], poly(ethylene glycol fumarate), and their copolymer, are also under investigation for diverse tissue engineering applications [224]. PPF undergoes bulk degradation by hydrolysis to produce fumaric acid which is a naturally occurring substance in the tricarboxylic acid cycle (Kreb's cycle), and propylene glycol [209].

3.4.3 Polyanhydrides

Polyanhydrides are surface-eroding polymers with low hydrolytic stability making them ideal candidates for drug delivery applications [225]. They possess a highly hydrophobic backbone that prevents water from penetrating into the scaffold interior, and a hydrolytically sensitive anhydride bond that confines the degradation to the surface, resulting in linear mass loss kinetics and zero-order drug release kinetics when used as drug delivery systems [210,226]. Their rapid degradation contributes to their poor mechanical properties prompting the incorporation of imide segments to create scaffolds from poly(anhydride-co-imide) [227], especially for orthopaedic applications [228]. Ibim et al. [229] studied the biocompatibility and osteocompatibility of poly(anhydride-co-imide) and found them to produce endosteal and cortical bone growth and a local tissue response similar to PLGA. They advocated the use of these polymers in weight-bearing orthopaedic applications. Burkoth et al. [230] used porogen leaching to create polyanhydride constructs which could be eventually filled with osteoblasts photoencapsulated in a hydrogel to potentially create a synthetic allograft for engineering bone. The degradation rate of polyanhydrides, which degrade by hydrolysis of the anhydride linkage, can be altered by making simple changes to the polymer backbone structure via a judicious choice of the diacid monomer [209]. Combining different amounts of these monomers could produce polymers with custom-designed degradation properties [214].

3.4.4 Poly(Ortho Esters)

Poly(ortho esters) (POEs) undergo surface erosion and the rate of degradation can be controlled by using diols having varying degrees of chain flexibility or with the incorporation of acidic and basic excipients [210]. Andriano et al. [231] found POE to possess better control over polymer mass loss with new tissue formation as well as better structural integrity relative to 50:50 PLGA. Also, bone mineral density in POE scaffolds was found to be 25% higher than PLGA scaffolds, although the amount of bone formed was inconsequential. Ng et al. [232] hypothesized that the appropriate choice of diols and their ratios could result in the formation of POEs that were viscous fluids in the 37°C–45°C range that converted to nondeformable highly viscous materials at or below 37°C. This could be of tremendous use in cases where slightly warmed materials could be injected into the desired site of injury or drug release, where they would eventually solidify at body temperature into, possibly, drug-releasing scaffolds. Incorporation of proteins or antigens could be achieved by simple mixing with the gently warmed polymer without the need of solvents or water. Kellomaki et al. [233] found the rate of hydrolysis of two POEs, as measured by the strength loss of the polymers, to be too rapid for load-bearing orthopaedic applications. Ng et al. [232] used POEs containing varying amounts of glycolic acid dimer segments in the polymer backbone to accurately control the erosion rate that proceeds by zero-order kinetics. This polymer, when placed in an aqueous environment would hydrolyze to produce glycolic acid which would catalyze hydrolysis of the ortho ester linkages of the polymer backbone. Thus, by varying the amount of acid segment in the polymer backbone one could finely tune the rate of degradation from a few days to several months [210].

3.4.5 Poly(Amino Acids) or Polycarbonates

Synthetic poly(amino acids) are very similar to naturally occurring proteins but possess low degradation rates, unfavorable mechanical properties, and immunogenicity [234]. Thus, amino acids have been used as monomeric building blocks in polymers lacking the conventional backbone structure present in peptides to overcome these drawbacks [234]. Tyrosine-derived polycarbonates are the most extensively studied from this group and possess a T_g in the range of 52°C–93°C and a decomposition temperature exceeding 290°C [234]. The backbone carbonate bond is hydrolyzed faster than the pendant chain ester bond, except under very acidic conditions ($pH \leq 3$) when the rates get interchanged due to acid-catalyzed hydrolysis of the ester bond [235]. Final degradation products of polycarbonates *in vitro* are desaminotyrosyl-tyrosine and alcohol while *in vivo* one can expect the former to enzymatically degrade into desaminotyrosine and L-tyrosine [234]. From a degradation–biocompatibility perspective, the tyrosine-derived polycarbonates were found to be similar to PLA when studied in a canine bone chamber model [236] and showed good potential as orthopaedic implant materials [237]. To decrease the hydrolytic stability of polycarbonates, the carbonyl oxygen was replaced by an imino group resulting in the production of polyiminocarbonates that had hydrolytically degradable fibers that retained the strength of polycarbonates [238]. Polycarbonates having an ethyl ester pendant group have shown to be quite osteoconductive, with good bone apposition and possessing adequate mechanical properties for load-bearing bone fixations [209]. Meechaisue et al. [239] used electrospinning to produce a mat of poly(DTE carbonate) fibers as tissue scaffolding material that supported the adhesion and propagation of three different cell lines.

3.4.6 Polyphosphazenes

Polyphosphazenes have an inorganic backbone of alternating phosphorus and nitrogen atoms that can be rendered hydrolytically unstable by substituting with appropriate organic side groups on the phosphorus atoms. Their good biocompatibility, synthetic flexibility, hydrolytic instability, nontoxic degradation products, ease of fabrication, and matrix permeability make them highly adaptable for tissue engineering [210], drug delivery [240], and gene delivery [241] applications. The pentavalency of phosphorus in polyphosphazenes provides active sites for attachment of drug molecules. The degradation products of these polymers are phosphates, ammonia, and the corresponding side groups, all of which are neutral and nontoxic [210]. Laurencin et al. [242] modulated cell growth and polymer degradation by varying the nature of the hydrolytically unstable side chains of the polyphosphazenes. They found the polymer to support osteoblast attachment and proliferation showing potential for skeletal tissue regeneration [243]. Ambrosio et al. [244] and Krogman et al. [245] designed blends of polyphosphazenes with PLGA to decrease the acidity of the degradation products of PLGA via the neutralizing effect produced by the degradation products of the polyphosphazene. Polyphosphazene nonwoven nanofiber meshes created by electrospinning were found to promote adhesion and proliferation of osteoblast-like cells [246]. Greish et al. [247] formed composites of HA and polyphosphazenes at physiologic temperatures via a dissolution–precipitation process that resulted in a mildly alkaline environment suitable for deprotonation of the acidic polyphosphazene and formation of calcium cross-links. Carampin et al. [248] used electrospinning to generate flat or tubular matrices of polyphosphazene comprising ultrathin fibers to mimic blood vessels. Neuromicrovascular endothelial cells formed a monolayer on the whole surface after 16 days of incubation, thereby demonstrating the feasibility of the polymer to form human tissues like vessels and cardiac valves.

3.4.7 Composites

As discussed earlier, scaffolds made of composites allow the tailoring of mechanical properties and resorption rates according to the specific needs of the implant site, as well as enhance bioactivity [5,14,21]. The use of composites is mostly in the arena of musculoskeletal tissue engineering, mainly bone, as that is where tailoring of mechanical properties is most crucial. The most commonly used composite combinations comprise HA, TCP, or bioactive glass particles or fibers used as fillers or coatings or both in PLA, PGA, or other resorbable polymers [14,21]. Zhang and Ma [249] incorporated HA into PLLA and PLGA to fabricate scaffolds for bone tissue engineering that had better osteoconductive properties as well as superior buffering capability and improved mechanical properties. Marra et al. [250] used blends of PLGA, PCL, and HA while Roether et al. [14] fabricated PDLLA foams coated with and impregnated by Bioglass as scaffolds for bone tissue engineering. Taboas et al. [40] created biphasic scaffolds with mechanically interdigitated PLA and sintered HA regions having 600 and 500 μm wide global pores, respectively.

3.5 CONCLUSIONS

Scaffolds form one of the most important components of a tissue engineering construct. Their functions, requirements, methods of fabrication, modifications, as well as commonly used synthetic scaffold materials have been discussed with the intention of impressing upon the reader the amount of research that is being done to create the ideal scaffold. One must appreciate that the requirements of different tissues in the body are unique and although most scaffold materials satisfy these requirements to varying degrees, there are some materials or combinations of materials that are better suited for specific applications. It would be extremely helpful, but difficult, to compare the best biomaterial candidates for different tissue applications by considering a common set of parameters and evaluation procedure so as to determine which one works the best. With newly emerging scaffold fabrication techniques like electrospinning as well as the continuous modification of existing methods, like cell and organ printing, along with the emergence of composite and hybrid materials as well as the benefits of adding nanoparticles/nanocomposites, the quest for finding the best scaffold seems to be within reach in the not too distant future.

REFERENCES

1. Langer, R. and J.P. Vacanti, Tissue engineering. *Science*, 1993, 260(5110): 920–926.
2. Muschler, G.F., C. Nakamoto, and L. Griffith, Engineering principles of clinical cell-based tissue engineering. *J. Bone Joint Surg*., 2004, 86-A(7): 1541–1558.
3. Agrawal, C.M. and R.B. Ray, Biodegradable polymeric scaffolds for musculoskeletal tissue engineering. *J. Biomed. Mater. Res*., 2001, 55(2): 141–150.
4. Hutmacher, D.W., T. Schantz, I. Zein, K.W. Ng, S.H. Teoh, and K.C. Tan, Mechanical properties and cell cultural response of polycaprolactone scaffolds designed and fabricated via fused deposition modeling. *J. Biomed. Mater. Res*., 2001, 55(2): 203–216.
5. Boccaccini, A.R. and J.J. Blaker, Bioactive composite materials for tissue engineering scaffolds. *Expert Rev. Med. Devices*, 2005, 2(3): 303–317.
6. Velema, J. and D. Kaplan, Biopolymer-based biomaterials as scaffolds for tissue engineering. *Advances in Biochemical Engineering/Biotechnology*, 2006, 102: 187–238.
7. Rickert, D., A. Lendlein, I. Peters, M.A. Moses, and R.-P. Franke, Biocompatibility testing of novel multifunctional polymeric biomaterials for tissue engineering applications in head and neck surgery: An overview. *Eur. Arch. Otorhinolaryngol*., 2006, 263: 215–222.
8. Hutmacher, D.W., Scaffolds in tissue engineering bone and cartilage. *Biomaterials*, 2000, 21(24): 2529–2543.
9. Vert, M., S.M. Li, G. Spenlehauer, and P. Guerin, Bioresorbability and biocompatibility of aliphatic polyesters. *J. Mater. Sci.: Mater. Med*., 1992, 3(6): 432–446.
10. Agrawal, C.M. and K.A. Athanasiou, Technique to control pH in vicinity of biodegrading PLA-PGA implants. *J. Biomed. Mater. Res*., 1997, 38(2): 105–114.
11. Hutmacher, D.W., A. Kirsch, K.L. Ackermann, and M.B. Hurzeler, A tissue engineered cell-occlusive device for hard tissue regeneration—a preliminary report. *Int. J. Periodontics Restorative Dent*., 2001, 21(1): 49–59.
12. Maquet, V., A.R. Boccaccini, L. Pravata, I. Notingher, and R. Jerome, Porous poly(α-hydroxyacid)/Bioglass® composite scaffolds for bone tissue engineering. I: Preperation and *in vitro* characterisation. *Biomaterials*, 2004, 25(18): 4185–4194.
13. Shikinami, Y. and M. Okuno, Bioresorbable devices made of forged composites of hydroxyapatite (HA) particles and poly-L-lactide (PLLA): Part I. Basic characteristics. *Biomaterials*, 1999, 20(9): 859–877.
14. Roether, J.A., J.E. Gough, A.R. Boccaccini, L.L. Hench, V. Maquet, and R. Jerome, Novel bioresorbable and bioactive composites based on bioactive glass and pollactide foams for bone tissue engineering. *J. Mater. Sci.: Mater. Med*., 2002, 13(12): 1207–1214.
15. Stamboulis, A., L.L. Hench, and A.R. Boccaccini, Mechanical properties of biodegradable polymer sutures coated with bioactive glass. *J. Mater. Sci.: Mater. Med*., 2002, 13(9): 843–848.
16. Rezwan, K., Q.Z. Chen, J.J. Blaker, and A.R. Boccaccini, Biodegradable and bioactive porous polymer/inorganic composite scaffolds for bone tissue engineering. *Biomaterials*, 2006, 27(18): 3413–3431.

17. Ang, K.C., K.F. Leong, C.K. Chua, and M. Chandrasekaran, Compressive properties and degradability of poly(ε-caprolactone)/hydroxyapatite composites under accelerated hydrolytic degradation. *J. Biomed. Mater. Res. A*, 2007, 80A(3): 655–660.
18. Freeman, J.W., M.D. Woods, and C.T. Laurencin, Tissue engineering of the anterior cruciate ligament using a braid-twist scaffold design. *J. Biomech*., 2007, 40(9): 2029–2036.
19. Webster, T.J. and E.S. Ahn, Nanostructured biomaterials for tissue engineering bone. *Adv. Biochem. Eng./Biotechnol.*, 2007, 103: 275–308.
20. Horch, R.A., N. Shahid, A.S. Mistry, M.D. Timmer, A.G. Mikos, and A.R. Barron, Nanoreinforcement of poly(propylene fumarate)-based networks with surface modified alumoxane nanoparticles for bone tissue engineering. *Biomacromolecules*, 2004, 5(5): 1990–1998.
21. Misra, S.K., S.P. Valappil, I. Roy, and A.R. Boccaccini, Polyhydroxyalkanoate (PHA)/inorganic phase composites for tissue engineering applications. *Biomacromolecules*, 2006, 7(8): 2249–2258.
22. Hench, L.L., Bioceramics. *J. Am. Ceram. Soc*., 1998, 81(7): 1705–1728.
23. Bilodeau, K. and D. Mantovani, Bioreactors for tissue engineering: Focus on mechanical constraints. A comparative review. *Tissue Eng*., 2006, 12(8): 2367–2383.
24. Androjna, C., R.K. Spragg, and K.A. Derwin, Mechanical conditioning of cell-seeded small intestine submucosa: A potential tissue-engineering strategy for tendon repair. *Tissue Eng*., 2007, 13(2): 233–243.
25. Mahmoudifar, N. and P.M. Doran, Tissue engineering of human cartilage in bioreactors using single and composite cell-seeded scaffolds. *Biotechnol. Bioeng*., 2005, 91(3): 338–355.
26. Darling, E.M. and K.A. Athanasiou, Articular cartilage bioreactors and bioprocesses. *Tissue Eng*., 2003, 9(1): 9–26.
27. Jeong, S.I., J.H. Kwon, J.I. Lim, S.-W. Cho, Y. Jung, W.J. Sung, S.H. Kim, Y.H. Kim, Y.M. Lee, B.-S. Kim, C.Y. Choi, and S.-J. Kim, Mechano-active tissue engineering of vascular smooth muscle using pulsatile bioreactors and elastic PLCL scaffolds. *Biomaterials*, 2005, 26(12): 1405–1411.
28. Shachar, M. and S. Cohen, Cardiac tissue engineering, ex-vivo: Design principles in biomaterials and bioreactors. *Heart Fail. Rev*., 2003, 8(3): 271–276.
29. Akhyari, P., P.W.M. Fedak, R.D. Weisel, T.-Y.J. Lee, S. Verma, D.A.G. Mickle, and R.-K. Li, Mechanical stretch regimen enhances the formation of bioengineered autologous cardiac muscle grafts. *Circulation*, 2002, 106(12 (Supplement I)): I-137–I-142.
30. Yarlagadda, P.K.D.V., M. Chandrasekharan, and J.Y.M. Shyan, Recent advances and current developments in tissue scaffolding. *Biomed. Mater. Eng*., 2005, 15(3): 159–177.
31. Moroni, L., J.A.A. Hendriks, R. Schotel, J.R.D. Wijn, and C.A.V. Blitterswijk, Design of biphasic polymeric 3-dimensional fiber deposited scaffolds for cartilage tissue engineering applications. *Tissue Eng*., 2007, 13(2): 361–371.
32. Liu, X., Y. Won, and P. Ma, Porogen-induced surface modification of nano-fibrous poly(L-lactic acid) scaffolds for tissue engineering. *Biomaterials*, 2006, 27(21): 3980–3987.
33. Li, J., A. Beaussart, Y. Chen, and A.F. Mak, Transfer of apatite coating from porogens to scaffolds: Uniform apatite coating within porous poly(DL-lactic-co-glycolic acid) scaffold *in vitro*. *J. Biomed. Mater. Res. A*, 2007, 80(1): 226–233.
34. Cai, K., K. Yao, Z. Yang, and X. Li, Surface modification of three-dimensional poly(D,L-lactic acid) scaffolds with baicalin: A histological study. *Acta Biomater*., 2007, 3(4): 597–605.
35. Koegler, W.S. and L.G. Griffith, Osteoblast response to PLGA tissue engineering scaffolds with PEO modified surface chemistries and demonstration of patterned cell response. *Biomaterials*, 2004, 25(14): 2819–2830.
36. Tiaw, K.S., S.W. Goh, M. Hong, Z. Wang, B. Lan, and S.H. Teoh, Laser surface modification of poly (ε-caprolactone) (PCL) membrane for tissue engineering applications. *Biomaterials*, 2005, 26(7): 763–769.
37. Nakayama, Y., S. Nishi, H. Ishibashi-Ueda, and T. Matsuda, Surface microarchitectural design in biomedical applications: *In vivo* analysis of tissue ingrowth in excimer laser-directed micropored scaffold for cardiovascular tissue engineering. *J. Biomed. Mater. Res*., 2000, 51(3): 520–528.
38. Karande, T.S., J.L. Ong, and C.M. Agrawal, Diffusion in musculoskeletal tissue engineering scaffolds: Design issues related to porosity, permeability, architecture and nutrient mixing. *Ann. Biomed. Eng*., 2004, 32(12): 1728–1743.
39. Yang, S., K.-F. Leong, Z. Du, and C.-K. Chua, The design of scaffolds for use in tissue engineering. Part II. Rapid prototyping techniques. *Tissue Eng*., 2002, 8(1): 1–11.

40. Taboas, J.M., R.D. Maddox, P.H. Krebsbach, and S.J. Hollister, Indirect solid free form fabrication of local and global porous, biomimetic and composite 3D polymer–ceramic scaffolds. *Biomaterials*, 2003, 24(1): 181–194.
41. Ishaug-Riley, S., G.M. Crane, M.J. Miller, A.W. Yasko, M.J. Yaszemski, and A.G. Mikos, Bone formation by three-dimensional stromal osteoblast culture in biodegradable polymer scaffolds. *J. Biomed. Mater. Res.*, 1997, 36(1): 17–28.
42. Sosnowski, S., P. Wozniak, and M. Lewandowska-Szumiel, Polyester scaffolds with bimodal pore size distribution of tissue engineering. *Macromol. Biosci.*, 2006, 6(6): 425–434.
43. Draghi, L., S. Resta, M.G. Pirozzolo, and M.C. Tanzi, Microspheres leaching for scaffold porosity control. *J. Mater. Sci.: Mater. Med.*, 2005, 16(12): 1093–1097.
44. Oh, S.H., I.K. Park, J.M. Kim, and J.H. Lee, *In vitro* and *in vivo* characteristics of PCL scaffolds with pore size gradient fabricated by a centrifugation method. *Biomaterials*, 2007, 28(9): 1664–1671.
45. Woodfield, T.B., C.A.V. Blitterswijk, J.D. Wijn, T.J. Sims, A.P. Hollander, and J. Riesle, Polymer scaffolds fabricated with pore-size gradients as a model for studying the zonal organization within tissue-engineered cartilage constructs. *Tissue Eng.*, 2005, 11(9–10): 1297–1311.
46. Harley, B.A., A.Z. Hastings, I.V. Yannas, and A. Sannino, Fabricating tubular scaffolds with a radial pore size gradient by a spinning technique. *Biomaterials*, 2006, 27(6): 866–874.
47. Mooney, D.J., K. McNamara, D. Hern, J.P. Vacanti, and R. Langer, Stablized polyglycolic acid fibre-based tubes for tissue engineering. *Biomaterials*, 1996, 17(2): 115–124.
48. Vacanti, J.P., M.A. Morse, W.M. Saltzman, A.J. Domb, A. Perez-Atayde, R. Langer, C.L. Mazzoni, and C. Breuer, Selective cell transplantation using bioabsorbable artificial polymers as matrices. *J. Pediatr. Surg.*, 1988, 23: 3–9.
49. Freed, L.E., G. Vunjak-Novakovic, R.J. Biron, D.B. Eagles, D.C. Lesnoy, S.K. Barlow, and R. Langer, Biodegradable polymer scaffolds for tissue engineering. *Biotechnology (NY)*, 1994, 12(7): 689–693.
50. Goldstein, A.S., G. Zhu, G.E. Morris, R.K. Meszlenyi, and A.G. Mikos, Effect of osteoblastic culture conditions on the structure of poly(DL-lactic-co-glycolic acid) foam scaffolds. *Tissue Eng.*, 1999, 5(5): 421–433.
51. Agrawal, C.M., J.S. McKinney, D. Lanctot, and K.A. Athanasiou, Effects of fluid flow on the *in vitro* degradation kinetics of biodegradable scaffolds for tissue engineering. *Biomaterials*, 2000, 21(23): 2443–2452.
52. Wu, L. and J. Ding, Effects of porosity and pore size on *in vitro* degradation of three-dimensional porous poly(D,L-lactide-co-glycolide) scaffolds for tissue engineering. *J. Biomed. Mater. Res. A*, 2005, 75(4): 767–777.
53. Lin, C.Y., N. Kikuchi, and S.J. Hollister, A novel method for biomaterial scaffold internal architecture design to match bone elastic properties with desired porosity. *J. Biomech.*, 2004, 37(5): 623–636.
54. Howk, D. and T.M. Chu, Design variables for mechanical properties of bone tissue scaffolds. *Biomed. Sci. Instrum.*, 2006, 42: 278–283.
55. Xie, J., M. Ihara, Y. Jung, I.K. Kwon, S.H. Kim, Y.H. Kim, and T. Matsuda, Mechano-active scaffold design based on microporous poly(L-lactide-co-epsilon-caprolactone) for articular cartilage tissue engineering: Dependence of porosity on compression force-applied mechanical behaviors. *Tissue Eng.*, 2006, 12(3): 449–458.
56. Moroni, L., J.R.D. Wijn, and C.A.V. Blitterswijk, 3D fiber-deposited scaffolds for tissue engineering: Influence of pores geometry and architecture on dynamic mechanical properties. *Biomaterials*, 2006, 27(7): 974–985.
57. Griffon, D.J., M.R. Sedighi, D.V. Schaeffer, J.A. Eurell, and A.L. Johnson, Chitosan scaffolds: Interconnective pore size and cartilage engineering. *Acta Biomater.*, 2006, 2(3): 313–320.
58. Lee, K.W., S. Wang, L. Lu, E. Jabbari, B.L. Currier, and M.J. Yaszemski, Fabrication and characterization of poly(propylene fumarate) scaffolds with controlled pore structures using 3-dimensional printing and injection molding. *Tissue Eng.*, 2006, 12(10): 2801–2811.
59. Suh, S.W., J.Y. Shin, J. Kim, J. Kim, C.H. Beak, D.-I. Kim, H. Kim, S.S. Jeon, and I.-W. Choo, Effect of different particles on cell proliferation in polymer scaffolds using a solvent-casting and particulate leaching technique. *ASAIO J.*, 2002, 48: 460–464.
60. Hou, Q., D.W. Grijpma, and J. Feijen, Porous polymeric structures for tissue engineering prepared by a coagulation, compression moulding and salt leaching technique. *Biomaterials*, 2003, 24: 1937–1947.

61. Murphy, W.L., R.G. Dennis, J.L. Kileny, and D.J. Mooney, Salt fusion: An approach to improve pore interconnectivity within tissue engineering scaffolds. *Tissue Eng*., 2002, 8(1): 43–52.
62. Gross, K.A. and L.M. Rodriguez-Lorenzo, Biodegradable composite scaffolds with an interconnected spherical network for bone tissue engineering. *Biomaterials*, 2004, 25(20): 4955–4962.
63. Hou, Q., D.W. Grijpma, and J. Feijen, Preparation of interconnected highly porous polymeric structures by a replication and freeze-drying process. *J. Biomed. Mater. Res*., 2003, 67(2): 732–740.
64. Darling, A.L. and W. Sun, 3D microtomographic characterization of precision extruded poly-ε-caprolactone scaffolds. *J. Biomed. Mater. Res. B: Appl. Biomater*., 2004, 70B(2): 311–317.
65. Wang, F., L. Shor, A. Darling, S. Khalil, W. Sun, S. Guceri, and A. Lau, Precision extruding deposition and characterization of cellular poly-ε-caprolactone tissue scaffolds. *Rapid Prototyp. J*., 2004, 10(1): 42–49.
66. Moore, M.J., E. Jabbari, E.L. Ritman, L. Lu, B.L. Currier, A.J. Windebank, and M.J. Yaszemski, Quantitative analysis of interconnectivity of porous biodegradable scaffolds with micro-computed tomography. *J. Biomed. Mater. Res. A*, 2004, 71(2): 258–267.
67. Li, S.H., J.R.D. Wijn, P. Layrolle, and K. de Groot, Synthesis of macroporous hydroxyapatite scaffolds for bone tissue engineering. *J. Biomed. Mater. Res*., 2002, 61: 109–120.
68. Agrawal, C.M., J.S. McKinney, D. Huang, and K.A. Athanasiou, The use of the vibrating particle technique to fabricate highly porous and permeable biodegradable scaffolds, in *Synthetic Bioabsorbable Polymers for Implants*, ASTM STP 1396, C.M. Agrawal, J.E. Parr, and S.T. Lin, Editors. 2000, American Society for Testing and Materials: West Conshohocken, PA. pp. 99–114.
69. LeBaron, R.G. and K.A. Athanasiou, Ex vivo synthesis of articular cartilage. *Biomaterials*, 2000, 21: 2575–2587.
70. Athanasiou, K.A., J.P. Schmitz, and C.M. Agrawal, The effects of porosity on degradation of PLA-PGA implants. *Tissue Eng*., 1998, 4: 53–63.
71. Li, S., J.R.D. Wijn, J. Li, P. Layrolle, and K. de Groot, Macroporous biphasic calcium phosphate scaffold with high permeability/porosity ratio. *Tissue Eng*., 2003, 9(3): 535–548.
72. Wang, H., J. Pieper, F. Peters, C.A. van Blitterswijk, and E.N. Lamme, Synthetic scaffold morphology controls human dermal connective tissue formation. *J. Biomed. Mater. Res. A*, 2005, 74(4): 523–532.
73. Hollister, S.J., C.Y. Lin, E. Saito, C.Y. Lin, R.D. Schek, J.M. Taboas, J.M. Williams, B. Partee, C.L. Flanagan, A. Diggs, E.N. Wilke, G.H. Van Lenthe, R. Muller, T. Wirtz, S. Das, S.E. Feinberg, and P.H. Krebsbach, Engineering craniofacial scaffolds. *Orthod. Craniofac. Res*., 2005, 8(3): 162–173.
74. Huang, Y.C., Y.Y. Huang, C.C. Huang, and H.C. Liu, Manufacture of porous polymer nerve conduits through a lyophilizing and wire-heating process. *J. Biomed. Mater. Res, B: Appl. Biomater*., 2005, 74(1): 659–664.
75. Spain, T.L., C.M. Agrawal, and K.A. Athanasiou, New technique to extend the useful life of a biodegradable cartilage implant. *Tissue Eng*., 1998, 4(4): 343–352.
76. Neves, A.A., N. Medcalf, M. Smith, and K.M. Brindle, Evaluation of engineered meniscal cartilage constructs based on different scaffold geometries using magnetic resonance imaging and spectroscopy. *Tissue Eng*., 2006, 12(1): 53–62.
77. Knackstedt, M.A., C.H. Arns, T.J. Senden, and K. Gross, Structure and properties of clinical coralline implants measured via 3D imaging and analysis. *Biomaterials*, 2006, 27(13): 2776–2786.
78. Tsang, V.L. and S.N. Bhatia, Fabrication of three-dimensional tissues. *Adv. Biochem. Eng./Biotechnol*., 2007, 103: 189–205.
79. Widmer, M.S., P.K. Gupta, L. Lu, R.K. Meszlenyi, G.R.D. Evans, K. Brandt, T. Savel, A. Gurlek, C.W.P. Jr, and A.G. Mikos, Manufacture of porous biodegradable polymer conduits by an extrusion process for guided tissue regeneration. *Biomaterials*, 1998, 19: 1945–1955.
80. Ma, P.X. and J.-W. Choi, Biodegradable polymer scaffolds with well-defined interconnected spherical pore network. *Tissue Eng*., 2001, 7(1): 23–33.
81. Liu, X. and P.X. Ma, Polymeric scaffolds for bone tissue engineering. *Ann. Biomed. Eng*., 2004, 32(3): 477–486.
82. Whang, K., T.K. Goldstick, and K.E. Healy, A biodegradable polymer scaffold for delivery of osteotropic factors. *Biomaterials*, 2000, 21: 2545–2551.
83. Li, J.P., J.R. de Wijn, C.A. van Blitterswijk, and K. de Groot, Porous Ti6Al4V scaffold directly fabricating by rapid prototyping: Preparation and *in vitro* experiment. *Biomaterials*, 2006, 27(8): 1223–1235.

84. Sodian, R., P. Fu, C. Lueders, D. Szymanski, C. Fritsche, M. Gutberlet, S.P. Hoerstrup, H. Hausmann, T. Lueth, and R. Hetzer, Tissue engineering of vascular conduits: Fabrication of custom-made scaffolds using rapid prototyping techniques. *Thorac. Cardiovasc. Surg.*, 2005, 53(3): 144–149.
85. Zein, I., D.W. Hutmacher, K.C. Tan, and S.H. Teoh, Fused deposition modeling of novel scaffold architectures for tissue engineering applications. *Biomaterials*, 2002, 23: 1169–1185.
86. Mooney, D.J., D.F. Baldwin, N.P. Suh, J.P. Vacanti, and R. Langer, Novel approach to fabricate porous sponges of poly(D,L-lactic-co-glycolic acid) without the use of organic solvents. *Biomaterials*, 1996, 17(14): 1417–1422.
87. Harris, L.D., B.S. Kim, and D.J. Mooney, Open pore biodegradable matrices formed with gas foaming. *J. Biomed. Mater. Res.*, 1998, 42(3): 396–402.
88. Maspero, F.A., K. Ruffieux, B. Muller, and E. Wintermantel, Resorbable defect analog PLGA scaffolds using CO_2 as solvent: Structural characterization. *J. Biomed. Mater. Res.*, 2002, 62: 89–98.
89. Cooper, A.I., Polymer synthesis and processing using supercritical carbon dioxide. *J. Mater. Chem.*, 2000, 10(2): 207–234.
90. Montjovent, M.O., L. Mathieu, B. Hinz, L.L. Applegate, P.E. Bourban, P.Y. Zambelli, J.A. Manson, and D.P. Pioletti, Biocompatibility of bioresorbable poly(L-lactic acid) composite scaffolds obtained by supercritical gas foaming with human fetal bone cells. *Tissue Eng.*, 2005, 11(11–12): 1640–1649.
91. Yoon, J.J., J.H. Kim, and T.G. Park, Dexamethasone-releasing biodegradable polymer scaffolds fabricated by a gas-foaming/salt-leaching method. *Biomaterials*, 2003, 24(13): 2323–2329.
92. Sheridan, M.H., L.D. Shea, M.C. Peters, and D.J. Mooney, Bioabsorbable polymer scaffolds for tissue engineering capable of sustained growth factor delivery. *J. Control. Release*, 2000, 64: 91–102.
93. Ennet, A.B., D. Kaigler, and D.J. Mooney, Temporally regulated delivery of VEGF *in vitro* and *in vivo*. *J. Biomed. Mater. Res. A*, 2006, 79(1): 176–184.
94. Kim, S.S., K.M. Ahn, M.S. Park, J.H. Lee, C.Y. Choi, and B.S. Kim, A poly(lactide-co-glycolide)/hydroxyapatite composite scaffold with enhanced osteoconductivity. *J. Biomed. Mater. Res. A*, 2007, 80(1): 206–215.
95. Nam, Y.S. and T.G. Park, Porous biodegradable polymeric scaffolds prepared by thermally induced phase separation. *J. Biomed. Mater. Res.*, 1999, 47: 8–17.
96. Whang, K., D.C. Tsai, E.K. Nam, M. Aitken, S.M. Sprague, P.K. Patel, and K.E. Healy, Ectopic bone formation via rhBMP-2 delivery from porous bioabsorbable polymer scaffolds. *J. Biomed. Mater. Res.*, 1998, 42(4): 491–499.
97. Moshfeghian, A., J. Tillman, and S.V. Madihally, Characterization of emulsified chitosan-PLGA matrices formed using controlled-rate freezing and lyophilization technique. *J. Biomed. Mater. Res. A*, 2006, 79(2): 418–430.
98. Nam, Y.S. and T.G. Park, Biodegradable polymeric microcellular foams by modified thermally induced phase separation method. *Biomaterials*, 1999, 20(19): 1783–1790.
99. Cao, Y., T.I. Croll, A.J. O'Connor, G.W. Stevens, and J. Cooper-White, Systematic selection of solvents for the fabrication of 3D combined macro- and microporous polymeric scaffolds for soft tissue engineering. *J. Biomater. Sci. Polym. Ed.*, 2006, 17(4): 369–402.
100. Guan, J., J.J. Stankus, and W.R. Wagner, Development of composite porous scaffolds based on collagen and biodegradable poly(ester urethane)urea. *Cell Transplantation*, 2006, 15(S1): S17–S27.
101. Gao, Y., W. Weng, K. Cheng, P. Du, G. Shen, G. Han, B. Guan, and W. Yan, Preparation, characterization and cytocompatibility of porous ACP/PLLA composites. *J. Biomed. Mater. Res. A*, 2006, 79(1): 193–200.
102. Blaker, J.J., V. Maquet, R. Jerome, A.R. Boccaccini, and S.N. Nazhat, Mechanical properties of highly porous PDLLA/Bioglass composite foams as scaffolds for bone tissue engineering. *Acta Biomater.*, 2005, 1(6): 643–652.
103. Rowland, A.S., S.A. Lim, D. Martin, and J.J. Cooper-White, Polyurethane/poly(lactic-co-glycolic) acid composite scaffolds fabricated by thermally induced phase separation. *Biomaterials*, 2007, 28(12): 2109–2121.
104. Helen, W., C.L. Merry, J.J. Blaker, and J.E. Gough, Three-dimensional culture of annulus fibrosus cells within PDLLA/Bioglass composite foam scaffolds: Assessment of cell attachment, proliferation and extracellular matrix production. *Biomaterials*, 2007, 28(11): 2010–2020.
105. Gong, Y., L. He, J. Li, Q. Zhou, Z. Ma, C. Gao, and J. Shen, Hydrogel-filled polylactide porous scaffolds for cartilage tissue engineering. *J. Biomed. Mater. Res. B, Appl. Biomater.*, 2007, 82(1): 192–204.

106. Mo, X., H.J. Weber, and S. Ramakrishna, PCL–PGLA composite tubular scaffold preparation and biocompatibility investigation. *Int. J. Artif. Organs*, 2006, 29(8): 790–799.
107. Cao, Y., G. Mitchell, A. Messina, L. Price, E. Thompson, A. Penington, W. Morrison, A. O'Connor, G. Stevens, and J. Cooper-White, The influence of architecture on degradation and tissue ingrowth into three-dimensional poly(lactic-co-glycolic acid) scaffolds *in vitro* and *in vivo*. *Biomaterials*, 2006, 27(14): 2854–2864.
108. Chun, K.W., K.C. Cho, S.H. Kim, J.H. Jeong, and T.G. Park, Controlled release of plasmid DNA from biodegradable scaffolds fabricated using a thermally-induced phase-separation method. *J. Biomater. Sci. Polym. Ed.*, 2004, 15(11): 1341–1353.
109. Qiu, Q.Q., P. Ducheyne, and P.S. Ayyaswamy, Fabrication, characterization and evaluation of bioceramic hollow microspheres used as microcarriers for 3-D bone tissue formation in rotating bioreactors. *Biomaterials*, 1999, 20(11): 989–1001.
110. Borden, M., M. Attawia, Y. Khan, and C.T. Laurencin, Tissue engineered microsphere-based matrices for bone repair: Design and evaluation. *Biomaterials*, 2002, 23(2): 551–559.
111. Borden, M., M. Attawia, and C.T. Laurencin, The sintered microsphere matrix for bone tissue engineering: *In vitro* osteoconductivity studies. *J. Biomed. Mater. Res.*, 2002, 61(3): 421–429.
112. Borden, M., S.F. El-Amin, M. Attawia, and C.T. Laurencin, Structural and human cellular assessment of a novel microsphere-based tissue engineered scaffold for bone repair. *Biomaterials*, 2003, 24(4): 597–609.
113. Borden, M., M. Attawia, Y. Khan, S.F. El-Amin, and C.T. Laurencin, Tissue-engineered bone formation *in vivo* using a novel sintered polymeric microsphere matrix. *J. Bone Joint Surg. Br.*, 2004, 86(8): 1200–1208.
114. Jiang, T., W.I. Abdel-Fattah, and C.T. Laurencin, *In vitro* evaluation of chitosan/poly(lactic acid-glycolic acid) sintered microsphere scaffolds for bone tissue engineering. *Biomaterials*, 2006, 27(28): 4894–4903.
115. Kofron, M.D., J.A.J. Cooper, S.G. Kumbar, and C.T. Laurencin, Novel tubular composite matrix for bone repair. *J. Biomed. Mater. Res. A*, 2007, 82(2): 415–425.
116. Leong, K.F., C.M. Cheah, and C.K. Chua, Solid freeform fabrication of three-dimensional scaffolds for engineering replacement tissues and organs. *Biomaterials*, 2003, 24(13): 2363–2378.
117. Sachlos, E. and J.T. Czernuska, Making tissue engineering scaffolds work. Review: The application of solid freeform fabrication technology to the production of tissue engineering scaffolds. *Eur. Cell. Mater.*, 2003, 5: 29–40.
118. Yang, S., K.-F. Leong, Z. Du, and C.-K. Chua, The design of scaffolds for use in tissue engineering. Part I. Traditional factors. *Tissue Eng.*, 2001, 7(6): 679–689.
119. Schantz, J.-T., A. Brandwood, D.W. Hutmacher, H.L. Khor, and K. Bittner, Osteogenic differentiation of mesenchymal progenitor cells in computer designed fibrin-polymer-ceramic scaffolds manufactured by fused deposition modeling. *J. Mater. Sci.: Mater. Med.*, 2005, 16(9): 807–819.
120. Cao, T., K.-H. Ho, and S.-H. Teoh, Scaffold design and *in vitro* study of osteochondral coculture in a three-dimensional porous polycaprolactone scaffold fabricated by fused deposition modeling. *Tissue Eng.*, 2003, 9 (Supplement 1(4)): S103–S112.
121. Rohner, D., D.W. Hutmacher, T.K. Cheng, M. Oberholzer, and B. Hammer, *In vivo* efficacy of bone-marrow-coated polycaprolactone scaffolds for the reconstruction of orbital defects in the pig. *J. Biomed. Mater. Res.*, 2003, 66B(2): 574–580.
122. Vozzi, G., C. Fliam, A. Ahluwalia, and S. Bhatia, Fabrication of PLGA scaffolds using soft lithography and microsyringe deposition. *Biomaterials*, 2003, 24(14): 2533–2540.
123. Wiria, F.E., K.F. Leong, C.K. Chua, and Y. Liu, Poly-epsilon-caprolactone/hydroxyapatite for tissue engineering scaffold fabrication via selective laser sintering. *Acta Biomater.*, 2007, 3(1): 1–12.
124. Williams, J.M., A. Adewunmi, R.M. Schek, C.L. Flanagan, P.H. Krebsbach, S.E. Feinberg, S.J. Hollister, and S. Das, Bone tissue engineering using polycaprolactone scaffolds fabricated via selective laser sintering. *Biomaterials*, 2005, 26(23): 4817–4827.
125. Tan, K.H., C.K. Chua, K.F. Leong, C.M. Cheah, W.S. Gui, W.S. Tan, and F.E. Wiria, Selective laser sintering of biocompatible polymers for applications in tissue engineering. *Biomed. Mater. Eng.*, 2005, 15(1–2): 113–124.
126. Chua, C.K., K.F. Leong, K.H. Tan, F.E. Wiria, and C.M. Cheah, Development of tissue scaffolds using selective laser sintering of polyvinyl alcohol/hydroxyapatite biocomposite for craniofacial and joint defects. *J. Mater. Sci.: Mater. Med.*, 2004, 15(10): 1113–1121.

127. Hao, L., M.M. Savalani, Y. Zhang, K.E. Tanner, and R.A. Harris, Selecetive laser sintering of hydroxyapatite reinforced polyethylene composites for bioactive implants and tissue scaffold development. *Proc. Inst. Mech. Eng. H, J. Eng. Med.*, 2006, 220(4): 521–531.
128. Cooke, M.N., J.P. Fisher, D. Dean, C. Rimnac, and A.G. Mikos, Use of stereolithography to manufacture critical-sized 3D biodegradable scaffolds for bone ingrowth. *J. Biomed. Mater. Res.*, 2003, 64B(2): 65–69.
129. Sodian, R., M. Loebe, A. Hein, D.P. Martin, S.P. Hoerstrup, E.V. Potapov, H. Hausmann, T. Lueth, and R. Hetzer, Application of stereolithography for scaffold fabrication for tissue engineered heart valves. *ASAIO J.*, 2002, 48(1): 12–16.
130. Arcaute, K., B.K. Mann, and R.B. Wicker, Stereolithography of three-dimensional bioactive poly (ethylene glycol) constructs with encapsulated cells. *Ann. Biomed. Eng.*, 2006, 34(9): 1429–1441.
131. Khalyfa, A., S. Vogt, J. Weisser, G. Grimm, A. Rechtenbach, W. Meyer, and M. Schnabelrauch, Development of a new calcium phosphate powder-binder system for the 3D printing of patient specific implants. *J. Mater. Sci.: Mater. Med.*, 2007, 18(5): 909–916.
132. Seitz, H., W. Rieder, S. Irsen, B. Leukers, and C. Tille, Three-dimensional printing of porous ceramic scaffolds for bone tissue engineering. *J. Biomed. Mater. Res. B, Appl. Biomater.*, 2005, 74(2): 782–788.
133. Giordano, R.A., B.M. Wu, S.W. Borland, L.G. Cima, E.M. Sachs, and M.J. Cima, Mechanical properties of dense polylactic acid structures fabricated by three dimensional printing. *J. Biomater. Sci. Polym. Ed.*, 1996, 8(1): 63–75.
134. Park, A., B. Wu, and L.G. Griffith, Integration of surface modification and 3D fabrication techniques to prepare patterned poly(L-lactide) substrates allowing regionally selective cell adhesion. *J. Biomater. Sci. Polym. Ed.*, 1998, 9(2): 89–110.
135. Sherwood, J.K., S.L. Riley, R. Palazzolo, S.C. Brown, D.C. Monkhouse, M. Coates, L.G. Griffith, L.K. Landeen, and A. Ratcliffe, A three-dimensional osteochondral composite scaffold for articular cartilage repair. *Biomaterials*, 2002, 23: 4739–4751.
136. Roy, T.D., J.L. Simon, J.L. Ricci, E.D. Rekow, V.P. Thompson, and J.R. Parsons, Performance of degradable composite bone repair products made via three-dimensional fabrication techniques. *J. Biomed. Mater. Res.*, 2003, 66A: 283–291.
137. Lee, M., J.C. Dunn, and B.M. Wu, Scaffold fabrication by indirect three-dimensional printing. *Biomaterials*, 2005, 26(20): 4281–4289.
138. Dellinger, J.G., J. Cesarano, and R.D. Jamison, Robotic deposition of model hydroxyapatite scaffolds with multiple architectures and multiscale porosity for bone tissue engineering. *J. Biomed. Mater. Res. A*, 2007, 82(2): 383–394.
139. Varghese, D., M. Deshpande, T. Xu, P. Kesari, S. Ohri, and T. Boland, Advances in tissue engineering: Cell printing. *J. Thorac. Cardiovasc. Surg.*, 2005, 129(2): 470–472.
140. Smith, C.M., A.L. Stone, R.L. Parkhill, R.L. Stewart, M.W. Simpkins, A.M. Kachurin, W.L. Warren, and S.K. Williams, Three-dimensional bioassembly tool for generating viable tissue-engineered constructs. *Tissue Eng.*, 2004, 10(9–10): 1566–1576.
141. Mironov, V., T. Boland, T. Trusk, G. Forgacs, and R.R. Markwald, Organ printing: Computer-aided jet-based 3D tissue engineering. *Trends Biotechnol.*, 2003, 21(4): 157–161.
142. Boland, T., V. Mironov, A. Gutowska, E.A. Roth, and R.R. Markwald, Cell and organ printing 2: Fusion of cell aggregates in three-dimensional gels. *Anat. Rec. A. Discov. Mol., Cell. Evol. Biol.*, 2003, 272(2): 497–502.
143. Sun, W., A. Darling, B. Starly, and J. Nam, Computer-aided tissue engineering: Overview, scope and challenges. *Biotechnol. Appl. Biochem.*, 2004, 39(Part 1): 29–47.
144. Hutmacher, D.W., M. Sittinger, and M.V. Risbund, Scaffold-based tissue engineering: Rationale for computer-aided design and solid free-form fabrication systems. *Trends Biotechnol.*, 2004, 22(7): 354–362.
145. Yeong, W.Y., C.K. Chua, K.F. Leong, and M. Chandrasekaran, Rapid protyping in tissue engineering: Challenges and potential. *Trends Biotechnol.*, 2004, 22(12): 643–652.
146. Bryant, S.J., C.R. Nuttelman, and K.S. Anseth, The effects of crosslinking density on cartilage formation in photocrosslinkable hydrogels. *Biomed. Sci. Instr.*, 1999, 35: 309–314.
147. Nguyen, K.T. and J.L. West, Photopolymerizable hydrogels for tissue engineering applications. *Biomaterials*, 2002, 23(22): 4307–4314.
148. Elisseeff, J., C. Puleo, F. Yang, and B. Sharma, Advances in skeletal tissue engineering with hydrogels. *Orthod. Craniofac. Res.*, 2005, 8(3): 150–161.

149. Anseth, K.S., C.N. Bowman, and L. Brannon-Peppas, Mechanical properties of hydrogels and their experimental determination. *Biomaterials*, 1996, 17(17): 1647–1657.
150. Ferruti, P., S. Bianchi, E. Ranucci, F. Chiellini, and V. Caruso, Novel poly(amido-amine)-based hydrogels as scaffolds for tissue engineering. *Macromol. Biosci.*, 2005, 5(7): 613–622.
151. Ferruti, P., S. Bianchi, E. Ranucci, F. Chiellini, and A.M. Piras, Novel agmatine-containing poly (amidoamine) hydrogels as scaffolds for tissue engineering. *Biomacromolecules*, 2005, 6(4): 2229–2235.
152. Mann, B.K., A.S. Gobin, A.T. Tsai, R.H. Schmedlen, and J.L. West, Smooth muscle cell growth in photopolymerized hydrogels with cell adhesive and proteolytically degradable domains: Synthetic ECM analogs for tissue engineering. *Biomaterials*, 2001, 22(22): 3045–3051.
153. Schmedlen, R.H., K.S. Masters, and J.L. West, Photocrosslinkable polyvinyl alcohol hydrogels that can be modified with cell adhesion peptides for use in tissue engineering. *Biomaterials*, 2002, 23(22): 4325–4332.
154. Burdick, J.A. and K.S. Anseth, Photoencapsulation of osteoblasts in injectable RGD-modified PEG hydrogels for bone tissue engineering. *Biomaterials*, 2002, 23(22): 4315–4323.
155. Sannino, A., P.A. Netti, M. Madaghiele, V. Coccoli, A. Luciani, A. Maffezzoli, and L. Nicolais, Synthesis and characterization of macroporous poly(ethylene glycol)-based hydrogels for tissue engineering applications. *J. Biomed. Mater. Res. A*, 2006, 79A(2): 229–236.
156. Martens, P.J., S.J. Bryant, and K.S. Anseth, Tailoring the degradation of hydrogels formed from multivinyl poly(ethylene glycol) and poly(vinyl alcohol) macromers for cartilage tissue engineering. *Biomacromolecules*, 2003, 4(2): 283–292.
157. Leach, J.B., K.A. Bivens, C.N. Collins, and C.E. Schmidt, Development of photocrosslinkable hyaluronic acid–polyethylene glycol–peptide composite hydrogels for soft tissue engineering. *J. Biomed. Mater. Res. A*, 2004, 70(1): 74–82.
158. Moffat, K.L. and K.G. Marra, Biodegradable poly(ethylene glycol) hydrogels crosslinked with genipin for tissue engineering applications. *J. Biomed. Mater. Res. B: Appl. Biomater.*, 2004, 71(1): 181–187.
159. Hiemstra, C., Z. Zhong, P.J. Dijkstra, and J. Feijen, PEG-PLA hydrogels by stereocomplexation for tissue engineering of cartilage. *J. Control. Release*, 2005, 101(1–3): 332–334.
160. Cascone, M.G., L. Lazzeri, E. Sparvoli, M. Scatena, L.P. Serino, and S. Danti, Morphological evaluation of bioartificial hydrogels as potential tissue engineering scaffolds. *J. Mater. Sci.: Mater. Med.*, 2004, 15(12): 1309–1313.
161. Wang, D.A., C.G. Williams, F. Yang, N. Cher, H. Lee, and J. Elisseeff, Bioresponsive phosphoester hydrogels for bone tissue engineering. *Tissue Eng.*, 2005, 11(1–2): 201–213.
162. Landers, R., U. Hubner, R. Schmelzeisen, and R. Mulhaupt, Rapid prototyping of scaffolds derived from thermoreversible hydrogels and tailored for applications in tissue engineering. *Biomaterials*, 2002, 23(23): 4437–4447.
163. Dhariwala, B., E. Hunt, and T. Boland, Rapid prototyping of tissue-engineering constructs, using photopolymerizable hydrogels and stereolithography. *Tissue Eng.*, 2004, 10(9–10): 1316–1322.
164. Hoffman, A., Hydrogels for biomedical applications. *Adv. Drug Deliv. Rev.*, 2002, 43(1): 3–12.
165. Drury, J.L. and D.J. Mooney, Hydrogels for tissue engineering: Scaffolds design variables and applications. *Biomaterials*, 2003, 24(24): 4337–4351.
166. Brandl, F., F. Sommer, and A. Goepferich, Rational design of hydrogels for tissue engineering: Impact of physical factors on cell behavior. *Biomaterials*, 2007, 28(2): 134–146.
167. Li, W.-J., C.T. Laurencin, E.J. Caterson, R.S. Tuan, and F.K. Ko, Electrospun nanofibrous structure: A novel scaffold for tissue engineering. *J. Biomed. Mater. Res.*, 2002, 60(4): 613–621.
168. Reneker, D.H. and I. Chun, Nanometre diameter fibres of polymer, produced by electrospinning. *Nanotechnology*, 1996, 7: 216–223.
169. Boland, E.D., G.E. Wnek, D.G. Simpson, K.J. Pawlowski, and G.L. Bowlin, Tailoring tissue engineering scaffolds using electrostatic processing techniques: A study of poly(glycolic acid) electrospinning. *J. Macromol. Sci. A, Pure Appl. Chem.*, 2001, 38(12): 1231–1243.
170. Yang, F., C.Y. Xu, M. Kotaki, S. Wang, and S. Ramakrishna, Characterization of neural stem cells on electrospun poly(L-lactic acid) nanofibrous scaffold. *J. Biomater. Sci. Polym. Ed.*, 2004, 15(12): 1483–1497.
171. Xin, X., M. Hussain, and J.J. Mao, Continuing differentiation of human mesenchymal stem cells and induced chondrogenic and osteogenic lineages in electrospun PLGA nanofiber scaffold. *Biomaterials*, 2007, 28(2): 316–325.

172. Kim, G. and W. Kim, Highly porous 3D nanofiber scaffold using an electrospinning technique. *J. Biomed. Mater. Res. B: Appl. Biomater.*, 2007, 81(1): 104–110.
173. Shin, M., H. Yoshimoto, and J.P. Vacanti, *In vivo* bone tissue engineering using mesenchymal stem cells on a novel electrospun nanofibrous scaffold. *Tissue Eng.*, 2004, 10(1–2): 33–41.
174. Li, W.-J., K.G. Danielson, P.G. Alexander, and R.S. Tuan, Biological response of chondrocytes cultured in three-dimensional nanofibrous poly(epsilon-caprolactone) scaffolds. *J. Biomed. Mater. Res.*, 2003, 67A(4): 1105–1114.
175. Yoshimoto, H., Y.M. Shin, H. Terai, and J.P. Vacanti, A biodegradable nanofiber scaffold by electrospinning and its potential for bone tissue engineering. *Biomaterials*, 2003, 24(12): 2077–2082.
176. Chen, M., P.K. Patra, S.B. Warner, and S. Bhowmick, Role of fiber diameter in adhesion and proliferation of NIH 3T3 fibroblast on electrospun polycaprolactone scaffolds. *Tissue Eng.*, 2007, 13(3): 579–587.
177. Deng, X.L., G. Sui, M.L. Zhao, G.Q. Chen, and X.P. Yang, Poly(L-lactic acid)/hydroxyapatite hybrid nanofibrous scaffolds prepared by electrospinning. *J. Biomater. Sci. Polym. Ed.*, 2007, 18(1): 117–130.
178. Meng, W., S.Y. Kim, J. Yuan, J.C. Kim, O.H. Kwon, N. Kawazoe, G. Chen, Y. Ito, and I.K. Kang, Electrospun PHBV/collagen composite nanofibrous scaffolds for tissue engineering. *J. Biomater. Sci. Polym. Ed.*, 2007, 18(1): 81–94.
179. Li, M., M.J. Mondrinos, X. Chen, M.R. Gandhi, F.K. Ko, and P.I. Lelkes, Co-elecetrospun poly(lactide-co-glycolide), gelatin, and elastin blends for tissue engineering scaffolds. *J. Biomed. Mater. Res. A*, 2006, 79(4): 963–973.
180. Pan, H., H. Jiang, and W. Chen, Interaction of dermal fibroblasts with electrospun composite polymer scaffolds prepared from dextran and poly lactide-co-glycolide. *Biomaterials*, 2006, 27(17): 3209–3220.
181. Townsend-Nicholson, A. and S.N. Jayasinghe, Cell electrospinning: A unique biotechnique for encapsulating living organisms for generating active biological microthreads/scaffolds. *Biomacromolecules*, 2006, 7(12): 3364–3369.
182. Thomas, V., M.V. Jose, S. Chowdhury, J.F. Sullivan, D.R. Dean, and Y.K. Vohra, Mechano-morphological studies of aligned nanofibrous scaffolds of polycaprolactone fabricated by electrospinning. *J. Biomater. Sci. Polym. Ed.*, 2006, 17(9): 969–984.
183. Li, W.J., R.L. Mauck, J.A. Cooper, X. Yuan, and R.S. Tuan, Engineering controllable anisotropy in electrospun biodegradable nanofibrous scaffolds for musculoskeletal tissue engineering. *J. Biomech.*, 2007, 40(8): 1686–1693.
184. Pham, Q.P., U. Sharma, and A.G. Mikos, Electrospun poly(epsilon-caprolactone) microfiber and multilayer nanofiber/microfiber scaffolds: Characterization of scaffolds and measurement of cellular infiltration. *Biomacromolecules*, 2006, 7(10): 2796–2805.
185. Li, W.J., J.A.J. Cooper, R.L. Mauck, and R.S. Tuan, Fabrication and characterization of six electrospun poly(alpha-hydroxy ester)-based fibrous scaffolds for tissue engineering applications. *Acta Biomater.*, 2006, 2(4): 377–385.
186. Moroni, L., R. Licht, J. de Boer, J.R. de Wijn, and C.A. van Blitterswijk, Fiber diameter and texture of electrospun PEOT/PBT scaffolds influence human mesenchymal stem cell proliferation and morphology, and the release of incorporated compounds. *Biomaterials*, 2006, 27(28): 4911–4922.
187. Vaz, C.M., S. van Tuijl, C.V. Bouten, and F.P. Baaijens, Design of scaffolds for blood vessel tissue engineering using a multi-layering electrospinning technique. *Acta Biomater.*, 2005, 1(5): 575–582.
188. Wutticharoenmongkol, P., N. Sanchavanakit, P. Pavasant, and P. Supaphol, Novel bone scaffolds of electrospun polycaprolactone fibers filled with nanoparticles. *J. Nanosci. Nanotechnol.*, 2006, 6(2): 514–522.
189. Lee, Y.H., J.H. Lee, I.G. An, C. Kim, D.S. Lee, Y.K. Lee, and J.D. Nam, Electrospun dual-porosity structure and biodegradation morphology of montmorillonite reinforced PLLA nanocomposite scaffolds. *Biomaterials*, 2005, 26(16): 3165–3172.
190. Thomas, V., S. Jagani, K. Johnson, M.V. Jose, D.R. Dean, Y.K. Vohra, and E. Nyairo, Electrospun bioactive nanocomposite scaffolds of polycaprolactone and nanohydroxyapatite for bone tissue engineering. *J. Nanosci. Nanotechnol.*, 2006, 6(2): 487–493.
191. van Lieshout, M.I., C.M. Vaz, M.C. Rutten, G.W. Peters, and F.P. Baaijens, Electrospinning versus knitting: Two scaffolds for tissue engineering of the aortic valve. *J. Biomater. Sci. Polym. Ed.*, 2006, 17(1–2): 77–89.
192. Zong, X., H. Bien, C.Y. Ching, L. Yin, D. Fang, B.S. Hsiao, B. Chu, and E. Entcheva, Electrospun fine-textured scaffolds for heart tissue constructs. *Biomaterials*, 2005, 26(26): 5330–5338.

193. Riboldi, S.A., M. Sampaolesi, P. Neuenschwander, G. Cossu, and S. Mantero, Electrospun degradable polyesterurethane membranes: Potential scaffolds for skeletal muscle tissue engineering. *Biomaterials*, 2005, 26(22): 4606–4615.
194. Kim, K., Y.K. Luu, C. Chang, D. Fang, B.S. Hsiao, B. Chu, and M. Hadjiargyrou, Incorporation and controlled release of a hydrophilic antibiotic using poly(lactide-co-glycolide)-based electrospun nanofibrous scaffolds. *J. Control. Release*, 2004, 98(1): 47–56.
195. Luu, Y.K., K. Kim, B.S. Hsiao, B. Chu, and M. Hadjiargyrou, Development of a nanostructured DNA delivery scaffold via electrospinning of PLGA and PLA-PEG block copolymers. *J. Control. Release*, 2003, 89(2): 341–353.
196. Nair, L.S., S. Bhattacharyya, and C.T. Laurencin, Development of novel tissue engineering scaffolds via electrospinning. *Expert Opin. Biol. Ther.*, 2004, 4(5): 659–668.
197. Boudriot, U., R. Dersch, A. Greiner, and J.H. Wendorff, Electrospinning approaches toward scaffold engineering—a brief overview. *Artif. Organs*, 2006, 30(10): 785–792.
198. Teo, W.E., W. He, and S. Ramakrishna, Electrospun scaffold tailored for tissue-specific extracellular matrix. *Biotechnol. J.*, 2006, 1(9): 918–929.
199. Smith, L.A. and P.X. Ma, Nano-fibrous scaffolds for tissue engineering. *Colloids Surf. B Biointerfaces*, 2004, 39(3): 125–131.
200. Nof, M. and L.D. Shea, Drug-releasing scaffolds fabricated from drug-loaded microspheres. *J. Biomed. Mater. Res.*, 2002, 59(2): 349–356.
201. Babensee, J.E., L.V. McIntire, and A.G. Mikos, Growth factor delivery for tissue engineering. *Pharm. Res.*, 2000, 17(5): 497–504.
202. Hu, Y., C. Zhang, S. Zhang, Z. Xiong, and J. Xu, Development of a porous poly(L-lactic acid)/hydroxyapatite/collagen scaffold as a BMP delivery system and its use in healing canine segmental bone defect. *J. Biomed. Mater. Res. A*, 2003, 67(2): 591–598.
203. Grondahl, L., A. Chandler-Temple, and M. Trau, Polymeric grafting of acrylic acid onto poly (3-hydroxybutyrate-co-3-hydroxyvalerate): Surface functionalization for tissue engineering applications. *Biomacromolecules*, 2005, 6(4): 2197–2203.
204. Ma, Z., C. Gao, Y. Gong, and J. Shen, Cartilage tissue engineering PLLA scaffold with surface immobilized collagen and basic fibroblast growth factor. *Biomaterials*, 2005, 26(11): 1253–1259.
205. Park, H., J.S. Temenoff, T.A. Holland, Y. Tabata, and A.G. Mikos, Delivery of TGF-α1 and chondrocytes via injectable, biodegradable hydrogels for cartilage tissue engineering applications. *Biomaterials*, 2005, 26(34): 7095–7103.
206. Lenza, R.F.S., W.L. Vasconcelos, J.R. Jones, and L.L. Hench, Surface-modified 3D scaffolds for tissue engineering. *J. Mater. Sci.: Mater. Med.*, 2002, 13(9): 837–842.
207. Chen, R.R. and D.J. Mooney, Polymeric growth factor delivery strategies for tissue engineering. *Pharm. Res.*, 2003, 20(8): 1103–1112.
208. Jang, J.-H., Z. Bengali, T.L. Houchin, and L.D. Shea, Surface adsorption of DNA to issue engineering scaffolds for efficient gene delivery. *J. Biomed. Mater. Res. A*, 2006, 77A(1): 50–58.
209. Gunatillake, P.A. and R. Adhikari, Biodegradable synthetic polymers for tissue engineering. *Eur. Cell. Mater.*, 2003, 20(5): 1–16.
210. Nair, L.S. and C.T. Laurencin, Polymers as biomaterials for tissue engineering and controlled drug delivery. *Adv. Biochem. Eng./Biotechnol.*, 2006, 102: 47–90.
211. Holland, T.A. and A.G. Mikos, Review: Biodegradable polymeric scaffolds. Improvements in bone tissue engineering through controlled drug delivery. *Adv. Biochem. Eng./Biotechnol.*, 2006, 102: 161–185.
212. Wang, S., L. Lu, and M.J. Yaszemski, Bone-tissue-engineering material poly(propylene fumarate): Correlation between molecular weight, chain dimensions, and physical properties. *Biomacromolecules*, 2006, 7(6): 1976–1982.
213. Peter, S.J., M.J. Yaszemski, L.J. Suggs, R.G. Payne, R. Langer, W.C. Hayes, M.R. Unroe, L.B. Alemany, P.S. Engel, and A.G. Mikos, Characterization of partially saturated poly(propylene fumarate) for orthopaedic application. *J. Biomater. Sci. Polym. Ed.*, 1997, 8(11): 893–904.
214. Temenoff, J.S. and A.G. Mikos, Injectable biodegradable materials for orthopaedic tissue engineering. *Biomaterials*, 2000, 21(23): 2405–2412.
215. Peter, S.J., P. Kim, A.W. Yasko, M.J. Yaszemski, and A.G. Mikos, Crosslinking characteristics of an injectable poly(propylene fumarate)/beta-tricalcium phosphate paste and mechanical properties of the crosslinked composite for use as a biodegradable bone cement. *J. Biomed. Mater. Res.*, 1999, 44(3): 314–321.

216. Peter, S.J., S.T. Miller, G. Zhu, A.W. Yasko, and A.G. Mikos, *In vivo* degradation of a poly(propylene fumarate)/beta-tricalcium phosphate injectable composite scaffold. *J. Biomed. Mater. Res.*, 1998, 41(1): 1–7.
217. Peter, S.J., L. Lu, D.J. Kim, and A.G. Mikos, Marrow stromal osteoblast function on a poly(propylene fumarate)/beta-tricalcium phosphate biodegradable orthopaedic composite. *Biomaterials*, 2000, 21(12): 1207–1213.
218. He, S., M.J. Yaszemski, A.W. Yasko, P.S. Engel, and A.G. Mikos, Injectable biodegradable polymer composites based on poly(propylene fumarate) crosslinked with poly(ethylene glycol)-dimethacrylate. *Biomaterials*, 2000, 21(23): 2389–2394.
219. Kempen, D.H., L. Lu, C. Kim, X. Zhu, W.J. Dhert, B.L. Currier, and M.J. Yaszemski, Controlled drug release from a novel injectable biodegradable microsphere/scaffold composite based on poly(propylene fumarate). *J. Biomed. Mater. Res. A*, 2006, 77(1): 103–111.
220. Payne, R.G., J.S. McGonigle, M.J. Yaszemski, A.W. Yasko, and A.G. Mikos, Development of an injectable, *in situ* crosslinkable, degradable polymeric carrier for osteogenic cell populations. Part 3. Proliferation and differentiation of encapsulated marrow stromal osteoblasts cultured on crosslinking poly(propylene fumarate). *Biomaterials*, 2002, 23(22): 4381–4387.
221. Lewandrowski, K.U., S.P. Bondre, D.L. Wise, and D.J. Trantolo, Enhanced bioactivity of a poly (propylene fumarate) bone graft substitute by augmentation with nano-hydroxyapatite. *Biomed. Mater. Eng.*, 2003, 13(2): 115–124.
222. Vehof, J.W.M., J.P. Fisher, D. Dean, J.-P.C.M.v.d. Waerden, P.H.M. Spauwen, A.G. Mikos, and J.A. Jansen, Bone formation in transforming growth factor beta-1-coated porous poly(propylene fumarate) scaffolds. *J. Biomed. Mater. Res.*, 2002, 60: 241–251.
223. Jabbari, E., S. Wang, L. Lu, J.A. Gruetzmacher, S. Ameenuddin, T.E. Hefferan, B.L. Currier, A.J. Windebank, and M.J. Yaszemski, Synthesis, material properties, and biocompatibility of a novel self-crosslinkable poly(caprolactone fumarate) as an injectable tissue engineering scaffold. *Biomacromolecules*, 2005, 6(5): 2503–2511.
224. Wang, S., L. Lu, J.A. Gruetzmacher, B.L. Currier, and M.J. Yaszemski, Synthesis and characterizations of biodegradable and crosslinkable poly(epsilon-caprolactone fumarate), poly(ethylene glycol fumarate), and their amphiphilic copolymer. *Biomaterials*, 2006, 27(6): 832–841.
225. Chasin, M., D. Lewis, and R. Langer, Polyanhydrides for controlled drug delivery. *Biopharm. Manuf.*, 1998, 1: 33–46.
226. Gopferich, A. and J. Tessmar, Polyanhydride degradation and erosion. *Adv. Drug Deliv. Rev.*, 2002, 54(7): 911–931.
227. Uhrich, K.E., S.E. Ibim, D.R. Larrier, R. Langer, and C.T. Laurencin, Chemical changes during *in vivo* degradation of poly(anhydride-imide) matrices. *Biomaterials*, 1998, 19(22): 2045–2050.
228. Attawia, M.A., K.E. Uhrich, E.A. Botchwey, R.S. Langer, and C.T. Laurencin, *In vitro* bone biocompatibility of poly(anhydride-co-imides) containing pyromellitylimidoalanine. *J. Orthop. Res.*, 1996, 14(3): 445–454.
229. Ibim, S.E., K.E. Uhrich, M. Attawia, V.R. Shastri, S.F. El-Amin, R. Bronson, R. Langer, and C.T. Laurencin, Preliminary *in vivo* report on the osteocompatibility of poly(anhydride-co-imides) evaluated in a tibial model. *J. Biomed. Mater. Res.*, 1998, 43(4): 374–379.
230. Burkoth, A.K., J.A. Burdick, and K.S. Anseth, Surface and bulk modifications to photocrosslinked polyanhydrides to control degradation behavior. *J. Biomed. Mater. Res.*, 2000, 51(3): 352–359.
231. Andriano, K.P., Y. Tabata, Y. Ikada, and J. Heller, *In vitro* and *in vivo* comparison of bulk and surface hydrolysis in absorbable polymer scaffolds for tissue engineering. *J. Biomed. Mater. Res.*, 1999, 48(5): 602–612.
232. Ng, S.Y., T. Vandamme, M.S. Taylor, and J. Heller, Synthesis and erosion studies of self-catalyzed poly (ortho ester)s. *Macromolecules*, 1997, 30(4): 770–772.
233. Kellomaki, M., J. Heller, and P. Tormala, Processing and properties of two different poly(ortho esters). *J. Mater. Sci.: Mater. Med.*, 2000, 11(6): 345–355.
234. Bourke, S.L. and J. Kohn, Polymers derived from the amino acid L-tyrosine: Polycarbonates, polyarylates and copolymers with poly(ethylene glycol). *Adv. Drug Deliv. Rev.*, 2003, 55(4): 447–466.
235. Tangpasuthadol, V., S.M. Pendharkar, R.C. Peterson, and J. Kohn, Hydrolytic degradation of tyrosine-derived polycarbonates, a class of new biomaterials. Part II: 3-yr study of polymeric devices. *Biomaterials*, 2000, 21(23): 2379–2387.

236. Choueka, J., J.L. Charvet, K.J. Koval, H. Alexander, K.S. James, K.A. Hooper, and J. Kohn, Canine bone response to tyrosine-derived polycarbonates and poly(L-lactic acid). *J. Biomed. Mater. Res.*, 1996, 31(1): 35–41.
237. Ertel, S.I., J. Kohn, M.C. Zimmerman, and J.R. Parsons, Evaluation of poly(DTH carbonate), a tyrosine-derived degradable polymer, for orthopaedic applications. *J. Biomed. Mater. Res.*, 1995, 29(11): 1337–1348.
238. Pulapura, S., C. Li, and J. Kohn, Structure–property relationships for the design of polyiminocarbonates. *Biomaterials*, 1990, 11(9): 666–678.
239. Meechaisue, C., R. Dubin, P. Supaphol, V.P. Hoven, and J. Kohn, Electrospun mat of tyrosine-derived polycarbonate fibers for potential use as tissue scaffolding material. *J. Biomater. Sci. Polym. Ed.*, 2006, 17(9): 1039–1056.
240. Lakshmi, S., D.S. Katti, and C.T. Laurencin, Biodegradable polyphosphazenes for drug delivery applications. *Adv. Drug Deliv. Rev.*, 2003, 55(4): 467–482.
241. Luten, J., J.H. van Steenis, R. van Someren, J. Kemmink, N.M. Schuurmans-Nieuwenbroek, G.A. Koning, D.J. Crommelin, C.F. van Nostrum, and W.E. Hennink, Water-soluble biodegradable cationic polyphosphazenes for gene delivery. *J. Control. Release*, 2003, 89(3): 483–497.
242. Laurencin, C.T., M.E. Norman, H.M. Elgendy, S.F. El-Amin, H.R. Allcock, S.R. Pucher, and A.A. Ambrosio, Use of polyphosphazenes for skeletal tissue regeneration. *J. Biomed. Mater. Res.*, 1993, 27(7): 963–973.
243. Laurencin, C.T., S.F. El-Amin, S.E. Ibim, D.A. Willoughby, M. Attawia, H.R. Allcock, and A.A. Ambrosio, A highly porous 3-dimensional polyphosphazene polymer matrix for skeletal tissue regeneration. *J. Biomed. Mater. Res.*, 1996, 30(2): 133–138.
244. Ambrosio, A.M., H.R. Allcock, D.S. Katti, and C.T. Laurencin, Degradable polyphosphazene/poly (alpha-hydroxyester) blends: Degradation studies. *Biomaterials*, 2002, 23(7): 1667–1672.
245. Krogman, N.R., A. Singh, L.S. Nair, C.T. Laurencin, and H.R. Allcock, Miscibility of bioerodible polyphosphazene/poly(lactide-co-glycolide) blends. *Biomacromolecules*, 2007, 8(4): 1306–1312.
246. Nair, L.S., S. Bhattacharyya, J.D. Bender, Y.E. Greish, P.W. Brown, H.R. Allcock, and C.T. Laurencin, Fabrication and optimization of methylphenoxy substituted polyphosphazene nanofibers for biomedical applications. *Biomacromolecules*, 2004, 5(6): 2212–2220.
247. Greish, Y.E., J.D. Bender, S. Lakshmi, P.W. Brown, H.R. Allcock, and C.T. Laurencin, Formation of hydroxyapatite-polyphosphazene polymer composites at physiologic temperature. *J. Biomed. Mater. Res. A*, 2006, 77(2): 416–425.
248. Carampin, P., M.T. Conconi, S. Lora, A.M. Menti, S. Baiguera, S. Bellini, C. Gandhi, and P.P. Parnigotto, Electrospun polyphosphazene nanofibers for *in vitro* rat endothelial cells proliferation. *J. Biomed. Mater. Res. A*, 2007, 80(3): 661–668.
249. Zhang, R. and P.X. Ma, Poly(α-hydroxyl acids)/hydroxyapatite porous composites for bone-tissue engineering. I. Preparation and morphology. *J. Biomed. Mater. Res.*, 1999, 44(4): 446–455.
250. Marra, K.G., J.W. Szem, P.N. Kumta, P.A. DiMilla, and L.E. Weiss, *In vitro* analysis of biodegradable polymer blend/hydroxyapatite composites for bone tissue engineering. *J. Biomed. Mater. Res.*, 1999, 47(3): 324–335.

4 Microfabrication Techniques in Scaffold Development

Christopher J. Bettinger, Jeffrey T. Borenstein, and Robert Langer

CONTENTS

4.1 INTRODUCTION

The advent of the semiconductor industry throughout the twentieth century has provided the secondary benefit of advanced manufacturing processes including the increased precision of the microfabrication of modern engineering materials such as metals and polymers. Advances in microfabrication techniques including micromachining and photolithography-aided processes such as dry and wet etching, metal deposition, and thin film growth have led to the ability to engineer systems and materials with well-defined features on the micron and submicron scale. Other more advanced techniques such as electron-beam (e-beam) photolithography and nanomanipulation have enabled the fabrication of structures with nanometer-scale precision. Although initially designed to support the rapid development demands associated with the integrated circuits industry, the application of micro- and nanometer-scale fabrication techniques has demonstrated invaluable utility in the design and development of engineered systems for biological and bioengineering applications.

The increasing precision of microfabricated devices has systematically enabled the manipulation of biological systems on a wide variety of length scales ranging from whole tissues and organs of centimeters in length, to individual cells on the micron scale, to individual biomolecules on the nanometer scale. One particularly promising application lies in the field of tissue engineering, a subset of bioengineering where micro- and nanometer systems have, and will continue to be of paramount importance. The application of these microfabricated systems has proven to be useful not only in the development and engineering of next-generation tissue engineered systems but also in aiding basic research efforts. Instrumentation that operates at the micron and submicron length scale is able to probe cells and tissues at biologically relevant length scales, which has led to the elucidation of some of the fundamental parameters of the cellular microenvironment that influence various fundamental biological processes such as differentiation [1], migration [2], embryonic development [3], and apoptosis [4]. Applying microfabricated systems toward the study and engineering of cell–matrix interactions is also of extreme importance. While the chemistry and biology of cell–matrix interactions has been studied extensively, the topography of this interface also plays an important role in regulating cell functions. The extracellular matrix (ECM) is composed of numerous structural proteins that are known to contain features at length scales from millimeters to nanometers. While larger structures are designed to provide macroscopic support, nanoscale features within the ECM are known to provide cues that influence essential cell functions such as proliferation, migration, spreading, contractility, tension, and traction forces. Numerous synthetic systems with a variety of submicron scale feature sizes and geometries have been used to study the behavior of cells in response to substrates rich in topographical cues [5].

In addition to well-defined substrates, cells have also been known to respond to randomly oriented topography such as nanometer-scale roughness. Topographic features on the order of 1 μm or smaller can influence cell functions that may be critical for tissue engineered systems including cell attachment, morphology, and directed migration, which are all critical cellular processes for controlling cell phenotype in cell-scaffold constructs. Cell alignment, for example, has been shown to play an important role in developing stronger tissues in the cases of smooth muscle cells, skeletal muscle, and fibroblasts. Topography has also been shown to influence the genotype and phenotype [6] including the upregulation of fibronectin mRNA levels in fibroblasts [7]. The generalized reaction of cell to topography has been extensively reviewed elsewhere [5,8,9] including other chapters within this text. Understanding of the fundamental processes related to cell–cell and cell–matrix interactions based on chemical, physical, topographical, and spatial microenvironmental cues is only one key aspect for the rational design of tissue engineering systems. It is also important to be able to apply these scientific discoveries through engineering both materials and material fabrication techniques for to further realize the potential for use in tissue engineering and regenerative medicine.

The evolution of increasingly sophisticated microfabricated systems will continue to play an important role in the advancement of tissue engineering and regenerative medicine. Although the current paradigm for the design and fabrication of tissue engineering scaffolds is biomimicry, advances in genomics, cell biology, and developmental biology will mandate proportional advances in the toolset for tissue engineering scaffold fabrication including the ability to control the spatial and temporal cellular environments on a micron and submicron level. There currently exists a wide range of "top–down" and "bottom–up" processes that have been developed to meet the corresponding increase in demand of micron and nanometer-scale precision in engineering biomaterials for tissue and organ regeneration systems [10,11]. However, there are multitudes of engineering challenges that remain to be addressed including the efficient incorporation of cells into scaffolds with sufficient spatial precision and the translation of the fundamentally two-dimensional photolithographic-based processes into three dimensions. This chapter focuses on the design and fabrication of tissue engineering scaffolds with micron and submicron scale features. The current state of the art in materials and materials processing for tissue engineering is surveyed with a specific focus at the interface of these disciplines with micro- and nanofabrication techniques.

4.2 APPLICATION OF TRADITIONAL MICROFABRICATION TECHNIQUES

4.2.1 Replica Molding of Biomaterials

Replica molding (RM) is the general term that encompasses a wide range of molding processes including hot embossing, solvent casting, and injection molding. Although RM has traditionally been used to create molded structures in a variety of engineering materials [12], the RM of biomaterials is the most straightforward technique toward creating microfabricated structures for tissue engineering applications. Photolithography remains a keystone technology in the fabrication of molds for use in RM processes. The advantages of RM include (1) feature resolutions down to 30 nm using polymeric materials, (2) the rapid and scalable production of microstructures across large surface areas, and (3) the inherent simplicity of the process. However, these advantages also result in limitations, including the two-dimensional nature of molds for RM using photolithographic processes. RM of inorganic materials such as poly(di-methylsiloxane) (PDMS) has been used extensively for biomedical applications including biosensors and microfluidic networks for cell culture [13–16]. However, the RM of biomaterials requires the adaptation of traditional processes to accommodate the chemical and physical properties of both synthetic and natural biomaterials.

Synthetic biodegradable polymers and natural biomaterials alike can be cast onto microfabricated molds to produce structures on substrates with feature resolutions as small as 30 nm. Synthetic thermoplastic biopolymers such as poly(ε-caprolactone) (PCL), poly(L-lactic acid) (PLA), and poly (L-lactic-*co*-glycolic acid) (PLGA) have been processed in this manner for various biomedical applications including tissue engineering scaffolds and devices for controlled drug release [10,17]. Melt casting, solvent casting, injection molding, and hot embossing can all be employed to attain well-defined feature geometries in polycrystalline or amorphous thermoplastic synthetic biomaterials. Similar processes can be adapted for fabricating tissue engineering scaffolds with conductive polymers for potential in nerve regeneration applications [18,19]. Thermoset biomaterials including crosslinked elastomeric networks such as poly(glycerol-*co*-sebacic acid) (PGS) [20] and poly(1,8-octanediol-*co*-citric acid) (POC) [21] require that the material be set into a given shape when initially molded followed by a chemical or physical crosslinking process. Therefore, processing of such materials may require the use of a sacrificial mold release layer consisting of a biologically benign material to prevent adhesion of these aggressive materials to the mold used during the final polymerization. Dilute sucrose solutions, which are typically used to prevent flocculation and coagulation in microparticle formulations, can be used to create thin films for aid in mold release of films while maintaining the fidelity of submicron sized features [22].

RM has also been used to produce micron scale and submicron scale features in natural biomaterials including structural proteins such as collagen, Matrigel, and sugar-based molecules including agarose [23–26]. A method reported by Tang et al. is also capable of incorporating cells within microfabricated gel constructs produced by micromolding. However, there are significantly more stringent processing limitations imparted on the processing of natural proteins, especially those that also incorporate cells within the structures. The potential for loss of function via protein denaturation limits processing conditions to generally low temperature processes using aqueous conditions. The inability to use elevated temperatures or organic solvents limits the suite of potential techniques, however, there are potential chemical methods that can be implemented including chemical crosslinking through established bioconjugation methods such as 1-ethyl-3-[3-dimehtyl-aminopropyl]carbodiimide hydrochloride (EDC) and *N*-hydroxysuccinimide (NHS), an established chemical route for bioconjugation of amines to carboxylates. Photopolymerization is another method for fabrication of replica-molded polymer sheets using mild processing conditions such as low temperatures and aqueous solvents.

4.2.2 Micromolding and Nanomolding of Biomaterials for Scaffold Topography

The ECM contains nanometer-scale features, which provide cues that mediate essential cell functions such as proliferation, migration, cytoskeleton remodeling, and differentiation. Synthetic systems with various submicron scale feature sizes and geometries have been used to study the behavior of cells in response to substrates rich in nanometer-scale topographical cues [5]. Cells also have been reported to respond to randomly oriented topography such as nanometer-scale roughness in addition to well-defined substrates with submicron features. Topographic features on the order of 1 μm or smaller in size and pitch can influence a number of cell functions including cell attachment, adhesion, morphology, and migration. Topography has also been shown to influence the gene profile [6] including the upregulation of fibronectin mRNA levels in fibroblasts [7]. The generalized reaction of cell to topography has been extensively reviewed elsewhere [5,8,9] including other sections within this text.

Given the demonstrated utility of being able to control cell function using nanotopography, this technology may also continue to play an important role in scaffold development as nanotopography is applied to natural and synthetic biomaterials. As previously mentioned, RM can reproduce features down to 30 nm, which is more than sufficient for inducing alterations in cell function. RM processes have been adapted to fabricate substrate nanotopography using a variety of biomaterials [23,27,28]. Some biomaterials require the addition of a sacrificial release layers, which promote mold release [22]. The materials used in mold release agents should be nontoxic, inexpensive, and soluble in aqueous agents to facilitate dissolution and release under mild conditions.

Although RM is a rapid, scalable, and facile method for producing ordered nanotopographic substrates, there are other processes available for producing nanotopographic features, which will be reviewed later in this chapter. One of the obvious limitations with nanotopographic substrates using traditional RM is the restriction to engineering principally two-dimensional systems. This restriction is not limiting in the design and fabrication of tissue engineered systems that are primarily two-dimensional structures such as the epidermis and epithelium of various organs. For example, laser micromachined wafers were used to fabricate PDMS masters for RM of collagen I sponges in work by Pins et al. These sponges contained features on the order of 40–310 μm, which were designed to mimic the structures of native epidermis. RM microfabricated sponges formed a basal lamina analogue when seeded with human epidermal keratinocytes. Another potential application for two-dimensional systems is in the retina of the eye. Corneal epithelium has been shown to respond to well-defined ling-grating substrate nanotopography with features as small as 40 nm using silicon [29]. However, further work must be performed in novel materials to allow for the potential for implantation. Two-dimensional RM could be expanded to other applications where the predominant structure is monolayer in nature. For example, RM of substrates using suitable biomaterials could also be envisioned for vascular implant applications [30], where the orientation of cells may prove to be important in inducing the appropriate phenotype. Two-dimensional sheets could be formed into tubes for applications in peripheral nerve regeneration as well [31].

The engineering of vital organs with complex microarchitecture requires the expansion from a two-dimensional platform. To address these unmet needs, tissue engineering strategies and scaffold fabrication techniques are continually moving toward development of three-dimensional systems. Subsequent nanotopography fabrication methods will also eventually have to expand to three-dimensional systems to accommodate advanced tissue engineering scaffold systems. One route toward pursuing this end is eliminating the use of photolithographic processing and instead utilizing physical properties of materials to produce cell-reactive nanotopographic materials [32]. These techniques will be described in detail elsewhere in this chapter.

4.2.3 Micromolding and Bonding of Biomaterials for Microfluidic Tissue Engineering Scaffolds

RM has been interfaced with nondegradable polymeric materials for the fabrication of microfluidic devices [33]. The development of microfluidics for biomedical engineering applications is extensive and permeates fields of study such as drug discovery and high-throughput screening. However, the appropriate processes must be tailored to establish efficient processes that utilize biodegradable materials. RM of biomaterials has proven to be useful in a variety of biomedical applications including drug delivery. RM coupled with appropriate lamination processes has led to the nascent subfield of microfluidic scaffold fabrication. This branch of tissue engineering is designed to address the issue of facile nutrient transport and waste removal within cell-seeded scaffolds. There have been a wide range of both synthetic and natural biomaterials, each of which require material-specific processes, that have been pursued to meet this demand for microfluidic, biodegradable scaffolds.

4.2.3.1 Microfluidic Scaffolds Using Synthetic Polymers

The primary advantage of synthetic polymers is the wide range of properties and processes that are available for use in design and fabrication. PLGA is among the first biomaterials to be used in biodegradable microfluidic systems [34]. Using a modified hot embossing technique, PLGA layers are molded with micron scale features using a PDMS mold. Heating PLGA to temperatures close to the melting point combined with appropriate amounts of mechanical pressure resulted in the fusion of polymer at the interface. Multiple PLGA layers are then laminated together using this precise thermal fusion bonding process to form enclosed microfluidic channels. Solvent bonding and embossing processes have also been developed for microfluidic scaffolds comprised of PLA, PGA, and corresponding PLGA copolymers [35,36]. RM and lamination processes have also been developed for the fabrication of cell-seeded microfluidic scaffolds from elastomeric, thermoset biodegradable polymers. Single layer and multilayer constructs have been fabricated using PGS, which have been used to support long-term perfused cell culture within the lumen of the microchannels [37]. These systems have been shown to support multiple cell types including both endothelial cells and hepatocytes.

4.2.3.2 Microfluidic Scaffolds Using Natural Biomaterials

Natural biomaterials offer many advantages over synthetic polymers as a material platform for microfluidic scaffolds. Natural biomaterials are typically composed proteins, which can usually support rapid cell attachment and have desirable degradation kinetics including long degradation times and nontoxic by-products. Similar to synthetic polymers, the fabrication of microfluidic scaffolds using natural biomaterials requires the adaptation of innovative soft-lithography techniques for molding and lamination of protein films. Microfluidic devices have been fabricated from gelatin, a protein composed primarily of collagen, and seeded with normal murine mammary epithelial cells (NMuMG). The observed cell morphology of NMuMGs in gelatin-based microfluidic device culture and static culture was similar. Although the motivation for fabricating these devices was to create a more native microenvironment for in vitro cell culture assays, the materials-specific processes developed for this purpose could also be used for the development of advanced scaffolds. Another natural biomaterial used for the fabrication of microfluidic scaffolds is silk fibroin, a 391 kDa protein that is produced by the silkworm *Bombyx mori*. Silk fibroin is a biodegradable material [38] that has shown potential for use as a biomaterial for many biomedical applications [39] including use as a scaffold biomaterial [40,41]. Silk fibroin, like many natural proteins, can be reconstituted into aqueous solutions. Aqueous silk fibroin solutions were used to produce films with micron and nanometer-scale features by replica molding on negative PDMS masters [23]. Delaminated silk fibroin films were bonded to form microfluidic devices (Figure 4.1). Hepatocytes were seeded and cultured under perfused for up to 5 days, while maintaining morphology and albumin production of cells grown in static culture.

Microfluidic devices fabricated from gelatin and silk fibroin have been demonstrated to support cell culture within the lumen of the devices. The relative lack of permeability of liquids within these biomaterials enables the precise control of liquids within these device modalities. Sequestering fluid to the lumen alone provides sufficient nutrient supply and waste removal for cells seeded within the microchannels of the devices. However, one of the promising applications of microfluidic scaffolds is the use of convection-aided mass transfer for large-volume tissue engineering scaffolds. The realization of this concept requires the fabrication of microfluidic scaffolds using highly

FIGURE 4.1 (See color insert following page 206.) Silk fibroin-based microfluidic devices. Replica molding can be used in conjunction with solvent-casting of aqueous silk fibroin solutions on microfabricated elastomeric masters. Replica-molded silk fibroin films produce high-fidelity features including (A) nanoscale posts with minimum widths of approximately 400 nm and (B) micron scale fluidic channels, which were used in subsequent experiments. (Scale bars are 5 μm and 500 μm in (A) and (B), respectively.) (C, D) SEM micrographs of the cross sections of microfabricated devices demonstrate retention of feature geometries in thin films with microchannel widths of approximately (C) 240 μm and (D) 90 μm. (Scale bars are 200 μm and 10 μm in (C) and (D), respectively.) (E, F) Patent microfluidic devices are demonstrated by fluorescent micrographs of devices perfused with rhodamine solution. Retention of the perfusate within the microchannels suggests robust bonding at the interface. (Scale bars are 500 μm and 50 μm in (E) and (F), respectively.) (From Bettinger, C.J. et al., *Adv. Mater.*, 19, 2847, 2007. With permission.)

water-permeable biomaterials. Toward this end, microfluidic devices have been fabricated using alginate [25], a naturally occurring polysaccharide that has been used extensively in tissue engineering applications [42,43]. One particularly useful property of alginate-based biomaterials is the ability to reversable crosslink this material using calcium-containing solutions to control the material properties [44]. Alginate microfluidic scaffolds were fabricated by replica-molding alginate films followed by bonding by calcium ion-induced crosslinking. The channels are sealed via chemical crosslinking by first chelating calcium from each of the faces to remove crosslinks, laminating the hydrogel slabs together, and then inducing chemical linkages via the addition of calcium chloride. Perfusion of the network (Figure 4.2A) with solutions of both low and high molecular weight compounds results in a variety of possibilities of transient concentration profiles throughout the hydrogel network (Figure 4.2B, C). Parameters such as flow geometry, volumetric flow rate, solute concentration, molecular weight of solute, and crosslinking density of hydrogel can be adjusted to control the spatial and temporal coordinates of concentration. The utility of such a system lies in the ability to potentially perfuse the ambient hydrogel network, which would presumably be housing seeded cells, to enhance nutrient supply and waste removal. This application would be especially useful for highly metabolically active cells such as hepatocytes. The development of convection-aided diffusion of biomolecules through microfluidic networks using natural biomaterials is a key development in advancing the field of tissue engineered scaffolds that are able to exceed spatial limitations that are defined by the diffusion limits. Rapid diffusion through

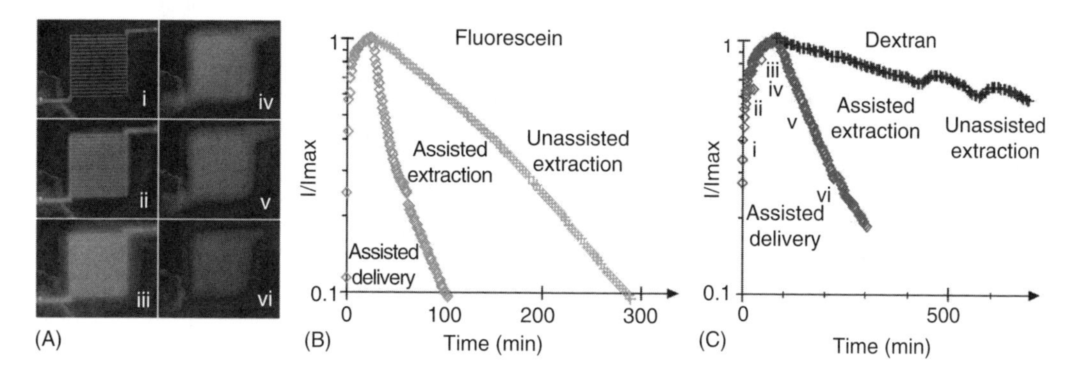

FIGURE 4.2 (See color insert following page 206.) Characterization of mass transfer in a microfluidic biomaterial (μFBM) fabricated from alginate microfluidic devices. (A) Fluorescence micrographs of a μFBM during assisted delivery (i–iii) and assisted extraction (iv–vi) of RITC-dextran. Assisted delivery refers to the operational mode of the device in which the solute is perfused through the microchannels and dissolves throughout the network. Assisted extraction refers to the operational mode where solute loaded within the alignate network is removed by perfusion of the microchannel network with an aqueous solution. These modes correspond to the cases of nutrient supply and waste removal, respectively. The rate of diffusion is primarily a function of the molecular weight of the solute, the crosslinking density of the network, the concentration of the solute in the perfused solution, and the flow rate of the perfused medium. (B, C) Temporal evolution of the normalized total intensity from fluorescence images, such as those in (A), during delivery and extraction of solutes (fluorescein, MW = 376 Da; RITC-dextran, MW = 70 kDa). Diamonds represent intensities during assisted delivery and assisted extraction. Crosses represent intensities during unassisted extraction from the same initial condition (i.e., achieved by delivery via the channels) as the assisted extraction experiment. The starting time of the unassisted evacuation has been shifted to match that of the assisted extraction. The dimensions of the construct as follows; the height of the gel $H = 0.29$ cm, the lateral dimension of the gel $L = 1$ cm, the height of the microchannels $h = 200$ μm, the width of the channels is $w = 100$ μm. The linear flow velocities of the fluid on the exterior of the construct is $u_r = 1$ cm/s while the velocity of the fluid within the microchannels is $u_c = 0.6$ cm/s. The solute concentrations are $c_0 = 20$ μmol/L for fluorescein and 10 μmol/L for RITC-dextran. (Reproduced from Cabodi, M. et al., *J. Amer. Chem. Soc.*, 127, 13788, 2005. With permission.)

networks will support the growth of encapsulated cells throughout the network. Microfluidic networks have also been fabricated using agarose [45], a naturally occurring unbranched polysaccharide that is present in the cell wall of some microorganisms. Like alginate, agarose is a natural polymer that can be cast into permeable gels. Microfluidic networks have been fabricated using a replica-molding processes combined with physical lamination techniques. This mild, aqueous processing is an enabling feature that facilitates encapsulation of cells with adequate viability. Agarose microfluidic networks were fabricated with encapsulated murine hepatocytes (AML-12) and perfused with medium. The viability of cells in perfused networks was 58% while viabilities of static networks remained below 20% at 3 days post seeding. The work outlined in this section, as well as other work not mentioned here, demonstrates the potential benefit of utilizing perfused microfluidic scaffolds as a tool for the next generation of tissue engineering scaffolds that are aimed at overcoming long-standing limitations.

4.2.3.3 Future of Microfluidic Systems in Tissue Engineering

The explosion of the use of microfluidics as engineering systems has largely focused on miniaturization of molecular analysis and genomics, portable devices for field-deployable biosensors, and interfacing fluid handling with microelectromechanical systems (MEMS) for improved automation [46]. However, the field of tissue engineering can also benefit significantly from advances in microfluidics technology, given that the appropriate biomaterials and biocompatible processes are interfaced with traditional microfabrication methods. The challenges that face the applications of microfluidics for advanced scaffold fabrication parallel those challenges faced by the microelectronic industry in the middle of the twentieth century. There are traditional issues that face any field of study based on microfabrication including increasing device yield and packaging, the term for seamlessly interfacing macroscopic instrumentation with devices that operate at micron length scales. Another essential cornerstone of developing the next generation of microfluidic scaffolds is the design, synthesis, and validation of novel materials and processing capabilities. The implementation of PDMS as a material for fabrication microfluidic devices [13] intended for analytical and diagnostic applications was essential because PDMS satisfied the material selection criteria set forth by the collective needs of the end user. The intermediate stage of the implementation of microfluidics for tissue engineering applications is the design and fabrication of complex microarchitecture for cell culture using traditional engineering materials, such as the aforementioned PDMS. For example, much work has been performed in creating designer microenvironments for hepatocytes within microfluidic devices [47,48]. These systems, though ultimately limited by their inorganic, nonbiodegradable material, are useful for the study of optimal conditions including microenvironmental effects such as shear stress, mass transport, and coculture. Lee et al. have moved beyond tranditional microfluidic geometries to fabricate a device that mimics the in vivo geometry and mass transport characteristics of a liver sinusoid [49]. Although these and other systems are not directly applicable to tissue engineering and organ regeneration applications, the technological developments surrounding the design, fabrication, cell culture techniques of these devices continue to lay the groundwork for future directly applicable tissue engineered systems. However, utilizing microfluidics for tissue engineering applications requires a more stringent set of material requirements. Materials used for microfluidic tissue engineering applications would ideally be resorbable, promote cell attachment, allow surface modification, and be amenable to facile processing. These unmet needs will continue to drive active research in synthetic biomaterial synthesis and development as well as natural biomaterial purification and characterization. In addition to material development, advances in materials processes must also be pursued to allow for efficient microfabrication strategies of novel materials. For example, advanced three-dimensional microfluidic systems using PDMS [50] must be expanded to facilitate similar strategies using biodegradable materials. In general, developing parallel strategies for novel, cell compatible biomaterial synthesis and processing will result in advanced microfluidic scaffolds to further scaffold development and ultimately tissue engineering.

4.2.4 Soft Lithography of Biomaterials

Photolithographic patterning processes lie at the cornerstone of micron scale systems as one of the most widely implemented tools for producing microfabricated structures. The principles of traditional photolithography have been translated directly to biomedical applications for the use of selective patterning of proteins, synthetic biomaterials, and cell-seeded biomaterials. One such technique involves the use of photoactivation to produce locally active regions that can readily immobilize proteins and other biomolecules. Though effective, alternatives to photografting of biomolecules would prove to be beneficial to promote rapid, scalable, inexpensive processing while retaining biological activity of the species. Microcontact printing (μCP), which was originally developed for patterning of chemical species [51], has since been modified for patterning biomolecules [52]. μCP and related processes are grouped together in a general technique known as soft lithography. Soft lithographic patterning of biological materials through various techniques has led to advanced tissue engineering scaffolds. A summary of the processes using soft lithography and related techniques is shown in Table 4.1.

4.2.4.1 Microcontact Printing

μCP is the direct deposition of molecules using a replica-molded, elastomeric stamp, typically fabrication from PDMS. Although μCP has been utilized for patterning molecules since the mid-1990s, this technique has only recently been applied to tissue engineering applications. μCP has typically been employed in this regard via the micropatterning of adhesion-promoting ECM proteins for selective adhesion of cells. μCP has been demonstrated on a variety of biomaterials including PLA, PLGA [53], and even human tissue [54]. Proteins patterned in this manner have been shown to control cell morphology, spreading, geometry [55], spatial arrangement, and relative orientation [56].

TABLE 4.1
Comparison of Techniques for Micropatterning of Biomaterials for Use in Tissue Engineering Scaffold Fabrication and Development

Technique	Patterning Modality	Compatible Materials	Minimum Feature Resolution	References
Replica molding (RM)	Silicon/PDMS master	Natural and synthetic polymers	30 nm	[13,22,23,26,34,37,143]
Microcontact printing (μCP)	PDMS stamp	Small molecules, polymers, proteins	35 nm (small molecules, proteins)	[52,144]
	Agarose stamp	Cells	10–15 μm (single cells)	[58]
Capillary force lithography (CFL)	PDMS stamp	Polymers, biopolymers	100 nm	[57,62,63]
Microfluidic patterning	3D PDMS microfluidics	Small molecules, proteins, cells	3–10 μm (proteins), 15–50 μm (cells)	[67,68,128–130]
Stencil micropatterning	PDMS stencil	Cells	100 μm	[65,66]
Dip-pen nanolithography (DPN)	AFM tip	Small molecules, proteins	10 nm	[86–89,145]

μCP of multiple proteins have also been demonstrated. Multistep protein patterning using μCP followed by backfilling with additional proteins in layer-by-layer techniques has led to the ability to precisely develop coculture systems with well-defined heterotypic cell–cell interactions [57]. The ability to control individual cell morphology, orientation, and cell–cell interactions may prove to be critical in engineering biomimetic microenvironments within tissue engineering scaffolds. Therefore, the advancement μCP printing has also been adopted for the use of direct cell patterning [58]. Stevens et al. demonstrated the use of replica-molded, elastomeric agarose stamps for the direct deposition of human osteoblasts on to porous hydroxyapatite scaffolds for applications in bone tissue engineering. This work highlights the benefits of using "cell-friendly" materials in μCP that allows for rapid, one-step patterning of mammalian cells. The subject of direct cell patterning will be further addressed in subsequent sections of this chapter.

4.2.4.2 Capillary Force Lithography and Other Soft Lithography Techniques

There are also a number of derivative techniques that are related to μCP designed to fabricated polymeric structures with high fidelity using elastomeric molds, stencils, and substrates. These techniques include micromolding in microcapillaries (MIMIC) [59], solvent-assisted microcontact molding (SAMIM) [60], and microtransfer molding (μTM) [61]. However, one technique that is especially useful for scaffold development is capillary force lithography (CFL), a soft lithographic technique that is complimentary to μCP [62]. Like μCP, CFL can be used for patterning structures with minimum feature sizes down to 100 nm in a variety of polymeric materials. CFL utilizes elastomeric molds for the sequestration of polymers in the voids of the mold, which results in microstructures when the mold is removed. CFL is especially useful in scaffold fabrication because of the large feature heights that can be achieved in additional to the large array of compatible biomaterials [63]. For example, this technique has been utilized for the micropatterning of repellant biomaterials such as poly(ethylene glycol) (PEG) and hyaluronic acid (HA), which, when used in combination with adhesion proteins (Figure 4.3), can be used to selectively pattern cells [64].

The prevalence of microfabrication and microtechnology laboratories and low-cost of materials has led to many variations of soft lithography. Soft lithography has expanded beyond simple stamping techniques for the direct patterning of proteins and cells. Folch et al. have developed microfabricated elastomeric stencils using PDMS for patterning of mammalian cells on two-dimensional substrates [65]. After overcoming technical challenges associated with the incompatible hydrophobic nature of PDMS and the aqueous environment for biomaterials, cells were patterned on a variety of substrate materials with various curvatures including gold, polystyrene, and collagen gels. Micropatterning using elastomeric stencils has been used in the coculture of hepatocytes with Kupffer cells to improve hepatocyte function in vitro [66]. Chiu et al. used complex microfluidic devices for the micropatterning of multiple proteins and cells [67]. This technique utilized complex three-dimensional microfluidic circuits that were mounted on the substrate to be patterned. Multiple protein solutions were then pumped throughout the network, selectively adsorbing to the substrate only at regions defined by the microfluidic structures. The structure can then be removed to reveal the final patterned substrate. This method had been previously developed by Delamarche et al. for the patterning of immunoglobulins [68].

4.2.4.3 Applications of Soft Lithography in Scaffold Development

Soft lithography will continue to play an important role in fundamental and applied biomedical engineering applications. For example, protein patterning has aided in elucidating fundamentals of cell biology through the use of designer microenvironments. Soft lithography will also serve to aid high-throughput screening assays for drug development or biomaterial deposition for cell differentiation studies [69,70]. Although soft lithography has been used to lay the groundwork for fundamental studies in cell biology, the direct application of soft lithography in tissue engineering through advanced scaffold development has been somewhat limited. The reason for this is primarily

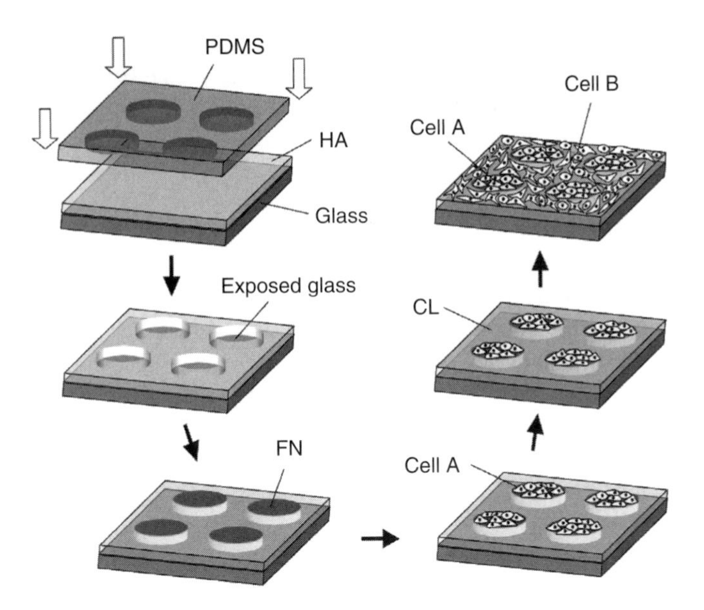

FIGURE 4.3 The scheme for fabrication of the coculture system using capillary force lithography and layer-by-layer deposition. Multiple types of soft lithographic techniques can be combined to form a variety of coculture systems. This schematic demonstrates one such example, in which capillary force lithography patterning of HA and layer-by-layer deposition of ECM proteins are combined to create a coculture of 3T3 fibroblasts and murine embryonic stem cells. Briefly, a few drops of HA solution were spun coated onto a glass slide, and a PDMS mold was immediately placed on the thin layer of HA. HA under the void space of the PDMS mold receded until the glass surface became exposed. The exposed region of a glass substrate was coated with fibronectin, where primary cells could be selectively adhered. Subsequently, the HA surface was complexed with collagen, allowing for the subsequent adhesion of secondary cells. This procedure is general and can be adapted for virtually any pair of cell types. (Reproduced from Fukuda, J. et al., *Biomaterials*, 27, 1479, 2006. With permission.)

due to the previously mentioned inherent two-dimensional nature of these techniques. As that paradigm for tissue engineering complex organs continues to evolve into three-dimensions, soft lithography will need to follow suit if it is to command a significant role in the next generation of scaffolds.

4.2.5 Electrodeposition of Biomaterials

Electrodeposition processes combined with photolithography have been used extensively in traditional microfabrication in a process termed LIGA. LIGA is an acronym for the German words *Lithographie*, *Galvanoformung*, *Abformung*, which translate into "lithography, electroplating, and molding" in English. LIGA is generally used in the fabrication of structures with feature sizes that fall between surface silicon micromachining and precision micromachining. LIGA is used to fabricated structures with lateral dimensions on the order of a few microns wide and several millimeters, which results in extremely high-aspect ratio structures. A wide variety of materials can be fabricated using LIGA including metals, metal alloys, plastics, and ceramics. LIGA and related processes for high-precision micromachining using traditional engineering materials have been reviewed extensively elsewhere [71]. There has been recent interest in the use of biodegradable eclectically conducting polymers in tissue engineering [72,73], with specific interest for applications in neuronal tissue regeneration [19]. LIGA presents a suitable method for the microfabrication of electrically conducting polymers for use as either neural prosthetics or scaffolds to help direct nerve growth for applications in peripheral nerve regeneration. In work by Lavan et al. poly(pyrrole)

(PPy), an electrically conducting polymer used extensively in biomedical applications [18,74], is electrodeposited on micropatterned gold islands to form micron scale structures of PPy on a silicon substrate [75]. This process is also amenable to electroforming three-dimensional structures with varying feature heights by designing the appropriate spacing of conductive features on the insulating substrate. The microfabrication of electrically conducting material could have a variety of applications including micropatterning and contact guidance of neurons. Although there are currently a wide variety of conducting polymers available for biomedical applications, there are significant limitations that preclude widespread use in tissue engineering including slow biodegradation rates, brittle mechanical properties, and poor cell attachment in some cases. These material deficiencies and the drive for implantable conducting biomaterials will also drive the development of novel conducting materials for a similar set of neuronal tissue engineering applications. As with other material processes, the drive materials development must be accompanied by a parallel pursuit of novel material processing capabilities.

4.2.6 Advanced Patterning Techniques

RM and soft lithography can collectively recapitulate features with resolutions that are suitable for cell and tissue engineering applications. There are other techniques that can reproduce submicron scale features without the need for photolithographic processes. Instead, these techniques employ other precision instrumentation to achieve micron and submicron scale feature resolution. However, like photolithographic processes, these technologies are primarily limited to patterning two-dimensional surfaces.

4.2.6.1 Laser Micromachining

Laser micromachining is a top–down process for rapid production of micron scale features. Laser ablation of polymers is typically performed using UV laser types such as excimer, argon-ion, fluorine, helium–cadmium, metal vapor, and nitrogen. Polymers are etched when the energy of the incident photoelectron is large enough to dissociate chemical bonds directly while imparting little thermal damage to the nonmachined regions. Microfabrication of polymers using laser ablation can be performed by either exposing the entire polymer substrate to UV irradiation through a photomask or by using a direct writing process. The minimum feature resolution of laser ablation is approximately 100 nm [76], which is an acceptable length scale for microfabricated structures for tissue engineering applications. The advantages of each method are scalability and elimination of the photolithographic patterns, respectively. Laser ablation has been used to fabricate a number of microsystems that would otherwise be difficult to produce using polymer RM techniques. For example, geometries such as through holes and trenches can be produced in biodegradable substrates with minimum feature sizes of approximately 10 μm in polymers such as poly(ether–ether–ketone) (PEEK) [77], poly(methyl methacrylate) (PMMA) [78], poly(vinyl alcohol) (PVA), PCL, PLA [79]. Laser ablation has also been used in combination with lamination techniques such as thermal or solvent lamination techniques to fabricate microfluidic prototypes [78,80]. Laser ablation, while proven directly useful for micromachining, can simultaneously functionalize surfaces that have undergone laser treatment. Laser irradiation can lead to the incorporation of nitrogen or oxygen molecules thereby creating functional groups such as amines or carboxylic acids. These functionalities can serve as precursors for surface modifications such as covalent linkages of peptides or non-bio-fouling agents. These surface modifications can also lead to a number of deleterious effects on materials used in microfluidics for certain applications. For example, it has been shown that laser ablation does not significantly impact the surface properties of PMMA, it does effect the electroosmotic mobility of polymer microchannels in materials such as poly(ethylene terephthalate glycol), poly(vinyl chloride), and poly(carbonate) [81]. As previously suggested, one potential limitation of laser ablation is the material selection, depending upon the intended final application. The ideal organic polymer for laser ablation is thermally and mechanically stable, yet susceptible to bond dissociation upon UV irradiation. Furthermore, the laser ablation process would ideally result in desirable surface modification to facilitate downstream material processing such as chemical modification, protein adsorption, or appropriate physical properties for use in microfluidics systems.

4.2.6.2 Colloidal Lithography

Nanoparticles offer an opportunity for exploitation in the pursuit of microfabrication of large surface areas without the need for a photomask or direct writing. Large numbers of nanoparticles with dimensions on the order of 50 nm can be made using a variety of solution-based methodologies such as emulsion phase separation. Laser ablating techniques can be used in combination with nanoparticles for the rapid fabrication of polymer substrates with arrays of nanoscale features with dimensions much smaller than the wavelength of light, which is a typical limitation using traditional photolithography. One approach, termed nanosphere lithography, uses an ordered template of nanometer-scale particles as a method to focus laser irradiation that is directed at the surface of the polymer substrate to create arrays of nanoscale pits [82]. A solution of silica nanoparticles is deposited onto the surface of the substrate. Upon evaporation of the solvent, surface tension effects largely outweigh the thermal energy (kT), which results in an energetically favored packing event. The result is spontaneous ordering of nanoparticles into a hexagonal close-packed configuration. A nearly identical method can be used to create ordered microscale sacrificial templates for use in porogen-leaching scaffold fabrication, which will be described in the future section on microscale self-assembly. Localized ablations is achieved as the ordered silica nanoparticles focus the UV irradiation, which can produce features such as pits as small as 30 nm in diameter. The size and geometry of the features can be varied by adjusting processing parameters including the diameter of the nanoparticles, wavelength of the laser, energy of radiation, and angle of incidence of the irradiation [83]. Nanosphere lithography provides a convenient method for creating large ordered arrays of feature dimensions that would otherwise be too slow, expensive, and in the case of features on the order of 10 nm, impossible using tradition photolithographic methods. While the set of potential feature geometries and sizes is inherently limited by the top–down nature of nanosphere lithography processes, the array of suitable materials is fundamentally similar to those used in laser ablation. Nonetheless, this method could serve to be useful in engineering polymeric substrates to study or enhance cell–matrix interactions by providing nanotopographic signals.

4.2.6.3 Dip-Pen Nanolithography

Dip-pen nanolithography (DPN) is a method that employs the use of atomic force microscopy instrumentation for patterning of single molecules [84] or multiple molecules [85] with nanometer-scale resolution. DPN has been adapted for the patterning of organic molecules including polymer nanowires, proteins [86], immunoglobulins [87], chemically modified oligonucleotides [88], and collagen I [89]. DPN was originally developed to achieve nanoscale patterning resolutions for the creation and modification of nanoscale devices. The evolution of DPN has provided some solutions to deficiencies experienced with the first generation of techniques such as more expansive suite of surface chemistries and improved control over resolution and printing [90]. However, there are some inherent, obtrusive limitations of using AFM as a tool for nanoscale lithography including costly equipment, poor scalability, and restrictions to printing on primarily two-dimensional surfaces. However, as with many previously mentioned microfabrication techniques, there exists a wide variety of tissue engineering applications for DPN, albeit a degenerate set when considering soft lithography techniques. Like aforementioned techniques, DPN could always be used for functionalizing two-dimensional scaffolds for added control in directing morphology, adhesion, migration, and cell–cell interactions.

4.3 FABRICATION OF THREE-DIMENSIONAL SCAFFOLDS

The motivation for scaffold design is typically focused on the ability to create scaffolds with predefined and precisely controlled unit cell geometries. Optimizing pore structures is useful in accommodating design parameters such as void percentage for cell seeding, bulk mechanical properties, and predicted concentrations of nutrients and waste throughout the scaffold. Controlling pore structure on a micron scale requires the recruitment of advanced three-dimensional fabrication techniques and computational topology design tools. The general field of rapid prototyping (RP), originally developed for rapid manufacturing techniques, forms three-dimensional objects with the help of custom fabrication hardware and computer-generated solid or surface models. Programs such as computer-aided design (CAD) coupled with the ability to fabricate arbitrary and complex three-dimensional structures through the use of RP and solid free-form fabrication (SFF) techniques have allowed the production of designer scaffolds with predefined microarchitecture. One of the major limitations in microfabricated scaffolds has been the inherent limitation of photolithography and surface sciences to two-dimensional platforms. Some photolithographic processes can be modified to create three-dimensional structures including the use of stepping techniques and the microscale origami. However, these techniques are costly and nonpractical for many applications including scaffold fabrication for tissue engineering. SFF techniques, through the implementation of controlled deposition using computer programs, do not require the use of multiple photomasks or an alignment step between layers. Systems that can freely incorporate controlled microfabrication processes in the "*z*-direction" have complete control over the microscale geometries and the relevant resulting macroscopic properties. Fabrication routes that control the fabrication in three dimensions have been devised to fabricate many classes of materials including polymers, composites, and ceramics (Figure 4.4). Of particular interest is controlling the porosity, mechanical moduli, and the interplay between these and other scaffold design parameters. For example, it may be beneficial to design a scaffold with mechanical moduli that matches the range of moduli found in either soft tissues (0.5–35 MPa) or hard tissues (10–1500 MPa). The following section outlines the current state of the art of SFF and outlines potential advancements and applications for use in scaffold fabrication.

4.3.1 Stereolithography and Selective Laser Sintering

Stereolithography (SLA) and selective laser sintering (SLS) are SFF techniques that use a laser-based curing device to photopolymerize liquid monomers or to sinter powered materials, respectively (Figure 4.4A,B). Both SLA and SLS processes start by converting a user-defined three-dimensional structure using CAD or similar computer software, which is then converted into a series of geometrically patterned two-dimensional "slices" of approximately 100 μm in thickness. A computer-controlled servo mechanism transmits information to the scanning laser and guides the location within the *x*–*y* plane, selectively bonding the material to form the structure. The build platform is then stepped down the same distance as the slice thickness and a new slice is scanned. SLA selectively polymerizes material from a vat of photopolymeric solution in a layer-by-layer process. SLA has been used to fabricate scaffolds from poly(propylene fumarate) and diethyl fumarate mixtures for hard tissues including the repair of bone defects [91]. PEG hydrogels that are end-modified with photoactive acrylate groups poly(ethylene glycol)-di methacrylate) (PEG-DMA) have been used extensively for photocrosslinkable systems for tissue engineering. PEG-DMA scaffolds incorporated with PLGA microparticles have also been fabricated for use in the tissue engineering of soft tissues [92]. Several research groups have also reported using digital micromirror devices (DMDs) to aid in SLA of various materials including polymers [93] and ceramics [94]. One previous limitation of SLA-based processes had been the inability to fabricate scaffolds that were large enough for clinical applications. This limitation has been overcome with the work of Cooke et al. as relevant length scales for repairing critical-sized boned defects can be

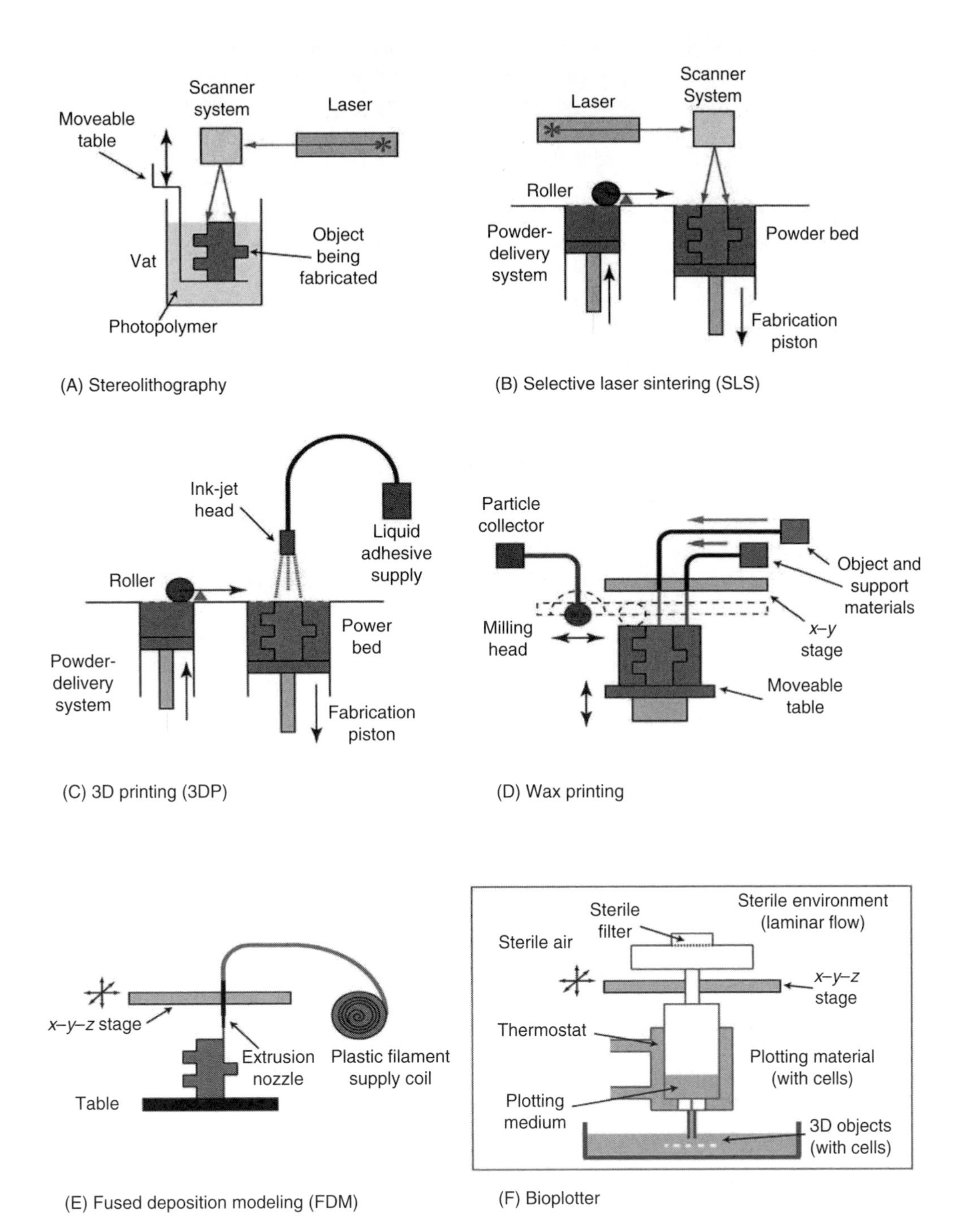

FIGURE 4.4 Schematics of SFF systems categorized by the processing technique. (A, B) Laser-based processing systems include the SLA system, which photopolyermerizes a liquid (A) and the SLS systems, which sinter powdered material (B). In each system, material is swept over a build platform that is lowered for each layer. (C, D) Printing-based systems, including 3D printing (C) and a wax printing machine (D). 3DP prints a chemical binder onto a powder bed. The wax-based system prints two types of wax material in sequence. (E, F) Nozzle-based systems. The fused deposition modeler prints a thin filament of material that is heated through a nozzle (E). The bioplotter prints material that is processed either thermally or chemically (F). (From Springer Science and Business Media and Worldwide and Guide to Rapid Prototyping Web site Copyright Castle Island Co., All rights reserved. With permission. [http://home.att.net/~castleisland/].)

fabricated using SLA. SLS, an SFF method that is operationally similar to SLA, selectively bonds material from a power bed using a carbon dioxide laser that is directed upon a power bed of material. The laser is scanned across the surface and sinters polymers or composites in a preprogrammed pattern for each layer. Low compaction forces during fabrication in SLS lead to highly porous structures. SLS has been used in the fabrication of scaffolds using calcium phosphate, PLA, PCL, and PEEK.

The spatial resolution and minimum feature size of SLA and SLS are governed by the laser spot size in the x–y plane and by slice thickness in the z-dimension. Laser spot sizes on the order of 250 μm are commonly attainable, while spot sizes as low as 70 μm have been achieved for modified systems. Feature resolution in the z-axis is theoretically limited by the precision of the mechanical stepping system that governs the slice thickness and typically is no smaller than 100 μm. In general, reducing feature resolution and slice thickness in SLA and SLS processes results in a dramatic decrease in the production speed. The current paradigm for tissue engineering only requires microfabrication techniques that are able to produce features that are on a similar order of magnitude of a few cells or approximately 50 μm. Nevertheless, there are potential engineering solutions to overcome problems of resolution should scaffold development require resolutions on the order of 1 μm. Variations of SLA processes could overcome the limitations due to slice thickness. Crosslinking of biomaterials via multiphoton excitation could also provide an efficient method of three-dimensional scaffold microfabrication. Multiphoton excitation crosslinking using UV has been applied to forming three-dimensional matrices of a wide range of polymers and bulk protein formulations such as collagen, laminin, fibronectin, polyacrylamide, bovine serum albumin, alkaline phosphatase, and various blends of the previous [95]. The distribution of intensity of irradiation in the z-axis theoretically could reduce the minimum feature sizes in this axis to approximately 20 μm or smaller. Remarkably, protein matrices fabricated using this method have been shown to remain active in this polymer matrix environment [96]. Further advancement of this technology could lead to the rapid and efficient fabrication of three-dimensional structures using a wide variety of biomaterials while simultaneously maintaining delicate conditions that can preserve biological activity.

4.3.2 Three-Dimensional Printing

Three-dimensional printing (3DP) is similar in practice to that of SLA and SLS in that a three-dimensional object is fabricated one two-dimensional slice at a time using a stepping system. One exception is that 3DP utilizes ink-jet printer technology to control the deposition of a chemical binder and selectively fuse material in a powder bed in order to create the object. Drug delivery and tissue engineering devices have been fabricated from poly(ethylene oxide) (PEO), PCL, and PLGA using 3DP. However, any biological material could be fabricated in principle using 3DP given the selection of an appropriate chemical binder. The resolution of objects created using 3DP varies upon the complexity of the object. Lines of 200 and 500 μm in width have been produced using 3DP of polymer solutions and pure solvents, respectively. Feature resolutions of approximately 1 mm are more likely for complex geometries. One significant limitation in the application of 3DP for scaffolds is the addition of the chemical binder. Even if the composition of the chemical binder itself is found to be nontoxic and biocompatible, the introduction of cytotoxic organic solvents such as chloroform or methylene chloride is undesirable. Postfabrication efforts to remove residual solvent such as vacuum drying are not completely effective; therefore the issue of cytotoxicity in 3DP-fabricated scaffolds remains.

4.3.3 Fused Deposition Modeling

Fused deposition modeling (FDM) is another method for SFF that has been employed in the fabrication of three-dimensional scaffold fabrication (Figure 4.4E). FDM uses a layer-by-layer deposition technique in which molten polymers or ceramics are extruded through nozzle with a small orifice, which merges with the material on the previously layer. The pattern for each layer is controlled by mechanical manipulation of the *x–y* position of the nozzle and can be different or arbitrary for each deposited layer. This technique has been applied to the production of three-dimensional scaffolds using polymers [PCL and high density polyethylene (HDPE)] and composites (PCL/hydroxyapatite). Fused deposition of ceramics has also been developed for fabrication of scaffolds from B-tricalcium phosphate. FDM processes can achieve pore sizes ranging from 160 to 700 μm with porosities ranging from 48% to 77%. PCL scaffolds fabricated via FDM have a compressive stiffness ranging from 4 to 77 MPa, which spans the mechanical stiffness for both soft and hard tissues [97]. One of the primary advantages of FDM and related processes is the high degree of precision that can be achieved in the *x–y* plane. However, control is the *z*-direction is limited and governed by the diameter of the extruded material through the nozzle. Additional limitations including pore anisotropy and the geometry of pore connectivity are substantially limited due to the continuous deposition process. The extrusion process limits the types of materials that can be processed and therefore has limited the wide applicability of FDM for scaffold fabrication. FDM is typically limited to synthetic thermoplastic polymers thereby eliminating many natural biomaterials or thermoset synthetic polymers.

4.3.4 Microsyringe Deposition

Microsyringe deposition is similar in practice to FDM in that a polymer is patterned on to a precision controlled surface using a continuous stream of material via layer-by-layer deposition in order to create three-dimensional structures [98]. Pressure assisted microsyringe-based deposition (PAM) uses compressed air to eject a solution of polymer in volatile solvent through a narrow capillary needle, which has a diameter between 10 and 20 μm. Control of the placement of solution in the *x–y* plane is controlled by a micropositioning system, which can achieve lateral precisions of 0.1 μm while the physical dimensions of the structures can range from 5 to 600 μm depending upon various processing parameters. PAM has been used for fabricate three-dimensional microfabricated scaffolds from PLGA, but could theoretically be expanded to any polymer, synthetic or natural, that is soluble in a volatile solvent. PAM presents several distinct advantages over SFF methods including dramatically improved spatial resolution and feature dimensions. Traditional SFF processes are known to have minimum resolutions on the order of 100 μm or larger and have principle limitations in minimum feature size. PAM also offers a convenient method toward fabrication of multiphase scaffolds with micron scale precision placement of multiple polymers. Low temperatures also allow potential integration systems for controlled release of proteins and other biomolecules to create favorable microenvironments for tissue regeneration.

4.3.5 Two-Photon Absorption

Two-photon absorption is a three-dimensional microfabricaiton technique that is able to produce features with resolutions of 120 nm, well below the diffraction limit of other photolithographic processes [99]. The materials used in this technique are acrylate-modified urethane resins, which are combined with photoinitiators to induce polymerization. These prepolymer resins are transparent to infrared, which enables deep penetration of irradiation and subsequent control over feature geometry deep within feature constructs. Although these techniques have been originally designed for use in MEMS systems, they have also been used in the three-dimensional scaffold development. Two-photon polymerization techniques have been used to fabricate scaffolds using ORMOCER, a proprietary inorganic–organic hybrid photopolymerizable material. A variety of cell types were seeded on these scaffolds including endothelial and neuroblastoma cells [100]. There are several obvious advantages of utilizing two-photon techniques for three-dimensional microfabrication including small feature resolutions and relatively rapid processing. However, there are distinct disadvantages including the cost of the equipment and the poor properties of materials that are currently available for use in two-photon systems. These disadvantages more than likely exist only because this technology has not fully matured in the commercial space. As two-photon technologies become more refined and adopted to various other application spaces such as tissue engineering and organ regeneration, the demand for affordable instrumentation and novel biomaterials that are suitable for these materials processing capabilities will be met.

4.3.6 Three-Dimensional Nanotopography

4.3.6.1 Electrospinning

Electrospinning is a convenient method for producing fibrous networks containing rich nanometer-scale texture with individual fiber diameters on the order of hundreds of nanometers. The electrospinning process draws a continuous narrow stream of material from a reservoir of polymer melt or solution to a collecting plate where the material accumulates, producing the fibrous mat. This is accomplished by inducing charge build-up on the surface of the solution through the application of strong voltages. When the voltage is sufficiently strong, the electrostatic potential overcomes the energy associated with surface tension of the bulk material at the orifice and the solution is accelerated toward the grounded collector. As the polymer solution is propelled to the collector, the solvent evaporates resulting in a continuous stream of ultrathin fibers. Electrospinning has been used to fabricate fibrous scaffolds using a variety of natural materials such as silk fibroin [101], collagen, polypeptides, and synthetic polymers such as PLGA, PCL, PVA, and PEO. Ceramics and composite fiber networks have also been produced using electrospinning techniques [102]. The diameters of individual fibers can range from approximately 30 nm to 1 μm and is varied by controlling the properties of the process such as polymer concentration, viscosity, conductivity, and applied voltage. Nanofibrous scaffolds produced by electrospinning typically result in a thin three-dimensional film of nonwoven mesh consisting of randomly oriented fibers. Mesh constructs can then be laminated together by thermal or solvent techniques, as previously mentioned. These processes produce bonds at the points of contact between the fibers and result in a crosslinked network. Nanofibrous systems with aligned fibers have also been synthesized by using a rotating drum as a collector [103] or, in the case of PLGA, by annealing the fibrous network after applying mechanical forces [104]. Fabricating topographically rich scaffolds with aligned nanofibers can exploit the contact guidance response in cells and lead to engineered cell functions, such as directed motility. Detailed processes and applications concerning electrospinning nanofibrous scaffolds have been reviewed extensively elsewhere [105] including more detailed sections of this text.

4.3.6.2 Colloidal Adsorption

Adsorption of colloids from solutions through dip coating is another economical method for imparting nanotopography on three-dimensional tissue engineering scaffolds [106]. The scaffold to be coated is simply incubated in a colloidal suspension of nanoparticles of specific size and composition, which are then adsorbed onto the scaffold surface. The composition of the nanoparticles can be resorbable polymers such as PLGA, bioinert metals such as gold, or ceramics such as silicon dioxide and titanium dioxide. Care must be made to ensure that the surface properties of the scaffold and colloid are compatible to ensure stable, irreversible adsorption onto the surface. This includes taking into account the relative surface energies and charge interactions in ionic medium. The advantages of this technique are somewhat obvious: colloidal adsorption is compatible with three-dimensional scaffolds with irregular shapes. Colloidal adsorption is inexpensive, rapid, and scalable. However, there are significant limitations including the limited functionality that can be imposed onto an existing scaffold. It is also difficult to obtain colloidal suspensions of nanoparticles that are monodisperse diameters, which negatively impacts the ability to form structures with long-range order. Nanoscale topography without long-range order is inadequate in terms of controlling cell functions at the level that is typically required for scaffold design. Colloidal adsorption is appropriate for effectively altering surface chemistry and topography in order to tune cell attachment, adhesion, and spreading on scaffolds.

4.3.7 Future of Solid Free-Form Fabrication for Tissue Engineering

SFF has been established as a convenient set of methods for use in three-dimensional scaffold fabrication of various materials including polymers, hydrogels, ceramics, and metallic biomaterials. One essential element to the continued use of RP in scaffold development is the ability not only to develop fabrication methods that are compatible with engineering materials, but also to modify fabrication techniques to enable the processing of biomaterials. For example, inkjet printing techniques have been used in the fabrication of collagen scaffolds [107]. Similar techniques must be optimized to enable the processing of other biomaterials with limited processing windows using RP.

Rational engineering design coupled with computer-controlled material processing techniques have allowed precision fabrication techniques that enable the control over the microarchitecture of three-dimensional porous scaffolds. In the near term, designing scaffolds using SFF techniques allows the possibility of indirectly controlling macroscopic properties such as the bulk mechanical properties and the permeability for nutrient transport considerations. Controlling the characteristics of pores on the micron scale such as pore geometry, connectivity, and porosity can lead to tissue-specific scaffold fabrication. SFF also allows the possibility of fabricating designer scaffolds for patient-specific organ regeneration applications. In general, the concept of made-to-order scaffolds is an appealing endeavor, which could serve currently unmet clinical needs such as scaffolds for reconstructive surgery. Although SFF has proven to be useful in engineering bulk properties through hierarchal design, future tissue engineering systems may require scaffolds with feature of resolution sizes that are smaller than that which are achievable with current SFF methods. Current design and fabrication strategies operate on a length scale of 100 μm or larger, which may be insufficiently precise to control cell-specific function. Consider the example of bone tissue engineering using a porous scaffold. Although optimal pore sizes between 200 and 600 μm have been suggested, empirical data for a number of systems suggests otherwise. Varying pore sizes between 500 and 1600 μm had no significant impact in bone growth of 3DP PLGA scaffolds seeded with osteocytes [108,109]. Hydroxy apatite scaffolds with pore sizes ranging from 400 to 1200 μm have been used in a mandibular defect model. Again, no significant difference in bone growth was detected across the range of pore sizes used in this study. These findings are consistent with the experimentally verified notion that the length scales of structures that govern many important cell functions such as adhesion, migration, proliferation, and differentiation are on the order of 10 μm or smaller. While the value of controlling macroscopic properties via hierarchal design should not be fully discounted, current SFF techniques as a stand alone for scaffold fabrication may not be sufficient for inducing appropriate biological responses. One option for fabricating scaffolds with nanoscale features using SFF is to simply improve the current resolution of the processes, which may prove to be too technically demanding or costly to justify future pursuit. Improving resolution to submicron scale precision would likely slow fabrication speed and would reduce the upper limit of scaffold volume. Therefore, more near-term advances in this field may stem from the integration of micron and submicron structures as a postfabrication modification for scaffolds produced by SFF such as electrospinning, colloidal adsorption, or phase separation and freeze-drying techniques that will be discussed in the following section.

4.4 MICROFABRICATION OF SCAFFOLDS USING PHYSICAL PROPERTIES

The behavior of a variety of cell types grown in three-dimensional microenvironments is well-known to differ vastly from cells grown on two-dimensional surfaces [110–112]. This widely understood phenomenon has prompted the drive to design and fabricate tissue engineering scaffolds that provide a suitable three-dimensional structure to support native cell function. The gravitation toward three-dimensional scaffolds has presented significant challenges in integrating control over spatial distribution and geometries of structures. Oftentimes, however, it may not be necessary to control specific structures on the micrometer length scale. Instead, designing the overall average scaffold geometry and characteristics can be designed by controlling parameters such as polymer composition in conjunction with physical processing parameters. Scaffold fabrication techniques has been accomplished using a variety of material processing approaches that engineer physio-chemical phenomenon. Some of these techniques are surveyed in the following section with more detailed treatments of these techniques found in other chapters within this text.

4.4.1 Microphase Separation

The characteristic phase separation behavior of polymer–solvent and polymer–polymer blends provides a useful means for rapidly fabricating porous tissue engineering scaffolds with ordered features on the micron and submicron length scale. The microphase morphology and characteristic length scale of features in polymer blends can be tuned by adjusting thermodynamic-dependent parameters such as the composition of the blend and kinetic parameters such as temperature, which govern the potential for spinodal decomposition and subsequent polymer demixing. The most commonly observed microphase morphologies are spherical, cylindrical, and lamellar, although numerous other classifications of morphologies have been observed in a variety of systems including block copolymer blends [113]. Modification of these bulk polymer blends via selective polymer etching or dissolution could lead to the development of tissue engineering scaffolds with short-range order. The phenomenon of polymer demixing and phase separation can also be adapted to the development of thin films as a method for chemical-induced topological modification. For example, partially compatible polymer blends spin coated into a thin film can produce topographically rich surfaces with structures of rounded pillar morphologies on the order of 100 nm in height and width [114]. Surfaces with polymer demixed nanotopography have been shown to affect the growth and spreading of human fibroblasts. Self-assembly of block copolymers can also be combined with nanopatterned substrates to achieve additional control over phase separation processes. Kim et al. present a method for obtaining highly uniform lamellae with variable amounts of long-range order using nanopatterned self-assembled monolayers on a two-dimensional substrate [115]. Further development of a similar process could lead to the development of integrating topographically rich surfaces with complex three-dimensional porous structures that enable the uniform presentation of topographical cues throughout the spatial extent of the scaffold. Electrospinning polymer blends provides another route for fabrication of nanometer-scale fibers with co-continuous phase separation by using ternary mixtures of PLA and poly(pyrrolidone) [116].

Freeze drying presents another possible route to producing scaffolds with micron scale features [117], although the control of spatial locations and distributions is reduced. A typical freeze-drying procedure begins with the addition of water to a solution of common synthetic biomaterials such as PLA or PLGA in organic solvent such as tetrahydrofuran or methylene chloride [118]. The solution is homogenized, poured into the appropriate mold, and then quenched rapidly at approximately −180°C using liquid nitrogen. This rapid quenching step leads to immediate phase separation within the scaffold, which is then freeze dried to remove the water and residual solvent. This process can produce scaffolds with pore sizes in the range of 15–35 μm. A similar method has been demonstrated in the production of porous PLA scaffolds with incorporated bioactive molecules that can provide cues for specific biological processes upon implantation [119]. Release kinetics of such systems suggest that slightly elevated temperatures in the range of 55°C–85°C and the use of organic solvents during preparation of these scaffolds has shown to slightly negatively impact the activity of the released molecules. In terms of engineering the microscale structure of these scaffolds, precise control of pore size, geometry, and interconnectivity is severely limited by the deferral to equilibrium processes in the preparation of such scaffolds. This leads to control of the averaged scaffold features, which can be manipulated through adjusting processing parameters.

4.4.2 Colloidal-Assisted Self-Assembly

Porogen leaching is a common approach to developing large, three-dimensional, porous scaffolds. In this established technique, the scaffold material is incorporated with a chemically or physically incompatible porogen. Upon scaffold fabrication via processes such as solvent evaporation or chemical and physical crosslinking, the porogen is selectively removed. One consequence of this method of fabrication technique is the high degree of porosity (typically in the range of 90% or higher), which is necessary to achieve an interconnected network for random porogen organization. Scaffolds with extremely high porosities are advantageous for tissue engineering metabolically active tissues by allowing for more rapid diffusion of nutrients and removal of waste products. High porosities also provide large surface areas per volume to allow cell attachment and proliferation. However, the mechanical properties of highly porous networks are severely compromised including a dramatic reduction in the relative stiffness of the porous and solid scaffolds, E_{cellular} and E_{solid}, respectively, with increasing porosity P. For the case of a cellular solid with randomly arranged porous features, this relationship is proportional to a geometrically defined constant through the following equation:

$$\frac{E_{\text{cellular}}}{E_{\text{solid}}} = C(1 - P)^2 \tag{4.1}$$

where C is an arbitrary constant. Reducing the porosity required for interconnection can be obtained through ordered assembly of uniform, micron scale, spherical, colloidal particles [120]. Physical templating through geometric constraints can produce scaffolds that exhibit dramatically increased stiffness and improved toughness, which has been show to effect cell behavior, when compared to randomly assembled porogen structures. This fabrication technique has been employed for the use of creating hydrogel networks with increased moduli using a porogen network consisting of ordered microspheres. PMMA microspheres of diameters ranging from 20 to 60 μm in a 30:70 water/ethanol mixture were deposited under physical perturbation to enhance characteristic hexagonal close-packed ordering. A solution of low molecular weight PEG-DMA modified with adhesion-promoting peptides was administered to the template and irradiated with UV to induce crosslinking of the scaffold. The PMMA porogen network was etched with acetic acid, leaving a porous hydrogel network that was shown to support the attachment and migration of NR6 fibroblasts. Using an ordered sacrificial pore structure has demonstrated an increase in stiffness by up to several orders of magnitude to the range of 10 kPa, which is comparable to that of soft tissues. In principle, this process could be applied to any number of polymer-porogen systems that exhibit physical properties that permit removal of the sacrificial phase such as relative selectively in dissolution, degradation, or differential melting points.

4.4.3 Supramolecular Self-Assembly

Ordered template assembly can also be achieved through self-assembly of biomolecules such as proteins or DNA. Self-assembly of peptide systems into supramolecular structures requires the rational incorporation of specific and selective interactions among the amino acid building blocks. The resulting structures can be tuned by manipulating amino acid sequences that provide the appropriate combinations of properties such as charge and residue hydrophilicity. Beta sheet structures can be produced by incorporating peptides with one hydrophobic surface and one hydrophilic surface in an aqueous solution. The hydrophobic surfaces become shielded while the charged hydrophilic residues remain in contact with the aqueous environment in an organization event that is similar in nature to in vivo protein folding. The hierarchal structure of this assembly process is evident by the ability to synthesize structures on the order of 30–100 nm by virtue of single peptide building blocks with lengths on the order of one nanometer. This technique has been utilized with natural amino acid sequences to produce a wide range of structures with nanometer-scale features including modified surfaces, filaments, and fibrils and peptide nanotubes for a variety of applications including tissue engineering and drug delivery [121].

Macromolecular structures can also assemble to produce stable networks of nanometer-scale fibers. Oligonucleotides can be engineered to form self-assembling DNA structures with features as large as 100 nm with 6 nm resolution [122,123]. In work by Rothemund, an assembly technique, termed "DNA origami" involves the combination of DNA "scaffolds," which provide the bulk components of the material, and DNA "staples," which induce self-assembly into a variety of preprogrammed structures. Self-assembling peptide networks have been used to support three-dimensional cell culture systems and feature fibers diameters on the order of 10 nm with pore sizes between 200 and 500 nm [124]. Reversible hydrogel structures have also been synthesized by engineering synthetic tri-block peptide structures [125]. Peptide constructs have been used to produce ordered peptide nanotubes, which may include other materials for various applications. In one approach, amyloid fiber peptides, the precursors to prions that can lead to Alzheimer's disease, can self-assembly to produce silver nanowires on the order of 100 nm in diameter [126]. Although the original intention of studying these processes was to aid in the assembly of inorganic materials such as metallic nanowires for applications in electronics [127], similar approaches could be applied to develop advanced tissue engineering constructs. One potential limitation of this class of materials for tissue engineering applications is the extremely small characteristic length scale of peptide-self-assembly. It may be necessary to fabricate tissue engineering constructs with feature definition that ranges over a wide range of length scales. Although nanoscale features can modulate biological function on the cellular level, controlling the structures of scaffolds on the length scale of hundreds of microns or larger may be required for some tissues to create biomimetic macroscopic geometries, improve mechanical strength, or optimize porosity connections to enhance nutrient transport and waste removal. Expanding the operable length scale of this technique could lead to the advancement in mesoscale assembly of larger structures with feature sizes on the order of tens of microns, which may result in a tissue engineering microenvironment that better supports spatially dependent cell processes such as migration, proliferation, and matrix remodeling. Combining template or self-assembly techniques with other scaffold fabrication techniques with much higher minimum resolutions may be a solution to controlling scaffold features and geometry across several orders of magnitude.

4.5 MICROFABRICATION OF CELL-SEEDED SCAFFOLDS

Coordinating the precise locations of cells placement of cells on scaffolds relative to is a promising method of capitalizing on specific cell–cell interactions that may lead to favorable conditions for tissue formation. However, three-dimensional scaffolds that are postseeded with cells cannot typically retain spatial segregation or control specific cell–cell interactions across multiple cell types. While photolithography-based cell-patterning techniques are a powerful method for controlling such interactions in a two-dimensional microenvironment, these methods rarely efficiently translate to three-dimensional systems. Advances in microfabrication of biomaterials and SFF have led to the potential to control the placement of polymers and cells with micron scale precision. Combining aspects of these existing technologies with a layer-by-layer approach is a promising approach to fabricating scaffolds that are preloaded with cells that are spatially controlled in arbitrary geometries. Typical scaffold fabrication techniques often employ harsh conditions such as the use of cytotoxic solvents or elevated temperatures. Hence, developing novel scaffold fabrication processes that are capable of maintaining viable cells is essential. This limitation also serves to drastically reduce the number of biomaterials that are suitable for such processes. Despite these challenges, a number of possibilities have been demonstrated using this approach to scaffold fabrication.

4.5.1 Layer-by-Layer Deposition

Microfluidic platforms provide a convenient method for controlling many aspects of the cellular microenvironment including spatially defined concentrations of nutrients, oxygen, and mechanical shear stresses as realized through fluid flow. Microfluidics can also be used to control spatial orientation through confinement within microchannels. One-step cell patterning in microfluidic devices has been used to create cellular arrays for diagnostic purposes. When used in combination with a layer-by-layer assembly strategy, micropatterning and microfluidics has proven useful for the controlled assembly of hierarchal constructs of multiple extracellular matrices and cell types [128]. A microfluidic patterned coculture system with three cell types has been demonstrated for potential use in vascular tissue engineering applications. Endothelial cells, smooth muscle cells, and fibroblasts can be each mixed with ECM protein solutions such as collagen, collagen–chitosan, and Matrigel and deposited sequentially [129,130] in a layer-by-layer method. The result is a three-dimensional layer of multiple cell types that are each contained predefined ECM proteins. First, silicon oxide surfaces were modified with 3-aminopropyltriethoxysilane using vapor deposition to enhance the adhesion of ECM proteins. PDMS-on-glass microfluidic devices were fabricated using the modified silicon oxide surfaces. The spatial layout of the tissues is defined by the layout of the microfluidic network. One selected cell–matrix solution is then flowed into the microfluidic network at 2°C at which point the system is heated to 37°C to induce gelation. Subsequent layers of cell–matrix systems can be added in a similar manner. After formation of the coculture structures, the PDMS layer is removed to allow for further characterization of cell function or matrix properties. The obvious advantage of this process is the use of cell-friendly conditions during fabrication such as the absence of toxic chemical such as solvents or photoinitiators and UV irradiation, which can improve viabilities of seeded cells. Tailoring specific cell–matrix combinations can be engineered to produce optimal conditions that improve the overall performance of cell-seeded scaffolds. Although the use of one microfluidic device limits the independent patterning of multiple cell types, this limitation could be potentially overcome by the use of multiple microfluidic patterning techniques, which has been demonstrated in other systems [131].

4.5.2 Photoencapsulation of Cell-Scaffold Constructs

Photomasks typically used for photolithography in traditional microfabrication processes can be used in combination with PEG-DMA solutions to pattern hydrogel structures with resolution of at least 10 μm. In one reported method, multiple steps of micropatterned photopolymerization processes can be coupled to produce three-dimensional cell–matrix structures with micron scale resolution [132]. Silicon dioxide substrates were first surface-functionalized with 3-(trimethoxysilyl)-propyl methacrylate (3-TPM) to allow for covalent bonding of hydrogels. Thin layers of cell-containing solutions of PEG-DMA hydrogel with model HepG2 cells were applied to the functionalized surface and exposed to UV irradiation through a photomask. The substrate is developed by washing in isotonic solution to remove the uncrosslinked regions. The substrate is then stepped in the z-direction by inserting a silicone spacer. This process can be repeated using arbitrary suspensions of cells in PEG-DMA solutions to create additional micropatterned layers. Photopolymerization of cell-encapsulated hydrogels in a layer-by-layer technique provides the flexibility of patterning cells layers in arbitrary geometries. PEG-DMA molecules modified with specific peptides that promote cell functions such as adhesion or migration can be integrated into the scaffold with ease. The result is a flexible hydrogel scaffold fabrication platform that can be used to control the microenvironment of multiple cell types while simultaneously governing the spatial configuration with micron scale precision. One drawback of such a system is the use of potentially cytotoxic photoinitiators in combination with UV irradiation as a means of creating microstructures. In the reported method, the viability of cells, as measured by MTT assay, is reduced in a dose-dependent manner as the time of UV exposure and concentration of acetophenone photoinitiator was varied between 0–60 sec and 0–3 mg/mL, respectively. UV irradiation can also cause irreversible DNA damage, which may impact the potential for use with stem cells and other progenitor cells. Another factor that may limit the utility of microfabricated cell-seeded scaffolds using photocrosslinkable hydrogels is the maintenance of pattern fidelity of small features with a high degree of precision. While the resolution of features greater than 200 μm is maintained at approximately 10% of the feature size, the minimum spatial resolution is greatly increased as the feature size is reduced below 100 μm. The actual dimensions of features can be increased up to 200% larger than the feature sizes on the photomask where the ratio is primarily dependent upon the exposure time. One mechanism that might explain the increase in feature size could be the inherent swelling nature of PEG hydrogels. Other possibilities may be the rapid migration of photoreactive species within the hydrogel during polymerization, which would lead to an overall deficiency in producing features smaller than 50 μm with tight resolution. Although this process has been adapted for the fabrication of PEG hydrogels, it could be expanded to other types of photoreactive polymeric species [133].

SFF techniques have demonstrated great promise in the design and fabrication of customizable tissue engineering scaffolds. Postseeded cell scaffolds, fabricated via SFF or other methods, limit the attachment of cells to the surface of the scaffold alone, which results in relatively low cell densities by volume. Combining aspects of SLA with cell–polymer solutions allows for the possibility of incorporating cells directly in the scaffold to achieve high cell densities. Such a system has been reported where Chinese hamster ovary (CHO) cells were photoencapsulated in PEG-DMA hydrogels using a modified SLA process [134]. The technical issues associated with this fabrication technique are simply the union of those with standard SLA processes and photoencapculation of living cells. Such challenges include attaining a reasonably small minimum feature size, producing tight resolutions, and maintaining high-cell viability of encapsulated cells. In the process developed by Dhariwala et al., minimum feature sizes of 150 μm in the z-direction and 250 μm in the x–y plane were achieved, which is similar to traditional SLA processes. As reported in previous photoencapsulation studies, the cell viability was highly dependent upon the concentration of photoinitiator.

4.5.3 Microscale Direct Deposition of Cells

The concept of direct deposition of cells is another promising prospective approach to obtaining patterns of cells in arbitrary geometries. Two methods for direct cell deposition using contact lithography have previously been mentioned in this treatment. These methods are (1) the use of an agarose stamp for the controlled direct deposition of cells on substrates [58] and (2) the use of an elastomeric stencil for patterned cellular deposition [65]. However, direct deposition of cells can be accomplished using other techniques as well.

4.5.3.1 Cell Spraying

The direct deposition of cells using a spraying technique in conjunction with a patterning mask is a convenient method for creating two-dimensional patterning of cells in succession, which results in a three-dimensional scaffold of micropatterned cells. This straightforward method aerosolizes a solution of cells using an off-the-shelf airbrush apparatus creating a suspension of liquid droplets containing cells, which is directed toward a surface of collagen gel [135]. Upon deposition of cells, the mask is removed leaving patterned cells on a collagen substrate with resolutions of approximately 100 μm. A second collagen gel (~400 μm in thickness) is then adsorbed to the surface followed by an additional cell spraying step. These sequences of steps can be repeated in succession to produce alternating layers of cells and ECM matrix proteins. Increased cytochrome P450 activity, an assay for liver-specific cell function, was observed when patterning NIH 3T3 fibroblasts in conjunction with plated primary hepatocytes in a short-term culture. The ability to create arbitrary geometries of patterned cells in a collagen sandwich is an appealing technology to preserve polarity and function of hepatocytes cultured in vitro [136]. The spraying process was shown not to dramatically affect cell viability or function of cultured human umbilical endothelial cells or primary hepatocytes. This approach, while similar in principle to other techniques for cell patterning including the use of micropatterned stamps or stencils, [65] provides additional functionality by combining patterned cell deposition with a layer-by-layer method, which controls the location of cells in the z-direction. Though collagen was used as the ECM protein in this specific example, any suitable ECM protein could be used in between cell layers, and can be optimized depending upon the specific tissue of interest. There are some disadvantages to this and similar techniques. Cell spraying requires the use of multiple masks for patterning cells in different geometries for each corresponding slice in the x–y plane. Also, the minimum resolution is still larger than the characteristic length scale for many microarchitectures found in the tissue of complex organs. For example, structures found in the lung alveoli, capillary beds, and liver sinusoids each have microarchitecture that would theoretically require 5–10 μm precision of individual cell placement.

4.5.3.2 Inkjet Printing of Cells

Another method of direct cell deposition employs the use of an off-the-shelf inkjet printer that uses a thermal ink deposition method, which is loaded with a suspension of CHO cells in a buffer solution. Cells are then deposited in a precise manner on to hydrogel-based substrates using a simple graphics program to define the location of the deposited cells in the x–y plane (Figure 4.4F). This process requires little modification to the commercially available printer and, despite the extreme stresses on cells that are associated with the printing process, viabilities of approximately 90% were maintained in the deposition of the robust CHO cell line as a proof-of-concept. The resolution of inkjet printing of cell suspensions was shown to be on the order of hundreds of microns, which is typical of most previously described SFF methods. Although this approach has demonstrated only two-dimensional patterned deposition of one cell type, this technology represents a flexible platform for future development of three-dimensional cell-seeded scaffolds.

4.5.3.3 Optical Methods for Noncontact Cell Deposition

The noncontact manipulation of cells through the use of single-beam gradient optical traps is an efficient method of controlling the spatial coordinates of a variety of cell types in suspension. This technology, also termed optical tweezers, has been demonstrated in the manipulation of a variety of biological components including cells, subcellular components, and microparticles coated with functional biomolecules. The use of high numerical aperture lenses leads to tight confinement of particles along the beam axis limiting the range of mobility in this direction, which ultimately leads to the drastic reduction in efficiency for depositing cells in this manner. By simply reducing the numerical aperture of the lens, micron- and submicron-sized particles can be directed along the beam path in a continuous stream and absorbed on to a target of interest. This general technique, termed laser-guided direct writing (LGDW) has been applied to a variety of organic and inorganic particles as well as mammalian cells [137]. During LGDW of cells, a laser emitting a near-infrared radiation is configured near a lens that is controlled using a three-axis micromanipulator and directed into a sterile chamber containing a suspension of cells and a target substrate at the bottom of the well. As cells randomly drift into the path of the laser, a gradient force dominates and brings the cell to the center of the beam. The laser beam then directs the cell along the beam path until it reaches the substrate. The position of the single cells or streams of cells can be directed by manipulating the position of the lens. This technique has been used for the direct deposition of embryonic chick spinal cord cells on a target substrate of untreated glass. One advantage of this technique is that cells can be deposited onto various substrates including collagen and Matrigel, which may lack the robust mechanical properties requisite for other soft lithography or patterning techniques. For example, human umbilical vascular endothelial cells (HUVECs) have been patterned

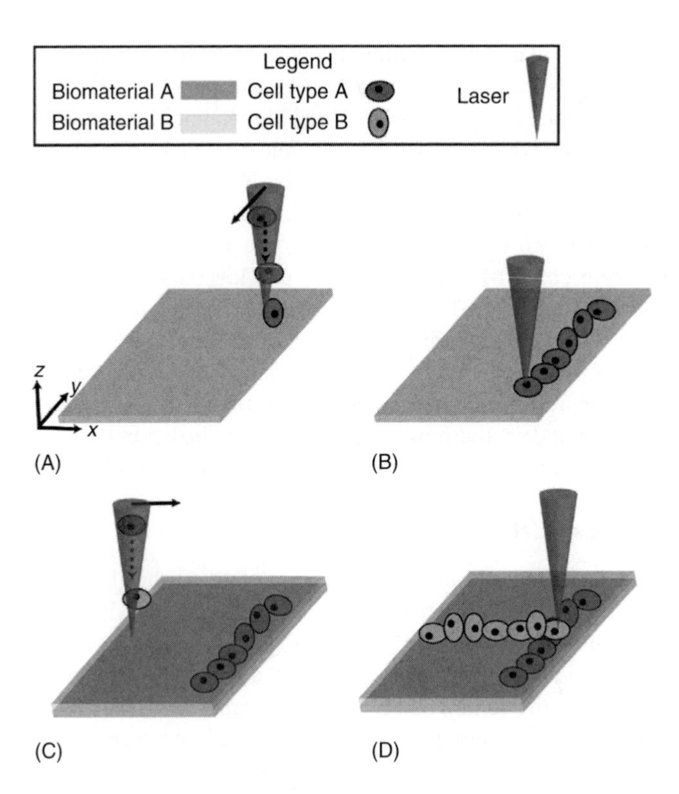

FIGURE 4.5 Schematic of LGDW for three-dimensional tissue engineering. (A) A suspension of cells is introduced on top of a biomaterial substrate in the presence of a laser that is moveable along the *x*–*y* axis. (B) As cells drift into the path of the laser, they are directed toward the substrate. The laser can then be moved to create individual lines of cells or form rasterized geometries. (C) Additional biomaterial layers can be adhered to the surface to create fresh substrates for (D) subsequent cell deposition of a different cell type, if desired. Repeating the steps of direct writing and biomaterial deposition can be used to create three-dimensional substrates, which may be potentially useful for a variety of tissue structures (see text).

directly on thin films of Matrigel, which is known to promote vascular organization [138,139]. This layer-by-layer technique can be combined with thin film deposition of ECM biomaterials to create three-dimensional structures (Figure 4.5). Multiple sequences of cell deposition of various types followed by the adsorption of a thin biomaterial layer can be cycled to create three-dimensional cell-seeded constructs. Multiple cell types can be deposited with ease, which can be used to produce cocultures to take advantage of heterotypic cell–cell interactions. In work by Nahmias et al., hepatocytes are patterned in combination with HUVECs to form nascent liver sinusoid structures. One potential drawback to large-scale implementation of this technology could be the slow speed of deposition. In the reported system, cells were deposited onto the surface at a rate of 2.5 cells per minute which suggests that the patterned deposition of 100–1000 cells would take on the order of hours to complete. This rate could theoretically be increased by increasing the concentration of cells within the suspension or by using an array of lasers to pattern multiple regions simultaneously.

4.6 FUTURE OF MICROFABRICATION TECHNOLOGY FOR SCAFFOLD DEVELOPMENT

The key to advancement of scaffold development is continually facilitating the interaction between at least these three key fields of study: (1) materials science, engineering, and processing, (2) cell biology, and (3) chemistry and materials synthesis. These and related fields will no doubt advance spontaneously as breakthrough developments in fields such as polymer synthesis, drug delivery, protein characterization, gene therapy, and stem cells continue at a frenetic rate. However, the interplay is essential in designing scaffolds that facilitate premeditated tissue regeneration responses. Cell biology and materials synthesis must be studied simultaneously to characterize cell–biomaterial interactions, which will ultimately serve as a basis for the design and synthesis of smart biomaterials. Materials synthesis and materials processing must move forward synergistically in order to develop materials processing strategies that are not only compatible with appropriate biomaterials, but are also able to accommodate technical and economic factors as well. Lastly, materials science and cell biology must be implemented in unison to study and eventually engineer the spatial and temporal microenvironments on a precise level is critical for inducing the desired cellular response, which is especially important in the field of tissue engineering and organ regeneration. One key thrust of the mutual development lies in the ability to create materials and systems on the nanometer length scale. The limitations of traditional photolithography have driven the implementation of other nanofabrication processes including nanoimprint lithography. Nanoimprint lithography processes are able to create reproducible features less than 30 nm [140]. Features on this length scale could be potentially utilized to fabricate synthetic basement membrane structures, which are known to contain features on a similar length scale [141,142]. Artificial cell substrates could be used to precisely interact with the filopodia and lamelopodia of cells, which could potentially be used to control cell functions such as migration and differentiation.

The current tissue engineering paradigm mandates that the driving force for scaffold design and fabrication is the desire to create a biomimetic system. This rudimentary albeit currently appropriate pursuit often manifests itself into designing biomimetic materials, processes, and cell-scaffold constructs that try to mimic properties of the tissue microenvironment such as cell phenotype, microarchitecture, mechanical properties, surface topography, and chemical cues. Biomimetic strategies for scaffold fabrication provide challenges that drive the overall improvement of microfabrication techniques for scaffold development and tissue engineering in general. These improvements may come in the form of newfound materials with improved tissue response or ease of processing, novel methods for controlling cell–cell or cell–biomaterial interactions, or new methods for the facile fabrication biomaterials into functional scaffolds that can promote regenerative processes across all hierarchal levels from proteins and cells to tissues and organs.

The primary role of microfabrication in scaffold development and tissue engineering is to function as a tool that can aid and assist in advancing individual fields of science and engineering that are either directly or indirectly related to tissue and organ regeneration. Furthermore, microfabrication techniques for biomaterials must also be viewed as a unifying medium for creating platforms for interdisciplinary research. For the most part, the instrumental role of microfabrication in tissue and organ regeneration has been permanently established only recently. However, the role of microfabrication technology in tissue engineering will become more prominent as it becomes more and more economically viable, ubiquitous, accessible to other scientific fields, and it continues to gain traction as a tool for both the fundamental and applied biological sciences.

REFERENCES

1. McBeath R, Pirone DM, Nelson CM, Bhadriraju K, and Chen CS. Cell shape, cytoskeletal tension, and RhoA regulate stem cell lineage commitment. *Dev Cell* 2004;6:483–495.
2. Jiang X, Bruzewicz DA, Wong AP, Piel M, and Whitesides GM. Directing cell migration with asymmetric micropatterns. *Proc Nat Acad Sci USA* 2005;102(4):975–978.
3. Lucchetta EM, Lee JH, Fu AL, Patel NH, and Ismagilov RF. Dynamics of Drosophila embryonic patterning network perturbed in space and time using microfluidics. *Nature* 2005;434:1134–1138.
4. Chen CS, Mrksich M, Huang S, Whitesides GM, and Ingber DE. Geometric control of cell life and death. *Science* 1997;276:1425–1428.

5. Flemming RG, Murphy CJ, Abrams GA, Goodman SL, and Nealey PF. Effects of synthetic micro- and nano-structured surfaces on cell behavior. *Biomaterials* 1999;20:573–588.
6. Dalby MJ, Riehle MO, Yarwood SJ, Wilkinson CDW, and Curtis ASG. Nucleus alignment and cell signaling in fibroblasts: Response to a micro-grooved topography. *Exp Cell Res* 2003;284:274–282.
7. Chou L, Firth JD, Uitto VJ, and Brunette DM. Substratum surface topography alters cell shape and regulates fibronectin mRNA level, mRNA stability, secretion and assembly in human fibroblasts. *J Cell Sci.* 1995;108:1563–1573.
8. Curtis ASG and Wilkinson CDW. Reactions of cells to topography. *J Biomater Sci Polym Ed* 1998; 9(12):1313–1329.
9. Curtis A and Wilkinson C. Topographical control of cells. *Biomaterials* 1997;18(24):1573–1581.
10. Armani DK and Liu C. Microfabrication technology for polycaprolactone, a biodegradable polymer. *J Micromech Microeng* 2000;10:80–84.
11. Chen SC and Lu Y. Micro- and nano-fabrication of biodegradable polymers. In: Mallapragada S, Narasimhan B (Eds.), *Handbook of Biodegradable Polymeric Materials and Their Applications*: American Scientific Publishers 2005:1–17.
12. Heckele M and Schomburg W. Review on micro molding of thermoplastic polymers. *J Micromech Microeng* 2004;14:R1–R14.
13. Duffy DC, McDonald JC, Schueller JA, and Whitesides GM. Rapid prototyping of microfluidic systems in poly(dimethylsiloxane). *Anal Chem* 1998;70(23):4974–4984.
14. Borenstein JT, Terai H, King KR, Weinberg EJ, Kaazempur-Mofrad MR, and Vacanti JP. Microfabrication technology for vascularized tissue engineering. *Biomed Microdevices* 2002;4(3):167–175.
15. Leclerc E, Sakai Y, and Fujii T. Microfluidic PDMS (polydimethylsiloxane) bioreactor for large-scale culture of hepatocytes. *Biotechnol Prog* 2004;20:750–755.
16. Leclerc E, Yasuyuki S, and Fujii T. Cell culture in 3-dimensional microfluidic structure of PDMS (polydimethylsiloxane). *Biomed Microdevices* 2003;5(2):109–114.
17. Richards-Grayson AC, Choi IS, Tyler BM, Wang PP, Brem H, Cima MJ, et al. Multi-pulse drug delivery from a resorbable polymeric microchip device. *Nat Mater* 2003;2:767–772.
18. Schmidt CE, Shastri VR, Vacanti JP, and Langer R. Stimulation of neurite outgrowth using an electrically onducting polymer. *Proc Nat Acad Sci USA* 1997;94:8948–8953.
19. George PM, Lyckman AW, LaVan DA, Hegde A, Leung Y, Avasare R, et al. Fabrication and biocompatibility of polypyrrole implants suitable for neural prosthetics. *Biomaterials* 2005;26:3511–3519.
20. Wang Y, Ameer GA, Sheppard BJ, and Langer R. A tough biodegradable elastomer. *Nat Biotechnol* 2002;20:602–606.
21. Yang J, Webb AR, and Ameer GA. Novel citric acid-based biodegradable elastomers for tissue engineering. *Adv Mater* 2004;16(6):511–515.
22. Bettinger CJ, Orrick B, Misra A, Langer R, and Borenstein JT. Microfabrication of poly(glycerol-sebacate) for contact guidance applications. *Biomaterials* 2006;27:2558–2565.
23. Bettinger CJ, Cyr KM, Matsumoto A, Langer R, Borenstein JT, and Kaplan DL. Silk fibroin microfluidic devices. *Adv Mater* 2007;19(19):2847–2850.
24. Paguirigan A and Beebe DJ. Gelatin based microfluidic devices for cell culture. *Lab Chip* 2005;6:407–413.
25. Cabodi M, Choi NW, Gleghorn JP, Lee CSD, Bonassar LJ, and Stroock AD. A microfluidic biomaterial. *J Am Chem Soc* 2005;127:13788–13789.
26. Tang MD, Golden AP, and Tien J. Molding of three-dimensional microstructures of gels. *J Am Chem Soc* 2003;125(43):12988–12989.
27. Owen GR, Jackson J, Chehroudi B, Burt H, and Brunette DM. A PLGA membrane controlling cell behaviour for promoting tissue regeneration. *Biomaterials* 2005;26:7447–7456.
28. Mills CA, Escarr J, Engel E, Martinez E, Errachid A, Bertomeu J, et al. Micro and nanostructuring of poly(ethylene-2,6-naphthalate) surfaces, for biomedical applications, using polymer replication techniques. *Nanotechnology* 2005;16(4):369.
29. Teixeira AI, Abrams GA, Bertics PJ, Murphy CJ, and Nealey PF. Epithelial contact guidance on well-defined micro- and nanostructured substrates. *J Cell Sci* 2003;116:1881–1892.
30. Sarkar S, Dadhania M, Rourke P, Desai TA, and Wong JY. Vascular tissue engineering: Microtextured scaffold templates to control organization of vascular smooth muscle cells and extracellular matrix. *Acta Biomater* 2005;1(1):93–100.

31. Belkas JS, Shoichet MS, and Midha R. Peripheral nerve regeneration through guidance tubes. *Neurol Res* 2004;26(2):151–160.
32. Jiang X, Takayama S, Qian X, Ostuni E, Wu H, Bowden N, et al. Controlling mammalian cell spreading and cytoskeletal arrangement with conveniently fabricated continuous wavy features on poly(dimethylsiloxane). *Langmuir* 2002;18:3273–80.
33. Metz S, Holzer R, and Renaud P. Polyimide-based microfluidic devices. *Lab Chip* 2001;1:29–34.
34. King KR, Wang CCJ, Kaazempur-Mofrad MR, Vacanti JP, and Borenstein JT. Biodegradable microfluidics. *Adv Mater* 2004;16(22):2007–2009.
35. Ryu W, Min SW, Hammerick KE, Vyakarnam M, Greco RS, Prinz FB, et al. The construction of three-dimensional micro-fluidic scaffolds of biodegradable polymers by solvent vapor based bonding of micro-molded layers. *Biomaterials* 2007;28(6):1174–1184.
36. Wang GJ, Chen CL, Hsu SH, and Chiang YL. Bio-MEMS fabricated artificial capillaries for tissue engineering. *Microsyst Technol* 2005;12:120–127.
37. Bettinger CJ, Weinberg EJ, Kulig KM, Vacanti JP, Wang Y, Borenstein JT, et al. Three-dimensional microfluidic tissue engineering scaffolds using a flexible biodegradable polymer. *Adv Mater* 2006;18:165–169.
38. Horan RL, Antle K, Collette AL, Wang Y, Huang J, Moreau JE, et al. In vitro degradation of silk fibroin. *Biomaterials* 2005;26:3385–3393.
39. Altman GH, Diaz F, Jakuba C, Calabro T, Horan RL, Chen J, et al. Silk-based biomaterials. *Biomaterials* 2003;24:401–416.
40. Sofia S, McCarthy MB, Gronowicz G, and Kaplan DL. Functionalized silk-based biomaterials for bone formation. *J Biomed Mater Res* 2001;54:139–148.
41. Karageorgiou V, Meinel L, Hofmann S, Malhotra A, Volloch V, and Kaplan D. Bone morphogenetic protein-2 decorated silk fibroin films induce osteogenic differentiation of human bone marrow stromal cells. *J Biomed Mater Res* 2004;71A(3):528–537.
42. Alsberg E, Anderson KW, Albeiruti A, Franceschi RT, and Mooney DJ. Cell-interactive alginate hydrogels for bone tissue engineering. *J Dent Res* 2001;80(11):2025–2029.
43. Kuo CK and Ma PX. Ionically crosslinked alginate hydrogels as scaffolds for tissue engineering: Part 1. Structure, gelation rate and mechanical properties. *Biomaterials* 2001;22(6):511–521.
44. Lee KY, Rowley JA, Eiselt P, Moy EM, Bouhadir KH, and Mooney DJ. Controlling mechanical and swelling properties of alginate hydrogels independently by cross-linker type and cross-linking density. *Macromolecules* 2000;33:4291–4294.
45. Ling Y, Rubin J, Deng Y, Huang C, Demirci U, Karp JM, et al. A cell-laden microfluidic hydrogel. *Lab Chip* 2007;7:756–762.
46. Whitesides GM. The origins and the future of microfluidics. *Nature* 2006;442:368–373.
47. Shito M, Kim NH, Baskaran H, Tilles AW, Tompkins RG, Yarmush ML, et al. In vitro and in vivo evaluation of albumin synthesis rate of porcine hepatoytes in a flat-plate bioreactor. *Artif Organs* 2001;25(7):571–578.
48. Park J, Berthiaume F, Mehmet T, Yarmush ML, and Tilles AW. Microfabricated grooved substrates as platforms for bioartificial liver reactors. *Biotechnol and Bioeng* 2005;90(5):632–644.
49. Lee PJ, Hung PJ, and Lee LP. An artificial liver sinusoid with a microfluidic endothelial-like barrier for primary hepatocyte culture. *Biotechnol and Bioeng* 2007;97(5):1340–1346.
50. Wu H, Odom TW, Chiu DT, and Whitesides GM. Fabrication of complex three-dimensional microchannel systems in PDMS. *J Am Chem Soc* 2003;125:554–559.
51. Xia Y and Whitesides GM. Soft lithography. *Angew Chem Int Edit* 1998;37(5):550–575.
52. Bernard A, Renault JP, Michel B, Bosshard HR, and Delamarche E. Microcontact printing of proteins. *Adv Mater* 2000;12(14):1067–1070.
53. Lin CC, Co CC, and Ho CC. Micropatterning proteins and cells on polylactic acid and poly(lactide-co-glycolide). *Biomaterials* 2005;26(17):3655–3662.
54. Lee CJ, Blumenkranz MS, Fishman HA, and Bent SF. Controlling Cell Adhesion on Human Tissue by Soft Lithography. *Langmuir* 2004;20(10):4155–4161.
55. Lehnert D, Wehrle-Haller B, David C, Weiland U, Ballestrem C, Imhof BA, et al. Cell behaviour on micropatterned substrata: Limits of extracellular matrix geometry for spreading and adhesion. *J Cell Sci* 2004;117:41–52.

56. McDevitt TC, Woodhouse KA, Hauschka SD, Murry CE, and Stayton PS. Spatially organized layers of cardiomyocytes on biodegradable polyurethane films for myocardial repair. *J Biomed Mater Res* 2003;66A(3):586–595.
57. Fukuda J, Khademhosseini A, Yeh J, Eng G, Cheng J, Farokhzad OC, et al. Micropatterned cell co-cultures using layer-by-layer deposition of extracellular matrix components. *Biomaterials* 2006; 27(8):1479–1486.
58. Stevens MM, Mayer M, Anderson DG, Weibel DB, Whitesides GM, and Langer R. Direct patterning of mammalian cells onto porous tissue engineering substrates using agarose stamps. *Biomaterials* 2005; 26(36):7636–7641.
59. Kim E, Xia Y, and Whitesides GM. Polymer microstructures formed by moulding in capillaries. *Nature* 1995;376:581–584.
60. Kim E, Xia Y, Zhao XM, and Whitesides GM. Solvent-assisted microcontact molding: A convenient method for fabricating three-dimensional structures on surfaces of polymers. *Adv Mater* 1997;9:651–654.
61. Zhao XM, Xia Y, and Whitesides GM. Fabrication of three-dimensional micro-structures: Microtransfer molding. *Adv Mater* 1996;8:837–840.
62. Suh KY, Kim YS, and Lee HH. Capillary force lithography. *Adv Mater* 2001;13(18):1386–1389.
63. Suh KY and Lee HH. Capillary force lithography: Large-area patterning, self-organization, and anisotropic dewetting. *Adv Funct Mater* 2002;12(6–7):405–413.
64. Suh KY, Seong J, Khademhosseini A, Laibinis PE, and Langer R. A simple soft lithographic route to fabrication of poly(ethylene glycol) microstructures for protein and cell patterning. *Biomaterials* 2001;25:557–563.
65. Folch A, Jo BH, Hurtado O, Beebe DJ, and Toner M. Microfabricated elastomeric stencils for micropatterning cell cultures. *J Biomed Mater Res* 2000;52:346–353.
66. Zinchenko YS, Schrum LW, Clemens M, and Coger RN. Hepatocyte and kupffer cells co-cultured on micropatterned surfaces to optimize hepatocyte function. *Tissue Eng* 2006;12(4):751–761.
67. Chiu DT, Jeon NL, Huang S, Kane RS, Wargo CJ, Choi IS, et al. Patterned deposition of cells and proteins onto surfaces by using three-dimensional microfluidic systems. *Proc Nat Acad Sci USA* 2000;12:2408–2413.
68. Delamarche E, Bernard A, Schmid H, Michel B, and Biebuyck H. Patterned delivery of immunoglobulins to surfaces using microfluidic networks. *Science* 1997;276(5313):779–781.
69. Flaim CJ, Chien S, and Bhatia SN. An extracellular matrix microarray for probing cellular differentiation. *Nat Methods* 2005;2(2):119–125.
70. Anderson DG, Levenberg S, and Langer R. Nanoliter-scale synthesis of arrayed biomaterials and application to human embryonic stem cells. *Nat Biotechnol* 2004;22:863–866.
71. Malek CK and Saile V. Applications of LIGA technology to precision manufacturing of high-aspect-ratio micro-components and -systems: A review. *Microelectron J* 2004;35(2):131–143.
72. Rivers TJ, Hudson TW, and Schmidt CE. Synthesis of a novel, biodegradable electrically conducting polymer for biomedical applications. *Adv Funct Mater* 2002;12(1):33–37.
73. Zelikin AN, Lynn DM, Farhadi J, Martin I, Shastri V, and Langer R. Erodible conducting polymers for potential biomedical applications. *Angew Chem Int Edit* 2002;41(1):141–144.
74. Gomez N and Schmidt CE. Nerve growth factor-immobilized polypyrrole: Bioactive electrically conducting polymer for enhanced neurite extension. *J Biomed Mater Res* 2007;81A(1):135–149.
75. Lavan DA, George PM, and Langer R. Simple, three-dimensional microfabrication of electrodeposited structures. *Angew Chem Int Edit* 2003;42(11):1262–1265.
76. Koch J, Fadeeva E, Engelbrecht M, Ruffert C, Gatzen HH, Stendorf AO, et al. Maskless nonlinear lithography with femtosecond laser pulses. *Appl Phys A-Mater* 2006;82:23–26.
77. Jensen MF, McCormack JE, Helbo B, Christensen LH, Christensen TR, and Geschke O. Rapid prototyping of polymer microsystems via excimer laser ablation of polymeric moulds. *Lab Chip* 2004;4:391–395.
78. Snakenborg D, Klank H, and Kutter JP. Microstructure fabrication with a CO_2 laser system. *J Micromech Microeng* 2004;14:182–189.
79. Kancharla VV and Chen S. Fabrication of biodegradable polymeric micro-devices using laser micromachining. *Biomed Microdevices* 2002;4(2):105–109.
80. Locascio LE, Ross DJ, Howell PB, Gaitan M. Fabrication of polymer microfluidic systems by hot embossing and laser ablation. *Methods Mol Biol* 2006;339:37–46.

81. Pugmire DL, Waddell EA, Haasch R, Tarlov MJ, and Locascio LE. Surface characterization of laser-ablated polymers used for microfluidics. *Anal Chem* 2002;74(4):871–878.
82. Lu Y, Theppakuttai S, and Chen SC. Marangoni effect in nanosphere-enhanced laser nanopatterning of silicon. *Appl Phys Lett* 2003;82(23):4143–4145.
83. Kosiorek A, Kandulski W, Glaczynska H, and Giersig M. Fabrication of nanoscale rings, dots, and rods by combining shadow nanosphere lithography and annealed polystyrene nanosphere masks. *Small* 2005;1(4):439–444.
84. Piner RD, Zhu J, Xu F, Hong S, and Mirkin CA. "Dip-pen" nanolithography. *Science* 1999;283 (5402):661–663.
85. Hong S, Zhu J, and Mirkin CA. Multiple ink nanolithography: Toward a multiple-pen nano-plotter. *Science* 1999;286(5439):523–525.
86. Noy A, Miller AE, Klare JE, Weeks BL, Woods BW, and DeYoreo JJ. Fabrication of luminescent nanostructures and polymer nanowires using dip-pen nanolithography. *Nano Lett* 2002;2(2):109–112.
87. Lee K-B, Park S-J, Mirkin CA, Smith JC, and Mrksich M. Protein nanoarrays generated by dip-pen nanolithography. *Science* 2002;295(5560):1702–1705.
88. Demers LM, Ginger DS, Park SJ, Li Z, Chung SW, and Mirkin CA. Direct patterning of modified oligonucleotides on metals and insulators by dip-pen nanolithography. *Science* 2002;296(5574): 1836–1838.
89. Wilson DL, Martin R, Hong S, Cronin-Golomb M, Mirkin CA, and Kaplan DL. Surface organization and nanopatterning of collagen by dip-pen nanolithography. *Proc Nat Acad Sci USA* 2001;98(24): 13660–13664.
90. Ginger DS, Zhang H, and Mirkin CA. The evolution of dip-pen nanolithography. *Angew Chem Int Ed* 2004;43,30.
91. Cooke MN, Fisher JP, Dean D, Rimnac C, and Mikos AG. Use of stereolithography to manufacture critical-sized 3D biodegradable scaffolds for bone ingrowth. *J Biomed Mater Res* 2002;64B:65–69.
92. Lu Y, Mapili G, Chen SC, and Roy K. Proceedings of the 30th annual meeting and exposition; 2003; Glasgow, Scotland.
93. Lu Y, Mapili G, Suhali G, Chen S, and Roy K. A digital micro-mirror device-based system for the microfabrication of complex, spatially patterned tissue engineering scaffolds. *J Biomed Mater Res* 2006;77A:396–405.
94. Chen W, Kirihara S, and Miyamoto Y. Fabrication of three-dimensional micro photonic crystals of resin-incorporating TiO_2 particles and their terahertz wave properties. *J Am Cer Soc* 2007;90(1):92–96.
95. Basu S and Campagnola PJ. Properties of crosslinked protein matrices for tissue engineering applications synthesized by multiphoton excitation. *J Biomed Mater Res* 2004;71A(2):359–368.
96. Basu S and Campagnola PJ. Enzymatic activity of alkaline phosphatase inside protein and polymer structures fabricated via multiphoton excitation. *Biomacromolecules* 2004;5(2):572–579.
97. Hollister SJ. Porous scaffold design for tissue engineering. *Nat Mater* 2005;4:518–524.
98. Vozzi G, Flaim C, Ahluwalia A, and Bhatia S. Fabrication of PLGA scaffolds using soft lithography and microsyringe deposition. *Biomaterials* 2003;24:2533–2540.
99. Kawata S, Sun H-B, Tanaka T, and Takada K. Finer features for functional microdevices. *Nature* 2001;412(6848):697–698.
100. Schlie S, Ngezahayo A, Ovsianikov A, Fabian T, Kolb HA, Haferkamp H, et al. Three-dimensional cell growth on structures fabricated from ORMOCER by two-photon polymerization technique. *J Biomater Appl* 2007;22:275–287.
101. Jin HJ, Fridrikh SV, Rutledge GC, and Kaplan DL. Electrospinning *Bombyx mori* silk with poly(ethylene oxide). *Biomacromolecules* 2002;3:1233–1239.
102. Li D, Wang Y, and Xia Y. Electrospinning of polymeric and ceramic nanofibers as uniaxially aligned arrays. *Nano Letters* 2003;3(8):1167–1171.
103. Xu CY, Inai R, Kotaki M, and Ramakrishna S. Aligned biodegradable nanofibrous structure: A potential scaffold for blood vessel engineering. *Biomaterials* 2004;25:877–886.
104. Zhong XH, Ran SF, Fang DF, Hsiao BS, and Chu B. Control of structure, morpohology, and property in electrospun poly(glycolide-co-lactide) non-woven membranes via post-draw treatments. *Polymer* 2003;44:4959.
105. Ma Z, Kotaki M, Inai R, and Ramakrishna S. Potential of nanofiber matrix as tissue engineering scaffolds. *Tissue Eng* 2005;11:101–109.

106. Wood MA. Colloidal lithography and current fabrication techniques producing in-plane nanotopography for biological applications *J R Soc Interface* 2007;4:1–17.
107. Yeong W-Y, Chua C-K, Leong K-F, Chandrasekaran M, and Lee M-W. Indirect fabrication of collagen scaffold based on inkjet printing technique. *Rapid Prototyping J* 2006;12(4):229–237.
108. Roy TD, Simon JL, Ricci JL, Rekow ED, Thompson VP, and Parsons JR. Performance of degradable composite bone repair products made via three-dimensional fabrication techniques. *J Biomed Mater Res A* 2003;66(2):283–291.
109. Simon JL, Roy TD, Parsons JR, Rekow ED, Thompson VP, Kemnitzer VP, et al. Engineered cellular response to scaffold architecture in a rabbit trephine defect. *J Biomed Mater Res A* 2003;66(2):275–282.
110. Cukierman E, Pankov R, and Yamada KM. Cell interactions with three-dimensional matrices. *Curr Opin Cell Biol* 2002;14:633–639.
111. Petersen OW, Ronnov-Jessen L, Howlett AR, and Bissell MJ. Interaction with basement membrane serves to rapidly distinguish growth and differentiation pattern of normal and malignant human breast epithelial cells. *Proc Nat Acad Sci USA* 1992;89:9064–9068.
112. Liu H and Roy K. Biomimetic three-dimensional cultures significantly increase hematopoietic differentiation efficacy of embryonic stem cells. *Tissue Eng* 2005;11(1–2):919–330.
113. Tang P, Qui F, Zhang H, and Yang Y. Morphology and phase diagram of complex block copolymers: ABC linear triblock copolymers. *Phys Rev E* 2004;69:031803, 1–8.
114. Dalby MJ, Childs S, Riehle MO, Johnstone H, Affrossman S, and Curtis ASG. Fibroblast reaction to island topography: Changes in cytoskeleton and morphology with time. *Biomaterials* 2003;24:927–935.
115. Kim SO, Solak HH, Stoykovich MP, Ferrier NJ, Pablo JJD, and Nealey PF. Epitaxial self-assembly of block copolymers on lithographically defined nanopatterned substrates. *Nature* 2003;424(6947):411–414.
116. Bognitzki M, Frese T, Steinhart M, Greiner A, and Wendorff JH. Preparation of fibers with nanoscaled morphologies: Electrospinning of polymer blends. *Polym Eng Sci* 2001;41(6):982–989.
117. Whang K, Thomas H, and Healy KE. A novel method to fabricate bioabsorbable scaffolds. *Polymer* 1995;36:837–841.
118. Yang F, Murugan R, Ramakrishna S, Wang X, Ma YX, and Wang S. Fabrication of nano-structured porous PLLA scaffold intended for nerve tissue engineering. *Biomaterials* 2004;25(10):1891–1900.
119. Lo H, Ponticiello MS, and Leong KW. Fabrication of conrolled release biodegradable foams by phase separation. *Tissue Eng* 1995;1:15–27.
120. Stachowiak AN, Bershteyn A, Tzatzalos E, and Irvine DJ. Bioactive hydrogels with and ordered cellular structure combine interconnected macroporosity and robust mechanical properties. *Adv Mater* 2005;14(4):399–403.
121. Zhang S. Emerging biological materials through molecular self-assembly. *Biotechnol Adv* 2002;20:321–339.
122. Park SH, Pistol C, Ahn SJ, Reif JH, Lebeck AR, Dwyer C, et al. Finite-size, fully addressable DNA tile lattices formed by hierarchical assembly procedures. *Angew Chem Int Ed* 2006:735–739.
123. Rothemund PWK. Folding DNA to create nanoscale shapes and patterns. *Nature* 2006;440:297–302.
124. Zhang S. Fabrication of novel biomaterials through molecular self-assembly. *Nat Biotechnol* 2003;21: 1171–1178.
125. Petka WA, Harden JL, McGrath KP, Wirtz D, and Tirrell DA. Reversible hydrogels from self-assembling artificial proteins. *Science* 1998;281(5375):389–392.
126. Reches M and Gazit E. Casting metal nanowires within discrete self-assembled peptide nanotubes. *Science* 2003;300:625–627.
127. Scheibel T, Parthasarathy R, Sawicki G, Lin XM, Jaeger H, and Lindquist SL. Conducting nanowires built by controlled self-assembly of amyloid fibers and selective metal deposition. *Proc Nat Acad Sci USA* 2003;100(8):4527–4532.
128. Tan JL, Liu W, Nelson CM, Raghavan S, and Chen CS. Simple approach to micropattern cells on common culture substrates by tuning substrate wettability. *Tissue Eng* 2004;10(5/6).
129. Tan W and Desai TA. Layer-by-layer microfluidics for biomimetic three-dimensional structures. *Biomaterials* 2004;24:1355–1364.
130. Tan W and Desai TA. Microscale multilayer cocultures for biomimetic blood vessels. *J Biomed Mater Res* 2005;72A:146–160.
131. Khademhosseini A, Yeh J, Eng G, Karp J, Kaji H, Borenstein J, et al. Cell docking inside microwells within reversibly sealed microfluidic channels for fabricating multiphenotype cell arrays. *Lab Chip* 2005;5:1380–1386.

132. Liu VA and Bhatia SN. Three-dimensional patterning of hydrogels containing living cells. *Biomed Microdevices* 2002;4(4):257–266.
133. Nijst CLE, Bruggeman JP, Karp JM, Ferreira L, Zumbuehl A, Bettinger CJ, et al. Synthesis and characterization of photocurable elastomers from poly(glycerol-co-sebacate). *Biomacromolecules* 2007;8(10):3067–3073.
134. Dhariwala B, Hunt E, and Boland T. Rapid prototyping of tissue-engineering constructs using photo-polymerizable hydrogels and stereolithography. *Tissue Eng* 2004;10(9/10):1316–1322.
135. Nahmias Y, Arneja A, Tower TT, Renn MJ, and Odde DJ. Cell patterning on biological gels via cell spraying through a mask. *Tissue Eng* 2005;11(5–6):701–708.
136. Moghe PV, Berthiaume F, Ezzell RM, Toner M, Tompkins RG, and Yarmush ML. Culture matrix configuration and composition in the maintenance of hepatocyte polarity and function. *Biomaterials* 1996;17(3):373–385.
137. Odde DJ and Renn MJ. Laser-guided direct writing of living cells. *Biotechnol Bioeng* 2000;67:312–318.
138. Nahmias Y, Schwartz RE, Verfaillie CM, and Odde DJ. Laser-guided direct writing for three-dimensional tissue engineering. *Biotechnol Bioeng* 2005;92(2):129–136.
139. Nahmias Y and Odde DJ. Micropatterning of living cells by laser-guided direct writing: Application to fabrication of hepatic-endothelial sinusoid-like structures. *Nat Protocols* 2006;1(5):2288–2296.
140. Chou SY, Krauss PR, and Renstrom PJ. Imprint of sub-25 nm vias and trenches in polymers. *Appl Phys Lett* 1995;67(21):3114.
141. Abrams GA, Goodman SL, Nealey PF, Franco M, and Murphy CJ. Nanoscale topography of the basementmembrane underlying the corneal epithelium of the rhesus macaque. *Cell Tissue Res* 2000;299:39–46.
142. Abrams GA, Schaus SS, Goodman SL, Nealey PF, and Murphy CJ. Nanoscale topography of the corneal epithelial basement membrane and Descemet's membrane of the human. *Cornea* 2000;19:57–64.
143. Xia Y, Kim E, Zhao X-M, Rogers JA, Prentiss M, and Whitesides GM. Complex optical surfaces formed by replica molding against elastomeric masters. *Science* 1996;273(5273):347–349.
144. Kumar A and Whitesides GM. Features of gold having micrometer to centimeter dimensions can be formed through a combination of stamping with an elastomeric stamp and an alkanethiol ink followed by chemical etching. *Appl Phys Lett* 1993;63(14):2002–2004.
145. Ginger DS, Zhang H, and Mirkin CA. The evolution of dip-pen nanolithography. *Angew Chem Int Edit* 2004;43(1):30–45.

Section II

Effect of Nanostructures on Cellular Responses and Tissue Regeneration

5 Cell Response to Nanoscale Features and Its Implications in Tissue Regeneration: An Orthopedic Perspective

Batur Ercan and Thomas Webster

CONTENTS

5.1 BONE PHYSIOLOGY

5.1.1 PROPERTIES AND STRUCTURE OF BONE

Bone is a living and growing tissue that forms the endoskeleton of vertebrates. The main functions of bone are protection of internal organs, support of physiological stresses and strains, storage of

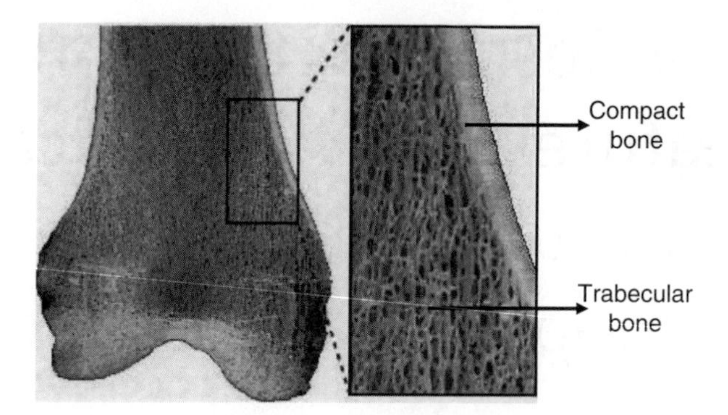

FIGURE 5.1 Compact and trabecular bone at the distal end of the femur. (From University of Glasgow, http://www.gla.ac.uk/ibls/fab/public/docs/xbone1x.html. With permission.)

calcium and phosphorus, hematopoiesis (or production of blood cells in the bone marrow), acid–base balance by absorbing or releasing alkaline salts to buffer the blood for excessive pH changes, and detoxification of the body by the removal of heavy metals from the blood [1].

Bone structure can be divided into two basic categories depending on its morphology: cortical (compact) and cancellous (trabecular). The characteristic placement of both of these bones can be seen in Figures 5.1 and 5.2 [2,3].

Accounting for 80% of the total mass of bone and containing 30% porosity, the smooth, rigid, and white outer layer of bone is cortical bone. It consists of closely packed osteons, matrix of the bone, lying parallel to the long axis of bone. At the center of these osteons, there is a central canal called a "Haversian canal" that contains blood vessels and nerves and is surrounded by concentric ring shaped matrices of cortical bone or lamellae. These lamellae contain bone cells, which are entrapped within spaces called lacunae. The connection between these lacunae and the Haversian canals is achieved by small canals called canaliculi [4].

In turn, trabecular bone is the inner part of bone. Accounting for 20% of the bone mass, its microstructure consists of large spaces between individual trabeculae (in the shape of plates), which leads to a high porosity content approaching 50%–90% [5] reducing the weight of the overall bone

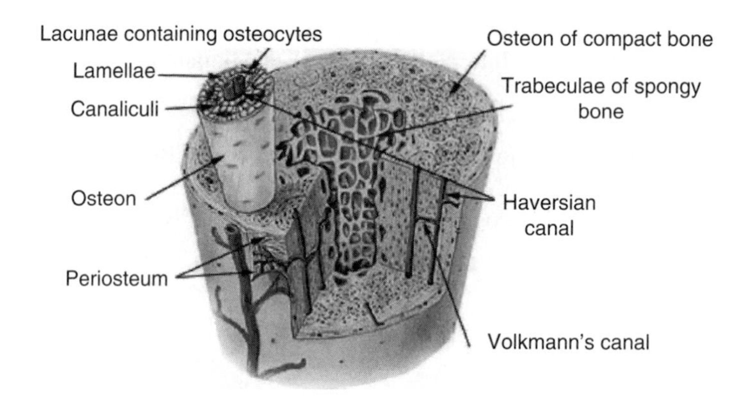

FIGURE 5.2 Compact and trabecular bones highlighting individual lamellae, osteons, and Haversian canals. Trabecular bone is the inner section of bone. Periosteum is a connective tissue membrane covering the outer surface of bones. Volkmann's canal connects osteons to each other and to the periosteum. (From Young, J.L., Fritz, A., Adamo, M.P. et al., U.S. National Cancer Institute's Surveillance, Epidemiology and End Results (SEER) Program, Structure of bone tissue, http://training.seer.cancer.gov/module_anatomy/unit3_2_bone_tissue.html. With permission.)

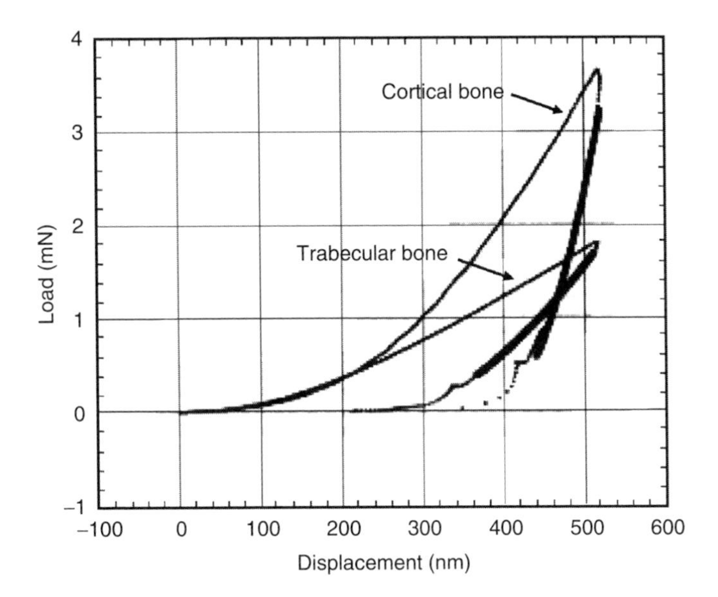

FIGURE 5.3 Force–displacement curves obtained by indentation tests (bold portion shows the elastic modulus of bone). Cortical bone has a higher elastic modulus than the trabecular bone. (From Zysset, P.K., Guo, X.E., Hoffler, C.E. et al., *J. Biomech.*, 32, 1005, 1999. With permission.)

structure. Although it seems like trabeculae are arranged randomly, they are organized to provide maximum strength. In addition, the trabecular bone is surrounded by bone marrow and it does not contain blood, muscles, or nerves [5]. Comparing both types of bones, compact bone has higher stiffness than cortical bone.

Figure 5.3 shows a load versus displacement graph of a human femur under nano-indentation, characterizing the differences in elastic moduli for both of these bone types [6]. It has been observed that a 20–39 year old human femur has a tensile strength of 124 MPa with a tensile elastic modulus of 17.6 GPa and a compressive strength of 170 MPa [7]. It is important to mention that the mechanical properties of bone drastically change with respect to the age, species, type (compact or spongy), porosity, anatomical differences, and health conditions of the individual. For example, the elastic modulus of human cadaver femur can decrease from 15.6 to 11.5 GPa for patients diagnosed with bone diseases [8]. In addition, bone is a very anisotropic structure, having different elastic moduli depending on the direction of loading. For instance, cortical bone tissue has a longitudinal elastic modulus of 17.4 GPa, transverse modulus of 9.6 GPa, bending modulus of 14.8 GPa, and a shear modulus of 3.5 GPa. Other than these properties, it is strain rate sensitive [9] and viscoelastic [10]. As a person reaches 40 years or more, bone mass of trabecular bone decreases by almost 50% and cortical bone mass decreases by approximately 25% due to the increased resorption of bone as one ages, leading to a decrease in strength and an increase in stiffness.

5.1.2 Micro- and Nano-Architecture of Bone

Bones can be classified in two categories depending on their micro-architecture: woven and lamellar bone. Woven bone contains coarse collagen fibers oriented in an irregular fashion. Due to this haphazard organization of collagen fibers, the mechanical properties of woven bone are more isotropic than those of lamellar bone. Primarily found in all fetal bones, woven bone is gradually replaced by lamellar bone as a person grows. Moreover, it is also found at rapidly growing areas, newly formed regions (such as healing of a fracture), as well as metaphyseal regions (or the growing region of bone).

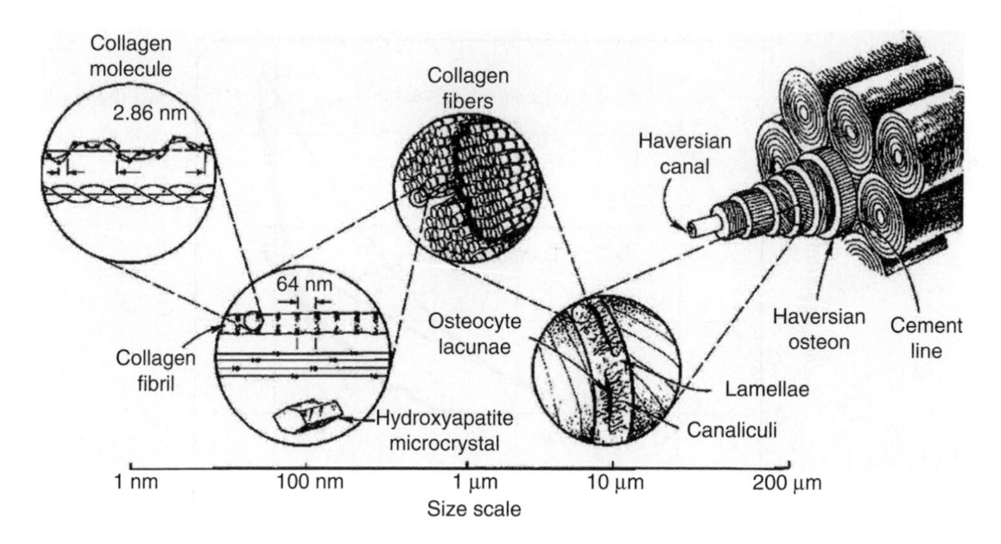

FIGURE 5.4 Hierarchical structure of human compact bone. The cement line is the outer boundary of an osteon. The pitch of the helical collagen molecule is 2.86 nm and the characteristic observed band on collagen fibrils is 64 nm. (From Lakers, R.S., *Nature*, 361, 511, 1993. With permission.)

However, the collagen fibers of lamellar bone are aligned in a parallel fashion to form lamellae. This parallel orientation of collagen fibers gives lamellar bone its anisotropic mechanical properties, having the highest strength parallel to the longitudinal axis of collagen fibers. Lamellar bone replaces maturing woven bone and it contains up to 100 times more mineralized matrix [7].

Regardless of whether it is woven or lamellar, bone is a nanostructured composite material comprising two major phases: an organic matrix phase (called an osteoid), mainly composed of proteins, and an inorganic phase that consists of salts deposited within the matrix. The matrix of the human femur is approximately 70% inorganic salts, 20% organic phase, and 10% water [7].

The main constituent of the organic phase is type I collagen (90% v/v). Type I collagen is synthesized by osteoblasts (or bone forming cells) generating a linear molecule 300 nm in length and 0.5 nm in diameter. These collagen molecules line up to create collagen fibrils, which come together to form collagen fibers further comprising the lamellae of the bone. This structural organization is seen in Figure 5.4 [11]. The remaining 10% of the organic phase of bone consists of noncollagenous proteins such as bone inductive proteins (i.e., osteonectin, osteopontin, osteocalcin), growth factors (e.g., insulin like growth factor), bone sialoprotein, bone proteoglycans, and proteolipids [7]. These proteins mediate functions of bone cell such as promoting, inhibiting, or regulating bone synthesis and resorption.

The inorganic phase is mainly composed of crystalline calcium hydroxyapatite (HA). Its chemical formula is $Ca_5(PO_4)_3(OH)$ (conventionally written as $Ca_{10}(PO_4)_6(OH)_2$ to emphasize the two molecules of HA within the unit cell) and with a hexagonal unit cell structure. The thickness and length of these crystals can range from 2 to 5 nm and 20 to 80 nm, respectively. The augmentation of these HA crystals is due to the maturity of the bone and the smaller crystals indicate that the bone matrix is new [5]. The stiffness of bone is mainly due to the presence of these HA crystals.

5.1.3 Bone Cells and Remodeling

There are three types of bone cells: osteoblasts, osteocytes, and osteoclasts. Due to the hierarchical structure of bone, all bone cells are naturally accustomed to interacting with both micro- and nanostructured entities. Their functions are described below.

5.1.3.1 Osteoblasts and Osteocytes

Osteoblasts and osteocytes (or differentiated osteoblasts) are bone forming cells and they differentiate from pluripotent mesenchymal progenitor cells [12]. Polarized with a Golgi apparatus, typical in highly secretory cells [12], osteoblasts secrete a bone matrix (osteoid) which further mineralizes to form bone. They also secrete organic proteins (i.e., type I collagen, osteopontin, osteocalcin, bone sialoprotein, etc.) and alkaline phosphatase (a dephosphorylation enzyme active in alkaline environments which is an indicator of bone turnover and bone cell function). These matrix proteins are then mineralized through the formation of HA crystals, the mechanism of which is not well understood to date.

When osteoblasts are entrapped within the bone matrix they have synthesized, they are called "osteocytes." Osteocytes are found inside lacunae, which are situated inside individual lamellae of the bone. They play an important role in the formation of the bone by secreting growth factors such as insulin like growth factor I and tissue growth factor β, which further control osteoblast functions by promoting their differentiation from immature osteoblasts (which do not secrete calcium) into mature osteoblasts (which secrete calcium) [13]. In vitro studies have shown that osteoblasts endure three different phases of growth after adhesion to an implant surface: proliferation and extracellular matrix (ECM) synthesis, ECM development and maturation, and ECM mineralization. These stages are outlined in Figure 5.5 [7].

5.1.3.2 Osteoclasts

Originating from the hematopoietic stem cell linage, osteoclasts are bone resorbing cells that are large and multinucleated (typically 6–8 nuclei) [12]. During bone resorption, they form ruffled cell membrane edges that increase the area of attachment. They secrete tartrate-resistant acid phosphatase and form Howship's lacunae (resorption peaks). They use carbonic anhydrase activity (CO_2 hydration) to lower the pH of the environment which increases the solubility of HA.

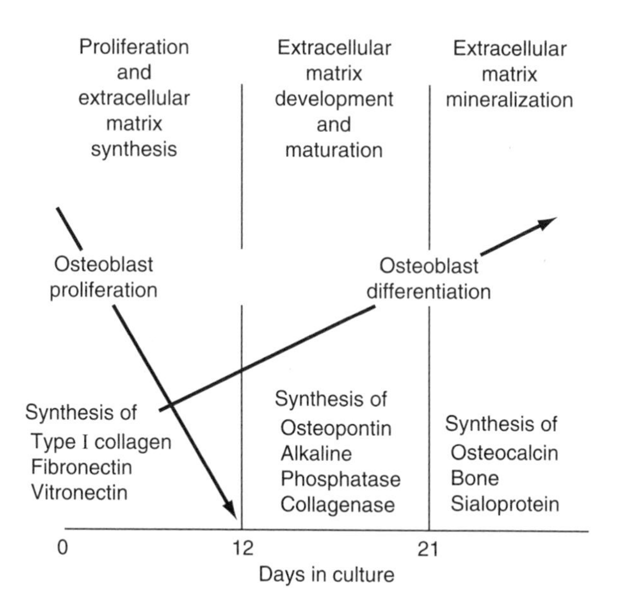

FIGURE 5.5 The differentiation phases that osteoblasts complete after adhesion to a new surface (i.e., an implanted biomaterial). These stages are proliferation and ECM synthesis, ECM development and maturation, and ECM mineralization. (From Webster, T.J., *Adv. Chem. Eng.*, 27, 125, 2001. With permission.)

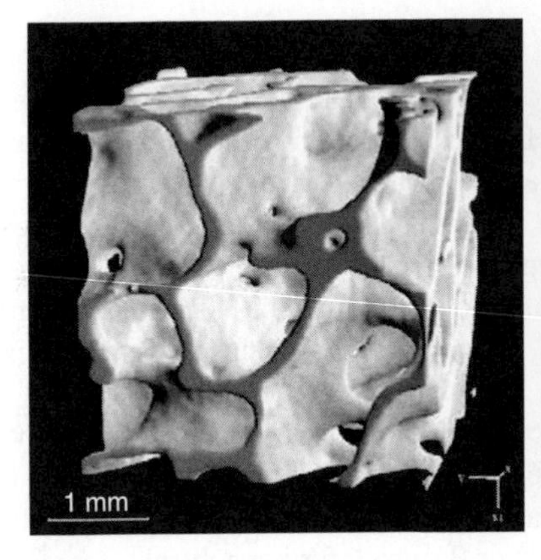

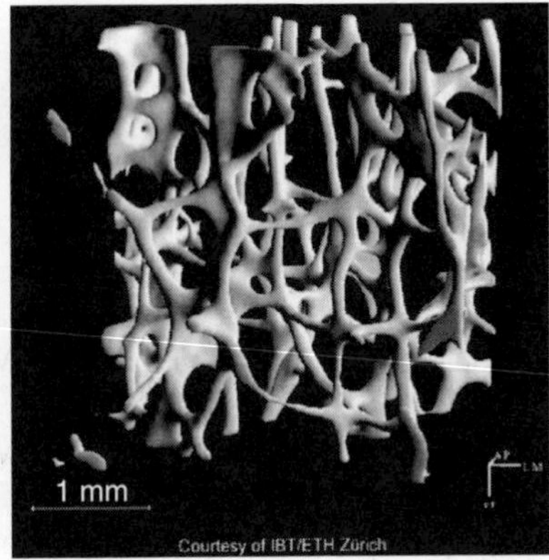

FIGURE 5.6 Micrographs of normal healthy bone (left) and osteoporotic bone (right). (From Rüegsegger P., *Calcified Tissue International*. 58(1), 24, 1996. With permission.)

Bones in the human body continuously undergo remodeling through the formation of the bone modeling unit (BMU). Both osteoblasts and osteoclasts act simultaneously toward remodeling bone which is, essentially, the replacement of the old bone tissue by new bone tissue. The activation of osteoclasts occurs in a couple of different ways, such as with the sensing of microcracks by osteocytes. Inactivated preosteoblasts activate preosteoclasts by a RANK ligand (i.e., a protein that activates the RANK receptor in preosteoclasts). This process causes preosteoclasts to unite and form multinuclear osteoclasts, which resorb bone for nearly 2 weeks. At the end of bone resorption, they undergo apoptosis (or cell suicide). The next stage in bone remodeling is the secretion of an osteoid by osteoblasts. It takes nearly 11 days for mineralization to start and ultimately 2 to 3 months for the formation of new mature bone [14]. The pace of this remodeling process undergoes some changes throughout the life span of a person, slowing down as the person ages. After the age of 40, bone formation cannot keep up with bone resorption, leading to a decrease in bone density [5]. One of the major bone diseases caused by the increase in bone resorption is osteoporosis. In Figure 5.6, the microstructure of healthy and osteoporotic bone is compared [15]. Patients with osteoporosis have weaker more fragile bones which are susceptible to fracturing. After a certain age, the only viable medical treatment to patients with a bone fracture is insertion of an orthopedic implant.

5.1.4 Cell Adhesion on Implant Surfaces

It has been argued that the fate of an implant is determined within the first seconds after implantation. When an implant is inserted, protein adsorption occurs within the first few seconds. This protein adsorption is crucial because it further mediates cell adhesion onto the implant surface. Mainly, the type of proteins adsorbed on implant surfaces and their conformation once adsorbed are the foremost parameters which control cell adhesion [16–18]. Initial protein adsorption is primarily controlled by properties of the implant surface such as surface topography, wettability, and chemistry [17,19–21]. Many studies have been conducted to investigate the effects of initial protein adsorption on cell adhesion to biomaterial surfaces. Two major proteins known to influence the attachment of bone cells to an implant surface are vitronectin and fibronectin.

Fibronectin is a somewhat large molecular weight (273,715 Da) [7] glycoprotein found in blood plasma and the ECM of bone [22]. Fibronectin promotes cell adhesion, migration, differentiation, and influences cell morphology. Figure 5.7 shows the structure of a fibronectin molecule. Vitronectin is also another blood plasma glycoprotein which is found in the ECM of many tissues [22]. Human

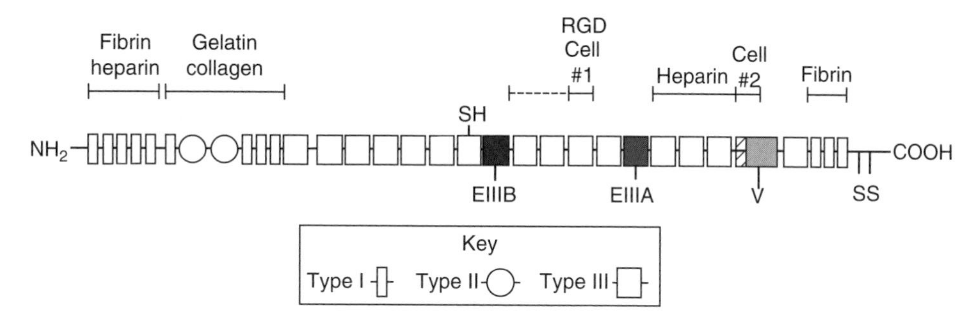

FIGURE 5.7 The structure of fibronectin. It is composed of three types of repeating subunits, type I, type II, and type III, being around 45, 60, and 90 amino acids, respectively. At three positions (EIIIB, EIIIA, and V) alternative splicing causes variations in structure. Splicing of V region is seen in humans where region V can vary from 0 to 120 amino acids. The two cell-binding sites are denoted in the figure as cell #1 and cell #2, where cell #1 includes the well-known RGD peptide sequence. Fibrin, gelatin, collagen, and heparin/heparin sulfate-binding sites are also indicated. (From Kreis, T. and Vale, R., *Guidebook to the Extracellular Matrix, Anchor, and Adhesion Proteins*, Oxford: Oxford University Press, 1999. With permission.)

vitronectin is found in two forms: one chain and two chain. The one-chain vitronectin has a 75 kDa band and the two-chain vitronectin dissociates into 65 and 10 kDa bands upon polyacrylamide gel electrophoresis [22]. Vitronectin plays an important role in many physiological processes (such as mediating the immune response, homeostasis, and blood coagulation). It is also very important in mediating cell adhesion and migration [7]. The structure of vitronectin is shown in Figure 5.8. Importantly, fibronectin and vitronectin (as well as all proteins) are natural nanostructured materials with which bone cells constantly interact.

These proteins are particularly important for bone cells. For example, surfaces with high hydrophilicity and roughness have been observed to increase vitronectin and fibronectin adhesion [18,23]. On titanium surfaces, in the absence of vitronectin, fibronectin, and fetal calf serum, no cell spreading was observed [18]. It was found that the attachment and spreading of bone-derived cells were reduced by 73%–83% only if vitronectin was removed from the cell culture medium [24].

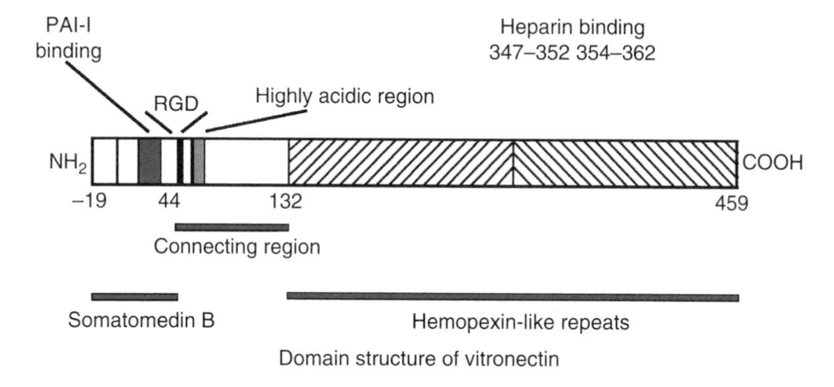

FIGURE 5.8 The structure of vitronectin. A somatomedin B domain (from 1 to 44 amino acid location at the N-terminus) contains an activator inhibitor-1 (PAI-1) binding site. This is followed by a highly acidic region, which contributes to the tertiary structure of the protein. A connecting region and a hemopexin-like region (i.e., a protein which binds to an iron containing heme group) are also present. An RGD domain is present immediately after the somatomedin B domain, between 45 and 47 amino acids. The two heparin-binding sites are between 347–352 and 354–362 amino acids. (From Kreis, T. and Vale, R., *Guidebook to the Extracellular Matrix, Anchor, and Adhesion Proteins*, Oxford: Oxford University Press, 1999. With permission.)

When only fibronectin was removed from the cell culture medium, the reduction in cell spreading and attachment on these surfaces was significantly less than when conducted with the vitronectin-depleted serum cell culture medium. When neither of those proteins were present in the medium, the bone-derived cells failed to attach [24]. In addition, cell adhesion strength was observed to increase linearly with adsorbed fibronectin surface density on a substrate in nonspecific fashion [25,26]. Both of these proteins have been shown to change conformation upon exposure to an extracellular environment. Temperature, pH, salt concentration, binding to surfaces (implant), and binding of bioactive molecules have been shown to provoke these conformational changes [5,22]. An important example to this phenomenon is the conformation change of vitronectin upon adsorption. Specifically, Garcia et al. showed that although the same amount of fibronectin was adsorbed onto different substrates, corresponding osteoblast-like cell adhesion can differ due to conformational changes in the adsorbed fibronectin [26]. For nanotechnology, similar importance of protein conformation on cell adhesion has been observed. Specifically, the investigation of nanophase alumina showed increased unfolding of vitronectin compared to its conventional (or micron grain size) counterpart [17]. Increased unfolding of vitronectin exposed more bioactive adhesive sites which further enhanced osteoblast adhesion on the nanophase alumina surfaces.

There are two main mechanisms of cell attachment onto proteins adsorbed onto an implant surface: through the specific peptide sequences (such as arginine–glycine–aspartic acid or RGD) or through the heparin-binding region as an adherent protein.

Some extracellular proteins (such as vitronectin, fibronectin, collagen, and laminin) contain RGD peptide sequences. Through this RGD peptide sequence, proteins attach to their respective cell membrane receptors called integrins. In Figure 5.9, cell attachment through an integrin receptor is shown [20].

Integrins are transmembrane glycoproteins which consist of two subunits, an α subunit and a β subunit. The α subunit is 1008–1152 amino acids long, with a cytoplasmic region of 22–32 and a

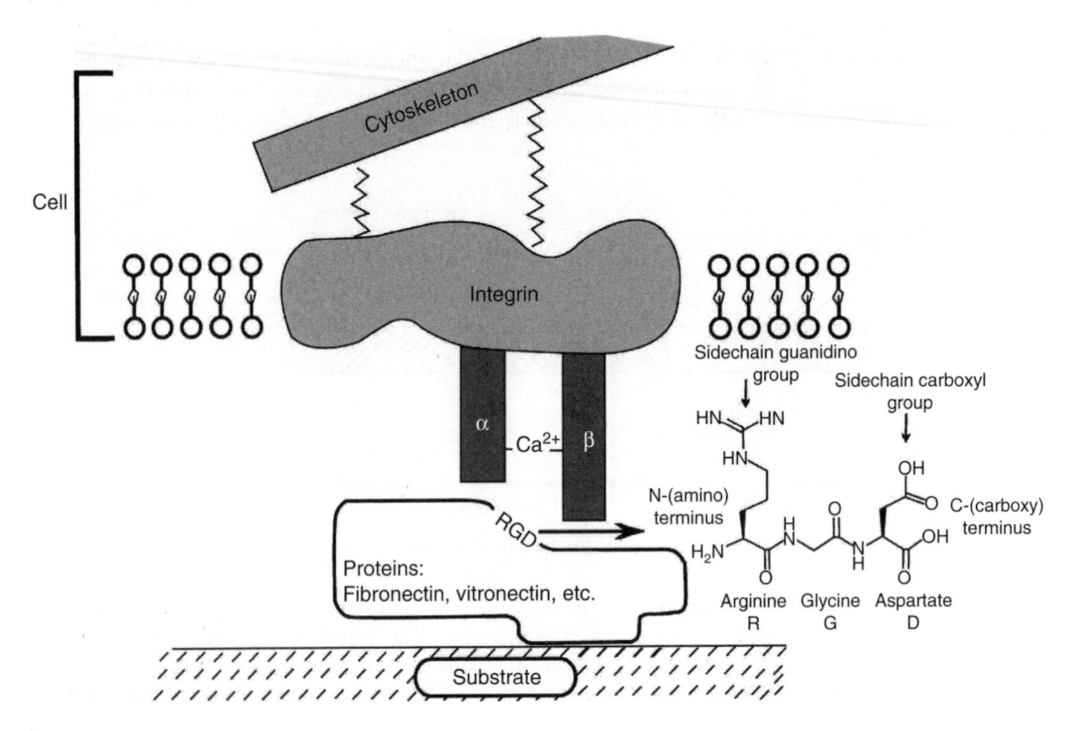

FIGURE 5.9 Cell attachment through an integrin receptor onto a substrate. (From Balasundaram, G. and Webster, T.J., *J. Mater. Chem.*, 16, 3737, 2006. With permission.)

transmembrane region of 20–29 amino acids [19]. The β subunit is 770 amino acids long with an intramembranous part of 26–29 amino acids and a cytoplasmic part of 20–50 amino acids [19]. The extracellular parts of these integrin subunits form a heterodimer. Osteoblasts have a wide range of integrin expression. These are α1, α2, α3, α4, α5, α6, αv, β1, β3, and β5. Different combinations of these subunits are possible and each combination has the potential to bind to one or more ligands, with some overlap in specificity where many integrins can bind to more than one protein and many proteins can be a ligand for more than one integrin [19]. This is most possibly due to the recognition of the RGD peptide sequence by many integrins. αvβ3 and αvβ5 integrins play a major role in attaching osteoblasts to vitronectin while the α5β1 integrin is used for fibronectin and the α2β1 integrin is used for collagen [3]. Internal forces between integrins and their corresponding ligands play an important role in mediating cellular functions. Studies show that the osteoblast adhesive force between the α5β1 integrin and a GRGDSP peptide sequence is 32 ± 2 pN [27]. Changes in these forces may further lead to changes in the patterning and the structure of the tissues. Moreover, the surface properties of the implant play an important role in integrin expression; chemical composition, surface topography, and wettability of the implant are major factors. For example, on a polystyrene substrate, the subunits expressed by integrin receptors of osteoblasts were α2, α4, α6, αv, β1, and β3 [28]. However, on rough CoCrMo alloy, α6 and β3 were not expressed on these cells [28]. For the case of a polished titanium alloy (specifically, titanium–aluminum–vanadium), the osteoblasts additionally expressed the α3 integrin [28].

Osteoblasts also adhered differently depending on surface topography. In an effort to mimic the natural structure of bone, nanotechnology has been used. Specifically, nanomaterials contain numerous biologically inspired nanometer surface features. Osteoblasts have been shown to adhere much more effectively on nanophase than currently used materials [29]. Clearly, a change in chemical composition, surface topography, and wettability all influence bone formation. When these parameters change, they change initial protein adsorption onto biomaterial surfaces, which further affects integrin expression, cell adhesion, and cell proliferation on these surfaces.

However, due to the fact that RGD peptide sequences promote osteoblast attachment, precoating substrates with this RGD peptide sequence has been widely investigated. Immobilizing surfaces with RGDS micropatterns was found to significantly enhance osteoblast and fibroblast adhesion [30]. Although this method shows promise in vitro (including increased cell spreading, proliferation, and differentiation), some problems have prevented the full usage of this technique in vivo. For example, the efficacy of RGD-binding integrins increases with the presence of other peptides in proteins adsorbed on the substrate (such as Pro-His-Ser-Arg-Asn (PHSRN), which binds to the α5β1 integrin of osteoblasts [19,31]). Additionally, the nature of the surgical technique seems to play an important role as well. Orthopedic implant surgeries are mechanically destructive to the weak adhesive forces of peptides immobilized onto implant surfaces. During surgery, the peptide coating can be severely damaged. Another reason for poor results in vivo can be that when a pre-RGD peptide coated implant is inserted in the body, other large proteins still adsorbed onto the implant surfaces, thus blocking the interaction of the RGD peptide sequence with the integrin receptor of the osteoblast. Due to the reasons stated above, researchers have been more interested in modifying the raw implant surface itself rather than using a peptide treatment process. As will be discussed, nanotechnology has introduced many different techniques to change surface properties to promote bone formation.

The second mechanism of osteoblast adhesion onto biomaterial surfaces is through cell membrane heparin sulfate proteoglycans. Attachment to the heparin-binding domains of proteins is electrostatic by nature. The heparin-binding domain contains some positively charged basic amino acids which interact with the negatively charged carboxylate and sulfate groups of the glycosaminoglycans found in the cell membrane [32]. It has been observed that blocking the heparin-binding sites of fibronectin with antibodies (such as platelet factor IV) inhibited 45% of osteoblast adhesion to fibronectin [33], and micropatterning the substrate surfaces with KRSR (a heparin sulfate-binding peptide) significantly increased osteoblast adhesion [30]. Interestingly,

KRSR immobilized on implant surfaces was found to decrease fibroblast (a fibrous tissue forming cell) adhesion. Therefore, the presence of heparin-binding sites in adsorbed proteins is also necessary to maximize the adhesion of osteoblasts onto biomaterials.

Lastly, integrins and heparin sulfate proteoglycans mediate the communication between ECM proteins and the cell, passing information from the ECM to the cell and vice versa. This process of intracellular signaling is very complex and still not clear. The first step is believed to be the autophosphorylation of focal adhesion kinase (FAK), followed by tyrosine phosphorylation and accumulation of other proteins [19]. These proteins further affect signal transduction and activate some transcription factors (such as activator protein-1 (AP-1)) through complex pathways which control cell proliferation and differentiation [5].

5.2 PROBLEMS WITH CURRENT IMPLANTS

Orthopedics is a multibillion dollar industry. In the United States alone, the total hospitalization costs for total hip replacements was $6.77 billion in the year 2003 [34]. In the year 2004, the number of total hip replacements was 234,000, which was twice the number of total hip replacement surgeries in 1991 [35]. Unfortunately, the number of revision hip replacements also doubled, from 23,000 to 46,000, between the years 1992 and 2003 [36]. The above statistics clearly show that current approaches fail to satisfy the needs for patients and improvements should drastically be made on the longevity of these implants.

In 2004, in the United States, the mean age of patients receiving total hip replacement surgeries was 66 [37], 33% of whom were under 65. When comparing the average life expectancy of a person at the age of 65 (17.9 years [5]) with the average lifetime of an implant (which is 12–15 years [38]), we can observe that the latter is relatively shorter; hence it is easy to understand the reason behind the high number of revision surgeries currently needed. This means that the majority of people having a total hip replacement surgery will require at least one revision surgery and for the case of young patients, even multiple surgeries might be necessary. In order to decrease the number of revision surgeries and to improve the quality of lives of many patients, the need for better and longer lasting implants is considerable.

Since the first known hip replacement which used ball and socket joints composed of ivory and nickel-plated steel screws to assist with fixation [39], implant failure has been a major problem in orthopedics. Improvements to implant design could only be conducted during trial and error methods [20] due to a lack of understanding of the main reasons which cause implant failure [5]. There are many reasons leading to the failure of an implant. Although they have not yet been perfectly understood or mastered to date, they can mainly be classified into two categories: mechanical factors (stress–strain imbalances, implant migration, and wear debris) and biological factors (foreign body reaction). To date, the typical trial and error process for developing better implants has relied on modifying implant chemistry (i.e., CoCrMo versus Ti versus stainless steel, etc.). Nanotechnology takes a different approach. It creates nanosurface features through a variety of means on any implant chemistry. It provides one material property (biologically inspired surface features) that can be used for any implant chemistry to promote bone growth. In this rest of this section, fundamentals concerning the hip implant/bone interface and how this bonding can fail through the use of traditional implants will be a primary focus since enhancing this bonding is a main advantage of nanotechnology [5].

5.2.1 Mechanical Factors

5.2.1.1 Stress–Strain Imbalance

The stress acting upon the tissue–implant interface depends on many factors (such as the design and the geometry of the implant, the presence of bone cement, which is used to bond the implant to the surrounding bone, and mechanical properties of the implant itself) [40]. When the distribution of stress and strain is uneven, the bone in the immediate vicinity of the implant undergoes nonphysiological levels of force which will further lead to aseptic loosening [41], bone resorption due to stress shielding, damage to the surrounding tissue, pain in the patient, and, eventually, implant failure. Experimental results using strain gauges and finite element analysis indicated the presence of a stress concentration is at the proximal (top) and distal (bottom) ends of a hip implant. By the use of finite element methods, it is possible to create the best design, which allows for the optimal load transfer, and reduce stress–strain concentrations at the implant–tissue interface by up to 70% [42]. Of course, an increase in bone growth at these sites will provide more support.

5.2.1.2 Implant Migration

Migration is the micromotion (0.1–1 mm) of an implant through time [43]. When this micromotion is excessive, the bone ingrowth into the implant surface will not occur [44]. In addition, implant migration and micromotion at the bone–prosthetic interface leads to debonding of the implant with the juxtaposed tissue which results in failure of the implant. It has been shown that acetabular cup migration is much higher for those cups which have failed than those which were successful. Due to these reasons, minimization of micromotion is one of the key issues concerning the orthopedic field.

The use of bone cement is the primary method to reduce implant migration and it has been shown to be effective by an order of magnitude in preventing micromotion [45]. Bone cement is basically a filler material, sitting in between the bone and the implant. The most commonly used bone cement material is poly(methyl methacrylate) (PMMA). Although the use of PMMA prevents micromotion, there are some inherent disadvantages of PMMA. For example, while removing bone to clear space for bone cement, the following disadvantages have been stated: necrosis of bone cells due to the exothermic high temperature (67°C–124°C) of PMMA polymerization (which further impairs local blood circulation [46]), leakage of unreacted monomers before polymerization, shrinkage during polymerization, large stiffness mismatch between the cement and the surrounding tissue, and the creation of extra interfaces (implant–cement, cement–bone) [47]. When bone cement is not used, the initial fixation of the implant to the bone becomes a significant problem. The bone should be fully bonded to the implant surface before any force is applied on the implant. Due to this reason, the early deposition of bone is vital in orthopedics. The implant materials should enhance bone formation and induce early bone deposition.

5.2.1.3 Wear

Wear is the generation of particles when two opposing surfaces under load move relative to each other, thus creating wear debris. Wear in implants not only causes the thinning of components with the creation of stress–strain imbalances and implant migration, but also produces wear debris which is released to the biological environment [48]. Wear can be due to mechanical factors (such as adhesion, abrasion, fatigue, etc.) or biological factors (such as corrosion of the implant surface by biological activities, etc.).

Wear debris causes osteolysis (or the localized resorption or dissolution of bone tissue) in the surrounding bone. For the wear debris to cause osteolysis, there is a practical wear threshold value which has been proposed: 0.05 mm/year [49]. When this threshold is exceeded, the occurrence of osteolysis is more likely. It is widely accepted that, as the wear rate decreases, so does the incidence of osteolysis. Some researchers contend that it is not the overall concentration of wear volume but it is the concentration of the wear volume within the critical size range (0.2–0.8 micron) that is important for a biological response [50].

The type of material, head radius, clearance, surface finish, and lubrication are all important factors which can increase the wear properties of an implant. Depending on these parameters, the wear properties of implants can vary significantly. For instance, metal on ultrahigh molecular weight polyethylene (UHMWPE) has a wear rate of 100–300 μm/year, ceramic on UHMWPE has a wear rate of 50–150 μm/year, metal on metal has a wear rate of 2–20 μm/year, and ceramic on ceramic has a wear rate of 2–30 μm/year [51]. However, despite their low wear rates, metal on metal implants have been criticized for having higher loosening rates, causing metal sensitivity which causes biological reactions to the alloy constituents producing nanometer scale wear debris [52,53]. Thus, just because a low wear rate is measured, biological factors of the associated wear debris must be taken into account.

5.2.2 Biological Factors

When an implant is inserted, some biological reactions take place near the implanted area. This is the body's response to this newly implanted foreign material. A successful implant promotes the adhesion of osteoblasts on the implant surface and the formation of new bone tissue, incorporating the implant into the body. However, if the body encapsulates the implant by the formation of soft fibrous tissue and tries to separate it as much as possible from surrounding bone, this is the sign of an unsuccessful implantation. The overall aim for orthopedic implants is to minimize the amount of fibrous tissue formation around the implant and to maximize new bone growth.

5.2.2.1 Host Response to Foreign Materials

After the implantation procedure, the body follows a sequence of local events during the healing response. These are acute inflammation, chronic inflammation, granulation tissue formation, foreign body reaction, and fibrosis. The steps and predominant cell types during these steps of wound healing are shown in Figure 5.10 [54].

Inflammation is the reaction of vascularized living tissue to local injury [54]. It activates a series of events which may heal tissue after implantation by recruiting parenchymal cells or cells intended to encapsulate the implant with fibrous tissue [55]. The initial inflammatory response is activated regardless of the type of biomaterial and the location of injury. Acute inflammation is a short-term (few days) response to the injury. Immediately after the surgery, the blood flow to the injury site increases. This implies an increased flow of proteins (some of them can promote anchorage dependent cell attachment), nutrition, immune cells, cells that can help recovery, mesenchymal stem cells, etc. The first cells to appear at the implantation site are white blood cells, mainly neutrophils. Afterward, neutrophils recruit monocytes to the inflammation area (where monocytes will further differentiate into macrophages which are well known to aid in implant material degradation). Neutrophils are then recruited to the site of inflammation by chemical mediators (chemotaxis) to phagocytose microorganisms and foreign materials [55]. They attach to the surfaces

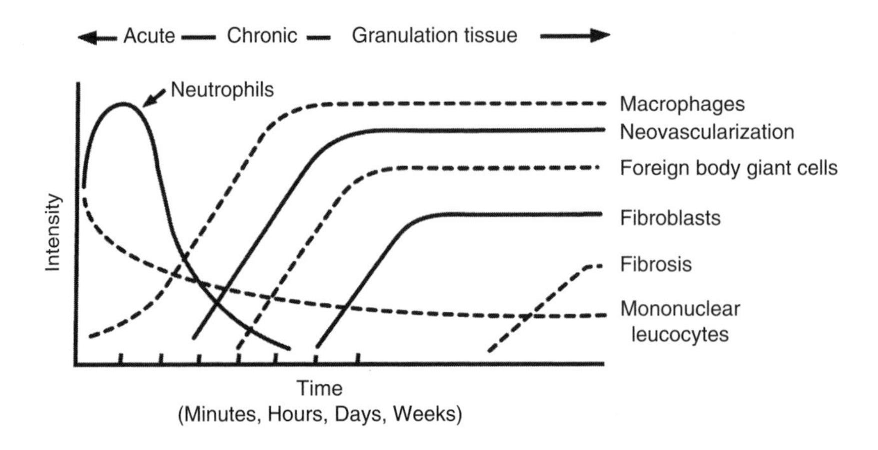

FIGURE 5.10 Time versus intensity graph for wound healing responses showing the predominant cell types at the inflammation site. (From Anderson, J.M., *Cardiovascular pathology*, 2(3), 33, 1993. With permission.)

of the biomaterial by adsorbed proteins, basically immunoglobulin G (IgG) and complement-activated fragment (C3b) through a process called "opsonization." Due to the relative size difference between orthopedic biomaterials and phagocytotic cells, cells cannot internalize the implant. This leads to a condition called "frustrated phagocytosis," which is the activation of phagocytic cells to produce extracellular products that attempt to degrade the biomaterial and, at the same time, recruit more cells to the implant site. Macrophages then secrete degradative agents (such as superoxides and free radicals) which severely damage both the juxtaposed tissue and possibly the implant.

Persisting inflammatory responses lead to chronic inflammation [55]. The main cell types observed during chronic inflammation are monocytes, macrophages, and lymphocytes. Neovascularization (formation of new blood vessels) also starts during this step of wound healing. Macrophages are the most important type of cells in chronic inflammation due to secretion of a great number of biologically active products such as: proteases, arachidonic acid metabolites, reactive oxygen metabolites, coagulation factors, and growth factors (which are important to recruit and promote fibroblasts functions).

The third step in the foreign body response is granulation tissue formation. Fibroblasts form granulation tissue. This tissue is the hallmark of the healing response. It is granular in appearance and contains many small blood vessels [5]. In addition, macrophages fuse together to form foreign body giant cells to attempt to phagocytose the foreign materials much more effectively. The amount of granulation tissue determines the extent of fibrosis [5].

The foreign body reaction, the fourth step in wound healing, contains foreign body giant cells and granulation tissue (which includes fibroblasts, capillaries, macrophages, etc.). Biomaterial surface properties are the key determining factor in this step. It has been shown that smooth surfaces (micron smooth) induce a foreign body reaction which is composed of macrophages one to two cells in thickness [55]. However, as the roughness (micron rough) increases, so does the foreign body reaction.

The last step in wound healing response is fibrosis, which is the fibrous tissue encapsulation of the implant. The successfulness of implantation generally depends on the proliferation capacity of the cells in the tissue. Tissues containing labile (proliferate throughout time) or stable (expanding) cells are less likely to enter the fibrosis step. However, if the cells comprising that tissue are stable (not growing, limited proliferation capacity), the chances of fibrous tissue formation is relatively higher. As stated before, fibrous encapsulation is not desired in bone tissue engineering applications because it leads to failure of the implant due to stress–strain imbalances. Quite simply, fibrous tissue cannot support physical stresses that bone can. The desired situation for an orthopedic implant is the recruitment of parenchymal cells, more specifically osteoblasts, around the implant as soon as possible.

5.3 EMERGING FIELD: NANOTECHNOLOGY

5.3.1 Introduction

Nanotechnology is the development of materials, devices, and systems exhibiting significantly altered and novel properties by gaining control at the atomic, molecular, and supramolecular level [56,57]. It has revolutionized numerous fields, including the orthopedic implant industry.

When the conventional dimensions of a material are decreased into nanoscale, some unusual changes in physiochemical properties result. This can be attributed to the size (i.e., size distribution), chemical composition (i.e., purity, crystallinity, electronic properties), surface structure (i.e., reactivity, surface groups), solubility, shape, and aggregation of nanometer compared to micron materials [58]. Although different experts define "nanomaterial" in different ways, the most commonly accepted definition is that a nanomaterial is one with a basic structural unit in the range of 1–100 nm.

Nanomedicine is a subset of nanotechnology and is the application of nanotechnology for disease diagnosis, treatment, or prevention. Although the discipline is at its infancy, it has great potential to change the methodology of medical sciences in the twenty-first century. For example, cancer cell detection [59], drug and gene delivery [60,61], tissue engineering [62], and bioimaging [63,64] will all drastically benefit from the advances in nanomedicine. Currently, only a few advantages of nanotechnology have been elucidated for bone tissue engineering applications which will be described below.

One of the main driving forces to explore nanomaterials in bone tissue engineering applications is that tissues in the human body are nanostructured, clearly opening the gates for numerous opportunities to improve medicine. For instance, as mentioned, HA molecules, the main constituent of the inorganic phase of the bone, are 2–5 nm thick and collagen type I, the main constituent of the organic phase of the bone, is 0.5 nm in diameter. This implies that bone cells are naturally accustomed to interacting with nanoscale features and surfaces. When implant surfaces are engineered to mimic the constituent components of bone, a better integration of the implant to the body can be expected. Additionally, nanoscale materials have more surface area, surface defects, number of atoms at the surface, and altered electron distributions, which changes their surface properties (i.e., reactivity) with respect to conventional materials, further effecting the interactions between the surface and proteins. Moreover, a decrease in grain or particle size into the nanometer regime would also alter mechanical, piezoelectric, and electrical properties. Briefly, advantages for the use of nanoscale materials in orthopedic applications can be classified into two broad categories: mechanical and biological advantages.

5.3.2 Differences in Mechanical Properties

5.3.2.1 Stress–Strain Imbalances and Implant Migration

Most of the ceramics and metals used in orthopedic applications cause stress shielding (one of the main reasons for bone resorption) due to the mismatch of mechanical properties between an implant and surrounding bone, causing aseptic loosening and migration of the implant. Hence, a more even stress–strain distribution and prevention of interfacial loosening is the main motivation for tailoring mechanical properties of bone implants. In particular, nanophase materials seem to bring great advantages in this respect.

For an orthopedic implant to give the best performance, it should mimic the elastic modulus, strength, and fracture toughness of natural bone. For a given crack size, as the fracture toughness of the implant increases, so does its strength, solving two problems simultaneously [65]. For orthopedic applications, enhancement of fracture toughness is generally succeeded by composite production methods. The use of a secondary phase for an improvement in fracture toughness can increase the energy to propagate a crack or more effectively dissipate the energy of a propagating crack through crack deflection, crack bridging, or phase transformation mechanism [65].

Researchers have focused on the use of HA for composite production, where HA is used either as a matrix or a reinforcing agent. Researchers have increased fracture toughness by adding reinforcing agents. For example, addition of partially stabilized zirconia (PSZ) as a reinforcing agent into the HA (conventional sized) matrix increases the fracture toughness from 1.73 to 2.3 $MPa \cdot m^{1/2}$, inducing greater toughness by the phase transformation of PSZ [66]. However, nanotechnology is not about new chemistries, it is about using a new size range.

The bending moduli attained through the use of nanoscale materials can much better mimic the bending moduli of the human femur (19 GPa) compared to micron conventional counterparts. For instance, conventional alumina has a bending modulus of 52 GPa but nanophase alumina has a bending modulus of 35 GPa, imitating the bone much more effectively [67].

Through the use of nano-HA as the matrix phase and 1.5% zirconia as the reinforcing agent, the strength of the composite increased from 183 to 243 MPa. The addition of a 4 wt% carbon nanotube (CNT) secondary face into HA increased the fracture toughness by 56% [68] and a 0.1 wt% sodium dodecyl sulfate functionalized CNT addition to poly(propylene fumarate) increased the compressive modulus of the polymer from 318 to 981 MPa due to crosslink formation [69]. Incorporation of 55 wt% nano-HA into polar polyamide (PA66) was found to be more effective than micro-HA incorporation, increasing the bending strength, tensile strength, and impact strength of the composite by 31.3%, 38.9%, and 68.0%, respectively [70]. The higher surface energy and surface activity of nano-HA crystals with respect to micro-HA crystals allowed for nano-HA to bind stronger to PA66, improving the mechanical properties.

When the strength of a ceramic material prepared by the powder compaction technique is considered, nanoceramics are calculated to be much stronger than their conventional sized counterparts. For example, the bending strength of conventional sized HA is 38–160 MPa and nano-HA is 183 MPa [71]. This improvement in strength is due to a minimized defect size through the use of nanopowders, which should be smaller than the powder dimensions if processed properly. However, due to large defects introduced during ceramic processing techniques, the advantage of increased strength through the use of nano-HA has not been realized to its full capacity. Improvements in processing techniques would increase nanoceramic strength and fracture toughness.

Due to the fact that bone is a composite of collagen proteins (polymer) and HA-based ceramics, the use of ceramics as a secondary phase inside polymers is a widely investigated approach [72]. The bending moduli of 40 and 50 wt% nanophase ceramic (alumina, titania, or HA)/PLA composites were found to be 2–4.5 times greater with respect to corresponding conventional ceramic/PLA composites [72]. For instance, the addition of 50 wt% alumina increased the bending modulus of pure PLA from 60 MPa to 3.5 GPa, much better emulating the bending modulus of the human femur, 19 GPa. Similarly, nanophase ceramic (titania or HA)/PMMA composites increased the bending modulus compared to conventional ceramic/PMMA composite formulations [72].

As discussed earlier, implant migration is a major problem in orthopedics. To prevent implant migration, bone cements are generally used. However, poor mechanical properties (especially fatigue properties) and adverse effects of PMMA necessitate some improvements not only in bone cement chemistries, but also in mechanical properties of bone cement. The agglomeration of radiopacifiers (which are generally 10 wt% ZrO_2 or $BaSO_4$ particles as x-ray contrast agents and reaction initiators) used in PMMA is one of the reasons for the poor mechanical properties of PMMA. Replacing micron sized radiospacifiers with nanosized radiopacifiers significantly increased tensile strength of the nanocomposite over that of the standard microcomposite bone cement due to a reduction of particle agglomerate size and increased resistance to crack propagation [73,74]. The addition of 5–10 wt% nano-HA powders into PMMA was found to increase tensile strength, Young's modulus, hardness, and toughness of PMMA. Incorporation of HA nanorods into a chitosan matrix, which has been proposed for bone cement, improves its strength and bending modulus, making a formulation which is twice as strong as that of PMMA. Moreover, the addition of HA can reduce water absorption, postponing the retention of mechanical properties of CS/HA composite under moisture conditions [75].

5.3.2.2 Wear

The detrimental effect of conventional sized wear debris has been investigated for a long time. Many studies have reported that conventional size wear debris (1–100 μm) decreased the proliferation of osteoblasts, increasing its detrimental effect as the particle sizes decrease [76], specifically emphasizing the detrimental effects of particles smaller than 1 μm [77,78]. However, in these research efforts, the effect of nanosize wear debris has not been emphasized. Whether nanosized wear debris is better or worse than micron sized wear debris is still a controversial topic many researchers are currently investigating. Specifically, some studies highlight the detrimental effects of nanosized wear debris by emphasizing their increased number, surface area, and solubility compared to conventional sized wear debris [79]. For example, nanophase titania and CNTs were found to induce morphological changes in neutrophils and decrease their survival rate [80]. Additionally, some reports indicated that on an equal-weight basis, CNTs are much more toxic than carbon black,

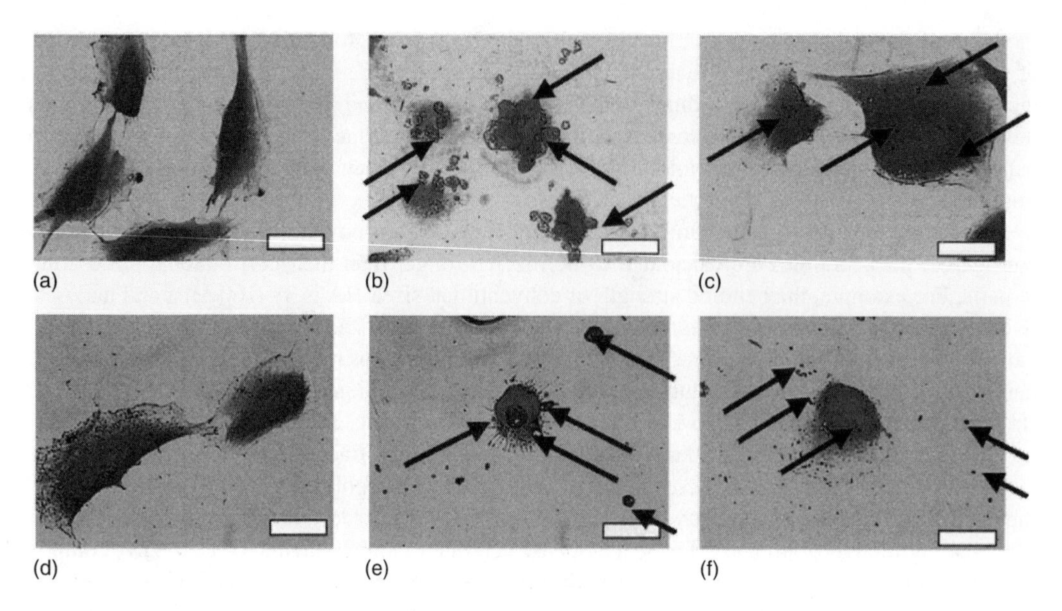

FIGURE 5.11 2 h (a, b, c) and 6 h (d, e, f) osteoblast morphology in the presence of alumina particles. (a, d) no particles (control), (b, e) conventional, and (c, f) nanophase alumina particles at concentrations of 100 μg/mL. Arrows indicate alumina particles. Bar = 10 μm. (From Gutwein, L.G. and Webster, T.J., *J. Nanoparticle Res.*, 4, 231, 2002. With permission.)

and can even be more toxic than quartz, which is considered a serious occupational health hazard for chronic inhalation exposure [81].

On the other hand, others have reported more favorable responses of nanoscale wear debris by showing that osteoblasts lived longer in the presence of nano compared to micron particulate ceramics [82]. In addition, a more well-spread morphology and increased cell proliferation was also observed in the presence of nanophase compared to micron particles, meaning a less adverse cell reaction to wear debris to nanophase ceramics [83]. In Figure 5.11, osteoblast morphologies in the presence of micro- or nanosized alumina particles are shown. These results indicated that by decreasing the particle size of alumina and titania to the nanometer regime, the adverse effects of wear debris to osteoblasts and chondrocytes (cartilage cells) can be minimized. Furthermore, direct contact toxicity studies showed less detrimental effects of carbon nanofibers on osteoblasts with respect to conventional sized carbon fibers [84]. The mechanism of better osteoblast health in the presence of nano compared to micron particles remains to be elucidated.

5.3.3 Differences in Biological Response and Bioactivity

5.3.3.1 Comparison for Wound Healing and Foreign Body Responses

As discussed in Section 5.2.2, wound healing is the natural response of body to the implant. Wound healing includes the recruitment of different cell types such as neutrophils, monocytes, macrophages, and foreign body giant cells. The occurrence of any of these cell types near the implant–tissue interface can be interpreted as a sign of wound healing and is undesirable due to an associated inhibition of bone formation.

Monocyte adhesion has been positively correlated to adsorbed fibrinogen on different polymeric surfaces. Preadsorbed proteins (such as IgG) have been found to significantly upregulate monocyte tumor necrosis factor-α release and procoagulant activity [85], none of which is desired on or around implants. Prevention of monocyte adhesion is the crucial step because monocytes differentiate into macrophages at the implant–tissue interface. It has been shown that smooth surfaces (micron smooth), induce a foreign body reaction which is composed of macrophages one to two cells in thickness. However, as the micron roughness increases, so does the foreign body reaction [55]. It is worth mentioning that these studies only investigated topographies in micron range, not nanorange. However, they provided evidence that surface topography of an implant plays an important role in the intensity of foreign body reactions.

Recent research by Liu-Snyder et al. showed significantly decreased macrophage adhesion and proliferation on nanophase alumina compared to conventional alumina after 12 and 24 h [86]. When macrophage adhesion diminishes, the detrimental effects of excessive inflammation would also decrease, this being another advantage of the use of nanomaterials as an orthopedic implant. Cytocompatibility assays on nano-HA crystals using human monocyte-derived macrophages showed no significant release of inflammatory cytokines (tumor necrosis factor alfa (TNF-α)) but an increase in the lactate dehydrogenase (LDH), an indicator of toxicity, at high nano-HA concentrations (100 million particle/0.5 million cells) was measured [87]. Additionally, the amount of the enzymes released from macrophages and foreign body giant cells during frustrated phagocytosis depends on the size of the biomaterial particle; the smaller the constituent particle size, the fewer enzymes will be released [55]. This release can be interpreted as when the grain or particle size of a material decreases to the nanosize, less detrimental inflammatory responses should be expected.

During the foreign body reaction, if frustrated phagocytosis occurs, the contents of the phagocytotic cells (degradative enzymes, superoxides, free oxygen radical, etc.) are released to the tissue–implant interface, whose effects are adverse for both the implant and juxtaposing tissue. However, once these free oxygen radicals are created during frustrated phagocytosis, one way to dispose of them is to attach these radicals to another surface. In this respect, the role of nanoscale materials is still under debate. It is known that nanoparticles attract free radicals at the cellular level [88] and, due to this characteristic, they have been proposed to be used as pharmacological agents as free-radical scavengers of oxygen. The use of nanophase implants can be advantageous for attracting free oxygen radicals on their surfaces, which would otherwise induce a more detrimental foreign body reaction. However, it is also clear that in addition to the attracting of free oxygen radicals, nanomaterials naturally generate reactive oxygen species [58]. Shrinkage in the size of nanomaterials generates structural defects and disrupts the electron configuration of materials, creating altered electrical properties which could establish specific surface groups that function as reactive sites. For example, in the case of oxygen superoxide radical creation, these reactive sites interact with an oxygen molecule and capturing of an electron creates a superoxide radical (O_2^-). Creation of a reactive oxygen species is so far the most widely criticized issue in nanomaterial toxicity because oxidative stress in the body can initiate inflammation and at high amounts; it can even lead to cell apoptosis [58]. Currently, there is not enough data on the toxicity effects of nanoscale materials; however, this is a very serious safety issue that should be extensively addressed before the widespread use of nanophase materials in biological applications.

Equally as interesting, another advantage of using nanoscale materials is its interaction with IgG. It has been shown that IgG adsorbs less on nanophase alumina with respect to conventional ceramics [17]. Less adsorption of IgG causes less neutrophil adhesion on nanophase ceramic surfaces, which is the first step in minimizing the foreign body reaction. Less IgG adsorption will also lead to less monocyte recruitment to the inflammation area, which will eventually result in less macrophage and less foreign body giant cell formation. Surfaces that strongly adsorb IgG promote long-term macrophage adhesion and they should be avoided [89]. Additionally, when less phagocytic cells are recruited to the inflammation area, there will be less secretion of free radicals and degradative enzymes, which are damaging to the cells near the inflammation area. These results reveal the promise of using nanophase materials to decrease the adverse effects of the foreign body reaction. If the intensity and duration of all five steps in wound healing (acute inflammation, chronic inflammation, granulation tissue formation, foreign body reaction, and fibrosis) can be diminished, which has already been partially fulfilled by use of nanophase ceramics, the patient can recover faster and the chances of implant success increases dramatically. Although it is hard to conclude the effects of using nanoscale materials on wound healing response just from protein adsorption as well as macrophage and neutrophil adhesion data alone, the results are promising.

Furthermore, the immune system response to nanophase and conventional sized materials is different. Immune system responses to an implant is not desired because it causes lysis of the tissue cells juxtapositioned to the implant and it activates the cells in the wound healing process through ligand–receptor mediated cellular adhesion, which eventually play an important role in implant failure. Basically, in order for the complement system (the branch of immune system that responds for recognition and elimination of foreign particles) to be activated, the thioester bond of the complement protein C4 is cleaved. A nanophase material cleaves this bond much less effectively than a conventional sized material, resulting theoretically in less complement activation and subsequent immune responses. One of the reasons for this behavior may be increased delocalization of electrons, which makes nanophase materials more stable.

Moreover, another function of the IgG antibody in the body is the activation of the immune system. When it adheres on the implant surfaces, it facilitates various aspects of the immune response. As stated above, nanophase materials promote less adsorption of IgG on their surfaces with respect to their conventional grain sized counterparts, further inducing less immune responses [17].

Besides these studies outlining the advantages of nanomaterials in minimizing the immune response, there have been studies on T cell interactions with nanomaterials. One of the major cell types in the immune response are T cells. When these cells do not respond to a biomaterial, there will be less of an immune system response. The T cell response to nanophase materials was found to be quite different than to conventional materials. On conventional titania, T cells were not very mobile, meaning they were strongly adherent. However, on nanophase titania, they were found to be much more mobile. These results demonstrated that nanophase materials may be recognized less by T cells than conventional materials [90].

After reducing wound healing and inflammation responses, a successful orthopedic material should enhance bone formation. There has been much research that nanoscale materials offer a great opportunity to fill this gap of enhancing bone formation where conventional materials have failed.

5.3.3.2 Comparison of Bioactivity

5.3.3.2.1 Metals

In most cases, the creation of nanostructured features on metals is achieved by chemical etching (such as by the use of H_2O_2) of conventional metals [91]. However, the etching process not only creates a nanotopography, but also alters the chemistry of the top surface, creating a thin metal oxide layer. Due to this, it is hard to determine if the changes in the cell behavior is a direct result of the nanotopography created or the change in chemistry of the surface. In this review and in nanotechnology in general, the main focus is to investigate what changes bone growth: chemistry or roughness. Powder processing techniques, more specifically, consolidation of nanometal powders without the use of heat, is one of the methods researchers have been using to investigate the effects of nanosurface topography changes on bone growth.

A comparison in osteoblast adhesion and mineral deposition between conventional metals (Ti, Ti6Al4V, and CoCrMo) traditionally used in orthopedics and their nanophase counterparts showed an increase in osteoblast functions on nanophase metals [92], the greatest mineral deposition occurring on nanophase CoCrMo [93]. Moreover, directed cell adhesion was observed on metal particle boundaries. This adhesion characteristic is shown in Figure 5.12 for nanophase and conventional Ti as an example. The fact that there are more grain boundaries on the surfaces of the nanophase Ti with respect to conventional Ti, and osteoblasts adhere at metal particle boundaries, could be an explanation as to why more bone growth occurs on nanometals [94]. Increases in roughness values of individual powders also proved to be crucial in mediating nanotopography of the samples and studies have demonstrated that as nanoroughness of the powders increase, so does adhesion [95]. In addition, nanophase Ti was more powerful at enhancing osteoblast adhesion than

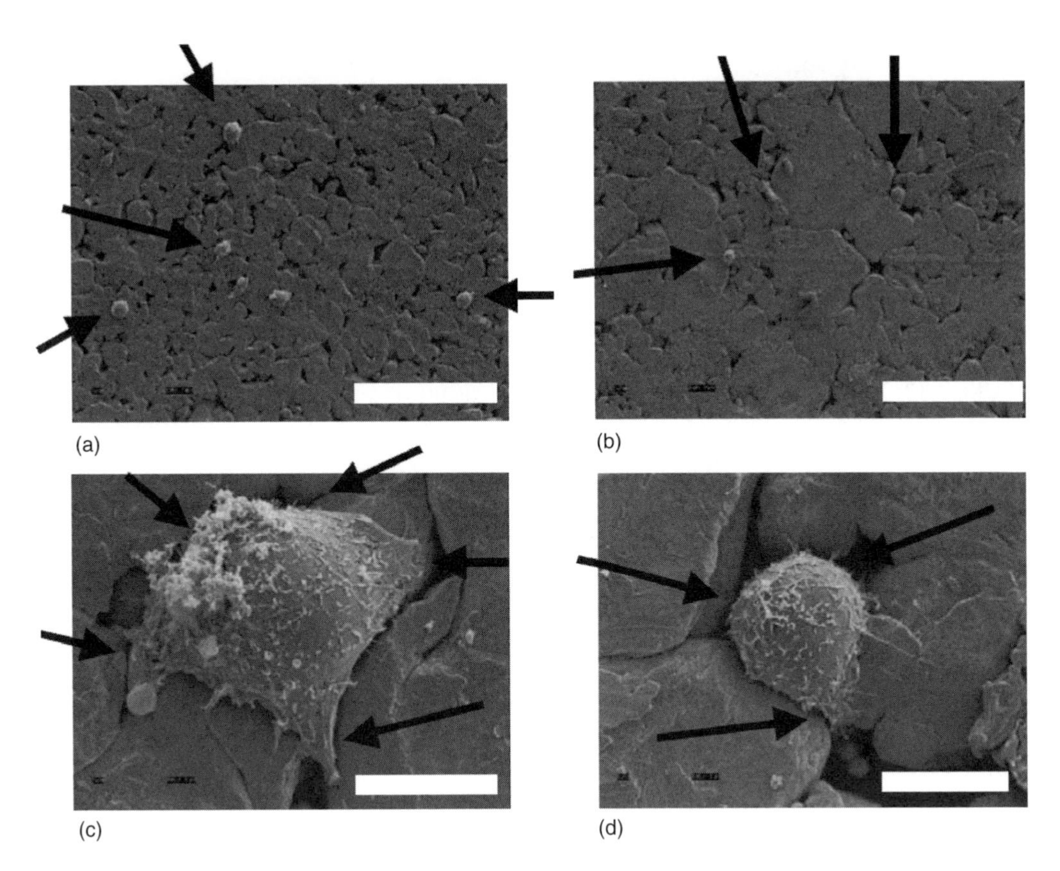

FIGURE 5.12 Scanning electron microscope (SEM) micrographs showing (a, c) osteoblast adhesion on nanophase Ti; (b, d) osteoblast adhesion on conventional Ti. Bar = 100 and 10 μm for low and high magnification images, respectively. Arrows indicate cells (low magnification) and area where cell protrusions are observed, specifically at particle boundaries (high magnification). (From Webster, T.J. and Ejiofor, J.U., *Biomaterials*, 25, 19, 4731, 2004. With permission.)

immobilizing peptides (such as the aforementioned RGD and KRSR) on the surfaces of conventional Ti [96].

Nanophase metalloids have also shown promise in orthopedic applications. In particular, many patients receiving implants do so due to the removal of cancerous bone. To help these patients, an anticancer chemistry, selenium (Se), was recently investigated as an orthopedic implant. The rationale behind the use of Se is that corrosion products could react with cancer cells or be used by surrounding tissues to prevent cancer reoccurrence. Osteoblast adhesion was found to be the highest on Se compacts composed of nanometer particles [97].

5.3.3.2.2 Ceramics

Most of the research in the nanophase bone tissue engineering field is on nanophase ceramics, including the first known report correlating increased bone cell function with decreased grain size [98]. All studies demonstrated increased in vitro osteoblast adhesion, proliferation, calcium deposition (an index of mineralization of bone matrix), and ECM protein synthesis (such as alkaline phosphatase) as the particulate size of a material decreases into the nanometer regime, in particular below 100 nm [17,29,99–104].

The most commonly investigated nanophase ceramics for orthopedic applications are HA, titania, and alumina. The main production method for nanophase ceramic orthopedic implants is powder consolidation followed by sintering. By manipulating the sintering time and temperature, it

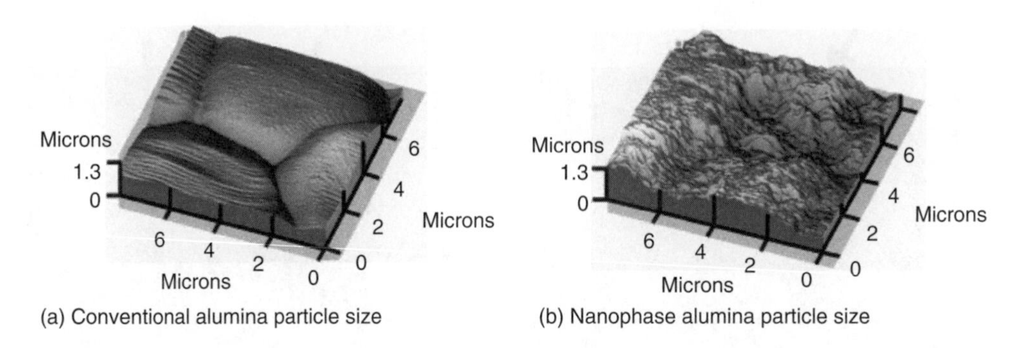

FIGURE 5.13 Atomic force microscope (AFM) images of (a) conventional alumina and (b) nanophase alumina. (From Webster, T.J., Siegel, R.W., and Bizios, R., *Biomaterials*, 20, 13, 1221, 1999. With permission.)

is possible to obtain desired grain sizes. Osteoblast adhesion experiments showed 30% and 40% increases on nanophase titania (grain size 32 nm) and alumina (grain size 46 nm), respectively, when compared with conventional ceramics [99]. The surface topography of nanophase and conventional alumina is depicted in Figure 5.13. Additionally, a step-function increase in the adhesion of osteoblasts was observed for alumina grain sizes between 49 and 67 nm and for titania between 32 and 56 nm. This critical grain size is important because ceramics with grain size below 100 nm were observed to promote other properties of ceramics such as electrical, catalytic, mechanical, etc. [58]. It also provides evidence that optimal surface structures for ceramics are below 60 nm in dimension.

Transferring nanotopographies of conventional and nanophase titania onto a model poly(lactic-co-glycolic acid) (PLGA) mold also increased osteoblast adhesion and proliferation. Figure 5.14 shows the surface topographies of nanophase and conventional titania together with corresponding PLGA molds. This finding proved that nanostructured surfaces of ceramics transferred to polymers

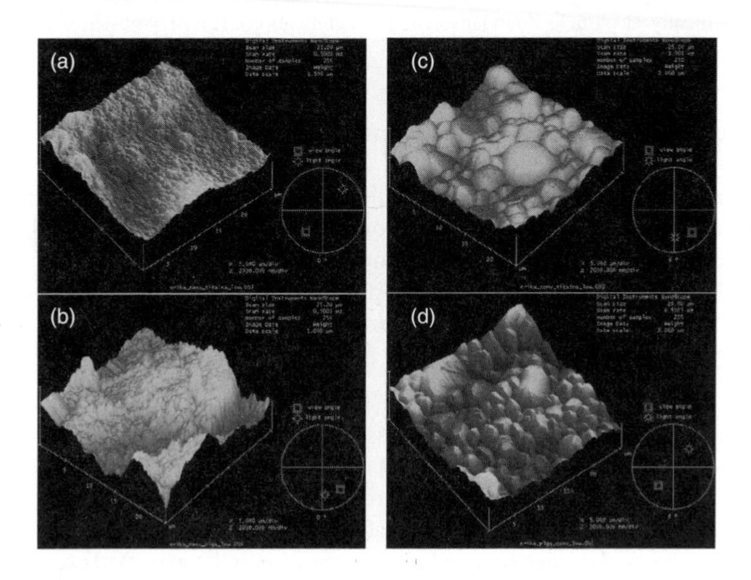

FIGURE 5.14 AFM images of (a) nanophase titania, (b) the PLGA mold of nanophase titania, (c) conventional titania, and (d) the PLGA mold of conventional titania. Root-mean-square values from AFM for (a), (b), (c), and (d) 5×5 μm^2 and 25×25 μm^2 scans were 29, 22, 13, and 12 nm, respectively. (From Palin, E., Liu, H., and Webster, T.J., *Nanotechnology*, 16, 1828, 2005. With permission.)

TABLE 5.1
Percent Increase in Alkaline Phosphatase and Extracellular Matrix Calcium Content for Osteoblasts Cultured on Nano Compared to Conventional Alumina, Titania, and HA after 28 Days

Chemistry	Grain Size (nm)		% Increase in Alkaline Phosphatase Synthesis on Nanophase Compared to Conventional	Increase in Extracellular Matrix Calcium on Nanophase Compared to Conventional
Al_2O_3	Conventional	167	36	4 times
	Nano	24		
TiO_2	Conventional	4250	22	6 times
	Nano	39		
HA	Conventional	179	37	2 times
	Nano	67		

Source: From Webster, T.J., Ergun, C., Doremus, R.H. et al., *Biomaterials*, 21, 17, 1803, 2000. With permission.
Note: HA = hydroxyapatite.

promote increased adhesion and proliferation of osteoblasts. The more the nanofeatures mimic those of natural bone, the further the increase in osteoblast adhesion and proliferation [105].

Moreover, calcium content in the ECM of osteoblasts cultured on nanophase alumina, titania, and HA was observed to be four, six, and two times greater than on respective conventional ceramics after 28 days respectively. Corresponding alkaline phosphatase synthesis was found to be 36%, 22%, and 37% higher on nanophase compared to conventional alumina, titania, and HA formulations. Additionally, osteoblast cell colony area (a measure of cell adhesion; the stronger the cells adhered, the less they migrated, occupying less surface area) decreased on alumina, titania, and HA when the grain size was reduced to nanoscale. Table 5.1 summarizes osteoblast activities on these three well-studied ceramics with nano and conventional grain sizes [100].

Another possible method to create nanostructured features on materials is anodization, which is the formation of an oxide nanopore or nanotube layer on the surface of metals by the use of electrochemical methods. Figure 5.15 shows the microstructure of an alumina membrane prepared by the anodization method. Anodized aluminum, titanium, and Ti6Al4V have been investigated for orthopedic applications [106–108]. Osteoblast cell adhesion, proliferation, and mineral deposition were all observed to increase on these surfaces. Moreover, vitronectin and fibronectin adsorption was observed to be higher on nanotubular structures formed by the anodization of Ti [109]. Interestingly, the pores created on anodized alumina were found to be the same size with filipodia (cell extensions of osteoblasts) where penetration of these filipodia inside the porous membrane is possible, causing a stronger cell adhesion by functioning as anchorage points for osteoblasts [108]. This behavior is shown in Figure 5.16.

The particle aspect ratio is also another parameter affecting the biological response of a nanomaterial. A nanofiber geometry can imitate the dimensions of the constituent phases of bone, HA crystals, and collagen fibers, with which bone cells are accustomed to interact. In vitro tests confirmed this hypothesis by showing increased osteoblast functions on nanofiber alumina (diameter is 2 nm and length >50 nm) with respect to a nanospherical alumina geometry [104,110]. A 29% increase in osteoblast proliferation and a 57% increase in calcium deposition

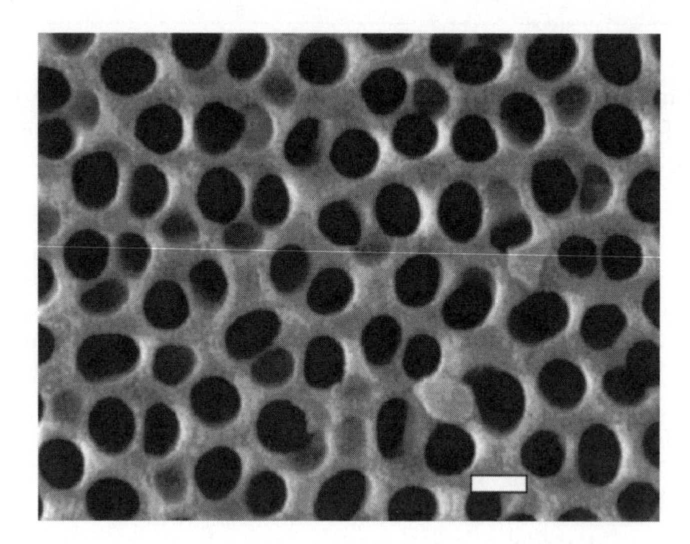

FIGURE 5.15 SEM images of 60 V alumina membranes showing porous structure. Scale bar = 100 nm. (From Swan, E.E.L., Popat, K.C., Grimes, C.A. et al., *J. Biomed. Mater. Res. A*, 72A, 3, 288, 2005. With permission.)

were observed on nanofiber alumina compared to other alumina surfaces [111]. Although the phase and chemistry of the materials were not the same in this last study, it showed great promise for increasing osteoblast functions upon aspect ratio changes of the nanoparticles.

Due to the presence of HA in bone, many studies have investigated HA and calcium phosphate ceramics for orthopedic applications [112–114]. Currently, HA is mainly used as a coating material on implant surfaces and the most widespread coating method is plasma spray deposition. However, it is hard to obtain nanograin sizes due to high temperatures and very rapid cooling during plasma spraying. Because of this, researchers are currently investigating new methods to coat implants with nano-HA [115,116]. Wet chemistry methods, where nano-HA is allowed to precipitate, have been proposed for coating surfaces with nano-HA. Although nanophase coatings of apatites can be applied on Ti successfully and promote bone formation in vivo [117], the coatings prepared by precipitation methods are not very controllable and are time consuming. IonTite, which is a high pressure based low temperature coating method, is also another proposed coating method. Investigations showed that IonTite deposited nanocrystalline HA coatings promote more osteoblast adhesion and calcium deposition compared to plasma sprayed commercially available HA coatings [113].

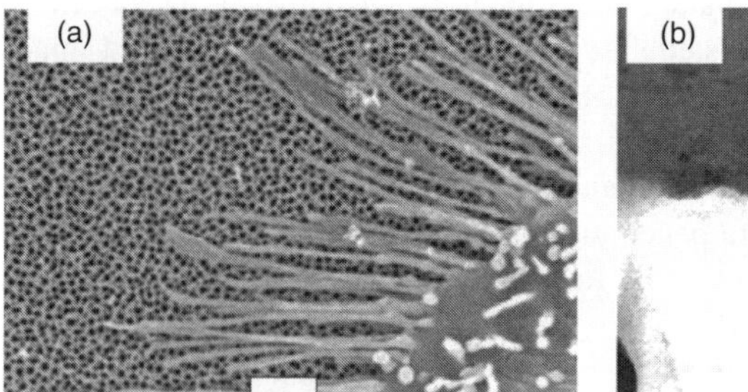

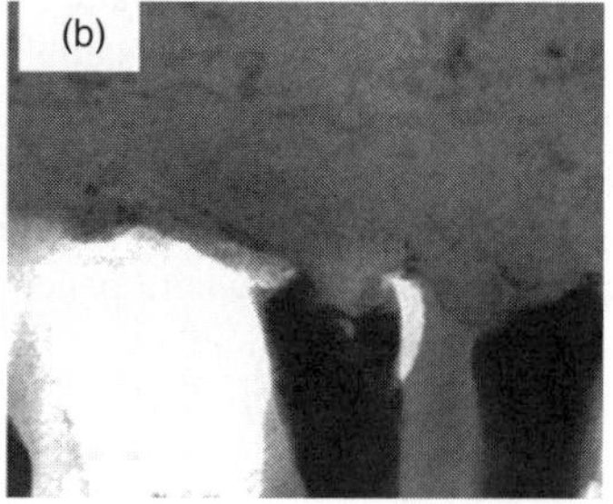

FIGURE 5.16 (a) SEM micrograph of an osteoblast on an alumina membrane and (b) transmission electron microscope (TEM) cross section showing penetration of filipodia into the alumina pores. Scale bar = 2 μm. (From Karlsson, M., Pålsgård, E., Wilshaw, P.R. et al., *Biomaterials*, 24(18), 3039, 2003. With permission.)

Equally as interesting, the latest results show that nanocrystalline HA and amorphous calcium phosphate can be functionalized with the cell adhesive RGD peptide sequence which has widely been studied on conventional sized counterparts. Decreasing the particulate size into the nanometer regime together with reducing the crystallinity of calcium phosphate ceramics was found to promote osteoblast adhesion to the same degree as functionalizing conventional HA with RGD [112]. Moreover, nano-amorphous calcium phosphate and nanocrystalline HA coated magnetic nanoparticles (Fe_3O_4) are currently being investigated as novel drug carrying systems which specifically attach to the osteoporotic bone by surface functionalized antibodies and can be further guided by an external magnet field [114].

It is worth mentioning that all the studies discussed so far have been conducted in vitro. The conditions in the human body are much more complex than what can be artificially created in a laboratory environment. For this reason, it is of paramount importance to test the efficacy of these in vitro results in vivo. A current study reported that the above-mentioned advantages of nanophase ceramics were also valid in vivo (Figures 5.17 and 5.18). For example, in this study porous tantalum scaffolds were coated with either nanophase or conventional sized HA particles. After 6 weeks of implementation into rat calvaria, much greater osseointegration was observed

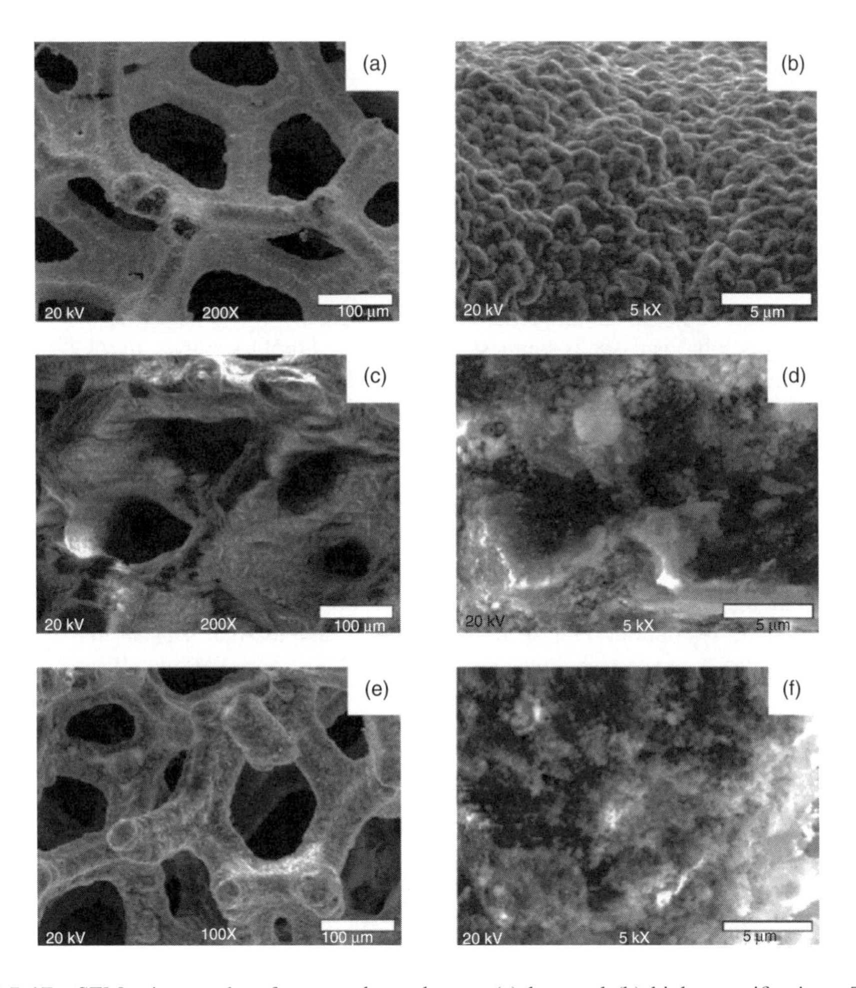

FIGURE 5.17 SEM micrographs of uncoated tantalum at (a) low and (b) high magnifications. Tantalum coatings of either micron grain size HA at (c) low and (d) high magnifications. Nanophase grain size HA at (e) low and (f) high magnifications. Scale bar = 100 and 5 μm, respectively. (From Sato, M., An, Y.H., Slamovich, E.B. et al., *Int. J. Nanomedicine*, 2008. With permission.)

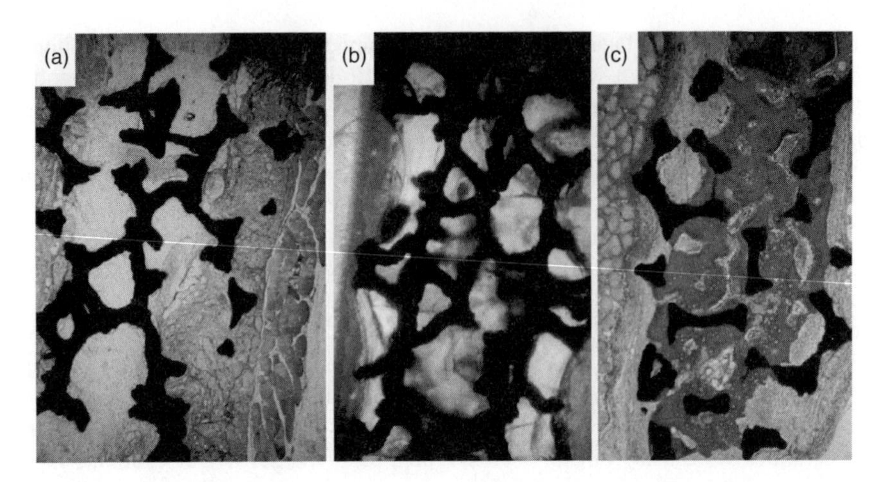

FIGURE 5.18 (See color insert following page 206.) Histology after 6 weeks implantation into rat calvaria of (a) uncoated tantalum and tantalum coatings, (b) micron grain size HA, and (c) nanophase grain size HA. Stain = Stevenel's blue and a counterstain of van Gieson's picrofuchsin. While little bone in growth is observed for (a) uncoated tantalum and (b) tantalum coated with micron grain size HA, much is observed for (c) tantalum coated with nanograin size HA. (From Sato, M., An, Y.H., Slamovich, E.B. et al., *Int. J. Nanomedicine*, 2008. With permission.)

for tantalum scaffolds which were coated with nanophase HA with respect to conventional HA coated tantalum scaffolds [118].

Most importantly, the competitive cell response is different on nanophase compared to conventional ceramics. Upon insertion of an orthopedic implant, as the wound healing process starts (which may end either with successful incorporation of the implant, parenchymal cell formation around the implant, or with failure of the implant), fibrous encapsulation and callus formation around the implant may begin. Fibrous tissue is synthesized mainly by fibroblasts; thus, fibroblast adhesion on an orthopedic implant is not desirable. It is also equally important for other competitive cells (such as endothelial cells) not to attach on the implant surfaces because their attachment means less space for osteoblasts to adhere. Studies have shown that there is less fibroblast and endothelial cell adhesion onto nanophase titania, alumina, and HA compared to conventional sized counterparts [29,119]. Moreover, osteoblast adhesion was more than 300% greater on these nanophase ceramics compared to fibroblast and endothelial cell adhesion [29]. In contrast to osteoblasts, fibroblast and endothelial cell adhesions were observed to decrease with an increase in nanoroughness [119].

The optimal implant should not only promote bone formation but also support bone remodeling by promoting bone resorption. This is important for many reasons such as the prevention of malnourishment due to excessive bone deposition and to control growth factor secretion by osteoclasts which in turn promotes bone deposition (meaning that bone growth occurs immediately after bone resorption). In a similar way to osteoblasts, osteoclast function was observed to increase on nanophase ceramics [119]. Tartrate-resistant acid phosphatase (TRAP) synthesis and the formation of resorption pits were greater on nanophase alumina and HA compared to their conventional sized counterparts. Increased solubility of nanophase ceramics upon exposure to TRAP was proposed to be the reason for this behavior [120]. Nanophase ceramics are well known to have increased solubility properties.

Nanophase ceramics have been also proposed for implantation due to their antibacterial properties. During surgery, bacteria (namely, *Staphylococcus epidermidis*) from the patient's own skin or mucosa enter the wound site, leading to an irreversible carbohydrate-based biofilm formation at the surface of the implant causing clinical failure. Due to this, an implant chemistry reducing *S. epidermidis* colonization is necessary. Nanophase ZnO, a well-known antibacterial agent, and nanophase TiO_2 decreased the adhesion of *S. epidermidis*, with nano-ZnO reducing colonization more than nano-TiO_2 [121]. Coupled with increased bone growth on these nanophase ceramics, such chemistries may have two key advantages for implantation: resist infection and increase bone growth.

Most importantly, all the above studies provided evidence of increased bone cell functions upon decreased grain sizes below 100 nm, adding another novel property of nanoscale materials to the field.

5.3.3.2.3 Polymers

As mentioned, the major component of the organic phase of bone is a polymeric-based material collagen. Additionally, the formation of biodegradable polymeric chemistries for bone tissue engineering is highly desirable. Researchers try to control and match the degradation rate of these polymers with bone growth rate so that as the polymer dissolves, the native tissue can penetrate inside the degraded part of the implant, reforming the new opened up space. Key nanophase polymer chemistries used in bone tissue engineering applications are poly(lactic acid) (PLA), poly(glycolic acid) (PGA), poly(lactic-co-glycolic acid) (PLGA), and polycaprolactone (PCL).

Techniques used to create nanotopographies in polymers are electron beam lithography, photolithography for creating an ordered topography, and polymer demixing, phase separation, colloidal lithography, chemical etching, and electrospinning for creating unordered surfaces (electrospinning can also be used to form ordered topographies of aligned fibers) [122].

Chemically treating microstructured polymers is a cheap and easy solution to create nanotopographies. Acidic structured polymers (such as PLGA and PCL) can be treated with a base, such as NaOH, while basic polymers (such as polyurethane (PU)) can be treated with an acid, such as HNO_3, to create nanometer surface features. Figure 5.19 shows nano-PLGA surfaces prepared by etching in 1N NaOH for 10 min [123]. Compared to conventional PLGA surfaces, nanorough PLGA showed increased osteoblast adhesion both on two-dimensional [124] and three-dimensional scaffolds [123]. Excitingly, endothelial cells' (which are initially accepted as competitive cells in bone tissue engineering applications) adhesion and proliferation were observed to decrease on chemically altered nanostructured surfaces [125]. Although not related to bone tissue engineering applications but nevertheless important to mention here, chondrocytes, bladder smooth muscle cells, and cardiac muscle cells were observed to adhere and proliferate much better on nanostructured PLGA surfaces created by this novel etching method [126–128].

Biodegradablility of polymers is also used for drug delivery purposes. In a recent research paper, phase separation and template leaching methods were used to produce a novel microporous and nanofibrous three-dimensional poly(L-lactic acid) scaffold, whose surface was immobilized with

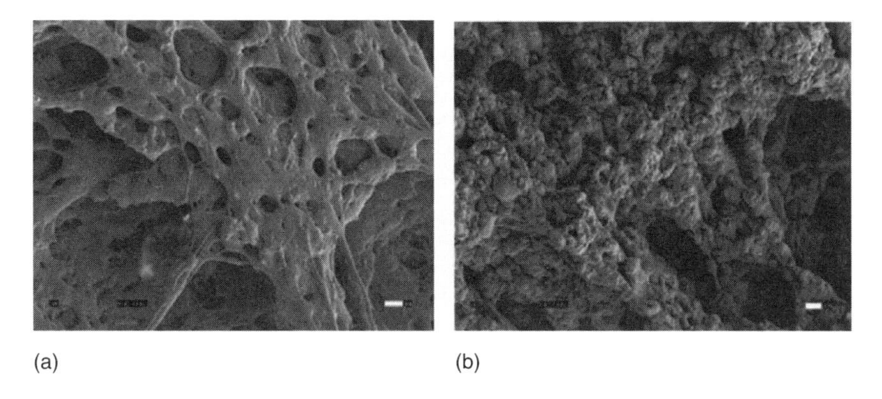

FIGURE 5.19 SEM micrographs showing (a) conventional PLGA and (b) nano-PLGA produced by salt etching method. Scale bar = 10 μm. (From Pattison, M.A., Wurster, S., Webster, T.J. et al., *Biomaterials*, 26, 15, 2491, 2005. With permission.)

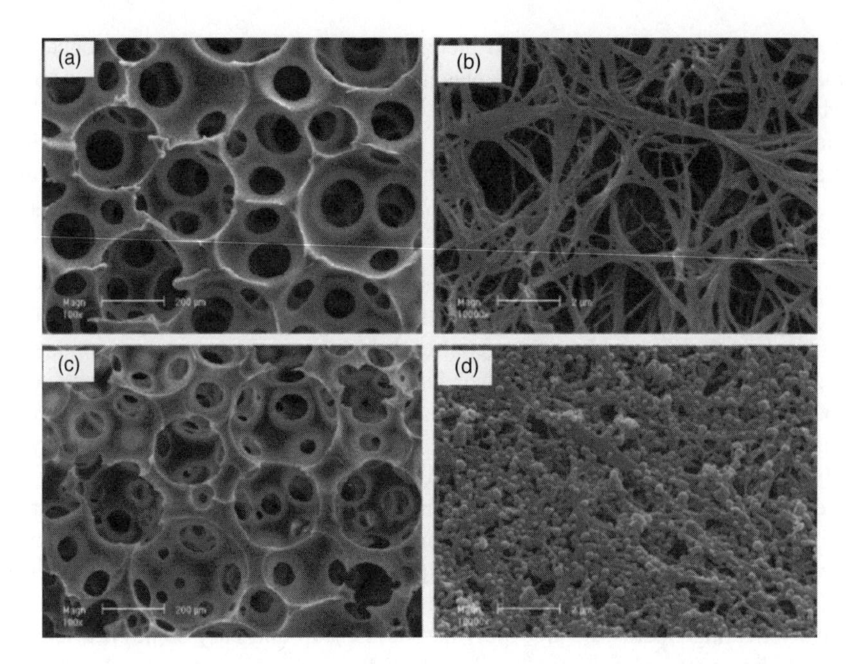

FIGURE 5.20 SEM micrographs of PLLA nanofibrous scaffolds (a, b) before and (c, d) after incorporation of rhBMP-7 in PLGA nanospheres. Scale bar = 200 μm (a, c) and 2 μm (b, d). (From Wei, G., Jin, Q., Giannobile, W.V. et al., *Biomaterials*, 28, 12, 2087, 2007. With permission.)

PLGA nanospheres. Recombinant human bone morphogenic protein 7 (rhBMP-7), which induces bone cell growth and function, was encapsulated within these PLGA nanospheres [129]. The microstructure of the novel template is depicted in Figure 5.20. Scaffolds with PLGA immobilized surfaces showed much more effective drug release with respect to scaffolds adsorbed with rhBMP-7, corresponding to considerably higher bone formation in vivo.

Electrospinning is another novel method to fabricate nanopolymers. Electrospinning is the injection of a polymer solution through a needle, where high voltage is applied to overcome the surface tension of the polymer solution causing initiation of a jet. Electrospun polymers can mimic the nanodimensions of extracellular proteins deposited by bone cells and can have porosities up to 90% with pore diameters reaching up to 100 μm [130]. Electrospun starch-based scaffolds having both nano- and micron sized fibers were shown to better mimic the ECM morphology of bone, increase osteoblast proliferation and alkaline phosphatase activity, and invoke a more well-spread osteoblast morphology with respect to scaffolds with only micron fibers [131]. It was also observed for the first time that utilizing the same polymeric scaffold, electrospun nano-PCL selectively differentiated human mesenchymal stem cells isolated into the osteogenic lineage [132]. Additionally, electrospun nano-PCL fibers were also shown to promote osteogenic differentiation of mesenchymal stem cells in vivo, showing embedded osteoblasts and mineralization throughout the nanopolymer scaffold [133].

5.3.3.2.4 Composites

Although both nanophase ceramics and polymers increase osteoblast functions, the inherent weaknesses they posses individually (i.e., low fracture toughness for ceramics and low strength for polymers) have led researchers to investigate different material composite combinations, where advantages of different material classes can be combined to fulfill the ultimate purpose of a better integrated implant.

Most of the research in the nanophase composites area is focused on polymer matrix composites with embedded nanophase ceramics due to their ability to mimic natural bone. One of the

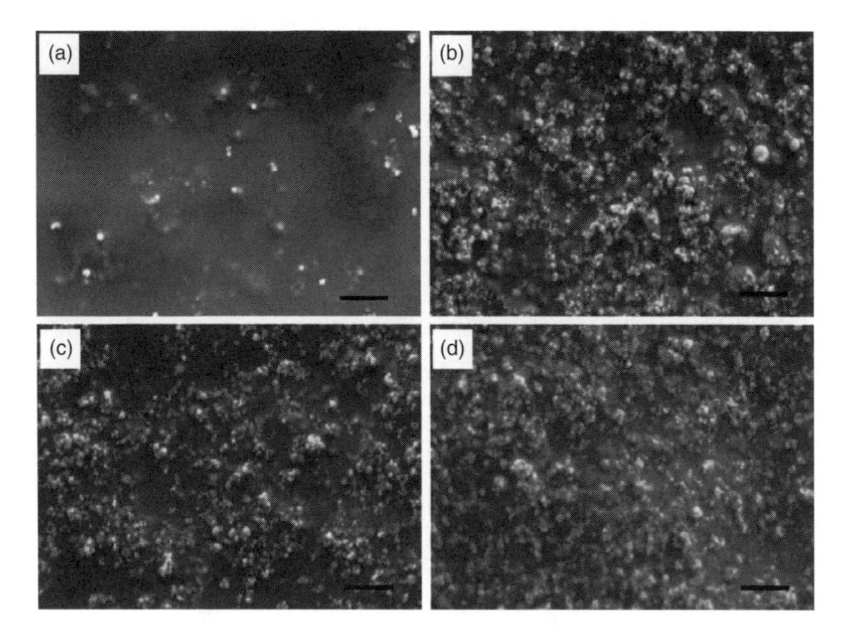

FIGURE 5.21 SEM micrographs of PLGA/titania composites sonicated for 10 min at (a) 25%, (b) 45%, (c) 45%, and (d) 70% of maximum sonication power, which is an indication of level of dispersion. Scale bar = 1 μm. (From Liu, H., Slamovich, E.B., and Webster, T.J., *Nanotechnology*, 16, 601, 2005. With permission.)

chemistries widely investigated is PLGA/titania. Addition of nanophase titania as a secondary phase inside biodegradable PLGA was found to improve osteoblast adhesion, alkaline phosphatase activity, and calcium deposition, which increased osteoblast function (investigated up to 30 wt% titania) [134]. Moreover, nanophase titania/nanophase PLGA composites showed the highest osteoblast adhesion with respect to other conventional titania/PLGA composite combinations [135]. In order to understand the effect of composite production parameters, specifically the dispersion power, on material properties and osteoblast behavior, different sonication powers were utilized when mixing the titania inside PLGA. The micrographs of the samples produced are shown in Figure 5.21. Results correlated increased dispersion of titania with decreased agglomeration together with increased osteoblast functions [136]. In fact, the highest amount of collagen synthesis (the major organic constituent in bone) was observed on samples sonicated at the highest power [137]. Moreover, the samples which mimic the nanoroughness of bone the best were shown to improve osteoblast adhesion and function the most [138].

Due to their unique electrical, mechanical, and geometric properties, carbon nanotubes and nanofibers have been of great interest in bone tissue engineering. The use of carbon nanofibers (CNFs) as a secondary phase in composite materials has been widely researched. PU/CNF composites increased osteoblast cell density as the CNF content in the composites increased. Excitingly, fibroblast density (which causes fibrous encapsulation of an orthopedic implant) decreased as the CNF content in PU/CNF composites increased [139]. Scanning electron microscopy (SEM) images of these composites are shown in Figure 5.22. These tendencies in cell adhesion observed with PU/CNF composites were also observed for PLGA/CNF composites [140]. Incorporation of 5 wt% CNFs inside the PLGA matrix doubled osteoblast adhesion with respect to pure PLGA compounds. In the case of polycarbonate urethane (PCU)/CNF composites, incorporation of 25 wt% CNFs improved osteoblast adhesion more than three times and reduced fibroblast adhesion by nearly 35% [141]. Additionally, the presence of a pyrolytic outer surface layer around the CNFs, which lowers the surface energy of the fibers, was also effective in promoting osteoblast adhesion [141]. Although

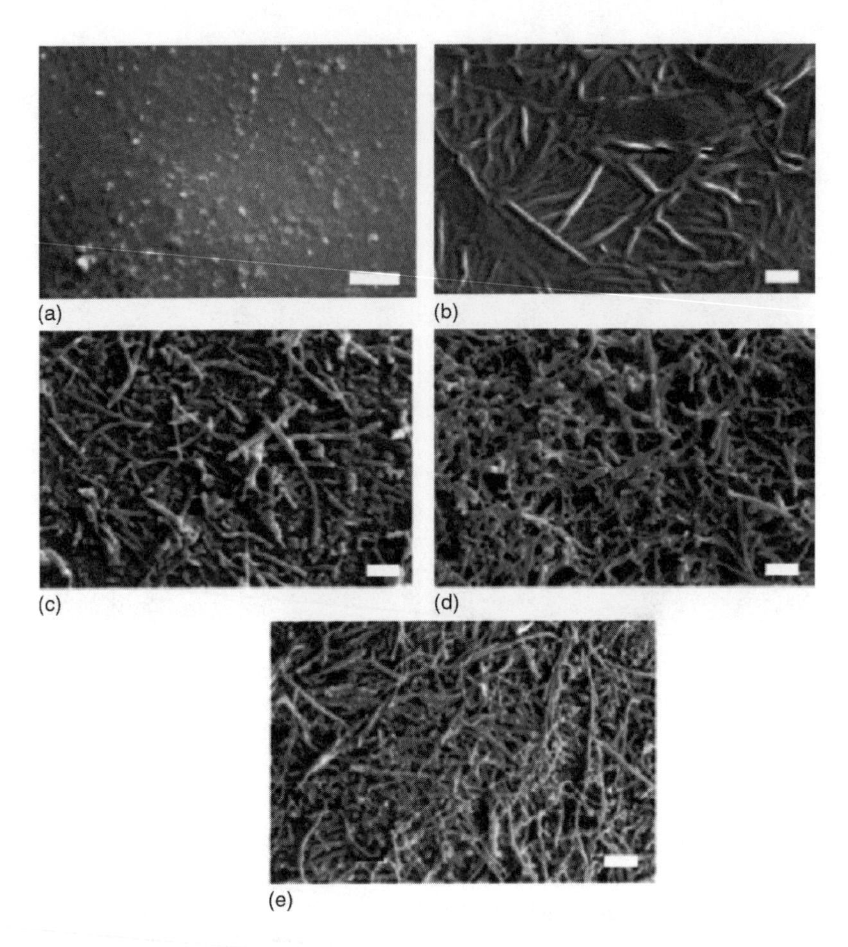

FIGURE 5.22 SEM images of PU/CNF composites with (a) 100:0, (b) 98:2, (c) 90:10, (d) 75:25, and (e) 0:100 PU/CN wt%. Scale bar = 1 μm. (From Webster, T.J., Waid, M.C., McKenzie, J.L. et al., *Nanotechnology*, 15, 48, 2004. With permission.)

not bone tissue engineering related, astrocyte (glial scar tissue forming cells in nervous system) adhesion, proliferation, and function also decreased on CNFs compared to conventional sized carbon fibers [142].

The use of HA as a secondary phase of a composite formulation is also common in bone tissue engineering [143]. This is because a nano-HA/collagen composite material has both organic and inorganic constituents resembling bone, mimicking not only its chemistry but also its topography. When such composites are cocultured in vitro with osteoblasts, it is possible to create a three-dimensional osteogenic cell/composite structure where new bone matrix synthesis can be achieved [144]. In vivo tests provide evidence that nano-HA/collagen composites biodegrade as mediated by foreign body giant cells and promote new bone formation [145]. In order to improve the poor mechanical properties of porous nano-HA/collagen composites, a secondary polymer, poly(lactic acid) can be incorporated [146]. In vitro studies showed osteoblast growth throughout the scaffold about 200–400 μm in depth within 12 days [147]. Results showed successful osteoblast adhesion, spreading, and proliferation after 1 week, giving promising results both in vitro and in vivo [146]. In addition, nano-HA can also be incorporated into poly(lactic acid) by the use of a thermally induced phase separation technique which can create a highly porous (90%) three-dimensional composite [148]. The addition of nano-HA has been found to increase protein adsorption on these scaffolds when compared to the conventional sized HA particles [148].

Addition of a ceramic secondary phase also improved the mechanical properties of polymer scaffolds. This has led researchers to investigate ceramic/ceramic composites. PSZ/HA composites, where zirconia has long been known to improve toughness by undergoing stress-induced transformation toughening [149], can be used as a secondary phase to improve the fracture toughness of HA [150]. It was observed that for the concentrations tested (10, 25, 40 wt% PSZ), as the grain size decreased to the nanoscale, osteoblast adhesion increased. Moreover, HA was used to improve the biocompatibility of ZrO_2–Al_2O_3 nanocomposites, where a 30 vol% addition gave the optimal osteoblast response and mechanical property combination [151].

5.3.3.2.5 Novel Materials

Compared to conventional sized carbon fibers, CNFs possess enhanced mechanical (i.e., stronger, have higher strength to weight ratios), electrical (conductivity), and surface (ability to be functionalized) properties [152], many of which are desirable for bone tissue engineering applications. Moreover, the increased aspect ratio of carbon nanofibers/tubes can be used to better mimic the constituents of bone. The main use of carbon nanofibers/tubes is as a secondary phase in a composite formulation, which was discussed in Section 5.3.3.2.4. However, additional research exists elucidating the interactions of only carbon nanofibers [84,110,152–154] and carbon nanotubes [155–159] scaffolds with living cells, in particular osteoblasts.

Such investigations demonstrated increased osteoblast adhesion [110], proliferation, and function (i.e., alkaline phosphatase synthesis and extracellular calcium deposition) on CNF compacts with respect to conventional sized carbon fibers [152]. This was further explained by molding conventional and nanosized carbon fiber compacts onto a model PLGA surface, so that only the effects of the CNF surface features were replicated. Increased osteoblast adhesion onto CNF polymer molds increased compared to conventional CNF polymer molds [160]. The microstructures of conventional carbon fibers and CNFs are depicted in Figure 5.23 [161].

In a recent study, CNF patterns were developed on a model polymer (PCU) by a novel imprinting method. Selective adhesion and alignment of osteoblasts were observed on CNF patterns. Results also showed direct deposition of calcium onto CNF patterns on PCU due to preferential adhesion of osteoblasts on CNFs rather than on PCU surfaces [153,154]. These results are promising toward mimicking the micro-alignment of calcium phosphate crystals naturally occurring in long bones of the body by CNF surface patterning. Figure 5.24 shows the CNF patterned surface and preferential adhesion of osteoblasts on CNF patterns [153]. The importance of this method is that if it can be combined with electrical stimulation, which is known to improve osteoblast function [162], it can result in a faster directional deposition of calcium on CNF surfaces.

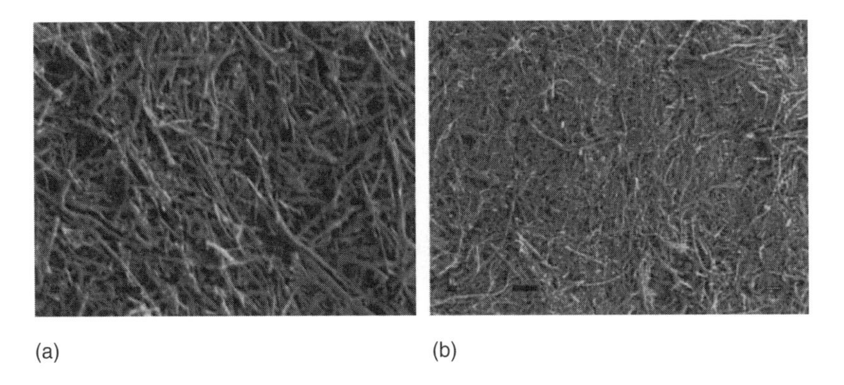

FIGURE 5.23 SEM images depicting (a) conventional carbon fibers (0.125 mm diameter) in comparison with (b) carbon nanofibers (60 nm diameter). Bars represent 1 μm. Bone cell function is enhanced on nanophase compared with conventional carbon fibers. (From Sato, M. and Webster, T.J., *Exp. Rev. Med. Dev.*, 1, 1, 105, 2004. With permission.)

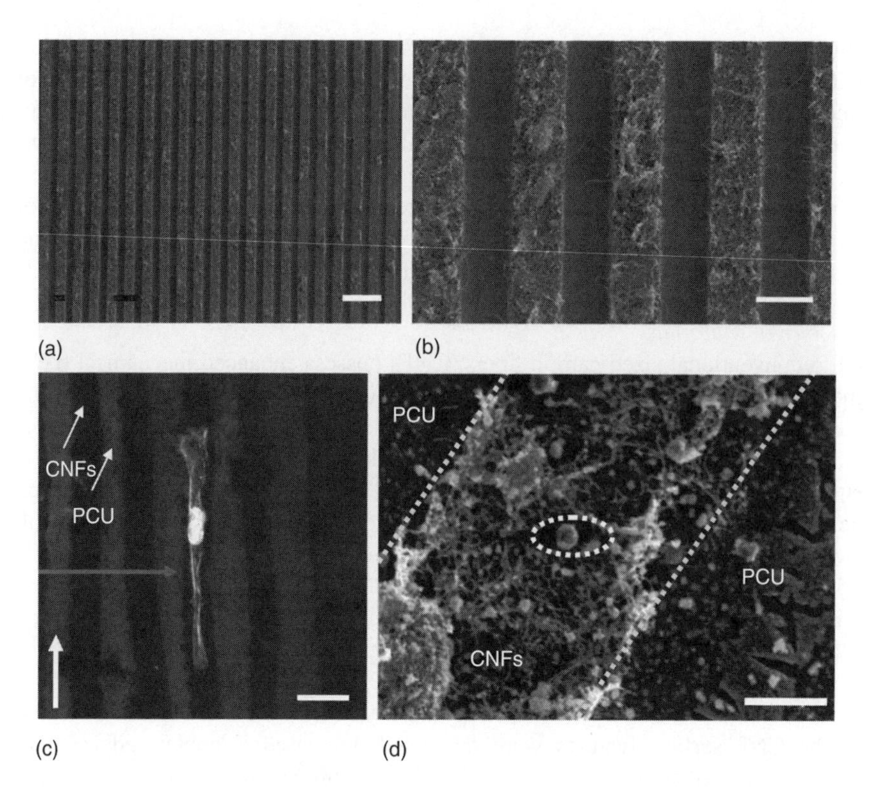

FIGURE 5.24 SEM micrographs of (a, b) CNF pattern on PCU in low and high magnification, respectively; (c) aligned actin in an osteoblast adhered selectively on CNF; (d) calcium phosphate crystal on CNF. Bars are (a) 100 μm, (b) 30 μm, (c) 20 μm, and (d) 10 μm. (From Khang, D., Sato, M., Price, R.L. et al., *Int. J. Nanomedicine*, 1, 1, 65, 2006. With permission.)

In fact, one study has used electric stimulation of osteoblasts on bone tissue engineering scaffolds of polylactic acid—carbon nanotube (80/20 wt%) composites. Although PLA is an insulative material, addition of 20% CNT makes it conductive, permitting the passage of an electrical current. When this composite was exposed to 10 μA alternating current for 6 h a day, a 46% increase in osteoblast proliferation was observed after 2 days and a 307% increase in calcium deposition was observed after 21 days with respect to nonelectrical stimulated samples, showing enhanced osteoblast functions [163].

Additionally, osteoblasts cultured on CNTs showed no decrease in membrane ion channel activity, which is necessary for the secretion of key bone-related proteins by osteoblasts, confirming the suitability of CNTs for bone tissue engineering applications [156]. Moreover, an extended osteoblast morphology was observed on CNTs [155]. This result has been confirmed by other researchers showing the highest proliferation rate on neutral charged single wall CNTs [159]. Most interesting, Tsuchiya et al. correlated osteoblast adhesion with carbon nanotube diameter. CNTs with a diameter of 10 nm showed more cell adhesion and TGF-β1 synthesis (an indication of normal cell growth) with respect to 200 nm diameter CNTs [157]. On the other hand, some investigations have emphasized contradicting results [158]. For example, osteoblasts were found to have a range of morphologies, some displayed reduced spreading and some showed reduced widening but increased elongation on CNT surfaces. The variation in osteoblast response to CNTs demonstrate that many parameters are playing a role in responses to these surfaces and not all CNTs are equal. In particular, the CNT production method (which further determines the micro- and nanostructure of the tubes), the presence of unreacted catalyst (generally, iron, nickel, or cobalt which are heavy metals), surface charge, surface chemistry, diameter, etc., all frequently change from one CNT to another.

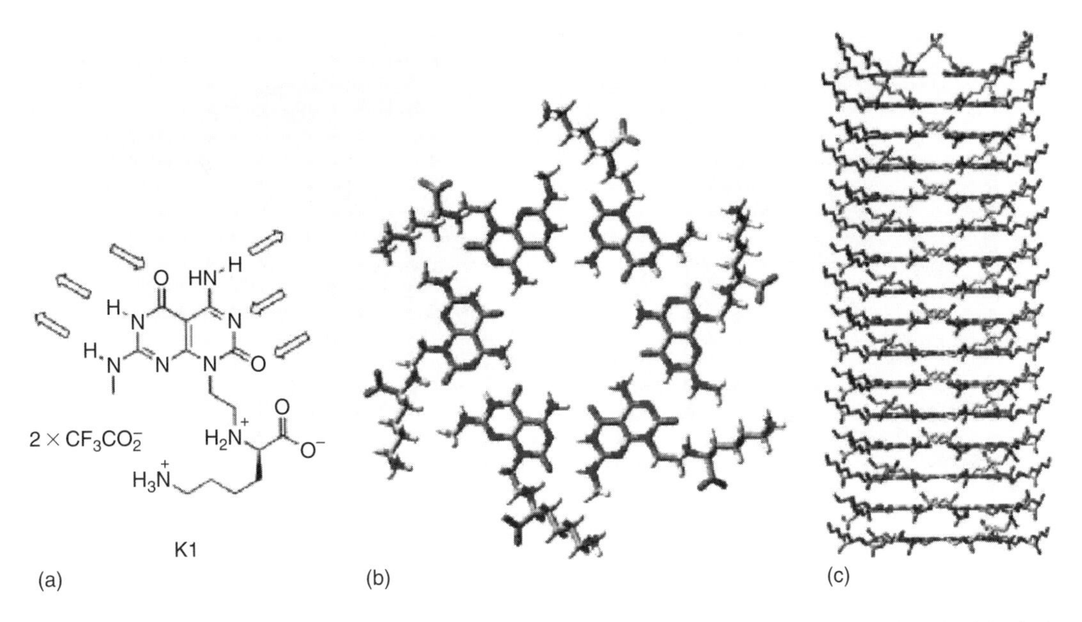

FIGURE 5.25 (a) Self-assembling module K1 undergoes spontaneous self-assembly under physiological conditions first into (b) a rosette super-macrocycle maintained by H-bonds and then into (c) a helicoidal stack of rosettes several micrometers long. (From Chun, A.L., Moralez, J.G., Webster, T.J. et al., *Biomaterials*, 26, 35, 7304, 2005. With permission.)

Biology offers inspiring models to nanotechnology. Since bone is a self-assembled collection of nanofibers [164], self-assembled nanostructures can give promising results. Helical rosette nanotubes (HRN), a self-assembled organic nanotube with dimensions that mimic the nanostructured constituent components of bone, are one of these novel formulations [165]. Figure 5.25 shows the organization of this nanotube. HRN are 3.5 nm in diameter and several microns in length. They consist of guanine and cytosine DNA base pairs which are obtained through an entropically driven self-assembly under physiological conditions (when added to water) or upon exposure to body temperatures. Titanium substrates coated with HRN showed increased osteoblast adhesion. Figure 5.26 shows titanium substrates coated with HRN [166]. The temperature driven (68°C) self-assembled HRN on titanium promotes osteoblast adhesion as much as uncoated titanium in the presence of serum proteins, showing the ability of HRN to act as a protein substitute.

Given these universal trends of increased osteoblast functions on numerous nanomaterials, it is intriguing to ask why osteoblasts behave so favorably on these novel nanophase materials. This question becomes even more interesting when asking why other cell types (such as fibroblasts) do not. This is explained in the next section.

5.3.3.3 Bone Cell Recognition on Conventional and Nanophase Materials

Understanding the reasons for promoted osteoblast behaviors on nanophase and conventional materials is the most critical step in future implant designs. In Section 5.1.3, bone cell attachment onto implant surfaces was introduced. In this section, the question "why nano?" will be investigated from the bone tissue engineering perspective.

As stated, changes in material properties manipulate protein adsorption characteristics, altering protein adsorption behavior which further influences cell adhesion onto biomaterial surfaces. When the grain size of a material is decreased into the nanoregime, some properties change such as increased electron delocalization, surface area, number of atoms at the surface, and surface defects, which are influential surface/protein interactions. Figure 5.27 depicts differences in atom positions for conventional compared to nanophase materials. The adsorption of major cell adhesive

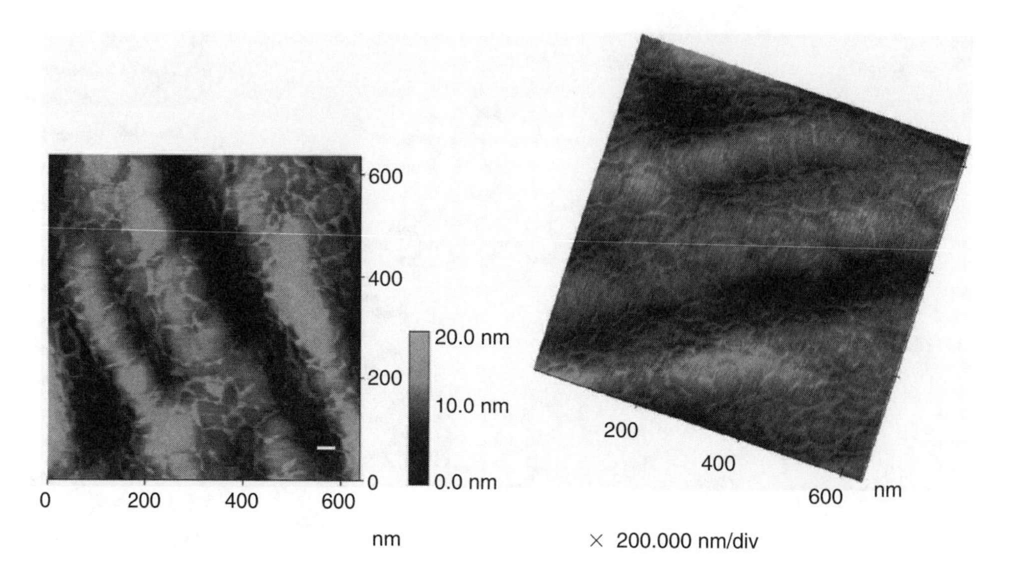

FIGURE 5.26 AFM micrographs of Ti substrates coated with 1 mg mL^{-1} HRN-K1 showing HRN formed networks on Ti surfaces. (From Chun, A.L., Moralez, J.G., Webster, T.J. et al., *Biomaterials*, 26, 35, 7304, 2005. With permission.)

proteins containing the RGD peptide sequence (vitronectin, fibronectin, collagen, and laminin) is affected due to these surface property changes and consequently adsorb onto the surfaces in a different fashion, specifically in higher amounts and in more optimal confirmations that increase the availability of cell-binding domains. When there are more bioactive RGD sequences available, more osteoblasts attach. Due to the fact that osteoblast proliferation and functions all depend on initial adhesion onto implant surfaces, the first step, osteoblast adhesion, is of high importance.

As emphasized many times here, nanophase materials increase osteoblast adhesion and proliferation. Adsorption of greater quantities of proteins, such as vitronectin, onto nanophase structures with respect to conventional materials is one of the reasons for this behavior. Research has shown that vitronectin adsorption is more than 30% greater for nanophase alumina compared to conventional alumina, correspondingly increasing cell adhesion [17]. Another reason for this behavior is decreased competitive protein adsorption (such as decreased apolipoprotein and albumin

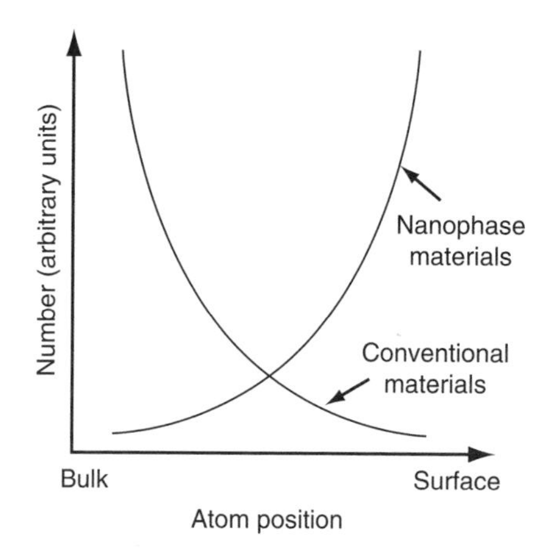

FIGURE 5.27 Graph of atom position-corresponding number of atoms. Higher number of atoms on surfaces of nanophase atoms is shown. (From Webster, T.J., *Adv. Chem. Eng.*, 27, 125, 2001. With permission.)

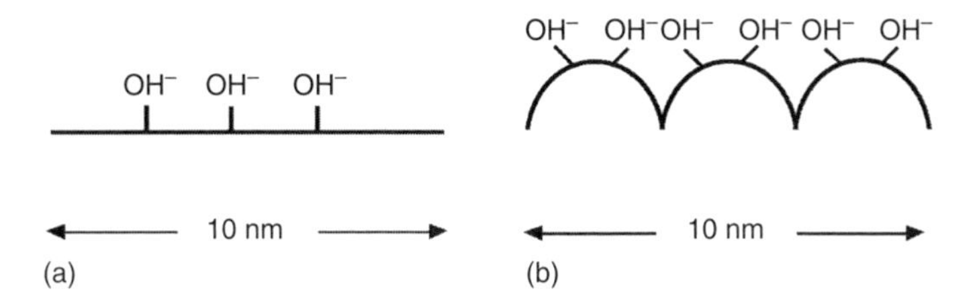

FIGURE 5.28 Schematics showing (a) conventional and (b) nanophase materials. (From Sato, M. and Webster, T.J., *Exp. Rev. Med. Dev.*, 1, 1, 105, 2004. With permission.)

adsorption on nanophase alumina [17]). When there is less competitive protein adsorption onto nanophase surfaces, there will be more available sites for proteins like vitronectin which favor osteoblast adhesion. Additionally, enhanced initial calcium adsorption onto hydrophilic surfaces is also a contributing factor, where a calcium-mediated mechanism has been proposed to be responsible for vitronectin adsorption. Most important of all, nanophase materials alter the conformation of the adsorbed proteins by promoting their unfolding. An example to this behavior is unfolding of vitronectin. Upon adsorption of vitronectin onto nanophase surfaces, the confirmation of the protein changes in such a way that cell adhesive epitopes (RGD and KRSR) are exposed, further enhancing interactions with osteoblast-membrane receptors.

In general, osteoblast functions increase on surfaces with higher nanoroughness (i.e., decreases in grain size or diameter of fibers to the nanoregime, increase in nanofiber content, etc.) [5]. One simplistic explanation for this is that the size of the proteins that mediate osteoblast adhesion is at the nanometer regime. A surface with nanotopography influences the availability of proteins cell adhesion to a greater extent than conventional surfaces which are nanosmooth. More scientifically speaking, the reason for increased osteoblast adhesion is increased electron delocalization for nanoscale materials. In the same linear space, there is more surface area for nanophase materials. This leads to greater delocalization of electrons. To illustrate this, MgO has less acidic OH^- groups on the surface due to a much higher proportion of edge sites for nanophase MgO, causing delocalization of electrons. The schematic of electron delocalization is shown in Figure 5.28. High electron delocalization in nanophase materials creates more electronegative surfaces compared to conventional materials, which increases electrostatic attractions between surfaces and proteins, drastically changing the way the surfaces interact with proteins. Increased adsorption of proteins on nanophase surfaces is due to this change in electrostatic attractions [29].

Wettability is also another parameter that changes as the material grain size decreases to nanoregime. Generally, as the grain size decreases, hydrophilicity increases. An example to this is nanophase alumina, showing three times smaller aqueous contact angles when the grain size is decreased from 167 to 24 nm [99]. Increased surface hydrophilicity increases protein adsorption, which further increases cell adhesion onto biomaterials. It has been shown that the adsorption of vitronectin, which is crucial in osteoblast adhesion, is increased on nanophase ceramics with greater wettability.

5.4 FUTURE DIRECTIONS

Although preliminary attempts to use nanoscale materials in bone tissue engineering seem encouraging, significant questions still remain unanswered. The toxicity of nanoparticles still needs to be addressed both from a commercial-scale manufacturing and usage as an implant point of view. For example, small nanoparticles can penetrate inside the human body through pores and the health effects of these particles are not yet known. Now that significantly increased bone growth has been observed on numerous nanomaterials, it is time to fully investigate their toxicity during manufacturing and/or implantation.

REFERENCES

1. Marieb, E.N. 1998. *Human Anatomy & Physiology, 4th ed.* Menlo Park, California: Benjamin/Cummings Science Publishing.
2. University of Glasgow. http://www.gla.ac.uk/ibls/fab/public/docs/xbone1x.html
3. Young, J.L., Fritz, A., Adamo, M.P. et al. U.S. National Cancer Institute's Surveillance, Epidemiology and End Results (SEER) Program, Structure of bone tissue, http://training.seer.cancer.gov/module_anatomy/unit3_2_bone_tissue.html
4. Young, J.L., Fritz, A., Adamo, M.P. et al. U.S. National Cancer Institute's Surveillance, Epidemiology and End Results (SEER) Program. http://www.web-books.com/eLibrary/Medicine/Physiology/Skeletal/Skeletal.htm
5. Price, R.L. 2004. Evaluation of nanofiber carbon and alumina for orthopedic/dental applications, PhD dissertation, Purdue University.
6. Zysset, P.K., Guo, X.E., Hoffler, C.E. et al. 1999. Elastic modulus and hardness of cortical and trabecular bone lamellae measured by nano-indentation in the human femur. *Journal of Biomechanics*, 32: 1005–1012.
7. Webster, T.J. 2001. Nanophase ceramics: The future orthopedic and dental implant material. *Advances in Chemical Engineering*, 27: 125–166.
8. Dickenson, R.P., Hutton, W.C., and Stott, J.R.R. 1981. The mechanical properties of bone in osteoporosis. *Journal of Bone and Joint Surgery. British Volume*, 63B(2): 233–238.
9. Linde, F., Norgaard, P., Hvid, I. et al. 1991. Mechanical properties of trabecular bone. Dependency on strain rate. *Journal of Biomechanics*, 24(9): 803–809.
10. Linde, F. 1994. Elastic and viscoelastic properties of trabecular bone by a compression testing approach. *Danish Medical Bulletin*, 41(2): 119–138.
11. Lakers, R.S. 1993. Materials with structural hierarchy. *Nature*, 361: 511–515.
12. Doll, B.A. 2005. *Basic Bone Biology and Tissue Engineering in Bone Tissue Engineering*, eds., Hollinger, J.O., Einhorn, T.A., Doll, B.A., Sfeir, C., pp. 3–90, CRC Press, LLC.
13. Kaplan, F.S., Hayes, W.C., Keaveny, T.M. et al. 1994. Form and function of bone. In: *Orthopaedic Basic Science*, ed., Simon, S.R., pp. 127–184. American Academy of Orthopedic Surgons, Rosemont, IL.
14. American Society for Bone and Mineral Research, Bone Curriculum, Bone Growth and Remodeling. 2007. http://depts.washington.edu/bonebio/ASBMRed/growth.html
15. Rüegsegger, P. 1996. "A microtomographic system for the nondestructive evaluation of bone architecture". *Calcified Tissue International*, 58(1):24–29.
16. Garcia, A.J., Ducheyne, P., and Boettiger, D. 1998. Effect of surface reaction stage on fibronectin-mediated adhesion of osteoblast-like cells to bioactive glass. *Journal of Biomedical Materials Research*, 40(1): 48–56.
17. Webster, T.J., Schadler, L.S., Siegel, R.W. et al. 2001. Mechanisms of enhanced osteoblast adhesion on nanophase alumina involve vitronectin. *Tissue Engineering*, 7(3): 291–301.
18. Degasne, I., Basle, M.F., Demais, V. et al. 1999. Effects of roughness, fibronectin and vitronectin on attachment, spreading, and proliferation of human osteoblast-like cells (saos-2) on titanium surfaces. *Calcified Tissue International*, 64: 499–507.
19. Siebers, M.C., Brugge, P.J., Walboomers, X.F. et al. 2005. Integrins as linker proteins between osteoblasts and bone replacing materials. A critical review. *Biomaterials*, 26(2): 137–146.
20. Balasundaram, G. and Webster, T.J. 2006. A perspective on nanophase materials for orthopedic implant applications. *Journal of Materials Chemistry*, 16: 3737–3745.
21. Christenson, E.M., Anseth, K.S., Beucken, J.J.J.P. et al. 2006. Nanobiomaterial applications in orthopedics. *Journal of Orthopaedic Research*, 25(1): 11–22.
22. Kreis, T. and Vale, R. 1999. *Guidebook to the Extracellular Matrix, Anchor, and Adhesion Proteins*, Oxford: Oxford University Press.
23. Lopes, M.A., Monteiro, F.J., Santos, J.D. et al. 1999. Hydrophobicity, surface tension, and zeta potential measurements of glass-reinforced hydroxyapatite composites. *Journal of Biomedical Materials Research*, 45(4): 370–375.
24. Howlett, C.R., Evans, D.M.M., Walsh, W.R. et al. 1994. Mechanism of initial attachment of cells derived from human bone to commonly used prosthetic materials during cell culture. *Biomaterials*, 15(3): 213–222.

25. Garcia, A.J., Ducheyne, P., and Boettigert, D. 1997. Cell adhesion strength increases linearly with adsorbed fibronectin surface density. *Tissue Engineering*, 3(2): 197–206.
26. Garcia, J.A., Duecheyne, P., and Boettiger, D. 1998. Effect of surface reaction stage on fibronectin-mediated adhesion of osteoblast-like cells to bioactive glass. *Journal of Biomedical Materials Research*, 40(1): 48–56.
27. Lehenkari, P.P. and Horton, M.A. 1999. Single integrin molecule adhesion forces in intact cells measured by atomic force microscopy. *Biochemical and Biophysical Research Communications*, 259: 645–650.
28. Sinha, R.K. and Tuan, R.S. 1996. Regulation of human osteoblast integrin expression by ortheopedic implant materials. *Bone*, 18: 451–457.
29. Webster, T.J., Ergun, C., Doremus, R.H. et al. 2000. Specific proteins mediate enhanced osteoblast adhesion on nanophase ceramics. *Journal of Biomedical Materials Research*, 51(3): 475–483.
30. Hasenbein, M.E., Andersen, T.T., and Bizios, R. 2002. Micropatterned surfaces modified with select peptides promote exclusive interactions with osteoblasts. *Biomaterials*, 23(19): 3937–3942.
31. Aota, S., Nomizu, M., and Yamada, K.M. 1994. The short amino acid sequence Pro-His-Ser-Arg-Asn in human fibronectin enhances cell-adhesive function. *Journal of Biological Chemistry*, 269: 24756–24761.
32. Rezania, A. and Healy, K.E. 1999. Biomimetic peptide surfaces that regulate adhesion, spreading, cytoskeletal organization, and mineralization of the matrix deposited by osteoblast-like cells. *Biotechnology Progress*, 15: 19–32.
33. Puleo, D.A. and Bizios, R. 1992. Mechanisms of fibronectin mediated attachment of osteoblasts to substrates in vitro. *Bone and Mineral*, 18: 215–226.
34. American Academy of Orthopedic Surgeons, Agency for Healthcare Research and Quality, Total hospitalization costs for hip replacements. http://www.aaos.org/Research/stats/Total%20Hospital%20Charges%20for%20Hip%20Replacement%20Chart.pdf
35. American Academy of Orthopedic Surgeons. Number of total hip replacement surgeries. National Hospital Discharge Survey, 1991–2004. http://www.aaos.org/Research/stats/Total%20Hip%20Replacement%20Chart.pdf
36. American Academy of Orthopedic Surgeons, Number revision hip replacement, procedures. National Hospital Discharge Survey, 1991–2004. http://www.aaos.org/Research/stats/Hip%20Revision%20Chart.pdf
37. American Academy of Orthopedic Surgeons, Facts on hip replacement. National Hospital Discharge Survey. 2004. http://www.aaos.org/Research/stats/Hip%20Facts.pdf
38. Dowson, D. 2001. New joints for the millennium: Wear control in total replacement hip joints. *Proceedings of the Institute of Mechanical Engineers. Part H: Journal of Engineering in Medicine*, 215(4): 335–358.
39. Hastings, G.W. 1980. Biomedical engineering and materials for orthopaedic implants. *Journal of Physics E: Scientific Instruments*, 13: 599–607.
40. Akay, M. and Aslan, N. 1996. Numerical and experimental stress analysis of a polymeric composite hip joint prosthesis. *Journal of Biomedical Materials Research*, 31(2): 167–182.
41. Jacob, H.A.C., Bereiter, H.H., and Buergi, M.L. 2007. Design aspects and clinical performance of the thrust plate hip prosthesis. *Proceedings of the Institution of Mechanical Engineers, Part H: Journal of Engineering in Medicine*, 221(1): 29–37.
42. Huiskes, R. and Boeklagen, R. 1989. Mathematical shape optimization of hip prosthesis design. *Journal of Biomechanics*, 22(8–9): 793–804.
43. Nunn, D., Freeman, M.A., Hill, P.F. et al. 1989. The measurement of migration of the acetabular component of hip prostheses. *Journal of Bone and Joint Surgery. British Volume*, 71B(4): 629–631.
44. Burke, D.W., O'Connor, D.O., Zalenski, E.B. et al. 1991. Micromotion of cemented and uncemented femoral components. *The Journal of Bone and Joint Surgery*, 73B(1): 33–37.
45. Charnley, J. and Kettlewell, J. 1965. The elimination of slip between prosthesis and femur. *The Journal of Bone and Joint Surgery*, 47B(1): 56–60.
46. Mjoberg, B. 1986. Loosening of the cemented hip prosthesis. The importance of heat injury. *Acta Orthopaedica Scandinavica.Supplementum*, 221: 1–40.
47. Gladius, L. 1997. Properties of acrylic bone cement: state of the art review. *Journal of Biomedical Materials Research*, 38(2): 155–182.

48. Schmalzried, T.P. and Callaghan, J.J. 1999. Current concepts review wear in total hip and knee replacements. *The Journal of Bone and Joint Surgery*, 81A(1): 115–136.
49. Dumbleton, J.H., Manley, M.T., and Edidin, A.A. 2002. A literature review of the association between wear rate and osteolysis in total hip arthroplasty. *The Journal of Arthroplasty*, 17(5): 649–661.
50. Ingham, E. and Fisher, J. 2000. Biological reactions to wear debris in total joint replacement. *Proceedings of the Institution of Mechanical Engineers, Part H: Journal of Engineering in Medicine*, 214(1): 21–37.
51. Dowson, D. 2001. New joints for the millennium: Wear control in total replacement hip joints. *Proceedings of the Institution of Mechanical Engineers*, 215(4): 335–358.
52. Dumbleton, J.H. and Mantley, M.T. 2005. Metal-on-metal total hip replacement. What does the literature say? *The Journal of Arthroplasty*, 20(2): 174–188.
53. Buscher, R., Tager, G., Dudzinski, W. et al. 2005. Subsurface microstructure of metal-on-metal hip joints and its relationship to wear particle generation. *Journal of Biomedical Materials Research Part B: Applied Biomaterials*, 72B(1): 206–214.
54. Anderson, J.M., 1993. Mechanisms of inflammation and infection with implanted devices. *Cardiovascular pathology*, 2(3):33–41.
55. Anderson, J.M., Gristina, A.G., Hanson, S.R. et al. 1996. Host reactions to biomaterials and their evaluation. In: *Biomaterials Science: An Introduction to Materials in Medicine*, eds. Ratner, B.D., Hoffman, A.S., Schoen, A.S., and Lemons, J.E., pp. 165–214, San Diego: Academic Press, Inc.
56. Webster, T.J. 2003. Improved bone tissue engineering materials. *American Ceramic Society Bulletin*, 82 (6): 23–28.
57. Roco, M.C. 2003. Nanotechnology: Convergence with modern biology and medicine. *Current Opinion in Biotechnology*, 14: 337–346.
58. Nel, A., Xia, T., Madler, L. et al. 2006. Toxic potential of materials at the nanolevel. *Science*, 311: 622–627.
59. Ferrar, M. 2005. Cancer nanotechnology: Opportunities and challenges. *Nature Reviews Cancer*, 5(3): 161–171.
60. Lu, Y. and Chen, S.C. 2004. Micro and nano-fabrication of biodegradable polymers for drug delivery. *Advanced Drug Delivery Reviews*, 56(11): 1621–1633.
61. Cohen, H., Levy, R.J., Gao, J. et al. 2000. Sustained delivery and expression of DNA encapsulated in polymeric nanoparticles. *Gene Therapy*, 7: 1896–1905.
62. Smith, L.A. and Ma, P.X. 2004. Nano-fibrous scaffolds for tissue engineering. *Colloids and Surfaces B: Biointerfaces*, 39: 125–131.
63. Kobayashi, H. and Brechbiel, M.W. 2005. Nano-sized MRI contrast agents with dendrimer cores. *Advanced Drug Delivery Reviews*, 57: 2271–2286.
64. Thalhammer, S. and Heckl, W.M. 2004. Nanotechnology and medicine. *4th IEEE Conference on Nanotechnology*, 577–579.
65. Webster, T.J. and Ahn, E.A. 2006. Nanostructured biomaterials for tissue engineering bone. *Advances in Biochemical Engineering/Biotechnology*, 103: 275–308.
66. Ioku, Y., Yoshimura, M., and Somiya, S. 1990. Microstructure and mechanical properties of hydroxyapatite ceramics with zirconia dispersion prepared by post-sintering. *Biomaterials*, 11(1): 57–61.
67. Webster, T.J., Siegel, R.W., and Bizios, R. 2001. Enhanced surface and mechanical properties of nanophase ceramics to achieve orthopaedic/dental implant efficacy. *Key Engineering Materials*, 192–195: 321–324.
68. Balania, K., Andersonb, R., Lahaa, T. et al. 2007. Plasma-sprayed carbon nanotube reinforced hydroxyapatite coatings and their interaction with human osteoblasts in vitro. *Biomaterials*, 28(4): 618–624.
69. Shi, X., Hudson, J.L., Spicer, P.P. et al. 2006. Injectable nanocomposites of single-walled carbon canotubes and biodegradable polymers for bone tissue engineering. *Biomacromolecules*, 7(7): 2237–2242.
70. Wang, X., Li, Y., Wei, J. et al. 2002. Development of biomimetic nano-hydroxyapatite/poly(hexamethylene adipamide) composites. *Biomaterials*, 23(24): 4787–4791.
71. Ahn, E.S., Gleason, N.J., and Ying, J.Y. 2005. The effect of zirconia reinforcing agents on the microstructure and mechanical properties of hydroxyapatite-based nanocomposites. *Journal of the American Ceramic Society*, 88(12): 3374–3379.
72. McManus, A.J., Doremus, R.H., Siegel, R.W. et al. 2005. Evaluation of cytocompatibility and bending modulus of nanoceramic/polymer composites. *Journal of Biomedical Materials Research Part A*, 72(1): 98–106.

73. Turell, M. and Bellare, A. 2007. A USAXS study of dispersion of barium sulfate particles in polymethylmethacrylate bone cement, http://aps.anl.gov/
74. Gomoll, A.H., Bellare, A., Fitz, W. et al. 1999. Nano-composite poly(methyl-methacrylate) bone cement, *MRS Fall Meeting*; *Nanophase and Nanocomposite Materials II*: 399–404.
75. Hu, Q., Li, B., Wang, M. et al. 2004. Preparation and characterization of biodegradable chitosan/hydroxyapatite nanocomposite rods via in situ hybridization: A potential material as internal fixation of bone fracture. *Biomaterials*, 25(5): 779–785.
76. Martínez, M.E., Medina, S., Campo, M.T.D. et al. 1998. Effect of polyethylene particles on human osteoblastic cell growth. *Biomaterials*, 19(1–3): 183–187.
77. Vermes, C., Chandrasekaran, R., Jacobs, J.J. et al. 2001. The effects of particulate wear debris, cytokines, and growth factors on the functions of MG-63 osteoblasts. *The Journal of Bone and Joint Surgery*, 83A (2): 201–211.
78. Shanbhag, A.S., Jacobs, J.J., Glant, T.T. et al. 1994. Composition and morphology of wear debris in failed uncemented total hip replacement. *Journal of Bone and Joint Surgery. British Volume*, 76B(1): 60–67.
79. Papageorgiou, I., Brown, C., Schins, R. et al. 2007. The effect of nano- and micron-sized particles of cobalt-chromium alloy on human fibroblasts in vitro. *Biomaterials*, 28(19): 2946–2958.
80. Tamura, K., Takashi, N., Akasaka, T. et al. 2004. Effects of micro/nano particle size on cell function and morphology. *Key Engineering Materials*, 254–256: 919–922.
81. Lam, C.W., James, J.T., McCluskey, R. et al. 2004. Pulmonary toxicity of single-wall carbon nanotubes in mice 7 and 90 days after intratracheal instillation. *Toxicology Science*, 77: 126–134.
82. Gutwein, L.G. and Webster, T.J. 2005. Increased viable osteoblasts density in the presence of nanophase compared to conventional alumina and titania particles. *Biomaterials*, 25(18): 4175–4183.
83. Gutwein, L.G. and Webster, T.J. 2002. Osteoblast and chondrocyte proliferation in the presence of alumina and titania nanoparticles. *Journal of Nanoparticle Research*, 4: 231–238.
84. Price, R.L., Haberstroh, K.M., and Webster, T.J. 2004. Improved osteoblast viability in the presence of smaller nanometer dimensioned carbon fibres. *Nanotechnology*, 15: 892–900.
85. Shen, M. and Horbett, T.A. 2001. The effects of surface chemistry and adsorbed proteins on monocyte/macrophage adhesion to chemically modified polystyrene surfaces. *Journal of Biomedical Materials Research*, 57(3): 336–345.
86. Liv-Synder, P., Khang, D., and Webster, T.J., Reduced responses of macrophages to nanophase alumina, Unpublished work.
87. Huang, J., Best, S.M., and Bonfield, W. 2004. In vitro assessment of the biological response to nano-sized hydroxyapatite. *Journal of Materials Science in Medicine*, 15: 441–445.
88. Rzigalinski, B.A., Danilisen, I., Strawn, E.T. et al. 2007. Nanoparticles for cell engineering—A radical concept. In: *Nanotechnologies for the Life Sciences, Tissue, Cell and Organ Engineering*, ed., Kumar, C.S.S.R., Volume 9, pp. 361–387, Wiley-VCH.
89. Jenney, C.R. and Anderson, J.M. 2000. Adsorbed serum proteins responsible for surface dependent human macrophage behavior. *Journal of Biomedical Materials Research*, 49(4): 435–447.
90. Khang, D., Webster, T.J., Unpublished work.
91. Oliveira, P.T.D. and Nanci, A. 2004. Nanotexturing of titanium-based surfaces upregulates expression of bone sialoprotein and osteopontin by culture osteogenic cells. *Biomaterials*, 25(3): 403–413.
92. Ward, B.C. and Webster, T.J. 2007. Increased functions of osteoblasts on nanophase metals. *Materials Science and Engineering C*, 27: 575–578.
93. Ward, B.C. and Webster, T.J. 2006. The effect of nanotopography on calcium and phosphorus deposition on metallic materials in vitro. *Biomaterials*, 27(16): 3064–3074.
94. Webster, T.J. and Ejiofor, J.U. 2004. Increased osteoblast adhesion on nanophase metals: Ti, Ti6Al4V, and CoCrMo. *Biomaterials*, 25(19): 4731–4739.
95. Ejiofor, J.U. and Webster, T.J. 2004. Bone cell adhesion on titanium implants with nanascale surface features. *International Journal of Powder Metallurgy*, 40(2): 43–54.
96. Balasundaram, G. and Webster, T.J. 2007. Increased osteoblast adhesion on nanograined Ti modified with KRSR. *Journal of Biomedical Research Part A*, 80(3): 602–611.
97. Perla, V. and Webster, T.J. 2005. Better osteoblast adhesion on nanoparticulate selenium—A promising orthopedic implant material. *Journal of Biomedical Materials Research Part A*, 75A(2): 356–364.

98. Webster, T.J., Siegel, R.W., and Bizios, R. 1998. An in vitro evaluation of nanophase alumina for orthopaedic/dental applications, *Bioceramics 11: Proceedings of the 11th International Symposium on Ceramics in Medicine*, eds. LeGeros, R.Z., LeGeros, J.P., pp. 273–276, New York: World Scientific.
99. Webster, T.J., Siegel, R.W., and Bizios, R. 1999. Osteoblast adhesion on nanophase ceramics. *Biomaterials*, 20(13): 1221–1227.
100. Webster, T.J., Ergun, C., Doremus, R.H. et al. 2000. Enhanced functions of osteoblasts on nanophase ceramics. *Biomaterials*, 21(17): 1803–1810.
101. Webster, T.J., Siegel, R.W., and Bizios, R. 1999. Design and evaluation of nanophase alumina for orthopaedic/dental applications. *NanoStructured Materials*, 12: 983–986.
102. Webster, T.J., Siegel, R.W., and Bizios, R. 2001. Nanoceramic surface roughness enhances osteoblast and osteoclast functions for improved orthopaedic/dental implant efficacy. *Scripta Materialia*, 44: 1639–1642.
103. Price, R.L., Gutwein, L.G., Kaledin, L. et al. 2003. Osteoblast function on nanophase alumina materials: Influence of chemistry, phase, and topography. *Journal of Biomedical Materials Research*, 67: 1284–1293.
104. Webster, T.J., Hellenmeyer, E.L., and Price, R.L. 2005. Increased osteoblast functions on theta + delta nanofiber alumina. *Biomaterials*, 26(9): 953–960.
105. Palin, E., Liu, H., and Webster, T.J. 2005. Mimicking the nanofeatures of bone increases bone-forming cell adhesion and proliferation. *Nanotechnology*, 16: 1828–1835.
106. Swan, E.E.L., Popat, K.C., Grimes, C.A. et al. 2005. Fabrication and evaluation of nanoporous alumina membranes for osteoblast culture. *Journal of Biomedical Materials Research Part A*, 72A (3): 288–295.
107. Yao, C., Perla, V., McKenzie, J. et al. 2005. Anodized Ti and Ti6Al4V possessing nanometer surface features enhances osteoblast adhesion. *Journal of Biomedical Nanotechnology*, 1(1): 68–73.
108. Karlsson, M., Pålsgård, E., Wilshaw, P.R. et al. 2003. Initial in vitro interaction of osteoblasts with nanoporous alumina. *Biomaterials*, 24(18): 3039–3046.
109. Yao, C. and Webster, T.J. 2006. Anodization: A promising nano-modification technique of titanium implants for orthopedic applications. *Journal of Nanoscience and Nanotechnology*, 6(9–10): 2682–2692.
110. Price, R.L., Haberstroh, K.M., and Webster, T.J. 2003. Enhanced functions of osteoblasts on nanostructured surfaces of carbon and alumina. *Medical and Biological Engineering and Computing*, 41: 372–375.
111. Price, R.L., Gutwein, L.G., Kaledin, L. et al. 2003. Osteoblast function on nanophase alumina materials: Influence of chemistry, phase and topography. *Journal of Biomedical Materials Research*, 67(4): 1284–1293.
112. Balasundaram, G., Sato, M., and Webster, T.J. 2006. Using hydroxyapatite nanoparticles and decreased crystallinity to promote osteoblast adhesion similar to functionalizing with RGD. *Biomaterials*, 27(14): 2798–2805.
113. Sato, M., Sambito, M.A., Aslani, A. et al. 2006. Increased osteoblast functions on undoped and yttrium-doped nanocrystalline hydroxyapatite coatings on titanium. *Biomaterials*, 27(11): 2358–2369.
114. Balasundaram, G. and Webster, T.J. 2006. Hydroxyapatite coated magnetic nanoparticles for the treatment of osteoporosis. Presented at the BMES Meeting, Chicago, IL.
115. Lee, S.H. and Kim, H.E. 2007. Nano-sized hydroxyapatite coatings on Ti substrate with TiO2 buffer layer by e-beam deposition. *Journal of the American Ceramic Society*, 90(1): 50–56.
116. Wang, Z.C., Chen, F., Huang, L.M. et al. 2005. Electrophoretic deposition and characterization of nano-sized hydroxyapatite particles. *Journal of Materials Science*, 40(18): 4955–4957.
117. Li, P., 2003. Biomimetic nano-apatite coating capable of promoting bone ingrowth. *Journal of Biomedical Materials Research Part A*, 66A(1): 79–85.
118. Sato, M., An, Y.H., Slamovich, E.B. et al. 2008. Increased osteointegration for tantalum scaffolds coated with nanophase compared to conventional hydroxyapatite. *International Journal of Nanomedicine*, 2008.
119. Webster, T.J., Siegel, R.W., and Bizios, R. 2001. Nanoceramic surface roughness enhances osteoblast and osteoclast functions for improved orthopaedic/dental implant efficacy. *Scripta Materialia*, 44: 1639–1642.
120. Webster, T.J., Ergun, C., Doremus, R.H. et al. 2001. Enhanced osteoclast-like cell functions on nanophase ceramics. *Biomaterials*, 22(11): 1327–1333.

121. Colon, G., Ward, B.C., and Webster, T.J. 2006. Increased osteoblast and decreased *Staphylococcus epidermidis* functions on nanophase ZnO and TiO2. *Journal of Biomedical Materials Research Part A*, 78A(3): 595–604.
122. Norman, J.J. and Desai, T.A. 2006. Methods for fabrication of nanoscale topography for tissue engineering scaffolds. *Annuals of Biomedical Engineering*, 34(1): 89–101.
123. Pattison, M.A., Wurster, S., Webster, T.J. et al. 2005. Three-dimensional, nano-structured PLGA scaffolds for bladder tissue replacement applications. *Biomaterials*, 26(15): 2491–2500.
124. Kay, S., Thapa, A., Haberstroh, K.M. et al. 2002. Nanostructured polymer/nanophase ceramic composites enhance osteoblast and chondrocyte adhesion. *Tissue Engineering*, 8(5): 753–761.
125. Miller, D.C., Thapa, A., Haberstroh, K.M. et al. 2004. Endothelial and vascular smooth muscle cell function on poly(lactic-co-glycolic acid) with nano-structured surface features. *Biomaterials*, 25(1): 53–61.
126. Park, G.E., Pattison, M.A., Park, K. et al. 2005. Accelerated chondrocyte functions on NaOH-treated PLGA scaffolds. *Biomaterials*, 26(16): 3075–3082.
127. Thapa, A., Miller, D.C., Webster, T.J. et al. 2003. Nano-structured polymers enhance bladder smooth muscle cell function. *Biomaterials*, 24(17): 2915–2926.
128. Miller, D.C., Thapa, A., Haberstroh, K.M. et al. 2002. Enhanced functions of cells on polymers with nanostructures surfaces. *Proceedings of the Second Joint EMBS/BMES Conference*, Houston, TX, 1: 755–756.
129. Wei, G., Jin, Q., Giannobile, W.V. et al. 2007. The enhancement of osteogenesis by nano-fibrous scaffolds incorporating rhBMP-7 nanospheres. *Biomaterials*, 28(12): 2087–2096.
130. Pham, Q.P., Sharma, U., and Mikos, A.G. 2006. Electrospinning of polymeric nanofibers for tissue engineering applications: A review. *Tissue Engineering*, 12(5): 1197–1211.
131. Tuzlakogu, K., Bolgen, N., Salgado, A.J. et al. 2005. Nano- and micro-fiber combined scaffolds: A new architecture for bone tissue engineering. *Journal of Materials Science: Materials in Medicine*, 16(12): 1099–1104.
132. Li, W.J., Tuli, R., Huang, X. et al. 2005. Multilineage differentiation of human mesenchymal stem cells in a three-dimensional nanofibrous scaffold. *Biomaterials*, 26(25): 5158–5166.
133. Shin, M., Yoshimoto, H., and Vacanti, J.P. 2004. In vivo bone tissue engineering using mesenchymal stem cells on a novel electrospun nanofibrous scaffold. *Tissue Engineering*, 10(1–2): 33–41.
134. Webster, T.J. and Smith, T.A. 2005. Increased osteoblast functions on PLGA composites containing nanophase titania. *Journal of Biomedical Materials Research Part A*, 74A(4): 677–686.
135. Kay, S., Thapa, A., Haberstroh, K.M. et al. 2002. Nanostructured polymer/nanophase ceramic composites enhance osteoblast and chondrocyte adhesion. *Tissue Engineering*, 8(5): 753–761.
136. Liu, H., Slamovich, E.B., and Webster, T.J. 2005. Increased osteoblast functions on (poly-lactic-co-glycolic acid) with highly dispersed nanophase titania. *Journal of Biomedical Nanotechnology*, 1(1): 83–89.
137. Liu, H., Slamovich, E.B., and Webster, T.J. 2005. Increased osteoblast functions on nanophase titania dispersed in poly-lactic-co-glycolic acid composites. *Nanotechnology*, 16: 601–608.
138. Liu, H., Slamovich, E.B., and Webster, T.J. 2006. Increased osteoblast functions among nanophase titania/poly(lactide-co-glycolide) composites of the highest nanometer surface roughness. *Journal of Biomedical Materials Research Part A*, 78A(4): 798–807.
139. Webster, T.J., Waid, M.C., McKenzie, J.L. et al. 2004. Nano-biotechnology: Carbon nanofibres as improved neural and orthopaedic implants. *Nanotechnology*, 15: 48–54.
140. Price, R.L., Haberstroh, K.M., and Webster, T.J. 2002. Increased adhesion on carbon nanofibers/polymer composite materials, *Proceedings of the Second Joint EMBS/BMES Conference*, Houston, TX, 625–626.
141. Price, R.L., Waid, M.C., Haberstroh, K.M. et al. 2003. Selective bone cell adhesion on formulations containing carbon nanofibers. *Biomaterials*, 24(11): 1877–1887.
142. McKenzie, J.L., Waid, M.C., Shi, R. et al. 2004. Decreased functions of astrocyte on carbon nanofiber materials. *Biomaterials*, 25(7–8): 1309–1317.
143. Murugan, R. and Ramakrishna, S. 2005. Development of nanocomposites for bone grafting. *Composites Science and Technology*, 65(15–16): 2385–2406.
144. Du, C., Cui, F.Z., Zhu, X.D. et al. 1999. Three-dimensional nano-HAp/collagen matrix loading with osteogenic cells in organ culture. *Journal of Biomedical Materials Research*, 44(4): 407–415.

145. Du, C., Cui, F.Z., Feng, Q.L. et al. 1998. Tissue response to nano-hydroxyapatite/collagen composite implants in marrow cavity. *Journal of Biomedical Materials Research*, 42(4): 540–548.
146. Liao, S.S., Cui, F.Z., Zhang, W. et al. 2004. Hierarchically biomimetic bone scaffold materials: Nano-HA/collagen/PLA composite. *Journal of Biomedical Materials Research Part B: Applied Biomaterials*, 69B(2): 158–165.
147. Liao, S.S., Cui, F.Z., and Zhu, Y. 2004. Osteoblasts adherence and migration through three dimensional porous mineralized collagen based composite: nHAC/PLA. *Journal of Bioactive and Compatible Polymers*, 19(2): 117–130.
148. Wei, G. and Ma, P.X. 2004. Structure and properties of nano-hydroxyapatite/polymer composite scaffolds for bone tissue engineering. *Biomaterials*, 25(19): 4749–4757.
149. Piconi, C. and Maccauro, G. 1999. Zirconia as a ceramic biomaterial. *Biomaterials*, 20(1): 1–25.
150. Evis, Z., Sato, M., and Webster, T.J. 2006. Increased osteoblast adhesion on nanograined hydroxyapatite and partially stabilized zirconia composites. *Journal of Biomedical Materials Research Part A*, 78A(3): 500–507.
151. Kong, Y.M.K., Bae, C.J., Lee, S.H. et al. 2005. Improvement in biocompatibility of ZrO_2–Al_2O_3 nano-composite by addition of HA. *Biomaterials*, 26(5): 509–517.
152. Elias, K.L., Price, R.L., and Webster, T.J. 2002. Enhanced functions of osteoblasts on nanometer diameter carbon fibers. *Biomaterials*, 23(15): 3279–3287.
153. Khang, D., Sato, M., Price, R.L. et al. 2006. Selective adhesion and mineral deposition by osteoblasts on carbon nanofibers patterns. *International Journal of Nanomedicine*, 1(1): 65–72.
154. Khang, D., Sato, M., and Webster, T.J. 2005. Directed osteoblast functions on micro-aligned patterns of carbon nanofibers on a polymer matrix. *Advanced Materials Science*, 3(10): 205–208.
155. Aoki, N., Yokoyama, A., Nodasaka, Y. et al. 2006. Strikingly extended morphology of cells grown on carbon nanotubes. *Chemistry Letters*, 25(5): 508–509.
156. Zanello, L.P. 2006. Electrical properties of osteoblasts cultured on carbon nanotubes. *Micro & Nano Letters*, 1(1): 19–22.
157. Tsuchiya, N., Sato, Y., Aoki, N. et al. 2007. Evaluation of multi-walled carbon nanotube scaffolds for osteoblast growth. *AIP Conference Proceedings*, 4th International Workshop on Water Dynamics, 898: 166–169.
158. George, J.H., Shaffer, M.S., and Stevens, M.M. 2006. Investigating the cellular response to nanofibrous materials by use of a multi-walled carbon nanotube model. *Journal of Experimental Nanoscience*, 1(1): 1–12.
159. Zanello, L.P., Zhao, B., Hu, H. et al. 2006. Bone cell proliferation on carbon nanotubes. *Nanoletters*, 6 (3): 562–567.
160. Price, R.L., Ellison, K., Haberstroh, K.M. et al. 2004. Nanometer surface roughness increases select osteoblast adhesion on carbon nanofibers compacts. *Journal of Biomedical Materials Research Part A*, 70A(1): 129–138.
161. Sato, M. and Webster, T.J. 2004. Nanobiotechnology: Implications for the future of nanotechnology in orthopedic applications. *Expert Review of Medical Devices*, 1(1): 105–114.
162. Wiesmann, H.P., Hartig, M., Stratmann, U. et al. 2001. Electrical stimulation influences mineral formation of osteoblast-like cells in vitro. *Biochimica et Biophysica Acta—Molecular Cell Research*, 1538(1): 28–37.
163. Supronowicz, P.R., Ajayan, P.M., Ullmann, K.R. et al. 2001. Novel current-conducting composite substrates for exposing osteoblasts to alternating current stimulation. *Journal of Biomedical Materials Research*, 59(3): 499–506.
164. Thomas, J.B., Peppas, N.A., Sato, M. et al. 2006. Nanotechnology and biomaterials. In: *CRC Nanomaterials Handbook*, ed. Gogotsi, Y., pp. 605–636, Boca Raton, CRC Press.
165. Chun, A.L., Moralez, J.G., Webster, T.J. et al. 2005. Helical rosette nanotubes: A biomimetic coating for orthopedics? *Biomaterials*, 26(35): 7304–7309.
166. Chun, A.L., Moralez, J.G., Fenniri, H. et al. 2004. Helical rosette nanotubes: A more effective orthopaedic implant material. *Nanotechnology*, 15: 234–239.

6 Surface Modification of Biomaterials at the Nanoscale: Biomimetic Scaffolds for Tissue Engineering Applications

Duron A. Lee and Cato T. Laurencin

CONTENTS

6.1 INTRODUCTION

A plethora of synthetic and natural biomaterials containing different properties are available for use in various clinical and biomedical applications [1]. Biomaterials have been used in many forms as prostheses, implant materials, matrices for guided tissue engineering, biosensors, microfluidic devices, and scaffolds, just to name a few [2]. Biomaterials often serve as substrates onto which cell populations can attach and migrate, function as cell delivery vehicles when combined and implanted with specific cell types, and used as drug carriers to promote specific cellular function in a localized region [3,4].

Most biomaterials used currently have suitable properties for the applications in which they are used. They are nontoxic, possess structural and mechanical stability, have favorable degradation properties, and are adequately biocompatible. While current biomaterials serve a valuable role in short-term biomedical applications, long-term and permanent implantation may pose significant safety, biomechanical, and inflammatory issues. The major challenge facing the development of the next generation of biomaterials is the need for more complex biological functionality and improved biocompatibility. In an attempt to accomplish these tasks, scientists and engineers have looked toward biology for ideas. Biological mimicry (or biomimetics) attempts to model biological properties or processes that naturally occur in the body toward developing biologically active materials with enhanced properties. Biomolecular signals can be incorporated directly onto a material surface or within the bulk material through the incorporation of biologically functional molecules derived from the biological microenvironment. Thus, a diverse range of biological functions can be built into materials through the incorporation of specific biomolecules including molecular ligands for engagement with cell surface receptors, enzymes to catalyze reactions, protein growth factors for site-specific delivery, or genetic material for cell transfections.

Biomimetic materials play a special role in tissue engineering strategies. This multidisciplinary approach to fabricating tissues in the laboratory uses a combination of engineered biomaterial scaffolds, cells, and biologically active molecules to accomplish this end. The biomaterial is crucial in the engineering of viable tissues for implantation and as such, recent attention has been focused on the development of biomimetic scaffolds for tissue engineering. The goal is to modify biomaterial surfaces to selectively interact with cell receptors through specific biomolecular recognition events to elicit specific biological or cellular functions (e.g., the promotion of normal wound healing or tissue regeneration) [5]. Engineering cell and tissue behavior at biomaterial surfaces at the nanoscale can be accomplished by chemically modifying surfaces with biologically active macromolecules (e.g., extracellular matrix-derived peptides) or physically modifying the surface topography at the nanoscale. Both parameters play an important role in the interaction between cells and a biomaterial surface [6].

We present some of the recent developments in biomimetic scaffold development by exploring different approaches to modify the surfaces of biomaterials. We highlight various methods to chemically and physically immobilize cell recognition motifs to material surfaces in an attempt to obtain controlled interactions between cells and synthetic substrates [1,2,7–13].

6.2 EXTRACELLULAR MATRIX

The extracellular matrix (ECM) is a complex structural entity surrounding and supporting cells. It is commonly referred to as the noncellular component of tissues. The ECM is composed of three major classes of biomolecules: (1) structural fibrillar proteins (e.g., collagen and elastin); (2) an amorphous interfibrillar matrix composed mainly of proteoglycans. These molecules form the complex, high molecular weight components of the ECM and are composed of a protein core to which long chains of repeating disaccharide units, termed of glycosaminoglycans (GAGs), are attached; and (3) specialized proteins (e.g., fibrillin, fibronectin, and laminin). ECMs are generally specialized for a particular function. The matrix may be mineralized to resist compression (e.g., bone) or dominated by tension resisting fibers (e.g., tendons).

The composition of the ECM proteins can have profound direct effects on cellular behavior as many of these protein molecules are highly tissue and cell specific. The effects of ECM molecules on cells are significant in conveying mechanical and chemical stimuli to influence cellular shape, spreading and growth; organization of the actin cytoskeleton; transcription activity and maintenance of cell differentiation [14–18]. The synergistic interaction between these matrix molecules and growth factors has also been shown to regulate cell behavior. Growth factors bind to the ECM through glycosaminoglycan side chains or protein cores, which increases growth factor stability and provides an appropriate cellular microenvironment for the regulation of cell proliferation and differentiation.

The importance of the ECM cannot be emphasized enough. The nanoscale interaction between cells and the ECM is essential for various basic biological systems [19–23]. Integration of components of the ECM with cell surface receptors and the consequent transduction of signals across the cell membrane serve to alter cell behavior and developmental fate [24,25]. Thus, disruptions of ECM can lead to uncontrolled cell growth and the developmental destruction of ECM during development can lead to tissue death.

6.3 CELLULAR ADHESION TO IMPLANTED BIOMATERIALS

Proteins make up the ECM of all native tissues and are present in all bodily fluids. While many are present in soluble form, others are present within the highly organized, cross-linked ECM of tissues [26]. Cells have a tremendous capacity to remodel their ECM environment [27] by removing proteins by proteolysis, adding proteins by synthesis and secretion, and cross-linking proteins by membrane-associated enzymes [28]. When biomaterials are implanted in vivo [29], nonspecific soluble and ECM protein absorption takes place immediately onto the surface [5]. A host of long-chain proteins arrange on the biomaterial surface in various orientations and three-dimensional (3-D) conformations ranging from native forms to denatured.

Cells interact indirectly with the biomaterial surface through the adsorbed layer of ECM proteins [30] and binding is mediated by interactions with cell surface receptors. Among them, the integrin superfamily is an important class of cell surface receptors that plays an important role in cellular adhesion to biomaterial surfaces. In addition to their major role as anchoring molecules, integrins are also involved in important cellular processes such as tissue organization, cell migration during embryogenesis, cellular differentiation, and in immune and inflammatory responses [31–33]. Integrins are heterodimers composed of two transmembrane subunits: a large α subunit containing several cation-binding sites and a smaller β subunit assembled together noncovalently in the presence of extracellular Ca^{2+}. There are several subfamilies of integrins, broadly defined by their β-subunit. With over 18 α and 8 β subunits, integrin molecules can assemble to form at least 24 different integrin heterodimers [34]. The composition of the αβ heterodimer also determines the specificity of the integrin toward various ECM protein ligands. Some of the combinations are highly specific, while others demonstrate promiscuity toward multiple ligands. Conversely, many ECM proteins serve as ligands for several integrin molecules (Figure 6.1) [35].

Integrins are involved in the cell anchoring process as well as the transduction of intracellular signals across the cell membrane [36,37] and involve several conformational changes of the integrin

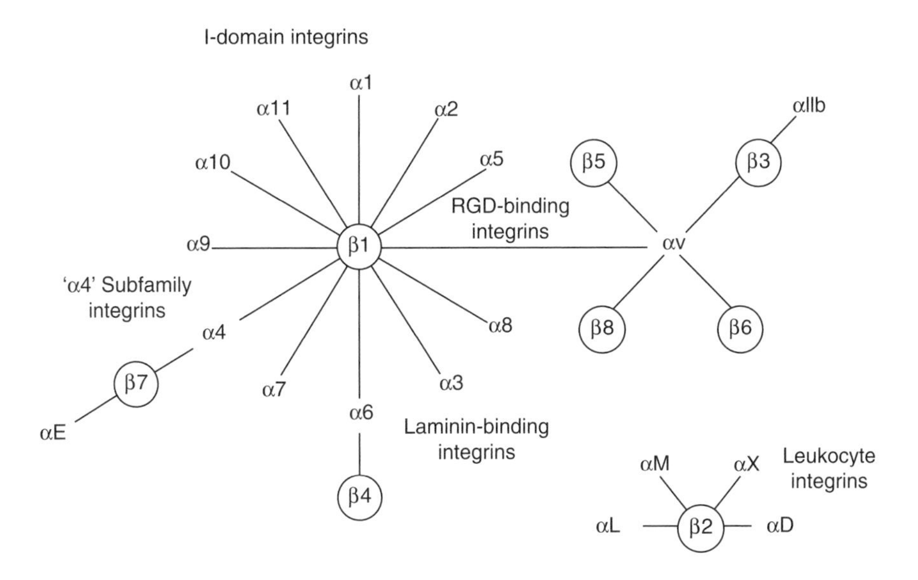

FIGURE 6.1 Integrin superfamily.

molecules [38]. A cascade of partially overlapping events characterizes integrin-mediated cell adhesion [39]. First, cells initially attach to ECM proteins through limited ligand binding that allows cells to withstand low shear forces. The cell cytoplasm then begins to spread across the substratum as cells flatten in morphology. Cytoskeletal proteins such as F-actin organize into microfilamentous bundles known as stress fibers. Lastly, clustered integrins in combination with other cytosolic, transmembrane, and membrane-associated molecules, form focal adhesion contacts between the cell and substratum [40–43]. The binding of the β-subunit to F-actin is mediated by actin-binding proteins (e.g., vinculin, talin, tension, α-actinin, paxillin), which serve to link ECM proteins to the actin cytoskeleton of cells. The cytosolic domain of the integrin β-subunit also participates in signal transduction by associating with a complex of signaling molecules comprised of kinases, phosphatases, and other adaptor proteins. Specifically, focal adhesion kinase is targeted to focal adhesions, where it associates with the cytoskeleton and is activated by autophosphorylation. Downstream signaling events occur as a result, leading to the activation of extracellular signal regulated kinases (ERK), mitogen-activated protein kinases (MAPK), and GTPases [36,44,45]. The end result translates into the regulation of differentiation, cytoskeletal organization, and other aspects of cellular behavior [46].

6.4 BIOMIMETIC SCAFFOLDS FOR TISSUE ENGINEERING

When native tissues or organs are severely damaged or injured, the ideal treatment is the replacement of the involved tissues with structurally and biologically functional ones. Current treatment strategies, including reconstruction surgeries and tissue and organ transplantation, can often result in inadequate and temporary results. In an attempt to address the limited alternatives currently used to replace tissue and organ structure and function loss, while finding feasible alternatives to partial and whole organ transplantation, new tissue strategies have been developed [47–50]. The term tissue engineering was first introduced at a National Science Foundation meeting in 1987 and was later defined as "an interdisciplinary field that applies the principles and methods of engineering and life sciences toward the fundamental understanding... and development of biological substitutes to restore, maintain or improve [human] tissue functions or a whole organ" [51–56]. Adding to the definition of tissue engineering is the "understanding the principles of tissue growth, and applying this to produce functional replacement tissue for clinical use" [57]. Laurencin et al. have further described tissue engineering as "the combination of biological, chemical, and engineering principles toward the repair, restoration, and replacement of tissues, using cells, scaffolds, and biologic factors alone or in combination" [54].

The underlying principle of tissue engineering is the guided application and control of cells, materials, and the microenvironment into which they are delivered [58]. In a tissue engineering approach, tissues or organs may be created in vivo, in vitro, or ex vivo and implanted into the patient [59]. For this to occur, there are generally three components employed in a tissue engineering approach to fabricating biological tissues: (1) viable, responsive cells; (2) a scaffold to support tissue formation; and (3) a growth-inducing stimulus. The site of tissue reconstruction must first contain a sufficient number of viable progenitor cells capable of producing the tissue of interest. If the number of cells is deficient, one must engineer the site to provide the necessary population of cells or provide some stimulus to recruit the necessary population of cells to the reconstruction site [60]. Secondly, the reconstruction site should be filled with a 3-D, porous scaffold that facilitates the attachment, migration, and proliferation of cells. It should also allow for the synthesis of ECM proteins and the ingrowth of tissue throughout the site. Thirdly, the cells in the graft site must receive the appropriate stimuli in the form of soluble, bioactive molecules (e.g., growth factors, cytokines, hormones) that will influence the cell phenotype and enhance the end result of tissue formation at the graft site [61,62].

Culturing isolated cells on a biomaterial scaffold, to be later transplanted into the target tissue for its regeneration [63], is a fundamental concept of tissue engineering. As such, the biomaterial plays an important role in most tissue engineering strategies [64]. Therefore, the importance of selecting the appropriate biomaterials that can serve as scaffolds for cellular attachment, proliferation, and differentiation has been emphasized in much of tissue engineering research [19]. There are a myriad of naturally occurring and synthetic materials that are currently used in clinical and biomedical applications, including metals, ceramics, polymers, bioglasses, and composite materials. Many biomaterials have been fabricated, synthesized, or processed into a variety of shapes and forms for specific applications that require them to interface with biological systems. While numerous synthetic materials are available for tissue engineering, one recurring problem that exists with implanted biomaterials is inadequate interaction between cells and the substrate. Poor cell–substrate interaction can lead to a number of problems including: in vivo fibrosis and implant encapsulation, foreign body reaction, inflammation, aseptic loosening, and local tissue resorption [65].

In an attempt to circumvent these problems, the development of biomaterials for tissue engineering applications has recently focused on the design of biomimetic materials. The design of biomimetic materials is an attempt to mimic the biological activity of the native extracellular microenvironment and create materials that are capable of eliciting or modulating specific biological responses and directing new tissue formation in vivo [66]. The ideal tissue engineering scaffold should therefore be one that closely mimics the natural environment of the ECM, both structurally and functionally. This biomimicry can be mediated by specific biomolecular recognition, which can be manipulated by altering the design parameters of the material itself [67–69], instead of being elicited by nonspecifically adsorbed ECM proteins.

The first step in the design of biomimetic materials is the introduction of a biomolecular component to the material. A wide variety of biological functions can be incorporated into materials including: cell recognition motifs to serve as ligands for engagement with cell surface receptors; enzymes to catalyze biological reactions; active biomolecules (e.g., growth factors, hormones, antibodies, pharmaceuticals) for site-specific delivery; and DNA and plasmid vectors for gene delivery and cell transfections [70,71]. Once incorporated into biomaterials, these bioactive molecules can directly influence a host of biological functions and cellular responses, such as cell adhesion, proliferation, and differentiation [72,73].

6.4.1 Proteins versus Peptides

Some of the earliest work on developing biomimetic surfaces involved the use of proteins naturally present in the ECM as coatings on material surfaces, such as fibronectin, vitronectin, collagen, and laminin [74]. The use of long-chain, ECM-derived proteins has several limitations, however. Proteins, in general, are large unstable molecules whose 3D conformation is held together by weak, noncovalent forces. Many are isolated from animal sources and although they are purified and processed before use, they may still elicit immunological responses and increase the risk of associated infections [75–78]. Upon substrate immobilization, large proteins with 3-D conformations will only have part of their structure in the proper orientation for cellular interaction due to their stochastic orientation on the material surface [79]. Long-chain proteins can also become randomly folded upon substrate adsorption such that cell receptor binding domains are not sterically available for proper cellular interaction. Material surface topography and chemistry (e.g., charge and hydrophobicity) can influence the orientation and conformation of the attached or adsorbed protein [80–82], thereby affecting its functionality. Improper protein orientation and conformational changes induced as a result from substrate immobilization can further lead to unwanted allergic reactions as sequestered antigenic epitopes present within the protein become available to elicit host inflammatory reactions. The native host environment can be another barrier in the development and the long-term survival of biomimetic materials using long-chain proteins. Protein loss can occur due to insolubility or desorption from the material surface [83] and subsequently eliminated from the body. Adsorbed proteins can also be destroyed by naturally occurring enzymes making long-term applications difficult. Inflammation and infection can further accelerate protein degradation. In addition to the overall low stability of these long-chain proteins and host inflammatory defenses, the protein immobilization process and subsequent storage could also impact their function.

The use of short peptide sequences as bioactive molecules offers many advantages. By presenting materials with cell recognition motifs as small, immobilized peptides, many of the shortfalls with using long-chain proteins as active biomolecules can be remedied. Shorter oligopeptide fragments are more cost-effective and highly stable during sterilization methods, heat treatment, pH changes, and storage [84–86]. The lack of 3-D conformation eliminates the concerns of protein denaturing, unsequestered antigenic epitopes, and unavailable cell binding sites. Compared to ECM proteins, which have many different cell recognition motifs, oligopeptides can be synthesized such that they contain a single motif. Thus, they can be made to selectively interact with one type of cell or adhesion receptor [1,87]. Additionally, because these small oligopeptide fragments have lower spatial requirements, they can be immobilized onto material surfaces with a higher surface density than larger proteins.

6.4.2 RGD-Derived Oligopeptide Fragments

Integrins role in cellular adhesion and signal transduction appears to be very important in regulating behavior within tissues indicate that cell adhesive proteins from the EMC and their interaction with integrins may thus be an important element in engineering biological tissues [88]. Further exploration of integrin binding to ECM proteins leads to the discovery of small signaling domains composed of several amino acid residues found within the primary structure of long-chain ECM proteins. This discovery transformed the approach to developing biomimetic scaffolds for tissue engineering. These oligopeptide sequences were found to react specifically with cell membrane receptors to elicit similar cellular effects as native ECM proteins. What resulted was the use of short ECM-derived, oligopeptide fragments for several biomimetic surface modification studies.

The most commonly used oligopeptide for biomimetic surface modification is the arginine–glycine–aspartic acid (RGD) tripeptide motif, which is the cell signaling domain derived from fibronectin and laminin (Figure 6.2). These adhesion-promoting, low-molecular weight oligopeptides are based on the primary structure of the cell receptor binding domains present within ECM proteins and have been shown to significantly modulate cellular behavior [89,90]. Several other RGD-containing sequences have been found in other ECM proteins. Todate, there are numerous short-chain, linear adhesive sequence motifs have been identified as bioactive, biomimetic molecules capable of promote cellular migration, adhesion, and proliferation.

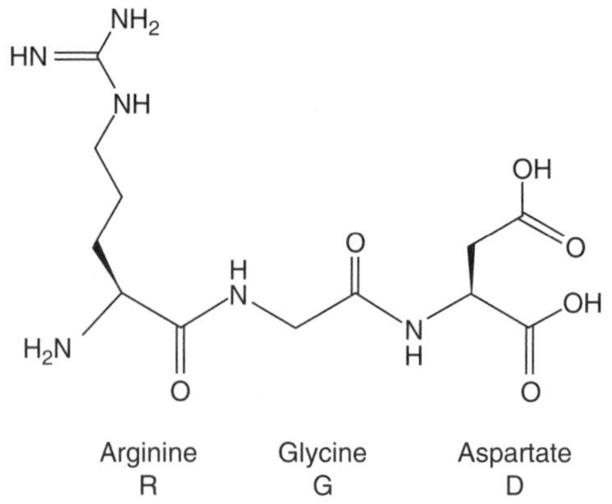

FIGURE 6.2 Structure of the RGD oligopeptide sequence.

RGD-containing oligopeptide molecules have quickly gained popularity as potential candidates for developing biomimetic scaffolds. Since these RGD adhesion motifs are found in multiple ECM proteins, a broad range of cell types is able to interact with these peptide sequences. The selective use of linear or cyclized RGD peptide sequences to endow bioactivity to materials has been demonstrated in several cell adhesion studies [30]. RGDs have been immobilized onto the surfaces of numerous biomaterial substrates (e.g., synthetic polymers, glasses, ceramics, silicon, and quartz [91–93], utilizing various peptide immobilization techniques and different cell types. The biological activity of these peptide-modified, biomimetic surfaces was evaluated and found to not only support the attachment of various cell types but also displayed receptor specificity, binding affinity and cellular response signaling similar to that elicited by the native protein [68,93]. Cellular behavior on RGD-modified surfaces also depends on the structure and conformation of RGD oligopeptide, as well as the density and arrangement of RGDs on the substrate surface. Some of the most commonly used RGD-containing oligopeptides for biomimetic materials include: RGD, RGDS, GRGD, YRGDS, and GRGDSP peptides [1]. Other cell specific, non-RGD derived oligopeptide sequences, such as YIGSR and IKVAV, VPGIG and FHRRIKA, derived from laminin, elastin, and heparin-binding domains, respectively, have also been immobilized onto various material substrates (Table 6.1) [94–98].

TABLE 6.1
Extracellular Membrane (ECM)-Derived Synthetic Oligopeptide Sequences Used for the Modification of Various Biomaterial Surfaces

ECM Molecule	Amino Acid Sequences	Materials	References
Fibronectin, vitronectin	RGD, RGDS, RGDT, RGDV, GRGD, GRGDS, GRGDY, GRGDF, PHSRN, PRRARV, YRGDG, YGRGD, GRGDSP, GRGDSY, GRGDVY, GRGDSPK, CGRGDSY, RGE, LDV, RGDTYRAY, REDV, KRSR	PET, PTFE	[86,110,129,218, 292]
		PCL, PLLA, PLGA	[107,146,293,294]
		PEG	[171,171,223,245,281, 295,296,297,298]
		PLA-*co*-Lys	[299,300]
		PS	[301]
		PCPU	[302,303]
		PU, PCU	[218,304,305,306,307]
		PE	[308]
		PMLG	[309]
		PEA	[119]
		PEG-PPF	[137,227,269]
		PU-PEG	[159]
		PHEMA	[129]
		PCEVE-PU	[310]
		PMMA	[213,311]
		PMMA-POEM	[126,312,313,314]
		FEP	[126]
		PVA	[160,315,316]
		PAN	[317]
		PEO/PPO/PEO	[85,109]
		DAS	[318]
		PPy	[206,319]
		p(NIPAAm)	[71]
		Dextran	[252]

(*continued*)

TABLE 6.1 (continued)
Extracellular Membrane (ECM)-Derived Synthetic Oligopeptide Sequences Used for the Modification of Various Biomaterial Surfaces

ECM Molecule	Amino Acid Sequences	Materials	References
		Silicone	[157]
		Glycophase glass	[93,94,128]
		Borosilicate glass	[320,321]
		Ti	[133,134,135,206]
		Si	[100]
Laminin	YIGSR, IKVAV	PEA	[199,320]
		PEP	[322]
Neural cell adhesion molecule	KHIFSDDSSE	Borosilicate glass	[320]
Collagen	DGEA, GFOGER	Borosilicate glass	[323]
		p(AAm-*co*-EG/AAc)	[324]
Bone sialoprotein	FHRRIKA, PRGDT	p(NIPAAm), p(AAm-*co*-EG/AAc)	[71,324]
		pAAm	[325,326]

Note: PET, poly(ethylene terephthalate); PTFE, poly(tetrafluoro ethylene); PCL, poly(caprolactone); PLLA, poly(L-lactic acid); PLGA, poly(lactic acid-*co*-glycolic acid); PEG, polyethylene glycol; PLA-*co*-Lys, poly(lactic acid-*co*-lysine); PS, polystyrene; PCPU, poly(carbonate urea)urethane; PU, poly(urethane); PCU, poly(carbonate urethane); PE, polyethylene; PMLG, poly(γ-methyl-L-glutamate); PEA, polyethylene acrylic acid copolymer; PEG-PPF, polyethylene glycol-poly(propylene fumarate); PU-PEG, PEG-modified poly(urathane); PHEMA, poly(hydroxyethyl methacrylate); PCEVE-PU, poly(chloroethylvinylether) urethane; PMMA, poly(methyl methacrylate); PMMA-POEM, poly(methyl methacrylate-*r*-polyoxyethylene methacrylate; FEP, poly(tetrafluoroethylene-*co*-hexafluoropropylene; PVA, polyvinyl alcohol; PAN, polyacrylonitrile; PEO/PPO/PEO, poly(ethylene oxide)-poly(propylene oxide) copolymer; DAS, dialdehyde starch; PPy, poly(pyrrole); P(NIPAAm), poly(*N*-isopropylacrylamide); Ti, titanium; Si, silicone; PEP, poly (ethylene propylene); p(AAm-*co*-EG/AAc), poly(acrylamide-*co*-ethylene glycol/acrylic acid); pAAm, poly(acrylamide).

6.4.3 Design Strategy for Fabricating Biomimetic Materials

Adhesion molecules were discovered to be crucial for the relationship between cells and ECM. The immobilization of small peptides containing the cell-specific adhesion domains of ECM proteins or growth factors to the base materials has proven to be a promising approach for creating biomimetic materials for artificial organs and devices, dental and medical implants, and tissue engineering applications. Incorporating biomolecular components to materials to make them biomimetic, through the immobilization of cell-specific binding motifs, can be achieved by a number of design strategies and techniques. One approach involves the incorporation of cell recognition, biomolecular motifs directly onto the material surface itself via chemical or physical modification schemes [30]. Another approach involves the incorporation of soluble, biologically active factors to the bulk material or encapsulated within the biomaterial [99], which can be released in situ to elicit a specific biological or cellular response or modulate new tissue formation [100–103]. These modified materials can be used in various applications in drug delivery and targeted tissue regeneration, among others [104]. These biologically active, biomimetic materials can play a crucial role in tissue engineering applications by serving a dual function: (1) as a scaffold to provide mechanical stability to the implant site and a substratum for cell anchorage and (2) as a bioactive, pharmacological agent that alters the course of tissue regeneration via the transmission of biological signals to cells in the surrounding tissue [105,106].

6.4.3.1 Physical Immobilization of Bioactive Peptides to Material Surfaces

Various strategies have been attempted to immobilize small biological motifs onto the surfaces of synthetic biomaterials devoid of active functional groups. Passive adsorption is a physical method for preparing surfaces with well-defined properties that do not rely on chemical processing [107]. Adsorption utilizes weak, nonspecific intermolecular interactions between the surface and the peptide species involved such as: hydrogen bonding, hydrophobic interactions, van der Waals forces, and weak valence electron interactions [105,108]. To obtain biomimetic materials, surfaces can be simply coated with biomimetic peptides or another material possessing active functional groups [e.g., poly(L-lysine), PLL] [1,107] that can be subsequently used to chemically react with oligopeptides. Materials can also be coated with hybrid molecules, such as PLL–RGD polymer–peptide molecules, which can be passively adsorbed to material surfaces or RGD-modified pluronics, which are block copolymers of ethylene oxide and propylene oxide (PEG/PPO) that can be coated onto substrate surfaces and immobilized by hydrophobic interactions [1,109].

Twedon et al. have demonstrated that surface-modified polyethylene terephthalate (PET) and polytetrafluoroethylene (PTFE), containing adsorbed RGD cell adhesive peptides to the material surfaces, accelerated healing and endothelial cell adhesion [110,111]. Reyes et al. adsorbed $\alpha_2\beta_1$ integrin-specific, collagen-derived oligopeptides to polystyrene and titanium surfaces. These type I collagen-mimetic peptides, containing the adhesion motif GFOGER, were found to activate focal adhesion kinase and ALP expression [112]. Additionally, these triple-helical peptides supported osteoblast gene expression and matrix mineralization similar to that induced by type I collagen [112] and improved in vivo peri-implant bone regeneration and osseointegration [113]. Collagen-derived GFOFER peptides have also been used for the modification of polystyrene surfaces by chemically cross-linking the peptides to a pre-adsorbed layer of albumin on the polymer surfaces using EDC. Biotinylated peptides containing a polyethylene oxide (PEO) linker were also immobilized to polystyrene surfaces via an adsorbed layer of avidin, a biotin-binding protein [114].

Groups have also employed the use of carrier molecules to assist in the adsorption and presentation of biomimetic peptides to material surfaces. Quirk et al. developed a simple way to present peptides to polymer surfaces by first attaching the peptide of interest to a long-chain, linear PLL polymers. This carrier molecule was used for the immobilization of the peptide sequences through simple adsorption to the PLA surface [115]. PLL-GRGDS polymer–peptide hybrid molecules have also been adsorbed onto PLA film surfaces [107,116], and PLAGA 3-D foam scaffolds [107], to which bovine aortic endothelial cells and human bone marrow derived cells demonstrated a marked increase in spreading and differentiation over the unmodified scaffolds. Poly (L-lysine)-*g*-poly(ethylene glycol) (PLL-*g*-PEG) molecules have been found to spontaneously absorb from aqueous solutions onto several biomaterial surfaces [117]. As such, RGD-conjugated, PLL-*g*-PEG polymers have been used as passive coatings on glass and polystyrene films and PLAGA microspheres [118].

While passive adsorption is an effective way to immobilize biomimetic peptides to the surface of materials, coatings only provide a transient modification of the material surface [1,119]. The inability to control peptide conformation and orientation upon substrate adsorption; peptide desorption, or wash-off; diffusion kinetics; and the inaccessibility to large molecules on the material surface are deficiencies with this method [120].

6.4.3.2 Chemical Immobilization of Bioactive Peptides to Material Surfaces

In order to fabricate biomimetic materials that can withstand long-term survival, stable immobilization of these biomolecular motifs to the substrate surface is crucial for the maintenance of bioactivity and ultimately proper functioning. Substrate-immobilized biomimetic oligopeptides should be able to withstand the contractile forces exerted by adhered cells upon the biomaterial surface during initial cellular attachment and resist internalization by cells [121,122]. The most extensively investigated approach to immobilize small, bioactive oligopeptide sequences is through covalent coupling directly to the biomaterial surface. Immobilization via chemical methods, however, can only be accomplished when surface reactive groups are present, which is not always the case with certain materials (e.g., polypyrole, poly(α-hydroxy acids)).

Thus, the covalent attachment of biological recognition molecules to the surfaces of some biomaterials is hampered by the lack of surface chemical reactivity [123]. To overcome this problem, researchers have utilized many different approaches to create functional groups on the surface of biomaterials to support the covalent attachment of biological recognition motifs.

6.4.3.2.1 Direct Covalent Immobilization of Bioactive Peptides

Much attention has been paid toward mimicking the ECM functions through the conjugation of cell-specific, biomimetic motifs to material surfaces. Introducing functional molecules such as to scaffolds surfaces in order to induce preferable host interactions, augment biological functions or enhance specific cellular responses [100] can be accomplished by covalent immobilization using chemical modification schemes. However, the presence of active surface functional groups is a necessary prerequisite for the immobilization of biomimetic molecules to biomaterial surfaces. In most cases, the immobilization of peptides to material surfaces relies on stable covalent bond formation between the nucleophilic N-terminus amine groups present in protein molecules and other activated functional groups on the solid material surface. Peptides have been immobilized on material surfaces by reacting the N-terminus amine group with carboxy, hydroxyl, epoxide, and aldehyde groups or other efficient leaving groups introduced by linkers, present on the material surface.

Carboxylic acid functional groups on biomaterial surfaces can be activated using a peptide-coupling reagent, such as a carbodiimide. Carbodiimides are heterobifunctional cross-linkers that mediate the formation of amide bonds from carboxylic acids (or phosphates) and amine groups. They have been used to generate peptide bonds from acid groups and amines through the formation of a stable *O*-urea derivative, which reacts readily with nucleophiles [124]. Carbodiimides are often used in the immobilization of peptides to material substrates as the water-soluble derivative, 1-ethyl-3-(3-dimethylaminopropyl)-carbodiimide (EDC) or as the organic-soluble derivative, *N,N′*-dicyclohexyl-carbodiimide (DCC).

Carbodiimide chemistry is a highly effective and widely used method to covalently immobilize biomimetic peptides onto various carboxylated materials via stable amide bonds. Unfortunately, this coupling method is not highly selective. Biomimetic peptides often contain reactive functional groups present within the constituent amino acid side chains (e.g., carboxyl and guanidine groups) that can lead to unwanted side reactions. *N*-Hydroxysuccinimide (NHS) is often used to assist the carbodiimide coupling by forming an active ester intermediate via condensation of the surface carboxylic acid group and NHS. The ester derivative is less prone to hydrolysis and can be prepared in advance and stored or used as an activated species in situ (e.g., in the presence of the amine nucleophile) without the risk of unwanted side reactions. The NHS-reactive ester intermediate is susceptible to nucleophilic attack by primary amines and results in the formation of stable amide bonds between the biomaterial surface and the N-terminus of the biomimetic peptide.

The immobilization of biomolecules to hydroxyl groups present on various biomaterial surfaces can be easily and directly accomplished with the use of highly reactive sulfonyl chlorides, such as 2,2,2-trifluoroethanesulfonyl chloride (tresyl chloride) [125]. Hydroxyl-containing surfaces can be preactivated with tresyl chloride [126–129] to yield sulfonated surfaces that can readily undergo nucleophilic attack by primary amines, thiols, and imidazole groups [130–132]. The hydroxyl groups of silica and agarose can be activated via reaction with tresyl chloride to form tresylated hydroxyl groups, which can be coupled to enzymes or peptides by direct reaction with the primary amine group [130,131]. Tresyl-activated surfaces can undergo a more rapid and efficient coupling of sulfhydryl-containing peptides than primary amines at physiological pH. Tresyl chloride treatment has been used to successfully immobilize fibronectin cell-adhesive proteins to titanium surfaces [133] through covalent and ionic interactions between oxygen of the terminal hydroxyl groups of titanium and the nitrogen group of fibronectin [134,135]. The immobilization of Polymyxin B (PMB), a cyclic polycationic peptide antibiotic, to a copolymer composed of ethyl acrylate and 2-hydroxyethyl methacrylate was successfully achieved using tresyl chloride, while preserving its antibacterial action against *E. coli* [136]. Alternatively, surface hydroxyl groups can be activated with *N,N′*-disuccinimidyl carbonate (DSC) or *p*-nitrophenyl chlorocarbonate (*p*-NPC) and coupled with peptides under aqueous conditions [137–140].

Aminated surfaces can be effectively immobilized with bioactive peptides by reacting the solid surface with homobifunctional linkers, such as glutaraldehyde, disuccinimidyl glutarate, or phenylene diisothiocyanate [141–143], via the N-terminus of the peptide. Carboxy-terminus immobilization can also occur via carbodiimide-mediated immobilization to the aminated surfaces. Surface amine groups can also be converted into other reactive species for immobilization reactions. Amine groups can be converted into carboxyl-containing surfaces by reacting with succinic anhydride or preactivated carboxyl groups using bis-activated moieties (e.g., disuccinimidyl tartrate or suberate, ethylene glycol-bis(succinimidyl succinate) or *N,N′*–disuccinimidyl carbonate [140,144,145]) and used for sulfonyl chloride-mediated peptide immobilization or other chemical immobilization strategies. Bistresyl-PEG was used in this manner as a linker molecule to immobilize RGD peptides to amine-functionalized surfaces for improved hepatocyte adhesion [145]. Aminated surfaces can also be converted into thiol-reactive surfaces for covalent attachment of bioactive molecules. Nanospheres were fabricated with PEG tethered chains on the surface bearing primary amino groups at the distal ends. The amino terminal groups were converted into thiol-reactive moieties by reacting nanospheres with *N*-succinimidyl-4-(*N*-maleimidomethyl)cyclohexane-1-carboxylate (SMCC) to produce maleimide-functional surfaces. Thiol-containing biomolecules, such as antihuman α-fetoprotein F_{ab}' fragment and cysteine-rich peptides, could then be covalently conjugated to the nanosphere surfaces via reaction of thiols with the surface maleimide groups (Figure 6.3) [147].

Chemoselective ligation is a more recent approach to chemically modify biomaterial surfaces that involve the selective covalent coupling of unique and mutually reactive functional groups under mild conditions. Selected pairs of functional groups are used to couple biomimetic peptides and other bioactive molecules to material surfaces via stable bonds without the need of activating agents or interfering with other functional groups [148]. These reactions are highly chemoselective and behave like molecular "velcro" [149]. Oxime ligation, a chemoselective reaction between an aldehyde group and a hydroxyl amino group, represents a site-specific method to link unprotected aminooxy-functionalized peptidic fragments to various marker molecules or material substrates. The high efficiency and selectivity of the aminooxy-aldehyde coupling reaction has been successfully demonstrated by attaching a variety of substances to proteins and immobilizing aminooxy-terminated RGD cyclopeptides to substrate surfaces [150,151]. The oxime ligation is compatible with most standard amino acid residues and, moreover, the oxime bond is known to be reasonably stable both in vitro and in vivo. Hydrazone ligation can be used to immobilize synthetic hydrazine-like peptides onto aldehyde- or α-oxoaldehyde-functionalized surfaces. Synthetic peptidoliposomes have been prepared via the chemoselective formation of α-oxohydrazone bonds between lysosomal-associated membrane protein (LAMP)-derived peptides and lipidic anchors. The in vitro functionality of the grafted-peptides was confirmed by selective AP-3 membrane protein recruitment to liposomes coupled with the synthetic cytoplasmic domain of LAMP [152,153]. Thiol-functionalized surfaces have been modified with bromoacetyl-containing RGD cyclopeptides [154], while thiol-containing RGD peptides have been linked to acrylic esters, acryl amides [155], and maleimide-functionalized surfaces [156,157]. Benzophenone and aromatic azide functionalized RGD peptides have also been developed as a versatile technique to photochemically immobilize RGDs on various surfaces (Figure 6.4) [158–161].

Biomaterials that lack naturally occurring surface functional groups are hampered by the limited chemical reactivity to support the covalent attachment of biomolecules. A variety of surface chemical and physical treatments have been used to overcome this problem and generate a host

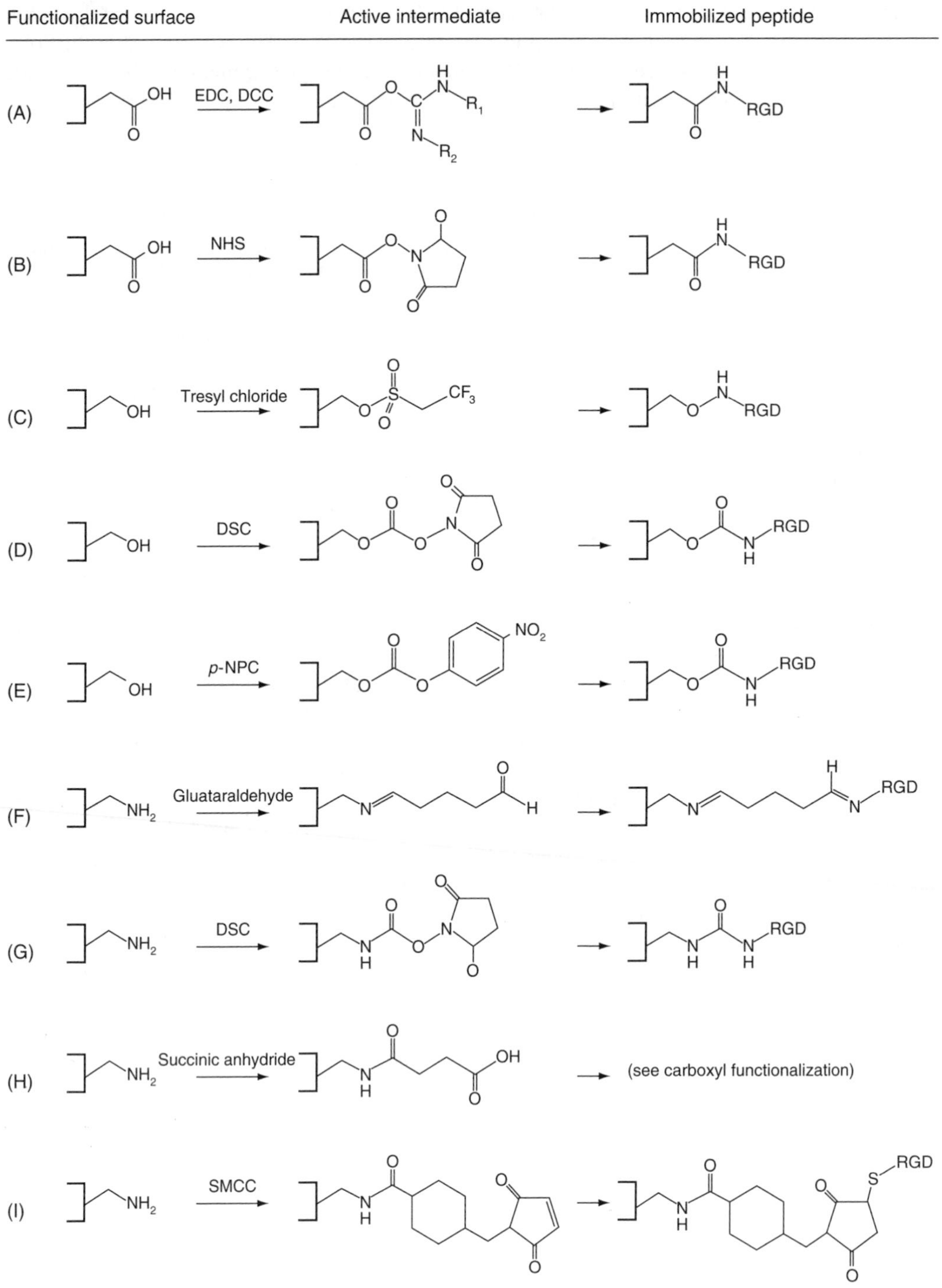

FIGURE 6.3 Direct chemical immobilization of bioactive peptides to reactive surfaces via terminal amine groups. (A) Carboxyl groups preactivated with carbodiimide (EDC, DCC) and (B) *N*-hydroxysuccinimide (NHS), (C) hydroxyl groups preactivated with tresyl chloride, (D) *N,N′*-disuccinimidyl carbonate (DSC) or (E) *p*-nitrophenyl chlorocarbonate (*p*-NPC). (F) Amine groups preactivated with glutaraldehyde and (G) converted into NHS-activated carboxyl groups by DSC, or (H) converted into carboxyl-containing surfaces with succinic anhydride. Amine groups can also be converted into maleimide-functionalized surfaces by SMCC for immobilization of thiol-containing peptides.

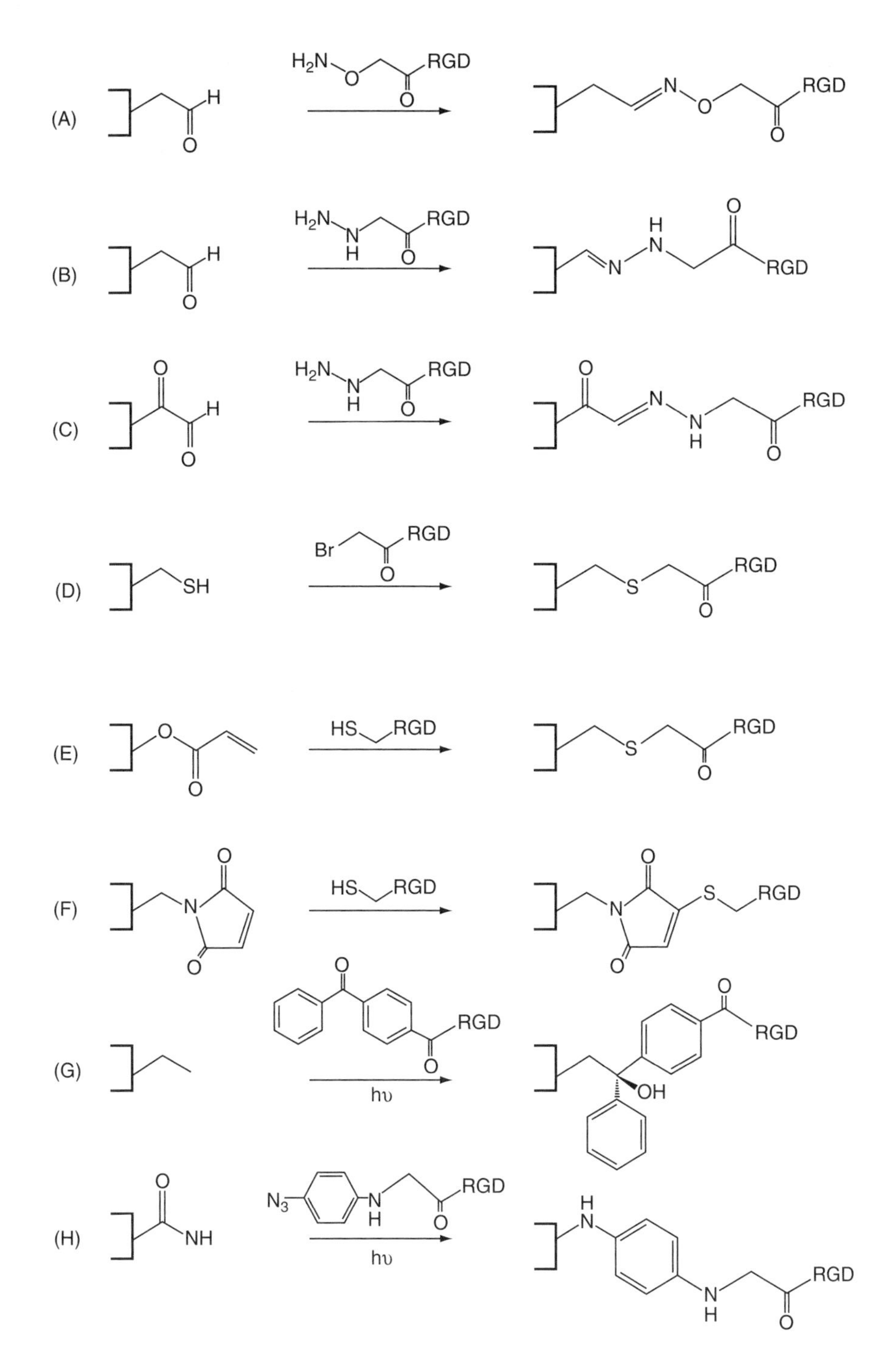

FIGURE 6.4 Chemoselective ligation of selective pairs of functional groups. (A) Oxime ligation of aldehyde surface and hydroxyl amino-RGD, (B) hydrazone ligation of aldehyde surface and hydrazine-RGD, (C) hydrazone ligation of oxo-aldehyde surface and hydrazine-RGD, (D) thiol surface and bromoacetyl-RGD, (E) acrylic ester surface and thiol-RGD, (F) maleimide surface and thiol-RGD, (G) polystyrene surface and benzophenone RGD, (H) PU-PEG surface and aromatic azide-RGD.

of biomimetic materials. Functional groups can be introduced to the material surface by chemical treatments, such as alkaline hydrolysis, aminolysis, and oxidation/reduction reactions. Hydroxyl groups, for example can be introduced to surfaces via alkaline hydrolysis, while aminolysis can be produced by reacting surfaces with diamines (e.g., hydrazine hydrate, ethylenediamine, and hexanediamine) resulting in amino-functionalized surfaces [162]. Surface-hydrolyzed poly(glycolide) sutures have been immobilized with bioactive molecules using carbodiimide chemistry [163]. Poly(L-lactide) microspheres have been hydrolyzed under alkaline conditions and surfaces immobilized with RGD peptides via carbodiimide coupling for use as chondrocyte microcarriers in dynamic culture conditions [164]. Poly(caprolactone) surfaces were aminolysed using hexanediane [165] and the aminated surfaces were covalently bound with laminin-derived sequences by carbodiimide chemistry [166]. Poly(tetrafluoroethylene-*co*-hexafluoropropylene) (FEP) films have been chemically modified with laminin-derived adhesive peptides by first chemically reducing the surfaces using sodium naphthalide. The resulting carbon–carbon double bonds were further modified to introduce hydroxyl groups via hydroboration/oxidation or carboxylic groups via oxidation. Hydroxyl- and carboxyl-functionalized surfaces were coupled with laminin adhesive peptides using tresyl chloride and demonstrated enhanced hippocampal neurite cell adhesion and extension [167]. Biomimetic chitosan films were fabricated by chemically treating with succinic anhydride to generate surface carboxyl groups, which were used for carbodiimide-mediated covalent immobilization of GRGDS peptides. GRGDS peptides were immobilized at a surface density of 10^{-9} mol/cm^2 and enhanced attachment, proliferation, migration, and differentiation of MC3T3-E1 cells [168]. Glass substrates have been modified with synthetic peptides derived from the neurite-outgrowth-promoting domain of the β2-chain of laminin. Peptides were successfully coupled to amine-functionalized surfaces using a heterobifunctional cross-linker and UV-photomasking was used to fabricate peptide patterns for neuronal cell adhesion and growth [169]. Aminophase glass slides were functionalized by incubation in aminopropyltriethoxysilane and used for carbodiimide-assisted immobilization of RGDS, VAPG and KQAGDV peptides [170,171]. RGD peptides have also been immobilized onto porous collagen matrices using periodate oxidation [172]. Other surface chemical reactions include silanization, chlorination, acylation, and quarternization reactions [123,173,174] to induce active functional groups on material surfaces for biomimetic peptide immobilization.

Plasma-treated surfaces have been used to introduce active functional groups to biomaterial surfaces for direct covalent immobilization of biomimetic molecules. Groups have used plasma treatments under a wide range of reacting gas types (ammonia, oxygen, nitrogen, hydrogen, ozone, etc.) to introduce various functional groups (e.g., carboxyl, hydroxyl, carbonyl, peroxyl, ether, and amine groups) to the surface of materials [175–179]. Hu et al. used ammonia gas plasma treatment to create active amine groups on the surfaces of poly(lactic acid) (PLA) polymer thin films and foam scaffolds. The introduced surface amine groups were coupled with glutarldehyde, followed by the cellular affinity motifs, poly(L-lysine) (PLL) and RGD-tripeptide. RGD-modified surfaces showed better OCP attachment than other films and increased alkaline phosphatase (ALPase) activity and calcium levels [178]. Poly(tetrafluoroethylene) fibers were surface modified using UV-activated ammonia plasma treatment to yield amine-functionalized groups for the coupling of laminin-derived cell adhesive peptides [179]. Radio frequency glow discharge has been used to modify polyurethane surfaces followed by coupling of RGDS peptides [180].

6.4.3.2.2 *Graft Polymerization*

Active functional groups can also be introduced to biomaterial surfaces by: synthesizing novel graft-copolymers with desired reactive groups to which biomimetic molecules can be immobilized [181]; creating polymer–peptide hybrid molecules that can be used as passive coatings or used to graft onto biomaterial surfaces in an attempt to guide cell attachment and development [182–186]; or combining materials together to form blends that possess covalent attachment sites [187,188]. End-grafting or in situ polymerization is a popular method for activating biomaterial surfaces by depositing thin polymer layers onto bulk material surfaces [189]. In this process, free radicals or active functional species are generated on a material surface to initiate the polymerization of a monomer species. The grafted polymer film is composed of a different composition than the underlying substrate and is covalently tethered to the material surface. Unique physicochemical and nanoscale topological properties can be engineered on substrate surfaces via the grafted polymer phase. This is an effective method for enhancing the chemical selectivity, modifying the surface topology and improving the biocompatibility and bioactivity of a material surface. Compared to other methods, graft polymerization can generally be performed under mild reaction conditions and low temperatures, and is relatively simple and inexpensive and can be used in a wide range of applications.

A variety of techniques have been employed to graft polymers onto material surfaces in an effort to make them more biologically active, including free radical, plasma deposition, photo-initiation, γ-irradiation, or wet chemistry methods [190]. Due to the large variety of well-suited reactive groups, functional monomers [191–197] and bioactive peptides available, the potential use of this process for fine-tuning surface properties for specific biomedical applications has attracted great interest from the biomaterials community [198]. Corona or plasma discharge is one of the most commonly used methods to introduce functional groups and bioactive molecules to material surfaces. In plasma-induced graft polymerization, material surfaces in are bombarded with energetic gaseous species (e.g., ions, electrons, free radicals, low energy photons, and excited molecular states [198]) whose energy is transferred and dissipated throughout the solid by a variety of chemical and physical processes. The result is functionalization of the material surface through the introduction of reactive molecular species (e.g., amine, peroxide, carboxyl groups). These plasma-treated, functionalized surfaces can subsequently be utilized as initiators for graft polymerization reactions for the covalent immobilization of biomimetic molecules [199–201].

Many groups have utilized the ability of combining surface treatment methods with graft polymerization schemes to modify the surfaces of various materials. Nakaoka et al. combined plasma treatment with graft polymerization to modify the surface of polyethylene (PE) films with biomimetic molecules. PE films were subjected to corona discharge to introduce peroxides onto the material surface, which initiated the graft polymerization of acrylic acid monomer on the PE surface. Bovine serum albumin, type I collagen, and RGDS peptides were immobilized on the film surfaces and the neural differentiation of midbrain cells were observed [202]. Photoinitiation reactions have been employed for the graft polymerization of RGD molecules onto material surfaces [203–205]. DeGiglio et al. electrochemically grafted poly-pyrole-3-acetic acid films onto titanium sheets, to which 4-fluoro-phenylalanine residues were immobilized using dicyclohexylcarbodiimide [206]. Chung et al. created biomimetic surfaces by first inducing nanoscale surface disturbances on poly(ε-caprolactone) (PCL) surfaces via solvent etching using a co-solvent system. Ethylene glycol moieties were embedded on the PCL surfaces, to which GRGD molecules were simultaneously photochemically grafted to form PCL-PEG-RGD surfaces. Modified surfaces supported human umbilical vein endothelial cells (HUVEC) adhesion and growth greater than unmodified surfaces showing promise in soft tissue engineering applications [207]. Biomimetic keratoprostheses were micropatterned by grafting di-amino-PEG onto poly(methyl methacrylate) (PMMA) surfaces using hydrolysis and carbodiimide chemistry or direct aminolysis. Functionalized grafted surfaces were subsequently immobilized with RGD peptides using a heterobifunctional cross-linker to couple peptides to the terminal amine groups on the grafted PEG layer via N-terminal cysteine sulfhydryl groups [208]. Hybrid molecules can be fabricated for use as coatings or in graft polymerization reactions for the fabrication of biomimetic materials. GRGD oligopeptides were photochemically grafted to *N*-succinimidyl-6-[4-azido-2-nitophenylamino]-hexanoate (SANPAH), which was subsequently grafted to chitosan surfaces by UV irradiation. The modified surfaces lead to improved adhesion and growth of human umbilical vein endothelial cells (HUVECs) [209].

Graft polymerization can result in producing specific surface properties for use in various applications to: improve cellular adhesion and spreading, enhance surface wettability, improve material biocompatibility, create non-fouling coatings and for surface functionalization and molecular immobilization. The modified surfaces are restricted to depths ranging from several hundred angstroms to tens of micrometers, without affecting the bulk properties of the material. As such, a variety of materials containing complex shapes can be treated. Graft polymerization can also be used to treat defined surface areas by using masks or resists [177,210].

6.4.3.3 Effects of Oligopeptide Surface Concentration and Distribution

Biomimetic scaffolds for tissue engineering can be fabricated for the purpose of regulating or controlling biological and cellular responses at the interface between biomaterials and the host environment. The in vitro and in vivo responses to these modified materials can be controlled not only by the type of biomimetic molecule immobilized to a biomaterial surface, but also by the ligand concentration, spatial distribution, orientation, and receptor-ligand affinity to these molecules (Figure 6.5) [30]. As such, these parameters are important considerations to make in designing surface-modified, biomimetic scaffolds for tissue engineering.

Many cellular processes within tissues are dependent upon the composition of the constituent ECM molecules. Thus, varying the concentration of available binding proteins can have profound

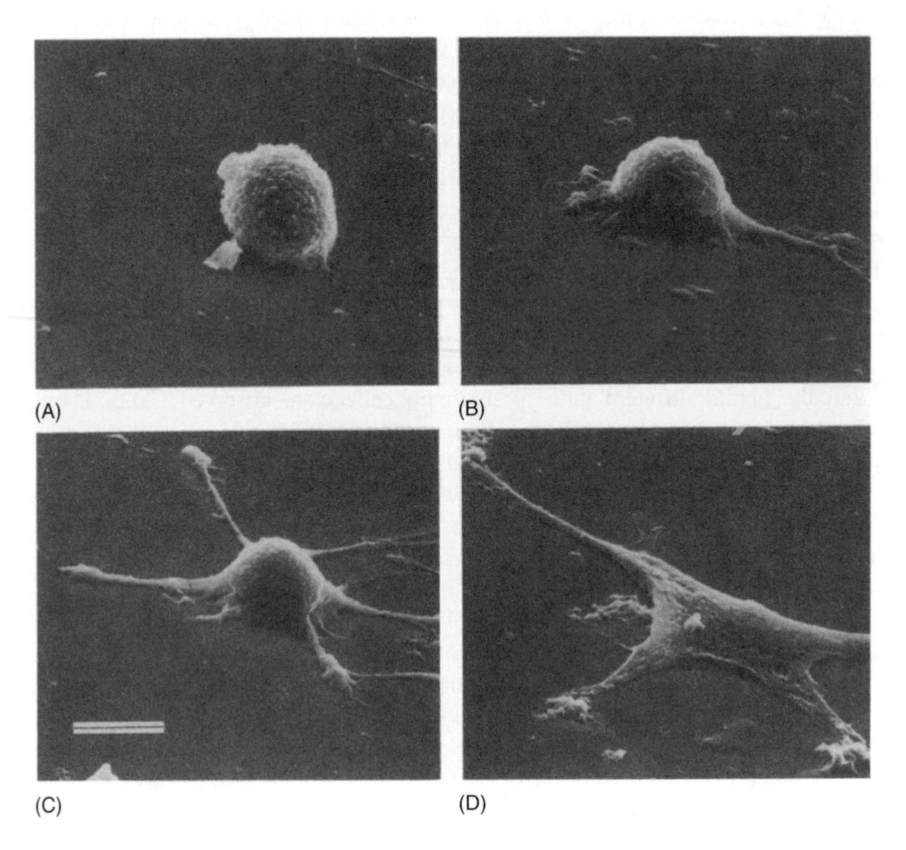

FIGURE 6.5 Scanning electron micrographs (SEMs) of adherent cells on substrates containing covalently grafted GRGDY indicating the classification into types I–IV. Cells adherent on substrates with varying surface concentration of peptide were scored according to the four morphological types represented here. (A) Type I, spheroid cells with no filapodial extensions, (B) type II, spheroid cells with one to two filapodial extensions, (C) type III, spheroid cells with greater than two filapodial extensions, (D) type IV, flattened morphology representative of well spread cells. Bar, 10 μm. (Reproduced from Massia, S.P. and Hubbell, J.A., *J.Cell Biol.*, 114, 1089, 1991. With permission.)

direct effects on cellular behavior. Early studies investigated the cellular response to various concentrations of cell adhesive ECM proteins adsorbed onto substrate surfaces. It was demonstrated that a minimum surface concentration of fibronectin and vitronectin, ranging from approximately 100 to 1300 fmol/cm^2, was needed to obtain sufficient cell spreading [93,211,212]. For surfaces immobilized with RGD peptides, cellular attachment and spreading appears to demonstrate a sigmoidal increase as a function of increasing RGD surface concentration [93,213–217]. A minimum surface density is therefore needed to elicit cellular responses. Massia et al. performed quantitative studies to examine the cellular response to GRGDY immobilized onto polymer-modified glass substrates via the N-terminal glycine residue [218] (Figure 6.5). A minimal amount of RGD peptide (1 fmol/cm^2) was found to be sufficient for fibroblast cell spreading, while at least 10 fmol/cm^2 was needed for $\alpha_v\beta_3$ integrin binding, focal contact formation, and cytoskeletal organization of F-actin stress fibers. At this surface concentration, approximately 60 RGD ligands/μm^2 are available for binding to cell surface integrins, demonstrating the biological potency of these synthetic, EMC-derived oligopeptides for use in biomimetic scaffold fabrication [189]. Drumheller et al. also examined the cellular response to RGD- and YIGSR-containing peptides grafted at various concentrations within cross-linked poly(acrylic acid) hydrogels via a PEG linker molecule [217,219]. Fibroblast cells required a peptide concentration of 12 pmol/cm^2 for cell spreading and 66 pmol/cm^2 for focal contact formation on these interpenetrating networks. While higher RGD surface densities are generally associated with enhanced cell adhesion and spreading, there is a point at which greater peptide immobilization may impede cellular responsiveness. Neff et al. demonstrated that maximum fibroblast proliferation occurred on polystyrene surfaces immobilized RGD peptides at an intermediate surface concentration of approximately 1.33 pmol/cm^2 [220]. Similarly, smooth muscle cell attachment and spreading was enhanced at intermediate peptide concentrations without limiting cell migration, proliferation, and matrix production [171]. Cell migration has also been shown to demonstrate a bi-phasic response to ligand density in vitro [221].

Cellular responsiveness to surface-modified biomaterials can also be influenced by the nanoscale spatial distribution of biomimetic peptides on the material surface. By distributing peptide ligands in a manner that can facilitate cell adhesion receptor aggregation, specific cellular responses may be enhanced [222]. A synthetic polymer-linking method was employed that allowed for variation of both the average surface density and the local spatial distribution of surface immobilized peptides [223]. YGRGD peptides were attached to PEG hydrogel-modified coverslips using star PEO tethers, composed of many PEO arms emanating from a central core. Peptide–polymer hybrid molecules were synthesized containing 1, 5, and 9 peptide ligands per star molecule and immobilized at a range of average ligand densities (1,000–200,000 ligands/μm^2) for cell culture. Maheshwari et al. observed that a significantly greater fraction of cells resisted higher shear stresses on clustered surfaces following centrifugation. Cell adhesion and migration on YGRGD-modified surfaces increased with average surface density and local cluster size. Cells also demonstrated well-formed stress fibers and focal contacts when peptide ligands were presented in clustered forms.

While nanoscale RGD clustering allows for the enhancement of specific cellular responses, the distance from which peptides are oriented from the substrate surface can also influence cell behavior. Cells may require RGD peptides to be immobilized with a minimum spacing between them and the anchoring substrate surface [94,224,225] such that they are flexible and experience minimal steric hindrance [30]. Short bionert polymer and amino acid chains (e.g., PEG, Gly–Gly–Gly–Gly, Pro–Val–Glu–Leu–Pro) have been used as spacers for peptide immobilization [226–229]. Hern et al. observed that surfaces modified with RGD peptides containing PEG spacers at a low surface density of 0.01 pmol/cm^2, promoted enhanced cell adhesion, while surfaces modified with RGDs without PEG spacers exhibited limited cell adhesion even at higher surface densities [230]. Platelet binding to poly(acrylonitrile) beads was shown to be enhanced by coupling up to 13 glycine residues to the N-terminus of a RGDF peptide [231]. Beer et al. suggested that the majority of RGD bonding sites could be reached by peptides that extend 11–32 Å from the substrate surface and most cell receptors could be reached with a spacer length of 46 Å. Craig et al. discovered that 12 amino acid residues were optimal for spacer length in novel oligopeptide coatings [232]. There are examples where no spacer molecule is needed for adequate cell adhesion as well as reduced cell adhesion when the spacer molecule is too long [233–237]. Nevertheless, the use of spacers in peptide immobilization may serve to: allow RGD peptides to access the integrin binding site located within the globular head of the molecule; facilitate the arrangement of peptides into spatial microdomains on substrate surfaces for enhanced cellular attachment; or compensate for substrate surface microscale roughness [1].

Receptor–ligand affinity and cell selectivity contributes to successful cell attachment and the modulation of cellular behavior on peptide-modified surfaces. Since each cell type has its own typical pattern of integrin expression, RGD peptides can be used to promote selective cell adhesion with high receptor affinity on modified surfaces [238]. Hirano et al. cultured five different cell lines on surfaces modified with RGD, RGDS, and RGDT peptides [239]. The tetrapeptides derived from fibronectin and collagen elicited greater cellular attachment, compared to the tripeptide sequences, due to greater integrin affinity toward the tetrapeptides molecules. Employing more affine RGD cyclopeptides in surface modification schemes can further enhance cellular attachment. Groups have observed increased cell attachment on surfaces modified with cyclic peptides compared to their respective linear conformations [240,241]. Some in vitro studies have demonstrated the feasibility of incorporating integrin specificity onto RGD-modified surfaces for selective cell adhesion [238]. Cyclo(RGDfV) peptides have demonstrated similar affinity toward $\alpha_5\beta_1$ and $\alpha_{IIb}\beta_3$ integrins; however, they have significantly higher affinity toward $\alpha_5\beta_3$ integrins [242]. Cyclic peptides G*PenGRGDSPC*A [243] and cyclo(RGDf(NMe)V) [244] can mediated selective $\alpha_V\beta_3$ integrin interactions, while linear RGD-derived peptides promote $\alpha_5\beta_1$ integrin-mediated cell adhesion [245–247]. These same linear RGD peptides, with lipophilic amino acid residues introduced next to aspartic acid (e.g., GRGD**F**, GRGD**Y**, GRGD**V**Y, GRGD**Y**PC), demonstrated higher affinity toward $\alpha_V\beta_3$ and $\alpha_{IIb}\beta_3$ integrins and selective cell attachment via these receptors [218,248–250]. Platelet agglutination can selectively proceed on GRGDF surfaces via $\alpha_{IIb}\beta_3$ integrins [251], while endothelial cells selectively adhere to GRGDVY surfaces via $\alpha_V\beta_3$ integrins [218,250]. RGDSPASSKP tethered surfaces demonstrated preferential adhesion to fibroblasts rather than endothelial cells via $\alpha_5\beta_1$ integrin selectivity [247]. Similarly, the more $\alpha_V\beta_3$ integrin-selective cyclo(RGDfK) surfaces promoted the adhesion of bone related cells and chondrocytes, while $\alpha_5\beta_1$ selective GRGDSP peptide functionalized surfaces enhanced fibroblast attachment and the $\alpha_5\beta_3$ selective cyclic G*PenGRGDSPC*A surfaces supported higher smooth muscle cell and endothelial cell densities [252].

6.5 BIOMIMETIC SCAFFOLDS FOR CARDIOVASCULAR TISSUE ENGINEERING

Autologous blood vessels are the replacement conduit of choice for small-diameter bypass, yet in more than 30% of these patients satisfactory vessels are absent or inadequate [253,254]. As an alternative, synthetic cardiovascular grafts can be used. However, they generally exhibit a lower patency rate due to thrombus formation and intimal occlusion upon implantation [255]. Thus, efforts toward developing vascular grafts with improved biocompatibility, mechanical compliance, and anti-thrombogenicity have been employed in a tissue engineering approach to fabricating cardiovascular tissues.

One approach to increase angiogenesis is to incorporate growth factors in material constructs [256–258], while another utilizes biological peptide fragments immobilized on substrate surfaces. Expanded tetrafluoroethylene was coated and immobilized with fibronectin and RGD-containing peptides for improved endothelial cell attachment [259]. RGD-containing sequences were incorporated into polyurethanes by chemical and photochemical modification, which demonstrated improved endothelial cell adhesion and proliferation [260–262]. Other peptides have been explored for more specific adhesion receptor-mediated interactions with endothelial cells. YIGSR peptide-immobilized poly(ethylene terephthalate) and glass surfaces were found to enhance endothelial cell attachment, spreading, and migration [263]. Fibronectin-derived REDV peptide was immobilized on surfaces and interacted with endothelial cell receptors to enhance attachment and spreading, but limited adhesion of vascular smooth muscle cells and platelets [94]. Biomimetic scaffolds were created by covalently immobilizing ephrin-A1 on the surface of poly(ethylene glycol)-diacrylate hydrogels by chemical modification and photopolymerization to examine their angiogenic properties. Ephrin-A1 immobilized on hydrogels was found to retain its capacity to stimulate endothelial cell adhesion in a dose-dependent manner as similar findings were observed on polystyrene culture wells pre-adsorbed with ephrin-A1 [264]. Human embryonic stem cells were encapsulated in a dextran-based hydrogel with or without immobilized regulatory factors: a tethered RGD peptide and microencapsulated vascular endothelial growth factor (VEGF). Results showed that the fraction of cells expressing the VEGF receptor, KDR/Flk-1, which is a vascular marker, increased up to 20-fold, as compared to spontaneously differentiated embryoid bodies (EBs) [265].

6.6 BIOMIMETIC SCAFFOLDS FOR ORTHOPAEDIC TISSUE ENGINEERING

The repair and replacement of damaged hard and soft tissues of the musculoskeletal system—such as bone, ligament, and tendon—are major clinical problems in the United States and around the world. The functional treatment of fracture nonunions and bone loss associated with trauma, cancer, and revision joint arthroplasty has become increasingly common for the orthopaedic surgeon and remains a significant challenge in the field of orthopaedic surgery. With nearly 30 million people in the United States sustaining some type of musculoskeletal injury annually, there are over 3 million orthopaedic procedures performed each year [266].

RGD peptides have been coupled to various materials to investigate its feasibility in bone tissue engineering applications. Sofia et al. and Chen et al. used silk-based scaffolds for the RGD coupling via carbodiimide chemistry for the attachment, spreading, proliferation, and differentiation of human Saos-2 osteoblast-like cells [267], fibroblasts and bone marrow stromal cells [268]. PLA and PLAGA have been coupled with RGD peptides using poly(L-lysine) and physical adsorption and demonstrated successful adhesion and growth of human osteoprogenitor cells [107]. RGD peptides have also been tethered to fumerate-based polyesters and have been shown to enhance rat mesenchymal stem cell responsiveness [137,269]. A rabbit model was used for the in vivo evaluation of cyclo(RGDfK)-coated PMMA implants for bone regeneration. Modified implants demonstrated enhanced and accelerated bone ingrowth, without interfacial fibrous tissue formation, compared to uncoated controls [213]. Karageorgiou et al. used carbodiimide chemistry to covalently immobilize BMP-2 onto silk fibroin films and cultured in the presence of human bone marrow stromal cells. Covalently coupled BMP-2 was retained on the silk surface at a significantly higher level that adsorbed surfaces and produced elevated alkaline phosphatase activity, calcium deposition, and higher osteogenic transcript levels [270].

Kardestuncer et al. demonstrated that RGD-modified silk fibroin could enhance the adhesion and proliferation of human tenocytes and supported their differentiation as evidenced by elevated transcript levels for decorin and Col-I [271], likely due to enhanced cell–cell interactions from increased cell density. RGD-modified silk fibers were also shown to support human bone marrow stromal cells and anterior cruciate ligament fibroblast adhesion, spreading, proliferation, and collagen matrix production in vitro [272].

Coating synthetic scaffolds with cyclic RGD peptides has been used to interact with specific integrin receptors on the cell surfaces. Integrin-expressing chondrocytes were cultivated on modified surfaces and binding specificity was attributed the immobilized RGD entity. Chondrocyte adhesion and morphology was found to change with increasing amounts of cyclic RGD peptides on the surface [273]. Human embryonic stem cells were encapsulated in poly(ethylene glycol)-diacrylate hydrogels modified with type I collagen, hylauronic acid or YRGDS-peptides and evaluated for chondrogenic activity. Cells encapsulated in RGD-modified hydrogels demonstrated neocartilage formation with basophilic extracellular matrix deposition within 3 weeks of culture and produced cartilage-specific gene up-regulation [274]. PLLA microspheres were surface modified with GRGDSPK peptides via carbodiimide chemistry and used as cell microcarriers for chondrocytes cultured in a flow intermittency bioreactor. Surface-modified microcarriers demonstrated 3D aggregation of cells on the RGD-modified microspheres after 7 days of culture with continued growth over 14 days [164]. However, PLLA and PLGA (50:50) thin films modified with a cellulose-binding domain (CBD)–RGD peptide proved not to be beneficial for in vitro neocartilage formation [275]. Hsu et al. fabricated a series of natural biodegradable materials, composed of chitosan–alginate–hyaluronate (C–A–H) complexes, into films and scaffolds and modified them with CBD-RGD peptides. Chondrocytes were cultivated in vitro on these materials and demonstrated enhanced cellular adhesion, proliferation and collagen and GAG synthesis, while rabbit knee cartilage defects were completely repaired by 6 months [276]. Meinel et al. evaluated RGD-modified silk, porous sponges for cartilage tissue engineering. Human MSC attachment, proliferation, and metabolic activity were markedly better on the slowly degrading silk scaffolds compared to their fast-degrading collagen sponge controls and supported continuous cartilage-like tissue formation [277].

6.7 BIOMIMETIC SCAFFOLDS FOR NEURAL TISSUE ENGINEERING

Paralysis due to nerve damage is a devastating injury that debilitates tens of thousands of Americans each year, and working with the nervous system has long been a challenge for doctors and scientists. Traditionally, tissue transplantation or peripheral nerve grafting is used to repair damaged or diseased regions in the CNS, but often encountered with donor shortage and immunological problems associated with infectious disease [278,279]. Poor regeneration of nerve tissue and rapid scar formation after a spinal cord injury deter endogenous repair of the damaged area. Hence, a strong need exists for innovative repair techniques, such as engineered nerve constructs that can span and regenerate the damaged spinal cord [280]. Tissue engineering is a burgeoning area in biomedical engineering to repair or replace the function of defective tissue. Tissue engineering has provided an alternative to conventional transplantation methods by using polymeric biomaterials with or without living precursor cells.

Surfaces modified with laminin-derived peptides, YIGSR and IKVAV, significantly improved P12 neural cell attachment [281]. Agarose gels derivatized with laminin-derived oligopeptide CDPGYIGSR (Cys–Asp–Pro–Gly–Tyr–Ile–Gly–Ser–Arg) guided neurite growth from dorsal root ganglia cells [282], while IKVAV derivatized gels significantly enhanced neurite outgrowth from PC12 cells [283]. Laminin-derived pentapeptides (YIGSR, IKVAV) were immobilized on fluorinated ethylene propylene films and demonstrated increased receptor-mediated NG108–15 cell attachment [284,285]. Similarly, poly(HEMA-*co*-AEMA) hydrogel modified with laminin-derived peptides guided cell adhesion and neurite outgrowth [286]. The enzymatic activity of factor XIIIa was employed to covalently incorporate bioactive peptides within fibrin during coagulation. Peptides derived from laminin (RGD, IKVAV, YIGSR, RNIAEIIKDI), and *N*-cadherin (HAV) alone and in combination were incorporated into fibrin gels and enhanced in vitro neurite extension and in vivo dorsal root ganglion axon regeneration [287]. Biological agents, such as hyaluronic acid and laminin-derived peptide fragments, have recently been used as doping agents during the electropolymerization of polypyrrole (PPy) and have been immobilized on the surfaces to interact locally with cells. Hyaluronic acid-modified surfaces promoted vascularization in vivo but did not promote PC-12 cell attachment in vitro [288], while the laminin fragment (CDPGYIGSR)-modified surfaces demonstrated in vitro preferential growth of neuroblastoma cells [289]. Furthermore, in vivo studies demonstrated that electrodes surface modified with PPy/CDPGYIGSR showed a significant increase in local neurofilament staining 1 week after implantation [290]. Stauffer et al. reported that laminin fragments, CDPGYIGSR and RNIAEIIKDI, were used as dopants in electropolymerization of the conducting polymer polypyrrole (PPy). Surfaces doped with a combination of the two peptides supported the highest neuronal density, while surfaces doped with RNIAEIIKDI had significantly longer primary neurites and less astrocyte adhesion than the common electrode material, gold [291].

6.8 CONCLUSIONS

The foremost challenge in developing a tissue engineered construct is the development of a synthetic microenvironment that can closely mimic the complex hierarchical micro- and nanoarchitecture of the ECM. Replicating the molecular-level spatial organization of biological cues, such as growth factors, cytokines enzymes, found in native tissue in vivo is also of particular interest in a biomimetic approach to engineering tissues in the laboratory. Researchers have demonstrated that a diversity of oligopeptide biomolecules exists that can be immobilized on materials provides useful and accessible options for surface modification schemes utilized to endow biomaterials with bioactive ligands, while maintaining biological activity. These strategies open up further options for selectively immobilize biomolecules to material surfaces in an effort directed toward encoding biological functions to materials ultimately related to directed cell and tissue regeneration.

REFERENCES

1. Hersel U, Dahmen C, and Kessler H. 2003. RGD modified polymers: Biomaterials for stimulated cell adhesion and beyond. *Biomaterials* 24:4385–4415.
2. Hubbell JA. 1995. Biomaterials in tissue engineering. *Biotechnology* 13:565–576.
3. Marler JJ, Upton J, Langer R, and Vacanti JP. 1998. Transplantation of cell in matrices for tissue engineering. *Adv. Drug Deliv. Rev.* 13:165–182.
4. Murphy WL and Mooney DJ. 1999. Controlled delivery of inductive proteins, plasmid DNA and cells from tissue engineering matrices. *J. Periodontal. Res.* 34:413–419.
5. Ratner BD and Bryant SJ. 2004. Biomaterials: Where we have been and where are we going. *Annu. Rev. Biomed. Eng.* 6:41–75.
6. Lao HK, Renard E, Linossier I, Langlois V, and Vallée-Rehel K. 2007. Modification of poly(3-hydroxy butyrate-co-3-hydroxyvalerate) film by chemical graft copolymerization. *Biomacromolecules* 8:416–423.
7. Langer R. 2000. Biomaterials in drug delivery and tissue engineering: One laboratory's experience. *Acc. Chem. Res.* 33:94–101.
8. Elbert DL and Hubbell JA. 1996. Surface treatments of polymers for biocompatibility. *Annu. Rev. Mater. Sci.* 26:365–394.
9. Ikada Y. 1994. Surface modification of polymers for medical applications. *Biomaterials* 15:725–736.
10. Drumheller PD, Herbert CB, and Hubbell JA. 1996. Bioactive peptides and surface design. *Bioprocess. Technol.* 23:273–310.
11. Lebaron RG and Athanasiou KA. 2000. Extracellular matrix cell adhesion peptides: Functional applications in orthopedic materials. *Tissue Eng.* 6:85–103.
12. Shakesheff KM, Cannizzaro SM, and Langer R. 1998. Creating biomimetic micro-environments with synthetic polymer-peptide hybrid molecules. *J. Biomat. Sci. Polym. Ed.* 9:507–518.
13. Hubbell JA. 1999. Bioactive biomaterials. *Curr. Opin. Biotechnol.* 10:123–129.
14. Mooney D, Hansen L, Vacanti J, Langer R, Farmer S, and Ingber D. 1992. Switching from differentiation to growth in hepatocytes: Control by extracellular matrix. *J. Cell. Physiol.* 151:497–505.
15. Blaschke RJ, Howlett AR, Desprez PY, Peterson OW, and Bissell MJ. 1994. Cell differentiation by extracellular matrix components. *Methods Enzymol.* 245:535–556.
16. Chen CS, Mrksich M, Huang S, Whitesides GM, and Ingber DE. 1997. Geometric control of cell life and death. *Science* 276:1425–1428.
17. Streuli CH, Schmidhauser C, Kobrin M, Bissell MJ, and Derynck R. 1993. Extracellular matrix regulates expression of the TGF-beta 1 gene. *J. Cell Biol.* 120:253–260.
18. Lelievre SA, Weaver VM, Nickerson JA, Larabell CA, Bhaumik A, Petersen OW, et al. 1998. Tissue phenotype depends on reciprocal interactions between the extracellular matrix and the structural organization of the nucleus. *Proc. Nat. Acad. Sci. USA* 95:14711–14716.

19. Masuko T, Iwasaki N, Yamane S, Funakoshi T, Majima T, Minami A, et al. 2005. Chitosan-RGDSGGC conjugate as a scaffold material for musculoskeletal tissue engineering. *Biomaterials* 26:5339–5347.
20. Hauschka SD and Konigsberg IR. 1966. The influence of collagen on the development of muscle colonies. *Proc. Natl. Acad. Sci. USA* 55:119–126.
21. Boudreau N, Myers C, and Bissell MJ. 1995. From lamini to lamin: Regulation of tissue-specific gene expression by the ECM. *Trends Cell. Biol.* 5:1–4.
22. Ingber D. 1991. Extracellular matrix and cell shape: Potential control points for inhibition of angiogenesis. *J. Cell Biochem.* 47:236–241.
23. Bissell MJ, Hall HG, and Parry G. 1982. How does the extracellular matrix direct gene expression? *J. Theor. Biol.* 99:31–68.
24. Borchiellini C, Coulon J, and Le Parco Y. 1996. The function of type IV collagen during drosophila muscle development. *Mech. Dev.* 58:179–191.
25. Murray MA, Fessler LI, and Palka J. 1995. Changing distributions of extracellular matrix components during early wing morphogenesis in Drosophila. *Dev. Biol.* 168:150–165.
26. Buck C and Horwitz AF. 1987. Cell surface receptors for extracellular matrix molecules. *Annu. Rev. Cell Bio.* 3:179–205.
27. Singer II, Scott S, Kawak DW, Kazazis DM, Gailit J, and Ruoslahti E. 1988. Cell surface distribution of fibronectin and vitronectin receptors depends on substrate composition and extracellular matrix accumulation. *J. Cell Biol.* 106:2171–2182.
28. Mosher DF. 1993. Assembly of fibronectin into extracellular matrix. *Curr. Opin. Struct. Biol.* 3:214–222.
29. Andeson JM. 2001. Biological responses to materials. *Annu. Rev. Mater. Res.* 31:81–110.
30. Shin H, Jo S, and Mikos AG. 2003. Biomimetic materials for tissue engineering. *Biomaterials* 24:4353–4364.
31. Ruoslahti E and Pierschbacher MD. 1987. New perspectives in cell adhesion: RGD and integrins. *Science* 238:491–497.
32. Albelda SM and Buck CA. 1990. Integrins and other cell adhesion molecules. *FASEB J.* 4:2868–2880.
33. Travis J. 1993. Biotech gets a grip on cell adhesion. *Science* 260:906–908.
34. van der Flier A and Sonnenberg A. 2001. Function and interactions of integrins. *Cell Tissue Res.* 305:285–298.
35. Plow EF, Haas TA, Zhang L, Loftus J, and Smith JW. 2000. Ligand binding to integrins. *J. Biol. Chem.* 275:21785–21788.
36. van der Flier A and Sonnenberg A. 2001. Function and interactions of integrins. *Cell Tissue Res.* 305:285–298.
37. Hynes RO. 1992. Integrins: Versatility, modulation, and signaling in cell adhesion. *Cell* 69:11–25.
38. Gottschalk K-E, Adams PD, Brunger AT, and Kessler H. 2002. Transmembrane signal transduction of the αIIbβ3 integrin. *Protein Sci.* 11:1800–1812.
39. Lebaron R. 2000. Biomaterials in drug delivery and tissue engineering. *Tissue Eng.* 6:85–103.
40. Zamir E and Geiger B. 2001. Molecular complexity and dynamics of cell-matrix adhesions. *J. Cell Sci.* 114:3583–3590.
41. Geiger B and Bershadsky A. 2001. Assembly and mechanosensory function of focal contacts. *Curr. Opin. Cell Biol.* 13:584–592.
42. Petit V and Thiery JP. 2000. Focal adhesions: Structure and dynamics. *Biol. Cell* 92:477–494.
43. Pande G. 2000. The role of membrane lipids in regulation of integrin functions. *Curr. Opin. Cell Biol.* 12:569–574.
44. Boudreau NJ and Jones PL. 1999. Extracellular matrix and integrin signalling: The shape of things to come. *Biochem. J.* 339:481–488.
45. Schwartz MA. 2001. Integrin signaling revisited. *Trends Cell. Biol.* 11:466–470.
46. Bilezikian JP, Raisz LG, and Rodan GA, (Eds.). 1996. In *Principles of Bone Biology*, p. 268. San Diego: Academic Press.
47. Brittberg M, Lindahl A, Nilsson A, Ohlsson C, Isaksson I, and Peterson L. 1994. Treatment of deep cartilage defects in the knee with autogenous chondrocyte transplantation. *New Engl. J. Med.* 331:889–895.
48. Bell E, Ivarsson B, and Merrill C. 1979. Production of tissue-like structure and contraction of collagen lattices by human fibroblasts of different proliferative potential in vitro. *Proc. Natl. Acad. Sci. USA.* 76:1274–1278.

49. Hansbrough JF, Christine D, and Hansbrough WB. 1992. Clinical trials of a living dermal tissue replacement placed beneath meshed, split-thickness skin graft on excised burn wounds. *J. Burn Care Rehabil.* 13:519–529.
50. Lundborg G, Gelberman RH, Longo FM, Dowell HC, and Varon S. 1982. In vivo regeneration of cut nerves encased in silicone tubes. *J. Neuropathol. Exp. Neurol.* 41:412–422.
51. Langer R and Vacanti JP. 1993. *Tissue Engr. Sci.* 260:920–926.
52. Langer R. 1997. Tissue engineering: A new field and its challenges. *Pharm. Res* 14:840–841.
53. Vacanti CA and Bonassar LJ. 1999. An overview of tissue engineered bone. *Clin. Orthop. Relat. Res.* 367S:S375–S381.
54. Laurencin CT, Attawia MA, Lu LQ, Borden MD, Lu HH, Gorum WJ, et al. 2001. Poly(lactide-co-glycolide)/hydroxyapatite delivery of BMP-2-producing cells: A regional gene therapy approach to bone regeneration. *Biomaterials* 22:1271–1277.
55. Ibarra C, Koski JA, and Warren RF. 2000. Tissue engineering meniscus: Cells and matrix. *Orthop. Clin. North Am.* 31:411–418.
56. Cima LG, Vacanti JP, Vacanti C, Ingber D, Mooney D, and Langer R. 1991. Tissue engineering by cell transplantation using degradable polymer substrates. *J. Biomech. Eng.* 113:143–151.
57. MacArthur BD and Oreffo ROC. 2005. Bridging the gap. *Nature* 433:19.
58. Vacanti CA and Vacanti JP. 2000. The science of tissue engineering. *Orthop. Clin. North Am.* 31:351–356.
59. Musgrave DS, Fu FH, and Huard J. 2002. Gene therapy and tissue engineering in orthopaedic surgery. *J. Am. Acad. Orthop. Surg.* 10:6–15.
60. Fleming JE, Cornell CN, and Muschler GF. 2000. Bone cells and matrices in orthopedic tissue engineering. *Orthop. Clin. North Am.* 31:357–374.
61. Laurencin CT, Attawia MA, Elgendy HE, and Herbert KM. 1996. Tissue engineered bone-regeneration using degradable polymers: The formation of mineralized matrices. *Bone* 19:93S–99S.
62. Attawia MA, Herbert KM, and Laurencin CT. 1995. Osteoblast-like cell adherence and migration through 3-dimensional porous polymer matrices. *Biochem. Biophys. Res. Commun.* 213:639–644.
63. Langer R and Vacanti JP. 1993. Tissue engineering. *Science* 260:920–926.
64. Hubbell JA. 1995. Biomaterials in tissue engineering. *Bio/Technology* 13:565–576.
65. Lee KB, Yoon KR, Woo SI, and Choi IS. 2003. Surface modification of poly(glycolic acid) (PGA) for biomedical applications. *J. Pharm. Sci.* 92:933–937.
66. Black J. 1999. *Biological Performance of Materials: Fundamentals of Biocompatibility*. New York: Marcel Dekker, Inc.
67. Sakiyama-Elbert SE and Hubbell JA. 2001. Functional biomaterials: Design of novel biomaterials. *Ann. Rev. Mater. Res.* 31:183–201.
68. Hubbell JA. 1999. Bioactive biomaterials. *Curr. Opin. Biotechnol.* 10:123–129.
69. Healy KE. 1999. Molecular engineering of materials for bioreactivity. *Curr. Opin. Solid State Mater. Sci.* 4:381–387.
70. Hoffman AS. 1992. Molecular bioengineering of biomaterials in the 1990s and beyond: A growing liaison of polymers with molecular biology. *Artificial Organs* 16:43–49.
71. Healy KE, Rezania A, and Stile RA. 1999. Designing biomaterials to direct biological responses. *Ann. NY Acad. Sci.* 875:24–35.
72. Ripamonti U. 1996. Osteoinduction in porous hydroxyapatite implanted in heterotopic sites of different animal models. *Biomaterials* 17:31–35.
73. Shin H, Jo S, and Mikos AG. 2003. Biomimetic materials for tissue engineering. *Biomaterials* 24:4353–4364.
74. Li JM, Menconi MJ, Wheeler HB, Rohrer MJ, Klassen VA, Ansell JE, et al. 1992. Precoating expanded polytetrafluoroethylene grafts alters production of endothelial cell-derived thrombomodulators. *J. Vasc. Surg.* 15:1010–1017.
75. Song E, Yeon KS, Chun T, Byun HJ, and Lee YM. 2006. Collagen scaffolds derived from a marine source and their biocompatibility. *Biomaterials* 27:2951–2961.
76. DeLustro F, Dasch J, Keefe J, and Ellingsworth L. 1990. Immune responses to allogeneic and xenogeneic implants of collagen and collagen derivatives. *Clin. Orthop. Relat. Res.* 263–279.
77. Schlosser M, Zippel R, Hoene A, Urban G, Ueberrueck T, Marusch F, et al. 2005. Antibody response to collagen after functional implantation of different polyester vascular prostheses in pigs. *J. Biomed. Mater. Res. A* 72:317–325.

78. Schwartzmann M. 2000. Use of collagen membranes for guided bone regeneration: A review. *Implant. Dent.* 9:63–66.
79. Elbert DL and Hubbell JA. 2001. Conjugate addition reactions combined with free-radical cross-linking for the design of materials for tissue engineering. *Biomacromolecules* 2:430–441.
80. Lhoest JB, Detrait E, van den Bosch, de Aguilar, and Bertrand P. 1998. Fibronectin adsorption, conformation, and orientation on polystyrene substrates studied by radiolabeling, XPS, and ToF SIMS. *J. Biomed. Mater. Res.* 41:95–103.
81. Hlady V and Buijs J. 1996. Protein adsorption on solid surfaces. *Curr. Opin. Biotechnol.* 7:72–77.
82. Underwood PA, Steele JG, and Dalton BA. 1993. Effects of polystyrene surface chemistry on the biological activity of solid phase fibronectin and vitronectin, analysed with monoclonal antibodies. *J. Cell Sci.* 104: 793–803.
83. Stivaktakis N, Nikou K, Panagi Z, Beletsi A, Leondiadis L, and Avgoustakis K. 2005. Immune responses in mice of beta-galactosidase adsorbed or encapsulated in poly(lactic acid) and poly(lactic-co-glycolic acid) microspheres. *J. Biomed. Mater. Res. A* 73:332–338.
84. Ito Y, Kajihara M, and Imanishi Y. 1991. Materials for enhancing cell adhesion by immobilization of cell-adhesive peptide. *J. Biomed. Mater. Res.* 25:1325–1337.
85. Neff JA, Caldwell KD, and Tresco PA. 1998. A novel method for surface modification to promote cell attachment to hydrophobic substrates. *J. Biomed. Mater. Res* 40:511–519.
86. Boxus T, Touillaux R, Dive G, and Marchand-Brynaert J. 1998. Synthesis and evaluation of RGD peptidomimetics aimed at surface bioderivatization of polymer substrates. *Bioorg. Med. Chem.* 6:1577–1595.
87. Ruoslahti E. 2003. The RGD story. A personal account. *Matrix Biol.* 22:459–465.
88. LeBaron RG and Athanasiou KA. 2000. Extracellular matrix cell adhesion peptides: Functional applications in orthopaedic materials. *Tissue Eng.* 6:85–103.
89. Pierschbacher MD and Ruoslahti E. 1984. Cell attachment activity of fibronectin can be duplicated by small synthetic fragments of the molecule. *Nature* 309:30–33.
90. Ruoslahti E and Pierschbacher MD. 1987. New perspectives in cell adhesion: RGD and integrins. *Science* 238:491–497.
91. Oosterom R, Ahmed TJ, Poulis JA, and Bersee HE. 2006. Adhesion performance of UHMWPE after different surface modification techniques. *Med. Eng. Phys.* 28:323–330.
92. Hoffmann K, Resch-Genger U, Mix R, and Friedrich JF. 2006. Fluorescence spectroscopic studies on plasma-chemically modified polymer surfaces with fluorophore-labeled functionalities. *J. Fluoresc.* 16:441–448.
93. Hubbell JA, Massia SP, and Drumheller PD. 1992. Surface-grafted cell-binding peptides in tissue engineering of the vascular graft. *Ann. N.Y. Acad. Sci.* 665:253–258.
94. Hubbell JA, Massia SP, Desai NP, and Drumheller PD. 1991. Endothelial cell-selective materials for tissue engineering in the vascular graft via a new receptor. *Biotechnology* 9:568–572.
95. Ranieri JP, Bellamkonda R, Bekos EJ, Vargo TG Jr, and Aebischer JAG. 1995. Neuronal cell attachment to fluorinated ethylene propylene films with covalently immobilized laminin oligopeptides YIGSR and IKVAV. II. *J. Biomed. Mater. Res.* 29:779–785.
96. Rezania A and Healy KE. 1999. Biomimetic peptide surfaces that regulate adhesion, spreading, cytoskeletal organization, and mineralization of the matrix deposited by osteoblast-like cells. *Biotechnol. Prog.* 15:19–32.
97. Bellamkonda R, Ranieri JP, and Aebischer P. 1995. Laminin oligopeptide derivatized agarose gels allow three dimensional neurite extension in vitro. *J. Neurosci. Res.* 41:501–509.
98. Schense JC, Bloch J, Aebischer P, and Hubbell JA. 2000. Enzymatic incorporation of bioactive peptides into fibrin matrices enhances neurite extension. *Nature Biotech.* 18:415–419.
99. Drumheller PD and Hubbell JA. 1994. Polymer networks with grafted cell adhesion peptides for highly biospecific cell adhesive substrates. *Anal. Biochem.* 222:380–388.
100. Davis DH, Giannoulis CS, Johnson RW, and Desai TA. 2003. Immobilization of RGD to silicon surfaces for enhanced cell adhesion and proliferation. *Biomaterials* 23:4019–4027.
101. Whitaker MJ, Quirk RA, Howdle SM, and Shakesheff KM. 2001. Growth factor release from tissue engineering scaffolds. *J. Pharm. Pharmacol.* 53:1427–1437.
102. Babensee JE, McIntire LV, and Mikos AG. 2000. Growth factor delivery for tissue engineering. *Pharm. Res.* 17:497–504.

103. Richardson TP, Murphy WL, and Mooney DJ. 2001. Polymeric delivery of proteins and plasmid DNA for tissue engineering and gene therapy. *Crit. Rev. Eukaryot. Gene Expr.* 11:47–58.
104. Hoffman AS. 1992. Molecular bioengineering of biomaterials in the 1990s and beyond: A growing liaison of polymers with molecular biology. *Artif. Organs* 16: 43–49.
105. Kouvroukoglou S, Dee KC, Bizios R, McIntire LV, and Zygourakis K. 2000. Endothelial cell migration on surfaces modified with immobilized adhesive peptides. *Biomaterials* 21:1725–1733.
106. Drumheller PD, Elbert DL, and Hubbell JA. 1994. Multifunctional poly(ethylene glycol) semi-interpenetrating polymer networks as highly selective adhesive substrates for bioadhesive peptide grafting. *Biotechnol. Bioeng.* 43:772–780.
107. Yang, Roach HI, Clarke NMP, Howdle SM, Quirk HR, Shakesheff KM, et al. 2001. Human osteoprogenitor growth and differentiation on synthetic biodegradable structures after surface modification. *Bone* 29:523–531.
108. Sanghvi AB, Miller KP, Belcher AM, and Schmidt CE. 2005. Biomaterials functionalization using a novel peptide that selectively binds to a conducting polymer. *Nat. Mater.* 4:496–502.
109. Neff JA, Tresco PA, and Caldwell KD. 1999. Surface modification for controlled studies of cell-ligand interactions. *Biomaterials* 20:2377–2393.
110. Tweden KS, Harasaki H, Jones M, Blevitt JM, Craig WS, Pierschbacher M, et al. 1995. Accelerated healing of cardiovascular textiles promoted by an RGD peptide. *J. Heart Valve Dis.* 4:S90–S97.
111. Olivieri MP and Tweden KS. 1999. Human serum albumin and fibrinogen interactions with an adsorbed RGD-containing peptide. *J. Biomed. Mater. Res.* 46:355–359.
112. Reyes CD and Garcia AJ. 2004. $\alpha 2\beta 1$ integrin-specific collagen-mimetic surface supporting osteoblastic differentiation. *J. Biomed. Mater. Res.* 69A:591–600.
113. Reyes CD, Petrie TA, Burns KL, Schwartz Z, and Garcia AJ. 2007. Biomolecular surface coating to enhance orthopaedic tissue healing and integration. *Biomaterials* 28:3228–3135.
114. Reyes CD and Garcia AJ. 2003. Engineering integrin-specific surface with triple helical collagen-mimetic peptide. *J. Biomed. Mater. Res.* 65A:511–523.
115. Quirk RA, Davies MC, Tendler SJB, and Shakesheff KM. 2000. Surface engineering of poly(lactic acid) by entrapment of modifying species. *Macromolecules* 33:158–260.
116. Quirk RA, Chan WC, Davies MC, Tendler SJB, and Shakesheff KM. 2001. Poly(L-lysine)-GRGDS as a biomimetic surface modifier for poly(lactic acid). *Biomaterials* 22: 865–872.
117. Elbert DL and Hubbell JA. 1998. Self-assembly and steric stabilization at heterogeneous, biological surfaces using adsorbing block copolymers. *Chem. Biol.* 5:177–183.
118. Faraasen S, Vörös J, Csúcs, Textor M, Merkle HP, and Walter E. 2003. Ligand-specific targeting of microspheres to phagocytes by surface modification with poly(L-lysine)-grafted poly(ethylene glycol) conjugate. *Pharm. Res.* 20:237–245.
119. Hirano Y, Okuno M, Hayashi T, Goto K, and Nakajima A. 1993. Cell attachment activities of surface immobilized oligopeptides RGD, RGDS, RGDV, RGDT, and YIGSR toward five cell lines. *J. Biomater. Sci. Polym. Ed.* 4:235–243.
120. Cha T, Guo A, Jun Y, and Zhu XY. 2004. Immobilization of oriented protein molecules on poly(ethylene glycol)-coated Si(111). *Proteomics* 4:1965–1976.
121. Pelham RJ and Wang YL. 1998. Cell locomotion and focal adhesions are regulated by the mechanical properties of the substrate. *Biol. Bull.* 194:348–350.
122. Castel S, Pagan R, Mitjans F, Piulats J, Goodman S, Jonczyk A, et al. 2001. RGD peptides and monoclonal antibodies, antagonists of avb3-integrin, enter the cells by independent endocytic pathways. *Lab. Invest.* 81:1615–1626.
123. Awenat KM, Davis PJ, Moloney MG, and Ebenezer W. 2005. A chemical method for the convenient surface functionalisation of polymers. *Chem. Commun.* 8:990–992.
124. Katz E, Schlereth DD, and Schmidt HL. 1994. Electrochemical study of pyrrolquinoline quinone covalently immobilized as monolayer onto a cystamine modified gold electrode. *J. Electroanal. Chem.* 367:59–70.
125. Jennissen HP. 1995. Cyanogen bromide and tresyl chloride chemistry revisited: The special reactivity of agarose as a chromatographic and biomaterial support for immobilizing novel chemical groups. *J. Mol. Recognit.* 8:116–124.
126. Banerjee P, Irvine DJ, Mayes AM, and Griffith LG. 2000. Polymer latexes for cell-resistant and cell-interactive surface. *J. Biomed. Mater. Res.* 50:331–339.

127. Tong YW and Shoichet MS. 1998. Peptide surface modification of poly(tetrafluoroethylene-co-hexafluoropropylene) enhances its interaction with central nervous system neurons. *J. Biomed. Mater. Res.* 42:85–95.
128. Massia SP and Hubbell JA. 1991. Human endothelial cell interactions with surface-coupled adhesion peptides on a nonadhesive glass substrate and two polymeric biomaterials. *J. Biomed. Mater. Res.* 25:223–242.
129. Massia SP and Hubbell JA. 1990. Covalently attached GRGD on polymer surfaces promotes biospecific adhesion of mammalian cells. *Ann. NY Acad. Sci.* 589:261–270.
130. Nilsson K and Mosbach K. 1981. Immobilization of enzymes and affinity ligands to various hydroxyl groups carrying supports using highly reactive sulfonyl chlorides. *Biochem. Biophys. Res. Comm.* 102:449–457.
131. Nilsson K and Mosbach K. 1987. Tresyl chloride-activated supports for enzyme immobilization. *Methods Enzymol.* 135:65–78.
132. Nilsson K and Mosbach K. 1984. Immobilization of ligands with organic sulfonyl chlorides. *Methods Enzymol.* 104:56–69.
133. Hayakwa T, Yoshinar M, and Nemoto K. 2003. Direct attachment of fibronectin to tresyl chloride-activated titanium. *J. Biomed. Mater. Res* 67A:684–688.
134. Hayakawa T, Yoshinari M, Nagai M, Yamamoto M, and Nemoto K. 2003. X-ray photoelectron spectroscopic studies of the reactivity of basic terminal OH of titanium towards tresyl chloride and fibronectin. *Biomed. Res.* 24:223–230.
135. Hayakwa T, Yoshinar M, and Nemoto K. 2005. Quartz-crystal microbalance-dissipation technique for the study of initial adsorption of fibronectin onto tresyl chloride-activated titanium. *J. Biomed. Mater. Res. Part B: Appl. Biomater.* 73B:271–276.
136. Tzorisa A, Hall EAH, Besselinkb GAJ, and Bergveld P. 2003. Testing the durability of polymyxin B immobilization on a polymer showing antimicrobial activity: A novel approach with the ion-step method. *Analytical Letters* 36:1781–1803.
137. Jo S, Shin H, and Mikos AG. 2001. Modification of oligo(poly(ethylene glycol) fumarate) macromere with a GRGD peptide for the preparation of functionalized polymer networks. *Biomacromolecules* 2:255–261.
138. Li JT, Carlsson J, Lin JN, and Caldwell KD. 1996. Chemical modification of surface active poly(ethylene oxide)-poly(propylene oxide) triblock copolymers. *Bioconjug. Chem.* 7:592–599.
139. Woerly S, Laroche G, Marchand R, Pato J, Subr V, and Ulbrich K. 1995. Intracerebral implantation of hydrogel-coupled adhesion peptides: Tissue reaction. *J. Neural Transplant Plast.* 5:245–255.
140. Morpurgo M, Bayer EA, and Wilchek M. 1999. *N*-hydroxysuccinimide carbonates and carbamates are useful reactive reagents for coupling ligands to lysines on proteins. *J. Biochem. Biophys. Methods* 38:17–28.
141. Benters R, Niemeyer NM, and Whorle D. 2001. Dendrimer-activated solid supports for nucleic acid and protein microarrays. *ChemBiochem.* 2:686–694.
142. Ynag M, Kong RYC, Kazmi N, and Leung AKC. 1994. Covalent immobilization of oligonucleotides on modified glass/silicon surfaces for solid-phase DNA hybridization and amplification. *Chem. Lett.* 27:257–258.
143. Pack SP, Kamisetty NK, Nonogawa M, Devarayapalli KC, Ohtani K,Yamada K, et al. 2007. Direct immobilization of DNA oligomers onto the amine-functionalized glass surface for DNA microarray fabrication through the activation-free reaction of oxanine. *Nucleic Acids Res.* 35:e110.
144. Kondoh A, Makino K, and Matsuda T. 1993. 2-Dimensional artificial extracellular-matrix-boadhesive peptide-immobilized surface design. *J. Appl. Polym. Sci.* 47:1983–1988.
145. Kim MR, Jeong JH, and Park TG. 2002. Swelling induced detachment of chondrocytes using RGD-modified poly(*N*-isopropylacryl-amide) hydrogel beads. *Biotechnol. Prog.* 18:495–500.
146. Carlisle ES, Mariappan MR, Nelson KD, Thomes BE, Timmons RB, Constantinescu A, et al. 2000. Enhancing hepatocyte adhesion by pulsed plasma deposition and poly-ethylene glycol coupling. *Tissue Eng.* 6:45–52.
147. Matsuya T, Tashiro S, Hoshino N, Shibata N, Nagasaki Y, and Kataoka K. 2003. A core–shell-type fluorescent nanosphere possessing reactive poly(ethylene glycol) tethered chains on the surface for zeptomole detection of protein in time-resolved fluorometric immunoassay. *Anal. Chem.* 75:6124–6132.
148. Tam JP, Yu Q, and Miao Z. 1999. Orthogonal ligation strategies for peptide and 154. protein. *Biopolymers* 51:311–332.

149. Camarero JA, Kwon Y, and Coleman MA. 2004. Chemoselective attachment of biologically active proteins to surfaces by expressed protein ligation and its application for "protein chip" fabrication. *J. Am. Chem. Soc.* 126:14730–14731.
150. Thumshirn G, Hersel U, Goodman SL, and Kessler H. 2003. Multimeric cyclic RGD peptides for tumor targeting: Solid-phase peptide synthesis and chemoselective oxime ligation. *Chemistry* 9:2717–2725.
151. Scheibler L, Dumy P, Boncheva M, Leufgen K, Mathieu H-J, Mutter M, et al. 1999. Functional molecular thin films: Topological templates for the chemoselective ligation of antigenic peptides to self-assembled monolayers. *Angew. Chem. Int. Ed.* 38:696–699.
152. Bourel-Bonnet L, Pécheur E-I, Grandjean C, Blanpain A, Baust T, Melnyk O, et al. 2005. Anchorage of synthetic peptides onto liposomes via hydrazone and r-oxo hydrazone bonds. Preliminary functional investigations. *Bioconjug. Chem.* 16:450–457.
153. Dubs P, Bourel-Bonnet L, Subra G, Blanpain A, Melnyk O, Pinel A-M, et al. 2007. Parallel synthesis of a lipopeptide library by hydrazone-based chemical ligation. *J. Combinatorial Chem.* 9:973–981.
154. Ivanov B, Grzesik W, and Robey FA. 1995. Synthesis and use of a new bromoacetyl-derivatized heterotrifunctional amino acid for conjugation of cyclic RGD-containing peptides derived from human bone sialoprotein. *Bioconj. Chem.* 6:269–277.
155. Fields GB, Lauer JL, Dori Y, Forns P, Yu y-C, and Tirrell M. 1998. Proteinline molecular architecture: Biomaterial applications for inducing cellular receptor binding and signal transduction. *Biopolymers* 47:143–151.
156. Houseman BT, Gawalt ES, and Mrksich M. 2003. Maleimide functionalized self-assembled monolayers for the preparation of peptide and carbohydrate biochips. *Langmuir* 19:1522–1531.
157. Lateef SS, Boateng S, Hartman TJ, Crot CA, Russell B, and Hanley L. 2002. GRGDSP peptide-bound silicone membranes withstand mechanical flexing in vivo and display enhanced fibroblast adhesion. *Biomaterials* 23:3159–3168.
158. Herbert CB, McLernon TL, Hypolite CL, Adams DN, Pikus L, Huang CC, et al. 1997. Micropatterning gradients and controlling surface densities of photoactivatable biomolecules on self-assembled monolayers of oligo(ethylene glycol) alkanethiolates. *Chem. Biol.* 4:731–737.
159. Lin YS, Wang SS, Chung TW, Wang YH, Chiou SH, Hsu JJ, et al. 2001. Growth of endothelial cells on different concentrations of Gly–Arg–Gly–Asp photochemically grafted in polyethylene glycol modified polyurethane. *Artif. Organs* 25:617–621.
160. Sugawara T and Matsuda T. 1995. Photochemical surface derivatization of a peptide containing Arg–Gly Asp (RGD). *J. Biomed. Mater. Res.* 29:1047–1052.
161. Chung T-Z, Lu Y-F, Wang H-Y, Chen W-P, Wang S-S, Lin Y-S, et al. 2003. Growth of human endothelial cells on different concentrations of Gly–Arg–Gly–Asp grafted chitosan surface. *Artificial Organs* 27:155–161.
162. Massia SP and Hubbell JA. 1992. Binding domains for cell surface proteoglycan-mediated adhesion. *J. Biol. Chem.* 267:10133–10141.
163. Lee K-B, Yoon KR, Woo SI, and Choi IS. 2003. Surface modification of poly(glycolic acid) for biomedical applications. *J. Pharm. Sci.* 92:933–937.
164. Chen R, Curran SJ, Curran JM, and Hunt JA. 2006. The use of poly(L-lactide) and RGD modified microspheres as cell carriers in a flow intermittency bioreactor for tissue engineering cartilage. *Biomaterials* 27:4453–4460.
165. Zhu Y, Gao C, Liu X, and Shen J. 2002. Surface modification of polycaprolactone membrane via aminolysis and biomacromolecules immobilization for promoting cytocompatibility of human endothelial cells. *Biomacromolecules* 3:1312–1319.
166. Santiage LY, Nowak RW, Rubin PJ, and Marra KG. 2006. Peptide-surface modification of poly (caprolactone) with laminin-derived sequences for adipose-derived stem cell applications. *Biomaterials* 27:2962–2969.
167. Tong YW and Shoichet MS. 1998. Peptide surface modification of poly(tetrafluoroethylene-co-hexafluoropropylene) enhances its interaction with central nervous system neurons. *J. Biomed. Mater. Res.* 42:85–95.
168. Li J, Yun H, Gong Y, Zhao N, and Zhang X. 2006. Investigation of MC3T3-E1 cell behavior on the surface of GRGDS-coupled chitosan. *Biomacromolecules* 7:1112–1123.
169. Matsuzawa M, Liesi P, and Knoll W. 1996. Chemically modifying glass surfaces to study substratum-guided neurite outgrowth in culture. *J. Neurosci. Methods* 69:189–196.

170. Mann BK, Tsai AT, Scott-Burden T, and West JL. 1999. Modification of surfaces with cell adhesion peptides alters extracellular matrix deposition. *Biomaterials* 20:2281–2286.
171. Mann BK and West JL. 2002. Cell adhesion peptides alter smooth muscle adhesion, proliferation, migration and matrix protein synthesis on modified surfaces and in polymer scaffolds. *J. Biomed. Mater. Res*. 60:86–93.
172. Ren D, Hou S, Wang H, Luo D, and Zhang L. 2006. Evaluation of RGD modification on collagen matrix. *Artif. Cells. Blood Substitutes Blood Technol. Biotechnol* 34:293–303.
173. McGrath MP, Sall ED, and Tremont SJ. 1995. Functionalization of polymers by metal-mediated processes *Chem. Rev*. 95:381–398.
174. Crist GT and Millan MJ. 1998. Conjugate addition of amino acid side chains to dyes containing alkynone, alkynoic ester and alkynoic amide linker arms. *Tetrahedron* 54:649–666.
175. Wan YQ, Qu X, Lu J, Zhu CF, Wan LJ, Yang JL, Bei J, and Wang S. 2004. Characterization of surface properties of poly(lactide-co-glycolide) after oxygen plasma treatment. *Biomaterials* 25:4777–4783.
176. Kim KS, Lee KH, Cho K, and Park CE. 2002. Surface modification of polysulfone ultrafiltration membrane by oxygen plasma treatment. *J. Membrane Sci*. 135–145.
177. Shen H, Hu X, Yang F, Bei J, and Wang S. 2007. Combining oxygen plasma treatment with anchorage of cationized gelatin for enhancing cell affinity of poly(lactide-co-glucolide). *Biomaterials* 28:4219–4230.
178. Hu Y, Shelley WR, Krajbich I, and Hollinger JO. 2002. Porous polymer scaffolds surface-modified with arginine-glycine-aspartic acid enhance bone cell attachment and differentiation in vitro. *J. Biomed. Mater. Res*. 64A:583–590.
179. Shaw D and Shoichet MS. 2003. Torard sponal cord injury repair strategies: Peptide surface modification of expanded poly(tetrafluoroethylene) fibers for guided neurite outgrowth in vitro. *J. Craniofac Sur*. 14:308–316.
180. Guan JJ, Sacks MS, Beckman EJ, and Wagner WRJ. 2002. Synthesis, characterization, and cytocompatibility of elastomeric, biodegradable poly(ester-urethane)ureas based on poly(caprolactone) and putrescine *J. Biomed. Mater. Res*. 61:493–503.
181. Hrkach J, Ou J, Lotan N, and Langer R. 1999. Poly(L-lactic acid-*co*-amino acid) graft copolymers. *J. Mater. Sci. Res*. 78:92–102.
182. Shakesheff K, Cannizzaro S, and Langer R. 1998. Creating biomimetic micro-environments with synthetic polymer-peptide hybrid molecules. *J. Biomater. Sci. Polym. Ed*. 9:507–518.
183. Quirk RA, Davies MC, Tendler SJB, and Shekesheff KM. 2000. Surface engineering of poly(lactic acid) by entrapment of modifying species. *Macromolecules* 33:158–260.
184. Quirk RA, Chan WC, Davies MC, Tendler SJB, and Shekesheff KM. 2001. Poly(L-lysine)-GRGDS as a biomimetic surface modifier for poly(lactic acid). *Biomaterials* 22:865–872.
185. Cannizzaro SM, Padera RF, Langer R, Rogers RA, Black FE, Davies MC, et al. 1998. A novel biotinylated degradable polymer for cell-interactive applications. *Biotechnol. Bioeng*. 58:529–535.
186. Demetriou M, Binkert C, Sukhu B, Tenenbaum HC, and Dennis JW. 1996. Fetuin/alpha2-HS glycoprotein is a transforming growth factor-beta type II receptor mimic and cytokine antagonist. *J. Biol. Chem*. 271:12755–12761.
187. Cook AD, Pajvani UB, Hrkach JS, Cannizzaro Sm, and Langer R. 1997. Colorimetric analysis of surface reactive amino groups on poly(lactic acid-co-lysine):poly(lactic acid) blends. *Biomaterials* 18:1417–1424.
188. Kubies D, Rypacek F, Kovarova J, and Lednicky F. 2000. Microdomain structure in polylactode-block-poly(ethylene oxide) copolymer films. *Biomaterials* 21:529:36.
189. Elbert DL and Hubbell JA. 1996. Surface treatments of polymers for biocompatibility. *Annu. Rev. Mater. Sci*. 26:365–394.
190. Yosomiya R, Morimoto K, Nakajima A, Ikada Y, and Suzuki T., (Eds.). 1990. Surface modification of matrix polymer for adhesion. In *Adhesion and Bonding in Composites*. 83–100. New York: Marcel Dekker, Inc.
191. Candan S. 2002. Radio frequency-induced plasma polymerization of allyl alcohol and 1-propanol. *Turk J. Chem*. 26:783–791.
192. Hamerli P, Weigel T, Groth T, and Paul D. 2003. Surface properties of and cell adhesion onto allylamine-plasma-coated polyethylenterephtalat membranes. *Biomaterials* 24:3989–3999.
193. Harsch A, Calderon JG, Timmons RB, and Gross GW. 2000. Pulsed plasma deposition of allylamine on polysiloxane: A stable surface for neuronal cell adhesion. *J. Neurosci. Meth*. 98:135–144.

194. Whittle JD, Short RD, Douglas CWI, and Davies J. 2000. Differences in the aging of allyl alcohol, acrylic acid, allylamine, and octa-1,7-diene plasma polymers as studied by x-ray photoelectron spectroscopy. *Chem. Mater.* 12:2664–2671.
195. Tarducci C, Schofield WCE, Badyal JPS, Brewer SA, and Willis C. 2001. Cyano-functionalized solid surfaces. *Chem. Mater.* 13:1800–1803.
196. Butoi CI, Mackie NM, Gamble LJ, Castner DG, Barnd J, Miller AM, et al. 2000. Deposition of highly ordered cf2-rich films using continuous wave and pulsed hexafluoropropylene oxide plasmas. *Chem. Mater.* 12:2014–2024.
197. Kelly JM, Short RD, and Alexander MR. 2003. Experimental evidence of a relationship between monomer plasma residence time and carboxyl group retention in acrylic acid plasma polymers. *Polymer* 44:3173–3176.
198. Watkins L, Bismarck A, Lee AF, Wilson D, and Wilson K. 2006. An XPS study of pulsed plasma polymerised allyl alcohol film growth on polyurethane. *Appl. Surf. Sci.* 252(23):8203–8211.
199. Ivanov VB, Behnisch J, Hollander A, Mehdorn F, and Zimmermann H. 1996. Determination of functional groups on polymer surfaces using fluorescence labeling. *Surf. Interface Anal* 24:257–262.
200. Davies J, Nunnerley CS, Brisley AC, Sunderland RF, Edwards JC, Kruger P, Kaes R, Paul AJ, and Hibbert S. 2000. Argon plasma treatment of polystyrene microtiter wells. Chemical and physical characterization by contact angle, ToF-SIMS, XPS and STM. *Colloid Surf A* 174:287–295.
201. Poncin-Epaillard F and Legeay G. 2003. Surface engineering of biomaterials with plasma techniques. *J. Biomater. Sci. Polym. Edn.* 14:1005–1028.
202. Nakaoka R, Tsuchiya T, and Nakamura A. 2003. Neural differentiation of midbrain cells on various protein-immobilized polyethylene films. *J. Biomed. Mater. Res.* 64A:439–446.
203. Thapa A, Webster J, and Haberstroh KM. 2003. Polymers with nanodimensioned surface features enhance bladders smooth muscle cell adhesion. *J. Biomed. Mater. Res.* 67A:1374–1383.
204. Chung TW, Lu YF, Wang SS, Lin YS, and Chu SH. 2002. Growth of human endothelial cells on photochemically grafted Gly-Arg-Gly-Asp (GRGD) chitosans. *Biomaterials* 23:4803–4809.
205. Calvert JW, Marra KG, Cook L, Kumta PN, DiMilla, and Weiss LE. 2000. Characterization of osteoblast-like behavior of culture bone marrow stromal cells on various polymer surfaces. *J. Biomed. Mater. Res.* 279–284.
206. DeGliglio E, Stefania C, Clavano C-D, Sabbatini L, Zambinin PG, Colucci S, et al. 2006. A new titanium biofunctionalized interface based on poly(pyrrole-3-acetic acid) coating: Proliferation of osteoblast-like cells and future perspectives. *J. Mater. Sci. Mater. Med.* 18:1781–1789.
207. Chung T-W, Yang M-G, Liu D-Z, Chen W-P, Pan C-I, and Wang S-S. 2005. Enhancing growth human endothelial cells on Arg–Gly–Asp (RGD) embedded poly(ε-caprolactone) (PCL) surface with nanometer scale of surface disturbance. *J. Biomed. Mater. Res.* 72A:213–219.
208. Patel S, Thkar RG, Wong J, McLeod SD, and Li S. 2006. Control of cell adhesion on poly(methyl methacrylate). *Biomaterials* 27:2890–2897.
209. Chung TW, Lu YF, Wang HY, Chen WP, Wang SS, Lin YS, et al. 2003. Growth of human endothelial cells on different concentrations of Gly–Arg–Gly–Asp grafted chitosan surface. *Artif. Organs* 27:155–161.
210. Konig U, Nitschke M, Pilz M, Simon F, Arnhold C, and Werner C. 2002. Stability and ageing of plasma treated poly(tetrafluoroethylene) surfaces. *Colloid Surf. B-Biointerf* 25:213–224.
211. Humphries MJ, Akiyama SK, Komoriya A, Olden K, and Yamada KM. 1986. Identification of an alternatively spliced site in human plasma fibronectin that mediates cell type-specific adhesion. *J. Cell Biol.* 103:2637–2647.
212. Underwood PA and Bennett FA. 1989. A comparison of the biological activities of the cell-adhesive proteins vitronectin and fibronectin. *J. Cell Sci.* 93:641–649.
213. Kantlehner M, Schaffner P, Finsinger D, Meyer J, Jonczyk A, Diefenbach B, et al. 2000. Surface coating with cyclic RGD peptides stimulates osteoblast adhesion and proliferation as well as bone formation. *Chembiochem* 1:107–114.
214. Jeschke B, Meyer J, Jonczyk A, Kessler H, Adamietz P, Meenen NM, et al. 2002. RGD-peptides for tissue engineering of articular cartilage. *Biomaterials* 23:3455–3463.
215. Danilov YN and Juliano RL. 1989. (Arg–Gly–Asp)n-albumin conjugates as a model substratum for integrin-mediated cell adhesion. *Exp. Cell Res.* 182:186–196.
216. Hern DL and Hubbell JA. 1998. Incorporation of adhesion peptides into nonadhesive hydrogels useful for tissue resurfacing. *J. Biomed. Mater. Res.* 39:266–276.

217. Drumheller PD and Hubbell JA. 1994. Polymer networks with grafted cell adhesion peptides for highly biospecific cell adhesive substrates. *Anal. Biochem.* 222:380–388.
218. Massia SP and Hubbell JA. 1991. An RGD spacing of 440 nm is sufficient for integrin αVβ3-mediated fibroblast spreading and 140 nm for focal contact and stress fiber formation. *J. Cell Biol.* 114:1089–1100.
219. Drumheller PD, Elbert DL, and Hubbell JA. 1994. Multifunctional poly(ethylene glycol) semi-interpenetrating polymer networks as highly selective adhesive substrates for bioadhesive peptide grafting. *Biotechnol. Bioeng.* 43:772–780.
220. Neff JA, Tresco PA, and Caldwell KD. 1999. Surface modification for controlled studies of cell–ligand interactions. *Biomaterials* 20:2377–2393.
221. Palecek SP, Loftus JC, Ginsberg MH, Luffenburger DA, and Horwitz AF. 1997. Integrin-ligand binding properties govern cell migration speed through cell-substratum adhesiveness. *Nature* 385:537–540.
222. Irvine DJ, Hue KA, Mayes AM, and Griffith LG. 2002. Simulations of cell-surface integrin binding to nanoscale-clustered adhesion ligands. *Biophys. J.* 82:120–132.
223. Maheshwari G, Brown G, Lauffenburger DA, Wells A, and Griffith LG. 2000. Cell adhesion and motility depend on nanoscale RGD clustering. *J. Cell Sci.* 113:1677–1686.
224. Craig WS, Cheng S, Mullen DG, Blevitt J, and Pierschbacher MD. 1995. Concept and progress in the development of RGD-containing peptide pharmaceuticals. *Biopolymers* 37:157–175.
225. Beer JH, Springer KT, and Coller BS. 1992. Immobilized Arg–Gly Asp (RGD) peptides of varying lengths as structural probes of the platelet glycoprotein IIb/IIIa receptor. *Blood* 79:117–128.
226. Hern DL and Hubbell JA. 1998. Incorporation of adhesion peptides into nonadhesive hydrogels useful for tissue resurfacing. *J. Biomed. Mater. Res.* 39:266–276.
227. Shin H, Jo S, and Mikos AG. 2002. Modulation of marrow stromal osteoblast adhesion on biomimetic oligo(poly(ethylene glycol) fumarate) hydrogels modified with Arg–Gly–Asp peptides and a poly(ethylene glycol) spacer. *J. Biomed. Mater. Res.* 61:169–179.
228. Loebsack A, Greene K, Wyatt S, Culberson C, Austin C, Beiler R, et al. 2001. In vivo characterization of a porous hydrogel material for use as a tissue bulking agent. *J. Biomed. Mater. Res.* 57:575–581.
229. Sakiyama-Elbert SE, Panitch A, Hubbell JA. 2001. Development of growth factor fusion proteins for cell-triggered drug delivery. *FASEB J.* 15:1300–1302.
230. Hern DL and Hubbell JA. 1998. Incorporation of adhesion peptides into nonadhesive hydrogels useful for tissue resurfacing. *J. Biomed. Mater. Res.* 39:266–276.
231. Beer HJ, Springer KT, and Coller BS. 1992. Immobilized Arg–Gly–Asp (RGD) peptides of varying lengths as structural probes of the platelet glycoprotein IIb/IIIa receptor. *Blood* 79:117–128.
232. Craig WS, Cheng S, Mullen DG, Blevitt J, and Pierschbacher MD. 1995. Concept and progress in the development of RGD-containing peptide pharmaceuticals. *Biopolymers* 37:157–175.
233. Elbert DL and Hubbell JA. 2001. Conjugate addition reactions combined with free-radical cross-linking for the design of materials for tissue engineering. *Biomacromolecules* 2:430–441.
234. Pelham RJ Jr, and Wang YL. 1998. Cell locomotion and focal adhesions are regulated by the mechanical properties of the substrate. *Biol. Bull.* 194:348–350.
235. Mammen M, Choi S-K, and Whitesides GM. 1998. Polyvalent interactions in biological systems: Implications for design and use of multivalent ligands and inhibitors. *Angew. Chem. Int. Ed.* 37:2755–2794.
236. Wong JY, Kuhl TL, Israelachvili JN, Mullah N, and Zalipsky S. 1997. Direct measurement of a tethered ligand-receptor interaction potential. *Science* 275:820–822.
237. Lo CM, Wang HB, Dembo M, and Wang YL. 2000. Cell movement is guided by the rigidity of the substrate. *Biophys. J.* 79:144–152.
238. Ruoslahti E. 2003. The RGD story. A personal account. *Matrix Biol.* 22:459–465.
239. Hirano Y, Okuno M, Hayashi T, Goto K, and Nakajima A. 1993. Cell-attachment activities of surface immobilized oligopeptides RGD, RGDS, RGDV, RGDT, and YIGSR toward five cell lines. *J. Biomater. Sci. Polym. Ed.* 4:235–243.
240. Delforge D, Gillon B, Art M, Dewelle J, Raes M, and Remacle J. 1998. Design of a synthetic adhesion protein by grafting RGD tailed cyclic peptides on bovine serum albumin. *Lett. Pept. Sci.* 5:87–91.
241. van der Pluijm G, Vloedgraven HJ, Ivanov B, Robey FA, Grzesik WJ, Robey PG, et al. 1996. Bone sialoprotein peptides are potent inhibitors of breast cancer cell adhesion to bone. *Cancer Res.* 56:1948–1955.
242. Pfaff M, Tangemann K, Muller B, Gurrath M, Muller G, Kessler H, et al. 1994. Selective recognition of cyclic RGD peptides of NMR defined conformation by αIIbβ3, αvβ3, and α5β1 integrins. *J. Biol. Chem.* 269:20233–20238.

243. Pierschbacher MD and Ruoslahti E. 1987. Influence of stereochemistry of the sequence Arg–Gly–Asp–Xaa on binding specificity in cell adhesion. *J. Biol. Chem.* 262:17294–17298.
244. Dechantsreiter MA, Planker E, Mathä B, Lohof E, Hölzemann G, Jonczyk A, et al. 1999. N-methylated cyclic RGD peptides as highly active and selective αvβ3 integrin antagonists. *J. Med. Chem.* 42:3033–3040.
245. Kao WJ and Hubbell JA. 1998. Murine macrophage behavior on peptide-grafted polyethyleneglycol-containing networks. *Biotechnol. Bioeng.* 59:2–9.
246. Pierschbacher MD and Ruoslahti E. 1987. Influence of stereochemistry of the sequence Arg–Gly–Asp–Xaa on binding specificity in cell adhesion. *J. Biol. Chem.* 262:17294–17298.
247. Kao WJ, Hubbell JA, and Anderson JM. 1999. Protein-mediated macrophage adhesion and activation on biomaterials: A model for modulating cell behavior. *J. Mater. Sci. Mater. Med.* 10:601–605.
248. Lin HB, Sun W, Mosher DF, Garciaecheverria C, Schaufelberger K, Lelkes PI, et al. 1994. Synthesis, surface, and cell adhesion properties of polyurethanes containing covalently grafted RGD-peptides. *J. Biomed. Mater. Res.* 28:329–342.
249. Beer JH, Springer KT, and Coller BS. 1992. Immobilized Arg–Gly Asp (RGD) peptides of varying lengths as structural probes of the platelet glycoprotein IIb/IIIa receptor. *Blood* 79:117–128.
250. Lin HB, Garciaecheverria C, Asakura S, Sun W, Mosher DF, and Cooper SL. 1992. Endothelial-cell adhesion on polyurethanes containing covalently attached RGD-peptides. *Biomaterials* 13:905–914.
251. Rezania A and Healy KE. 1999. Integrin subunits responsible for adhesion of human osteoblast-like cells to biomimetic peptide surfaces. *J. Orthop. Res.* 17:615–623; *Cell Res.* 182:186–196.
252. Massia SP and Stark J. 2001. Immobilized RGD peptides on surface grafted dextran promotes biospecific cell attachment. *J. Biomed. Mater. Res.* 56:390–399.
253. Sayers RD, Raptis S, Berce M, and Miller JH. 1998. Long-term results of femorotibial bypass with vein or polytetrafluoroethylene. *Br. J. Surg.* 85:934–938.
254. Larsen CC, Kligman F, Tang C, Kottke-Marchant K, and Marchant RE. 2007. A biomimetic peptide fluorosurfactant polymer for endothelialization of ePTFE with limited platelet adhesion. *Biomaterials* 28:3537–3548.
255. Ratcliffe A. 2000. Tissue engineering of vascular grafts. *Matix Biol.* 19:353–357.
256. Nillsen ST, Geutjes PJ, Wismans R, Schalkwijk J, Daamen WF, and van Kuppevelt TH. 2006. Increased angiogenesis in acellular scaffolds by combined release of FGF2 and VEGF. *J. Control Release* 116: e88–e90.
257. Chang Y, Lai PH, Wei HJ, Lin WW, Chen CH, Hwang SM, et al. 2007. Tissue regeneration observed in a basic fibroblast growth factor-loaded porous acellular bovine pericardium populated with mesenchymal stem cells. *J. Thorac. Cardiovasc. Surg.* 134:65–73.e1–4.
258. Cushing MC, Liao JT, Jaeggli MP, and Anseth KS. 2007. Material-based regulation of the myofibroblast phenotype. *Biomaterials* 28:3378–3387.
259. Walluscheck KP, Steinhoff G, Kelm S, and Haverich A. 1996. Improved endothelial cell attachment on ePTFE vascular grafts pretreated with synthetic RGD-containing peptides. *Eur. J. Vasc. Endovasc. Surg.* 12:321–330.
260. Lin HB, Garcia-Echeverria C, Asakura S, Sun W, Mosher DF, and Cooper SL. 1992. Endothelial cell adhesion on polyurethanes containing covalently attached RGD-peptides. *Biomaterials* 13:905–914.
261. Lin Y-S, Wang S-S, Chung T-W, Wang Y-H, Chiou S-H, Jen-Ji H, et al. 2001. Growth of endothelial cells on different concentrations of Gly–Arg–Gly–Asp photo-chemically grafted in polyethylene glycol modified polyurethane. *Artif. Organs* 25:617–621.
262. Guan J, Stankus JJ, and Wagner WR. 2007. Biodegradable elastomeric scaffolds with basic fibroblast growth factor release. *J. Control Release* 120:70–78.
263. Dee KC, Anderson TT, and Bizios R. 1994. Cell function on substrates containing immobilized bioactive peptides. *Mar. Res. Soc. Symp. Proc.* 331:115–119.
264. Moon JJ, Lee SH, and West JL. 2007. Synthetic biomimetic hydrogels incorporated with ephrin-A1 for therapeutic angiogenesis. *Biomacromolecules* 8:42–49.
265. Ferreira LS, Gerecht S, Fuller J, Shieh HF, Vunjak-Novakovic G, and Langer R. 2007. Bioactive hydrogel scaffolds for controllable vascular differentiation of human embryonic stem cells. *Biomaterials* 28:2706–2717.
266. Market Dynamics: Bone Substitutes and Growth Factors. 2002. Datamonitor.
267. Sofia S, McCarthy MB, Gronowicz G, and Kaplan DL. 2001. Functionalized silk-based biomaterials for bone formation, *J. Biomed. Mater. Res* 54:139–148.

268. Chen J, Altman GH, Karageorgiou V, Horan R, Collette A, and Volloch V.2003. Human bone marrow stromal cell and ligament fibroblast responses on RGD-modified silk fibers, *J. Biomed. Mater. Res. A.* 67:559–570.
269. Jo S, Engel PS, and Mikos AG. 2000. Synthesis of poly(ethylene glycol)-tethered poly(propylene fumarate) and its modification with GRGD peptides. *Polymer* 41:7595–7604.
270. Karageorgiou V, Meinel L, Hofmann S, Malhotra A, Volloch V, and Kaplan D. 2004. Bone morphogenetic protein-2 decorated silk fibroin films induce osteogenic differentiation of human bone marrow stromal cells, *J. Biomed. Mater. Res. A* 71:528–537.
271. Kardestuncer T, McCarthy MB, Karageorgiou V, Kaplan D, and Gronowicz G. 2006. RGD-tethered silk substrate stimulates the differentiation of human tendon cells. *Clin. Orthop. Relat. Res.* 448:234–239.
272. Chen J, Altman GH, Karageorgiou V, Horan R, Collette A, Volloch V, et al. 2003. Human bone marrow stromal cell and ligament fibroblast responses on RGD-modified silk fibers. *J. Biomed. Mater. Res. A.* 67:559–570.
273. Jeschke B, Meyer J, Jonczyk A, Kessler H, Adamietz P, Meenen NM, et al. 2002. RGD-peptides for tissue engineering of articular cartilage. *Biomaterials* 23:3455–3463.
274. Hwang NS, Varghese S, Zhang Z, and Elisseeff J. 2006. Chondrogenic differentiation of human embryonic stem cell-derived cells in arginine–glycine–aspartate-modified hydrogels. *Tissue Eng.* 12:2695–2706.
275. Hsu SH, Chang SH, Yen HJ, Whu SW, Tsai CL, and Chen DC. 2006. Evaluation of biodegradable polyesters modified by type II collagen and Arg–Gly–Asp as tissue engineering scaffolding materials for cartilage regeneration. *Artif. Organs* 30:42–55.
276. Hsu SH, Whu SW, Hsieh SC, Tsai CL, Chen DC, and Tan TS. 2004. Evaluation of chitosan-alginate-hyaluronate complexes modified by an RGD-containing protein as tissue-engineering scaffolds for cartilage regeneration. *Artif. Organs* 28:693–703.
277. Meinel L, Hofmann S, Karageorgiou V, Zichner L, Langer R, Kaplan D, et al. 2004. Engineering cartilage-like tissue using human mesenchymal stem cells and silk protein scaffolds. *Biotechnol. Bioeng.* 88:379–391.
278. Hudson T, Evans G, and Schmidt C. 2003. Engineering strategies for peripheral nerve repair. *Orthop. Clin. North Am.* 31:485–498.
279. Yang F, Murugan R, Ramakrishna S, Wang X, Ma Y-X, and Wang S. 2004. Fabrication of nano-structured porous PLLA scaffold intended for nerve tissue engineering. *Biomaterials* 25:1891–1900.
280. Park H, Cannizzaro C, Vunjak-Novakovic G, Langer R, Vacanti CA, and Farokhzad OC. 2007. Nanofabrication and microfabrication of functional materials for tissue engineering. *Tissue Eng.* 13:1867–1877.
281. Dai W, Belt J, and Saltzman WM. 1994. Cell-binding peptides conjugated to poly(ethylene glycol) promote neural cell aggregation. *Biotechnology* 12:797–801.
282. Borkenhgen M, Clemence J-F, Sigrist H, and Aebischer P. 1999. Three dimensional extracellular matrix engineering in the nervous system. *J. Biomed. Mater. Res* 40:392–400.
283. Bellamkonda R, Ranieri JP, and Aebischer P. 1995. Laminin oligopeptide derivatized agarose gels allow three dimensional neurite extension in vitro. *J. Neurosci. Res.* 41:501–509.
284. Ranieri JP, Bellamkonda R, Bekos EJ, Vargo TG, Gardella JA Jr, and Aebischer P. 1995. Neuronal cell attachment to fluorinated ethylene propylene films with covalently immobilized laminin oligopeptides YIGSR and IKVAV. II. *J. Biomed. Mater. Res.* 29:779–785.
285. Ranieri JP, Bellamkonda R, Bekos EJ, Gardella JA Jr, Mathieu HJ, Ruiz L, et al. 1994. Spatial control of neuronal cell attachment and differentiation on covalently patterned laminin oligopeptide substrates. *Int. J. Dev. Neurosci.* 12:725–735.
286. Yu TT and Shoichet MS. 2005. Guided cell adhesion and outgrowth in peptide-modified channels for neural tissue engineering. *Biomaterials* 26:1507.
287. Schense JC, Bloch J, Aebischer P, and Hubbell JA. 2000. Enzymatic incorporation of bioactive peptides into fibrin matrices enhances neurite extension. *Nature Biotech.* 18:415–419.
288. Collier JH, Camp JP, Hudson TW, and Schmidt CE. 2000. Synthesis and characterization of polypyrrole-hyaluronic acid composite biomaterials for tissue engineering applications. *J. Biomed. Mater. Res.* 50:574–584.
289. Cui X, Lee VA, Raphael Y, Wiler JA, Hetke JF, and Anderson DJ. 2001. Surface modification of neural recording electrodes with conducting polymer/biomolecule blends. *J. Biomed. Mater. Res.* 56:261–272.

290. Cui X, Wiler J, Dzaman M, Altschuler RA, and Martin DC. 2003. In vivo studies of polypyrrole/peptide coated neural probes. *Biomaterials* 24:777–787.
291. Stauffer WR and Cui XT. 2006. Polypyrrole doped with 2 peptide sequences from laminin. *Biomaterials* 27:2405–2413.
292. Marchand-Brynaert J, Detrait E, Noiset O, Boxus T, Schneider YJ, and Remacle C. 1999. Biological evaluation of RGD peptidomimetics, designed for the covalent derivatization of cell culture substrata, as potential promoters of cellular adhesion. *Biomaterials* 20:1773–1782.
293. Yamaoka T, Hotta Y, Kobayashi K, and Kimura Y. 1999. Synthesis and properties of malic acid-containing functional polymers. *Int. J. Biol. Macromol.* 25:265–271.
294. Eid K, Chen E, Griffith L, and Glowacki J. Effect of RGD coating on osteocompatibility of PLGA-polymer disks in a rat tibial wound. 2001. *J. Biomed. Mater. Res.* 57:224–231.
295. Kao WJ and Lee D. 2001. In vivo modulation of host response and macrophage behavior by polymer networks grafted with fibronectin-derived biomimetic oligopeptides: The role of RGD and PHSRN domains. *Biomaterials* 22:2901–2909.
296. Liu Y and Kao WJ. 2002. Human macrophage adhesion on fibronectin. The role of substratum and intracellular signaling kinases. *Cell Signal.* 14:145–152.
297. Kao WJ and Liu Y. 2001. Utilizing biomimetic oligopeptides to probe fibronectin–integrin binding and signaling in regulating macrophage function in vitro and in vivo. *Front. Biosci.* 6:D992–D999.
298. Kao WJ and Liu Y. 2002. Intracellular signaling involved in macrophage adhesion and FBGC formation as mediated by ligand–substrate interaction. *J. Biomed. Mater. Res.* 62:478–487.
299. Cook AD, Hrkach JS, Gao NN, Johnson IM, Pajvani UB, Cannizzaro SM, et al. 1997. Characterization and development of RGD-peptide-modified poly(lactic acid-co-lysine) as an interactive, resorbable biomaterial. *J. Biomed. Mater. Res.* 35:513–523.
300. Barrera DA, Zylstra E, Lansbury PT, and Langer R. 1993. Synthesis and RGD peptide modification of a new biodegradable copolymer poly(lactic acid-co-lysine). *J. Am. Chem. Soc.* 115:11010–11011.
301. McConachie A, Newman D, Tucci M, Puchett A, Tsao A, Hughs J, et al. 1999. The effect on bioadhesive polymers either freely in solution or covalently attached to a support on human macrophages. *Biomed. Sci. Instrum.* 35:45–50.
302. Gumusderelioglu M and Turkoglu H. 2002. Biomodification of non-woven polyester fabrics by insulin and RGD for use in serum-free cultivation of tissue cells. *Biomaterials* 23:3927–3935.
303. Krijgsman B, Seifalian AM, Salacinski HJ, Tai NR, Punshon G, Fuller BJ, et al. 2002. An assessment of covalent grafting of RGD peptides to the surface of a compliant poly(carbonate-urea)urethane vascular conduit versus conventional biological coatings: Its role in enhancing cellular retention. *Tissue Eng.* 8:673–680.
304. Yanagi M, Kishida A, Shimotakahara T, Matsumoto H, Nishijima H, Akashi M, et al. 1994. Experimental study of bioactive polyurethane sponge as an artificial trachea. *ASAIO J.* 40:M412–M418.
305. Lin HB, Zhao ZC, Garcia-Echeverria C, Rich DH, and Cooper SL. 1992. Synthesis of a novel polyurethane co-polymer containing covalently attached RGD peptide. *J. Biomater. Sci. Polym. Ed.* 3:217–227.
306. Lin HB, Garcia-Echeverria C, Asakura S, Sun W, Mosher DF, and Cooper SL. 1992. Endothelial cell adhesion on polyurethanes containing covalently attached RGD-peptides. *Biomaterials* 13:905–914.
307. Anderheiden D, Klee D, Höcker H, Heller B, Kirkpatrick CJ, and Mittermayer C. 1992. Surface modification of a biocompatible polymer based on polyurethane for artificial blood vessels *J. Mater. Sci.: Mater. Med.* 3:1–4.
308. Johnson R, Harrison D, Tucci M, Tsao A, Lemos M, Puckett A, et al. 1997. Fibrous capsule formation in response to ultrahigh molecular weight polyethylene treated with peptides that influence adhesion. *Biomed. Sci. Instrum.* 34:47–52.
309. Kugo K, Okuno M, Masuda K, Nishino J, Masuda H, and Iwatsuki M. 1994. Fibroblast attachment to Arg–Gly Asp peptide-immobilized poly(g-methyl l-glutamate). *J. Biomater. Sci. Polym. Ed.* 5: 325–337.
310. Sanchez M, Deffieux A, Bordenave L, Borquey C, and Fontanille M. 1994. Synthesis of hemocompatible materials. Part 1: Surface modification of polyurethanes based on poly(chloroalkylvinylether)s by RGD fragments. *Clin. Mater.* 15:253–258.
311. Schaffner P, Meyer J, Dard M, Wenz R, Nies B, Verrier S, et al. 1999. Induced tissue integration of bone implants by coating with bone selective RGD-peptides in vitro and in vivo studies. *J. Mater. Sci. Mater. Med.* 10:837–839.

312. Irvine DJ, Mayes AM, and Griffith LG. 2001. Nanoscale clustering of RGD peptides at surfaces using Comb polymers. 1. Synthesis and characterization of Comb thin films. *Biomacromolecules*. 2:85–94.
313. Irvine DJ, Ruzette A-VG, Mayes AM, and Griffith LG. 2001. Nanoscale clustering of RGD peptides at surfaces using comb polymers-2. Surface segregation of comb polymers in polylactide. *Biomacromolecules* 2:545–556.
314. Koo LY, Irvine DJ, Mayes AM, Lauffenburger DA, and Griffith LG. 2002. Co-regulation of cell adhesion by nanoscale RGD organization and mechanical stimulus. *J. Cell Sci.* 115:1423–1433.
315. Kondoh A, Makino K, and Matsuda T. 1993. 2-Dimentiona artificial extracellular-matrix bioadhesive peptide immobilized surface design. *J. Appl. Polym. Sci.* 47:1983–1988.
316. Kobayashi H and Ikada Y. 1991. Corneal cell adhesion and proliferation on hydrogel sheets bound with cell-adhesive proteins. *Curr. Eye Res.* 10:899–908.
317. Beer JH, Springer KT, and Coller BS. 1992. Immobilized Arg–Gly Asp (RGD) peptides of varying lengths as structural probes of the platelet glycoprotein IIb/IIIa receptor. *Blood* 79:117–128.
318. Holland J, Hersh L, Bryhan M, Onyiriuka E, and Ziegler L. 1996. Culture of human vascular endothelial cells on an RGD-containing synthetic peptide attached to a starch-coated polystyrene surface: Comparison with fibronectin-coated tissue grade polystyrene. *Biomaterials* 17:2147–2156.
319. De Giglio E, Sabbatini L, Colucci S, and Zambonin G. 2000. Synthesis, analytical characterization, and osteoblast adhesion properties on RGD-grafted polypyrrole coatings on titanium substrates. *J. Biomater. Sci. Polym. Ed.* 11:1073–1083.
320. Kam L, Shain W, Turner JN, and Bizios R. 2002. Selective adhesion of astrocytes to surfaces modified with immobilized peptides. *Biomaterials* 23:511–515.
321. Hasenbein ME, Andersen TT, and Bizios R. 2002. Micropatterned surfaces modified with select peptides promote exclusive interactions with osteoblasts. *Biomaterials* 23:3937–3942.
322. Ranieri JP, Bellamkonda R, Bekos EJ, Vargo TG Jr, and Aebischer JAG. 1995. Neuronal cell attachment to fluorinated ethylene propylene films with covalently immobilized laminin oligopeptides YIGSR and IKVAV. II. *J. Biomed. Mater. Res.* 40:371–377.
323. Mineur P, Guignandon A, Lambert ChA, Amblard M, Lapière ChM, and Nusgens BV. 2005. RGDS and DGEA-induced [Ca2+]i signalling in human dermal fibroblasts. *Biochim. Biophys. Acta.* 1746:28–37.
324. Harbers GM and Healy KE. 2005. The effect of ligand type and density on osteoblast adhesion, proliferation, and matrix mineralization. *J. Biomed. Mater. Res. A.* 75:855–869.
325. Graham KL, Zeng W, Takada Y, Jackson DC, and Coulson BS. 2004. Effects on rotavirus cell binding and infection of monomeric and polymeric peptides containing alpha2beta1 and alphaxbeta2 integrin ligand sequences. *J. Virol.* 78:11786–11797.
326. Mizuno M and Kuboki Y. 2001. Osteoblast-related gene expression of bone marrow cells during the osteoblastic differentiation induced by type I collagen. *J. Biochem. (Tokyo).* 129:133–138.

Section III

Promising Nanofabrication Techniques to Develop Scaffolds for Tissue Engineering

7 Electrospun Polymeric Nanofiber Scaffolds for Tissue Regeneration

Syam Prasad Nukavarapu, Sangamesh G. Kumbar, Jonathan G. Merrell, and Cato T. Laurencin

CONTENTS

7.1 INTRODUCTION

Electrospinning is a technique to generate submicron to nanometer scale fibers from polymer solutions or melts. It is also known as electrostatic spinning, which has some characteristics common with electrospraying and the traditional fiber drawing process. Electrospinning is a simple and facile technique to produce nonwoven, interconnected nanofiber mats that have potential applications in the field of engineering and medicine [1–5]. The recent surge in nanotechnology has adopted electrospinning as an elegant technique to produce nanofiber structures for biomedical applications such as wound healing, tissue engineering, and drug delivery [6–8].

In tissue engineering, scaffold supports initial cellular activities and disappears by the time the tissue formation is completed. It is desired that a scaffold by itself should mimic tissue in structure, provide required mechanical strength and the appropriate porosity for cellular infiltration and growth. Traditional scaffolding methodologies such as solvent casting and particulate leaching [9], gas foaming [10], freeze drying, [11] and 3D printing [12] lack the ability to form scaffolds that duplicate the native tissue structure. On the other hand, electrospining has demonstrated its ability to fabricate nanofiber scaffolds that closely mimic the native extracellular matrix (ECM) structure [13]. Further, the technique has exhibited its versatility to produce scaffolds with a range of mechanical properties (plastic to elastic), optimized porosity, and pore volume for tissue engineering applications [14,15].

Biodegradable synthetic polymers such as poly(D,L-lactide) (PLA), poly(D,L-lactide-*co*-glycolide) (PLAGA), polycaprolactone (PCL), polyphosphazenes etc., have been electrospun to obtain nanofiber scaffolds [15–18]. These nanofiber matrices have exhibited a close resemblance to collagen and elastin fibrils in the native ECM structure in terms of their size and fiber arrangement. Since they closely mimic the structure of the ECM, synthetic nanofiber scaffolds showed excellent cellular adhesion and growth behavior for their applications in tissue engineering [13,19]. Electrospinning was also used to create scaffolds of naturally occurring polymers such as collagen, elastin, fibrinogen, chitosan, alginate, hyaluronic acid, and silk. Such natural electrospun nanofiber scaffolds, because of their structural and compositional similarity with native ECM, attracted lot of attention for tissue regeneration applications [20]. Various polymer compositions such as synthetic–synthetic, synthetic–natural, and natural–natural polymer blends were also electrospun to produce new scaffolds with required bioactivity and mechanical properties suitable for vascular, dermal, neural, and cartilage tissue engineering [21–23]. Also efforts were on to create polymer–ceramic electrospun scaffolds for bone tissue engineering applications [24–26].

In addition to electrospinning, thermally induced phase separation (TIPS) and molecular self-assembly are the two other methods used for nanofiber scaffold fabrication [27,28]. However, TIPS can be applied only to crystalline polymers and thus far only used to poly(L-lactic acid) (PLLA) to fabricate nanofiber scaffolds. Further the molecular self-assembly suffers from the complex chemistry involved with the self assembling peptide synthesis. The other drawbacks of these two techniques include less control on process parameters and the inability to produce oriented nanofibers. In contrast, electrospinning is a viable process that has already been applied to various synthetic and natural polymers to derive scaffolds with appropriate physical and biochemical properties suitable to regenerate a tissue of interest.

7.2 ELECTROSPINNING: BASICS AND INSTRUMENTATION

7.2.1 Electrospinning Setup

Basic electrospinning experimental setup essentially consists of a polymer solution or melt delivery system, power source, and a collector or target. A syringe or a capillary tube supplies polymer solution or melt at a desired rate, and a high-power source will provide necessary electric potential to initiate a polymer jet from a pendent droplet. The grounded target or collector located at a certain distance from the polymer source will collect spun nanofibers. To provide a constant desirable polymer solution or melt flow it is advantageous to have a syringe pump in the electrospinning setup and thus all recent experimental setups use programmable pumps. Earlier experimental setups were based on gravity, wherein the syringe is positioned at a correct angle relative to the ground such that the polymer drips pump free at a desired rate. A positive terminal of the high-power source will be connected to the syringe needle and the other will be grounded to the target. Shape, size, and motion of the target determine the nanofiber alignment and thus depending on the application various targets or collectors were designed. Random nanofibers deposition was possible with stationary aluminum foil as a target while rotating drums or metal wheels were used to fabricate aligned and tubular nanofiber scaffolds. The distance from the syringe tip to the grounded target (called the tip-to-target distance or working distance) also controls the fiber diameters during electrospinning.

7.2.2 Spinning Process and Mechanisms

Various polymers, copolymers, polymer blends, and polymer–ceramic composites have been electrospun by dissolving them in suitable solvents like water, inorganic acids, organic solvents as well as solvent mixtures. Electrospinning process starts with the application of high electric potential ($\approx$1 kV/cm) to the pendant drop of a polymer solution. This electrical field induces charge separation and hence causes charge repulsion within the polymer droplet. A polymer jet is initiated when the opposing electrostatic force overcomes the surface tension of the polymer solution. Just before the jet formation, polymer droplet under the influence of electric field assumes cone shape with convex sides and a rounded tip, which is known as the Taylor cone. Once the jet is launched, the dominating electric force favors straight elongation of the polymer droplet. At the end of the straight elongation, the Columbic interaction within the jet and with external electric field manifest into an off-axial radial component of velocity that result in polymer jet bending and whipping. During the entire process, polymer jet under the influence of electrical force elongates thousands or even millions of times before the ultra thin solid fiber collects onto the target.

Majority of the studies in the literature optimized the key process parameters such as solution viscosity and applied electric field to obtain defect free nanofibers. Though the electrospinning process had been known for long time, understanding physics behind the process and a fundamental model to explain the process are still lacking. Visualizing the spinning process using high-speed cameras has recently led to few models to explain and predict the process accurately [29]. Yarin and Reneker proposed a comprehensive model and explained the experimental observations by introducing linear bending instabilities during the fiber elongation process [30]. Later, Rutledge and coworkers modeled the process to account for nonlinear whipping behavior of the jets observed in the process of nanofiber formation [31].

7.2.3 Background and Statistics

The electrospinning technique evolved on the basis of electrospraying that has been widely used for around 100 years. Both the processes require high-voltage to initiate a liquid jet. The electrified jet further breaks into fine droplets and particles in electrospraying [32], whereas a solution with high enough viscosity ends up forming continuous fiber in electrospinning. Some of the important mile stones in the process of electrospinning and chronological developments are summarized in Figure 7.1. Though electrospinning was reported in 1934, further development and the progress were relatively slow for the next 50 years. By year 2000, as part of the nanotechnology revolution, electrospinning technique attracted greater attention and more than 100 different polymers were electrospun for various technological applications. Laurencin and coworkers were the first one to report the feasibility of electrospun highly porous nanofiber scaffolds for biomedical applications [15]. Since nanofiber scaffolds combine the material biodegradability and fiber diameters in the range of native ECM fibers, they have generated lot of excitement for regenerating tissues such as skin, blood vessel, and bone. A literature survey from the SciFinder Scholar indicates an exponential growth in the number of publications on electrospun biodegradable scaffolds as readily seen from Figure 7.2. Such a rapid growth is an indicative of the popularity of the technique that may ensure various nanofiber-based scaffolds in the near future for tissue regeneration applications.

7.2.4 Relevance of Electrospun Scaffolds in Tissue Engineering

In tissues and organs, cells are surrounded by ECM and most of the cellular functions are mediated by the ECM in association with signaling molecules. The ECM comprises 10–100 nm

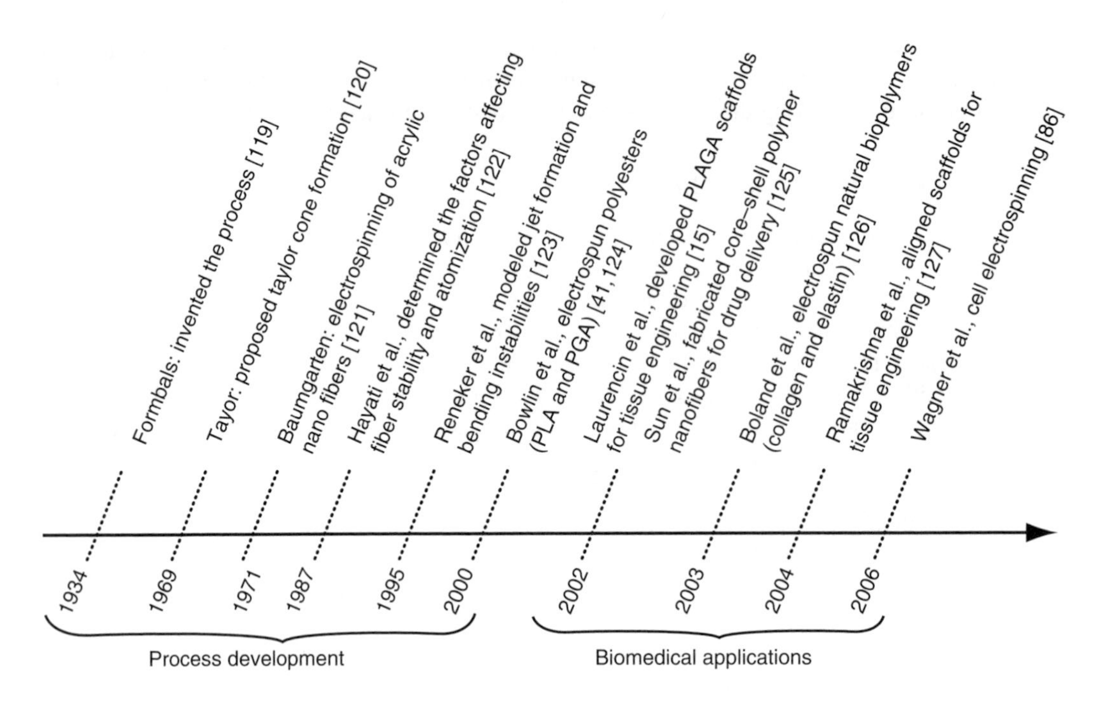

FIGURE 7.1 Chronological events in the process of electrospinning to fabricate nanofiber matrices for variety of applications. Electrospinning also known as electrostatic spinning is a century old technique; however, its potential in diversified high technology applications has been realized lately. Recently, electrospinning has emerged as a leading technique to create nanofeatured scaffolds for tissue regeneration applications.

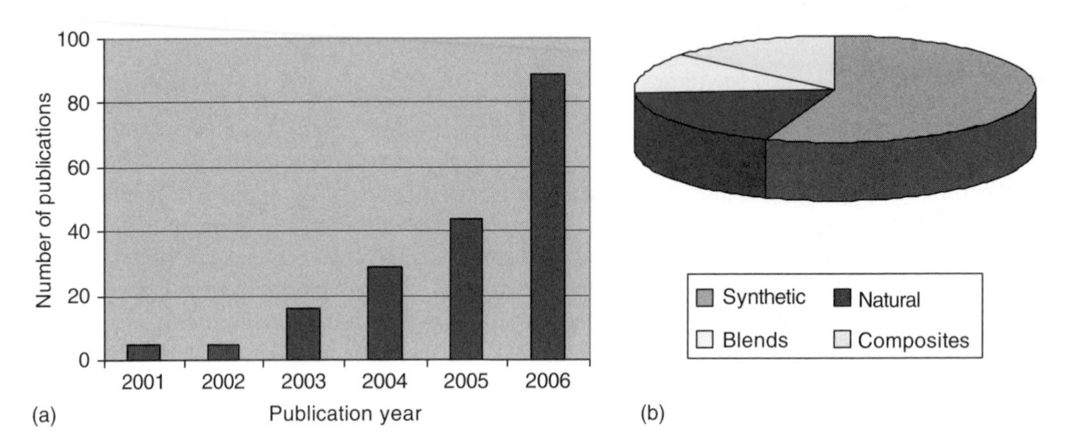

FIGURE 7.2 (See color insert following page 206.) Electrospinning statistics. (a) Number of papers in the literature based on a SciFinder Scholar search with terms "electrospinning, scaffolds, tissue engineering." This search also includes 77 publications reported in 2007 between January and October that are not shown in the graph. (b) These publications can be categorized into number of papers dealing with synthetic, natural, their blends and composite scaffolds as presented in the Pie chart. Majority of the publications dealt with the natural polymers; however, the recent efforts are towards the electrospinning of blends that combine the properties of naturals and synthetics to achieve improved mechanical properties and bioactivity.

sized fibrils that mainly composed of collagen, elastin, and fibronectin. Due to their close proximity and high surface area, the ECM nanofibrils provide an opportunity for the cells to interact with signaling molecules and hence control the cell functions and tissue properties. Engineering tissues require scaffolds that closely mimic the structure and function of the ECM [33]. Electrospinning is a versatile method to create biodegradable scaffolds with fiber sizes close to the ECM fibrils. Thus electrospun PLLA, PLAGA, PCL, and polyphosphazene nanofiber scaffolds provide topographical cues that result in improved cellular performance. In particular collagen, elastin, or their blends with synthetic polymers PLAGA or PCL mimic the ECM in structure and composition. These nanofiber scaffolds have shown improved cell adhesion, proliferation, differentiation, migration, and gene expression. The improved cellular response observed for electrospun scaffolds may translate into accelerated tissue healing *in vivo* and thus serve as better scaffolds for tissue engineering.

7.3 ELECTROSPINNING TO GENERATE NANOFIBROUS POLYMERIC SCAFFOLDS

7.3.1 Key Parameters Influencing Fiber Diameter and Morphology

Obtaining fibers with nanoscale diameters is easily accomplished by means of electrospinning. However, consistently obtaining fibers of desired diameter and morphology is more challenging. The most common morphological defects of nanofibers are beads that remain as small polymer aggregates in the nanofiber structure. Low polymer solution concentrations often produce more beads than fibers, and an optimum polymer concentration will produce bead-free nanofiber matrices as presented in Figure 7.3. Furthermore, for each polymer and solvent system these parameters need to be optimized since the polymer solution properties are different for each pair. Electrospinning is governed by several parameters and these parameters need to be optimized in order to produce fibers with the required diameter and morphology. These parameters can be grouped into three categories: polymer solution variables, controlled variables, and ambient variables.

7.3.1.1 Polymer Solution Variables

The most important parameter in electrospinning is polymer solution concentration. Polymer solutions of high concentrations are more viscous than low concentration solutions and can be readily electrospun into fibers. Low-viscosity solutions are more prone to produce beads (Figure 7.3a), and bead occurrences are reduced as concentration is increased (Figure 7.3c). For example, polystyrene (PS) solutions in tetrahydrofuran (THF) when electrospun produced only beads below 20 wt%, beads and fibers between 20–35 wt%, and only fibers at 35 wt% [34]. Concentrations exceeding a certain upper threshold, inherent and unique to each polymer, produce solutions that cannot be electrospun [35]. Although increase in viscosity produces fewer beads and tends to increase fiber diameter, this feature may or may not be desired. Solutions of PLAGA-*b*-PEG-NH_2 polymer in different solvents of equal concentration resulted in different solution viscosity [36] ranging from 220 to 1261cP and the resulting electrospun nanofiber matrices showed average fiber diameters from 449 to 761 nm, respectively.

Charge density of the polymer solution is another variable that can potentially affect the fiber size and morphology. Solvent composition and presence of the conducting moieties (salts, metallic particles, and carbon nanotubes) in the solution affect the charge density of the polymer solution. Addition of salts such as sodium chloride in small quantities (1 wt%) to PLA polymer solution resulted in bead-free nanofibers [35]. The increased solution conductivity is purported to result in greater charge density on the surface of the solution as it is pulled from the Taylor cone. The increased charge generates more force to stretch the polymer into fibers as it is pulled toward the target.

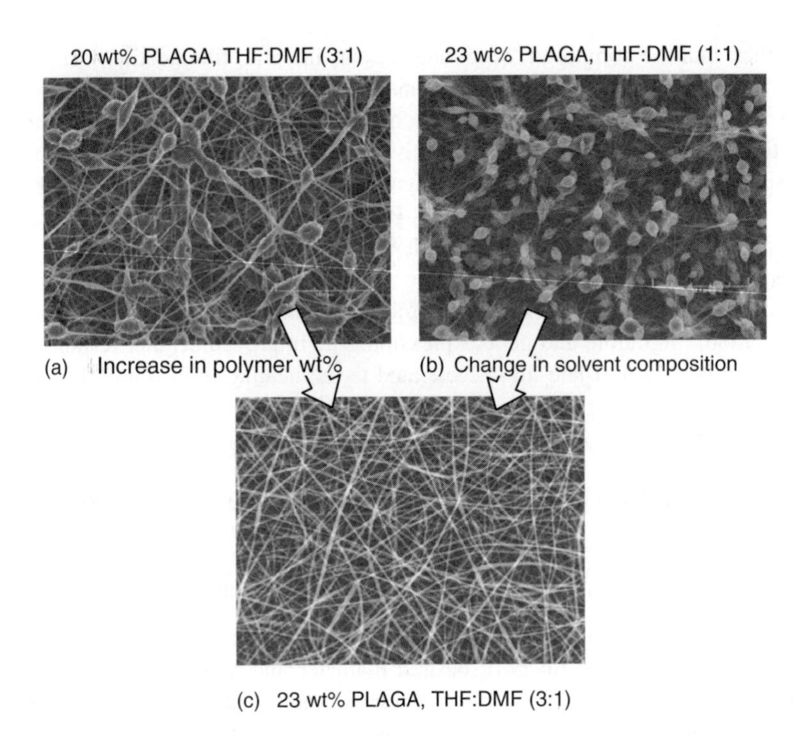

FIGURE 7.3 Scanning electron micrographs (SEMs) present morphology of the electrospun nanofiber matrices that resulted at different polymer concentrations and solvent compositions. The other spinning parameters were kept at a constant working distance of 20 cm, 20 kV applied potential, 18 gauge needle, and a flow rate of 2 mL/h. SEM of electrospun PLAGA 85/15 (0.7 dL/g) showing (a) bead and string formation at a polymer concentration of 20 wt% and THF:DMF (3:1) solvent, (b) more beads in the nanofiber structure at a concentration of 23 wt%, and THF:DMF (1:1), (c) defect free nanofibers at a concentration of 23 wt%, and THF:DMF (3:1) solvent. These observed morphological changes from (a) to (c) clearly demonstrate the effect of solution variables on the electrospun nanofiber morphology.

7.3.1.2 Controlled Variables

Controlled variables such as applied voltage, capillary diameter, and polymer solution rate need to be optimized to produce desired nanofibers. A minimal electrical potential called threshold voltage is necessary to provide sufficient force to extract the polymer fibers from the Taylor cone and pull them to the target. However, increased voltage can adversely affect the morphology of the fibers making them "rougher" in appearance. For instance, smooth poly(ethylene oxide) (PEO) electrospun fibers were produced at 5.5 kV and when the voltage was increased to 9.0 kV the nanofiber surface became coarse [37]. Increased voltage during electrospinning of PLA produced more beads and further bead morphology changed from being spindle-like to more spherical [35]. Such observed morphological changes as a result of applied voltages may be due to the position of the polymer jet initiation [37].

The diameter of the needle also has an effect on the size and morphology of the fibers formed. Larger needle diameters tend to produce fewer beads. Laurencin and coworkers previously shown that electrospinning of methylphenoxy substituted polyphosphazene (PNmPh) at a concentration of 3 wt% using a 25-guage needle produced fibers with spindle-like beads while the 18-guage needle spun bead-free fibers [18]. The distance from the needle tip to the grounded target has a significant effect on the electrospun fibers. Shorter distances tend to generate fewer beads. For instance, PCL nanofibers electrospun in Laurencin laboratories at a 10 cm working distance were bead free while at 20 and 30 cm contained lot of beads.

Polymer solution or melt feed rate affects the morphology and nanofiber fabrication rate. For instance, PLLA nanofibers electrospun at a higher feed rate of 75 μL/min produced large fibers with large beads [35]. While decrease in polymer feed rate reduced the fiber diameter and produced smaller diameter beads.

7.3.1.3 Ambient Variables

There are very few studies conducted to investigate the effects of ambient parameters such as temperature, humidity, and airflow on the electrospinning process. Most of the studies showed reduced fiber diameter when the electrospinning was carried out at elevated temperatures. This was attributed to the reduced viscosities of polymer solutions observed at higher temperatures [38]. Casper et al. studied the affect of humidity on the nanofiber surface morphology [39]. Humidity less than 25% did not affect the polystyrene fiber formation, whereas the fibers deposited at humidity above 30% showed pore formation on the nanofiber surface. Increase in humidity (30%–70%) caused higher number of pores with increased pore size and diameter. Though, the pore formation was not completely understood, this was mainly attributed to the evaporative cooling phenomenon that takes place when spinning is done in a highly humid environment [40]. Airflow can adversely affect electrospinning. As highly charged polymer chains are pulled toward the target, they spin in a large spiral loops before reaching the target. During electrospinning all the fibers carry the same charge and repel each other until they reach the grounded target. Airflow can disrupt the spiral pattern as the fibers move toward the target, and encourage interactions that would not normally occur. Thus it is always important to encase the electrospinning device using a nonconducting material to avoid the airflow and any possible temperature and humidity changes that affect the final fiber diameter and morphology.

7.3.2 Various Scaffold Materials

7.3.2.1 Synthetic Polymers

Synthetic biodegradable polymers have shown great promise as scaffold materials for tissue engineering. To date, researchers have investigated poly(α-esters), poly(ortho esters), poly(phosphazenes), and poly(anhydrides) as potential degradable materials for scaffolding applications. Among these, poly(α-esters) including poly(glycolic acid) (PGA), PLA, PLLA, PLAGA, and PCL have long history of use as synthetic biodegradable materials, and considered for electrospinning to generate nanofiber scaffolds for tissue engineering applications.

The electrospinning of PGA was accomplished by dissolving PGA in 1,1,1,3,3,3 hexafluoro-2-propanol (HFIP) at various concentrations to obtain fibers in the range of 0.2–1.2 μm [41]. PLA, which is more hydrophobic than PGA because of its methyl group, was easily dissolved in organic solvents to achieve electrospinning. The resulting fiber matrices were strong, noncompliant, and showed longer degradation times than PGA nanofiber matrices [42]. PLAGA, a copolymer of PGA and PLA with tunable mechanical and degradation properties, was electrospun to generate 3D nanofiber architectures with high porosity and mechanical strength to be used as scaffolds for tissue engineering [15]. Later, PLAGA (85/15) and PLAGA (50/50) were extensively used to create ECM mimicking nanofiber scaffolds with varied fiber diameter, porosity, mechanical strength, and fiber orientation for various tissues regenerative applications [43,44]. Polydioxanone, a highly flexible polymer with the degradation rate falling between PGA and PLA was also electrospun in HFIP to generate flexible scaffolds for soft tissue engineering [45]. PCL was also widely investigated for electrospinning to form nanofiber scaffolds with properties such as slow-degradation (1–2 years), and high modulus and elasticity [46]. Polyphosphazenes are yet another class of polymers with controlled physicochemical, mechanical, and degradation properties suitable for various biomedical applications. Laurencin laboratories for the first time electrospun several polyphosphazenes and the list of these polymers include poly[bis(p-methylphenoxy)phosphazene] (PNmPh), poly[bis(ethyl alanato)phosphazene] (PNEA), and their copolymer (PNEAmPh), and poly[bis(carboxylatophenoxy)phosphazene] (PCPP) [18,47,48]. These novel nanofiber scaffolds degraded into neutral and biocompatible degradation products and were suitable for bone tissue engineering applications [18,48]. Recently, polyhydroxybutyrate (PHB) and poly(hydroxybutyrate-*co*-valerate) (PHBV) were also electrospun to generate scaffolds for bone regeneration [49].

7.3.2.2 Natural Polymers

Polymers of natural origin are proven to be biocompatible and known to serve as better scaffold materials for tissue engineering. Such electrospun natural nanofiber matrices closely mimic the natural ECM in structure, composition, and function. Some of the natural materials considered for electrospinning were (1) ECM proteins such as collagen type I, II, III, and elastin, (2) serum proteins such as fibrinogen, hemoglobin, and myoglobin, and (3) naturally occurring polymers such as chitosan, alginate, hyaluronic acid, and silk.

Collagen is the principal component of the ECM in many tissues and hence chosen for electrospinning to generate tissue engineered scaffolds that mimic the structural, chemical, and biological properties of native collagen. Collagen type I and III were electro processed with an organic solvent HFIP and achieved fiber sizes 100 ± 40 nm and 250 ± 150 nm, respectively [50]. Elastin, another key component of the ECM, is insoluble in native form. Soluble forms such as elastin fragments (α-elastin, κ-elastin) or recombinant human tropoelastin were developed into nanofiber scaffolds that display elasticity and resilience for vascular tissue engineering applications [51]. Electrospinning of serum protein fibrinogen with HFIP and minimal essential medium (MEM) resulted fiber matrices containing 80–700 nm fiber diameter suitable for wound healing applications [14]. Globular proteins such as hemoglobin and myoglobin were also spun into nanofiber conduits for oxygen delivery induced wound healing applications [52]. In addition to these biopolymers, efforts were also made to produce nanofiber scaffolds from natural biodegradable materials such as chitosan, alginate, hyaluronic acid, and silk via electrospinning [53–55]. For convenience, natural polymers are often blended with biocompatible water soluble poly(vinyl alcohol) (PVA), poly (ethylene glycol) (PEG), and PEO to aid the fiber formation while maintaining the biocompatibility [56,57]. All these bio/natural polymer matrices were subjected to cross-linking reaction using cross-linking agents such as vapors of glutaraldehyde, formaldehyde, and many acrylate based photo cross-linkable groups to acquire required mechanical strength to be used as tissue engineering scaffolds.

7.3.2.3 Polymer–Polymer Blends

Nanofibers were fabricated from synthetic–synthetic, synthetic–natural, and natural–natural polymers by electrospinning. Such nanofiber scaffolds combine the properties of individual polymers and showed improved biocompatibility and mechanical support during the tissue regeneration. Some of the popular combinations that were electrospun into nanofiber scaffolds include the blends of PLA–PCL, PLAGA–dextran, PLAGA–gelatin–elastin, PCL–gelatin, PHBV–collagen, polydioxanone–elastin, polyaniline–gelatin, collagen–elastin, and gelatin–hyaluronic acid.

Nanofiber PLG, PLA, and PLAGA scaffolds show good mechanical strength however, lack extensibility. Addition of PCL to PLG and PLA improved extensibility (PGA to 400% and PLA to 200%) of the nanofiber matrices and led to scaffolds with the required strength as arterial grafts [58]. Though natural polymers are good candidate materials for tissue engineering, lack of mechanical strength and poor processability limit their use for scaffolding applications. Electrospinning natural polymers in combination with synthetic polymers have led to a new set of materials with improved biocompatibility and mechanical performance. For example, electrospinning of PLAGA–elastin–gelatin resulted in bioactive ECM mimicking scaffolds with optimal mechanical/elastic properties suitable for engineering soft tissues [59]. Blending gelatin with PCL improved the processability of gelatin for electrospinning and attained the required mechanical strength and elasticity without

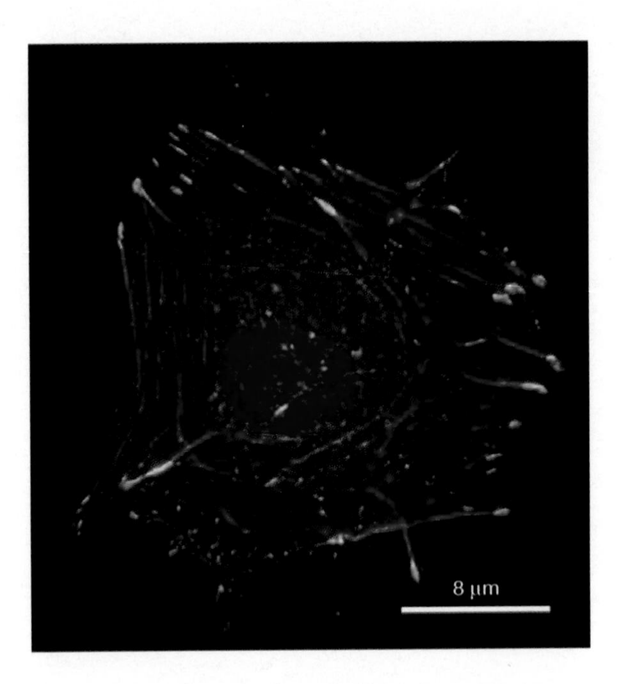

FIGURE 2.6 Fluorescence image of human umbilical cord vein endothelial cells (HUVECs) adhered on a two-dimensional (2D) glass surface in the presence of ECM proteins and proteoglycans. Actin filaments (red) and integrins are connected through vinculin proteins (green). Blue: nucleus-DAPI, green: antivinculin–FITC, red: Alexa Fluor 594 phalloidin-F-actin (Veiseh, M. unpublished observation).

FIGURE 4.1 Silk fibroin-based microfluidic devices. Replica molding can be used in conjunction with solvent-casting of aqueous silk fibroin solutions on microfabricated elastomeric masters. Replica-molded silk fibroin films produce high-fidelity features including (A) nanoscale posts with minimum widths of approximately 400 nm and (B) micron scale fluidic channels, which were used in subsequent experiments. (Scale bars are 5 μm and 500 μm in (A) and (B), respectively.) (C, D) SEM micrographs of the cross sections of microfabricated devices demonstrate retention of feature geometries in thin films with microchannel widths of approximately (C) 240 μm and (D) 90 μm. (Scale bars are 200 μm and 10 μm in (C) and (D), respectively.) (E, F) Patent microfluidic devices are demonstrated by fluorescent micrographs of devices perfused with rhodamine solution. Retention of the perfusate within the microchannels suggests robust bonding at the interface. (Scale bars are 500 μm and 50 μm in (E) and (F), respectively.) (From Bettinger, et al., *Adv. Mater.*, 19, 2847, 2007. With permission.)

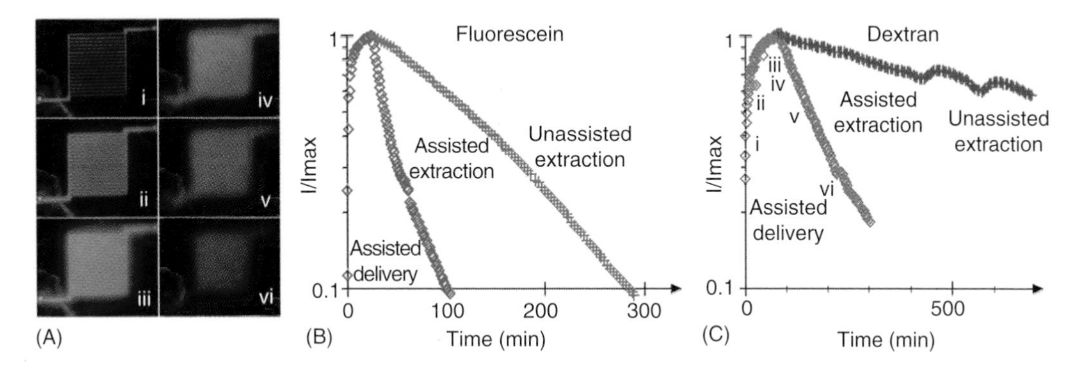

FIGURE 4.2 Characterization of mass transfer in a microfluidic biomaterial (μFBM) fabricated from alginate microfluidic devices. (A) Fluorescence micrographs of a μFBM during assisted delivery (i–iii) and assisted extraction (iv–vi) of RITC-dextran. Assisted delivery refers to the operational mode of the device in which the solute is perfused through the microchannels and dissolves throughout the network. Assisted extraction refers to the operational mode where solute loaded within the alignate network is removed by perfusion of the microchannel network with an aqueous solution. These modes correspond to the cases of nutrient supply and waste removal, respectively. The rate of diffusion is primarily a function of the molecular weight of the solute, the crosslinking density of the network, the concentration of the solute in the perfused solution, and the flow rate of the perfused medium. (B, C) Temporal evolution of the normalized total intensity from fluorescence images, such as those in (A), during delivery and extraction of solutes (fluorescein, MW = 376 Da; RITC-dextran, MW = 70 kDa). Diamonds represent intensities during assisted delivery and assisted extraction. Crosses represent intensities during unassisted extraction from the same initial condition (i.e., achieved by delivery via the channels) as the assisted extraction experiment. The starting time of the unassisted evacuation has been shifted to match that of the assisted extraction. The dimensions of the construct as follows; the height of the gel $H = 0.29$ cm, the lateral dimension of the gel $L = 1$ cm, the height of the microchannels $h = 200$ μm, the width of the channels is $w = 100$ μm. The linear flow velocities of the fluid on the exterior of the construct is $u_r = 1$ cm/s while the velocity of the fluid within the microchannels is $u_c = 0.6$ cm/s. The solute concentrations are $c_0 = 20$ μmol/L for fluorescein and 10 μmol/L for RITC-dextran. (From Cabodi, et al., *J. Amer. Chem. Soc.*, 127, 13788, 2005. With permission.)

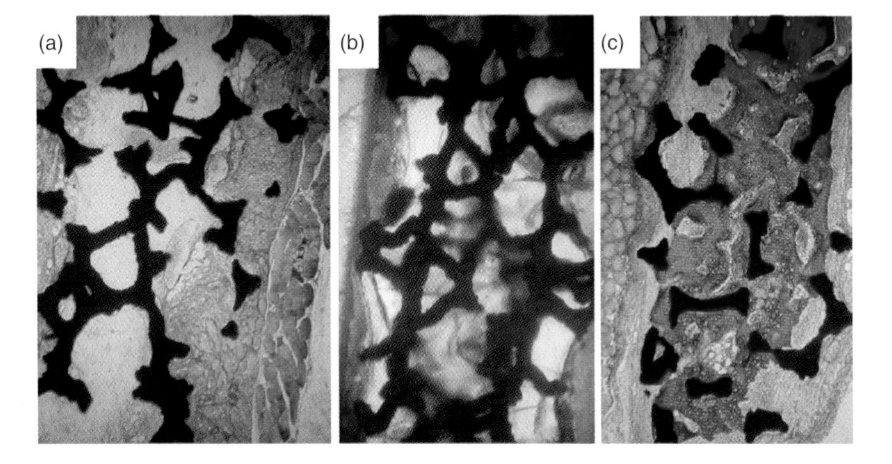

FIGURE 5.18 Histology after 6 weeks implantation into rat calvaria of (a) uncoated tantalum and tantalum coatings, (b) micron grain size HA, and (c) nanophase grain size HA. Stain = Stevenel's blue and a counterstain of van Gieson's picrofuchsin. While little bone in growth is observed for (a) uncoated tantalum and (b) tantalum coated with micron grain size HA, much is observed for (c) tantalum coated with nanograin size HA. (From Sato, M., An, Y.H., Slamovich, E.B., and Webster, T.J., *Int. J. Nanomedicine*, in press, 2008. With permission.)

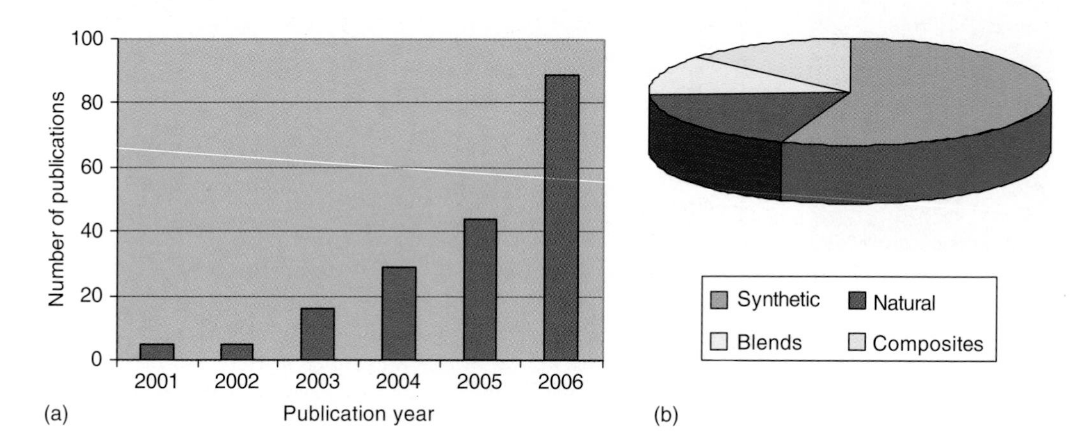

FIGURE 7.2 Electrospinning statistics. (a) Number of papers in the literature based on a SciFinder Scholar search with terms "electrospinning, scaffolds, tissue engineering." This search also includes 77 publications reported in 2007 between January and October that are not shown in the graph. (b) These publications can be categorized into number of papers dealing with synthetic, natural, their blends and composite scaffolds as presented in the Pie chart. Majority of the publications dealt with the natural polymers; however, the recent efforts are towards the electrospinning of blends that combine the properties of naturals and synthetics to achieve improved mechanical properties and bioactivity.

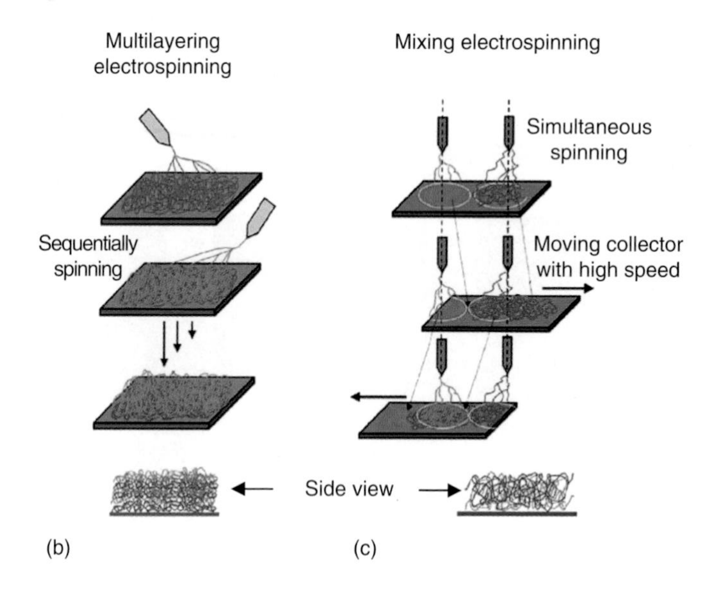

FIGURE 7.4 Various methods of electrospinning: (b) Multilayered electrospinning. (c) Mixing electrospinning. Multijet electrospinning provided opportunities to improve spinning process for obtaining large area uniform thick nanofiber mats, with production rate that match the industrial fiber production process. Native tissues are often multicomponent with layered or mixed arrangement. Multilayering or mixing electrospinning are two novel approaches to build scaffolds that duplicate the native tissue structure. (From Kidoaki, S., Kwon, I. K., and Matsuda, T., *Biomaterials*, 26, 37, 2005. With permission.)

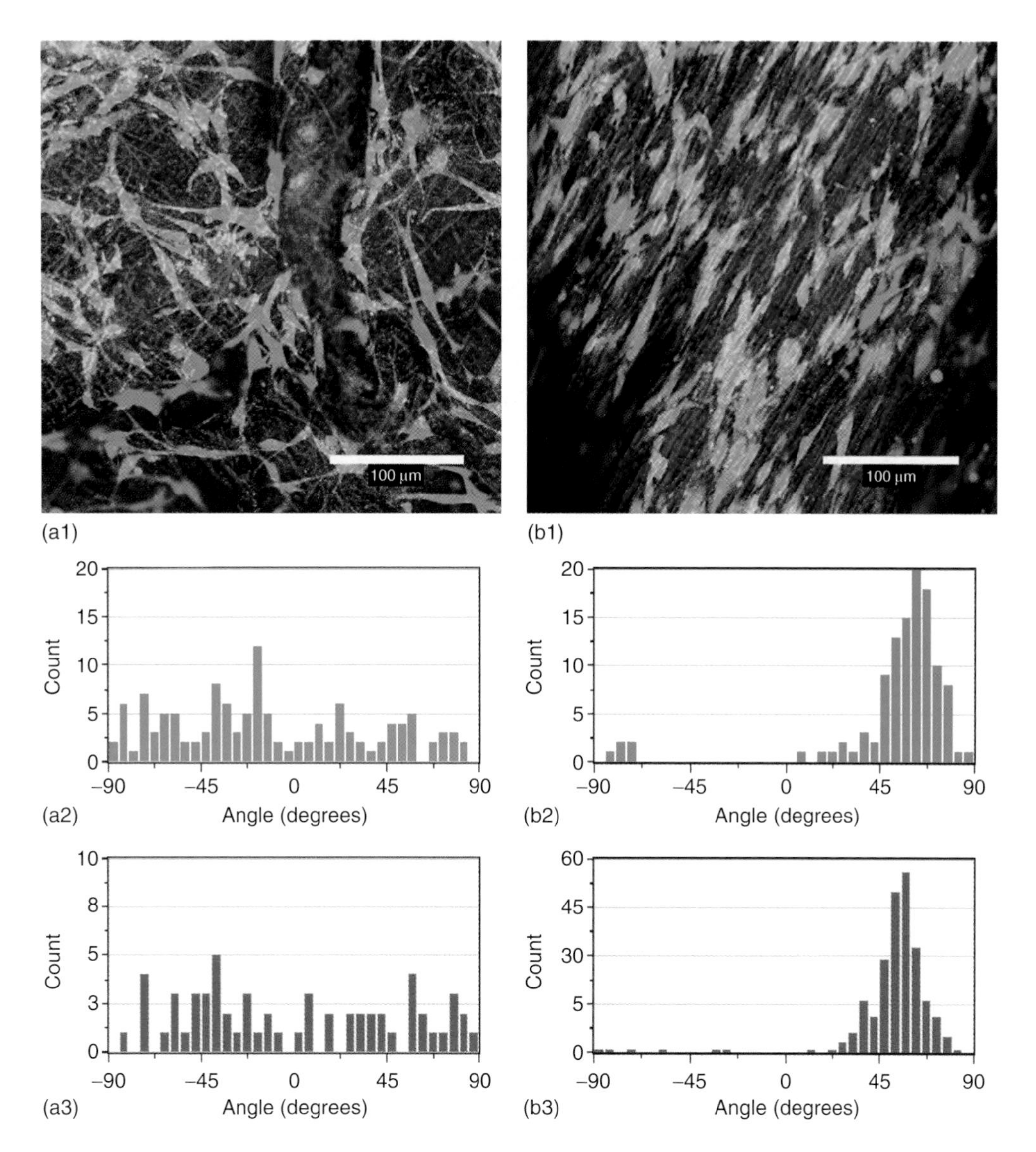

FIGURE 9.8 Superimposed and pseudocolored images of hMSCs (green) and nanofibers (red) in the (a1) random and (b1) oriented PCL nanofibrous scaffolds after 24 h cell culture. Histograms representing distributions of cell orientation angles with respect to the horizontal axis for hMSCs on (a2) random and (b2) oriented scaffolds. Histograms representing distributions of nanofiber orientation angles with respect to the horizontal for (a3) random and (b3) oriented scaffolds. Histograms in (a2) and (a3) were constructed based on measurements of 127 cells and 234 nanofibers, respectively, while 124 cells and 249 nanofibers were used to construct the histograms in (b2) and (b3), respectively.

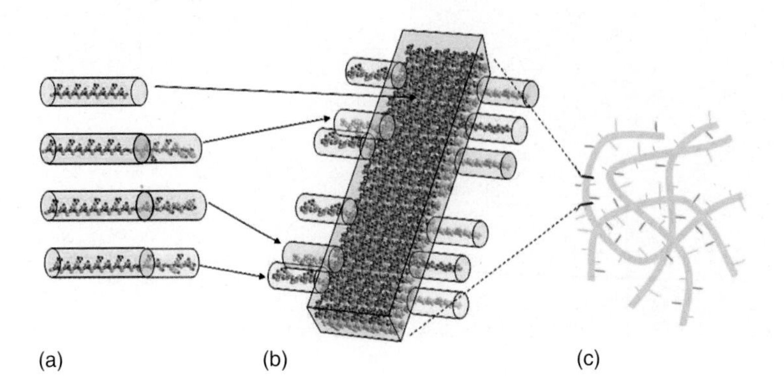

FIGURE 11.1 Schematic illustration of the designer self-assembling peptide scaffold. (a) Direct extension of the self-assembling peptide sequence by adding different functional motifs. Longer cylinders on the left side represent the self-assembling backbone and the shorter cylinders on the right side represent various functional peptide motifs. (b) Molecular model of a self-assembling peptide nanofiber with functional motifs flagging from both sides of the double β-sheet nanofibers. Either mono or multiple functional (or labeled) peptide motifs can be mixed at the same time. The density of these motifs can be easily adjusted by simply mixing them in various ratios, 1:1–1,000,000 or more before the assembling step. (c) They then will be part of the self-assembled scaffold.

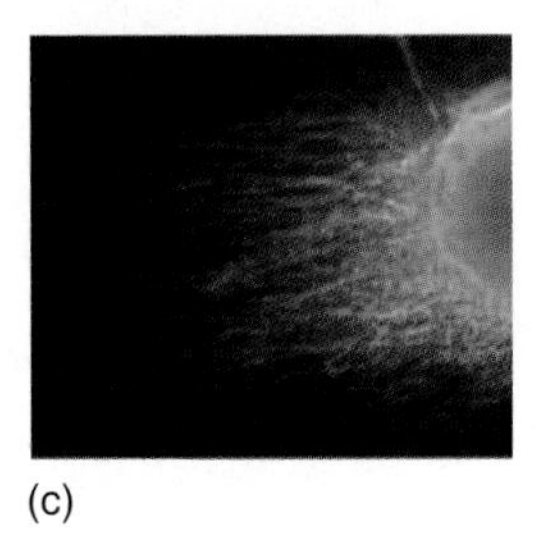

FIGURE 12.2 (c) Dorsal root ganglia from P3 pups cultured on aligned electrospun fibers.

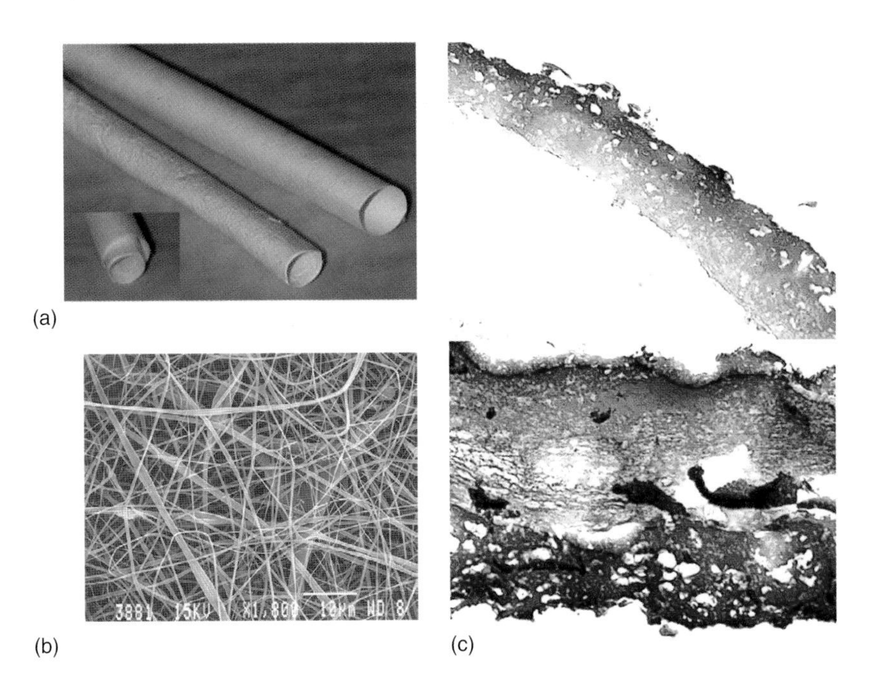

FIGURE 13.5 (a) Collagen electrospun scaffolds fabricated in 2 and 4 mm tubular structures for small-diameter vascular graft engineering. (b) An electrospun blend of collagen type I, type III, and elastin (in a 40:40:20 ratio) showing fibers with an average diameter of 490 ± 220 nm and random fiber orientation. (c) Top: Histological section demonstrating three defined layers of vessel wall architecture: intimal (dark purple, left), medial (light purple, center) and adventitial (dark purple, right). Bottom: Three-layered prosthetic electrospun scaffold after seeding with endothelial cells, vascular smooth muscle cells, and fibroblasts, demonstrating similar definition of layering as native tissue. (From Boland, E.D., et al., *Front Biosci.*, 9, 1422, 2004. With permission.)

chemical cross-linking [60]. Biodegradable conducting polyaniline contained gelatin nanofibers were also achieved by electrospinning to generate biocompatible conducting scaffolds for neural tissue engineering [23]. The nanofiber blend of collagen–elastin biopolymer scaffolds possesses resistance from rupture (from collagen) and irreversible deformation from the pulsatile blood flow (from elastin) to serve as an ideal scaffold for vascular tissue engineering [61].

7.3.2.4 Polymer–Ceramic Composites

Nanofibers loaded with nanoparticles present a composite nanofiber system that is known to increase mechanical strength and also the scaffold biofunctionality. Such particulate system consisting of hydroxyapatite (HA) and β-$Ca_3(PO_4)_2$ nanoparticles improves the osteocompatibility, while carbon nanotubes (CNTs) impart higher mechanical strength and electrical conductivity to the resulting scaffolds.

Electrospun triphasic nanofiber scaffold made of PCL–collagen I–HA (30 nm) simulates the bone in structure and composition. Scaffolds with an average fiber diameter 180 ± 50 nm (matches well with the native collagen bundle in ECM) and tensile modulus 2.2 GPa are promising candidates for bone tissue engineering [24]. Electro processed gelatin–HA resulted in biomimetic nanocomposite nanofiber scaffolds with HA composition gradient applicable for guided bone tissue engineering [62]. Biodegradable ethyl valinato substituted polyphosphazene (PNEV) was also electrospun with HA to create mechanically robust osteocompatable scaffolds with fiber diameters in the range of 100–800 nm [63]. Electrospun PLA nanofiber scaffolds containing multiwalled carbon nanotubes showed enhancement in mechanical properties and such electrically conducting scaffolds are suitable for neural tissue engineering applications [64]. Since CNTs are electrically conducting, the addition of CNTs resulted in an increased surface charge density and hence reduction in fiber diameter during electrospinning. Using various compositions of HA and CNTs, it is possible to improve mechanical strength and electrical conductivity of these scaffolds.

7.3.3 Various Methods

7.3.3.1 Solution and Melt Electrospinning

Electrospinning in general is applied to higher viscosity fluids. For polymers that have glass transition above room temperature, the high viscosity fluid state is only achieved either by dissolving them in a suitable solvent or processing at a temperature close to the melting point. The vast majority of research done to date has used solution electrospinning due to the fact that polymer solutions can be spun with little concern for ambient temperature and they readily produce very small fibers. Optimization of fiber diameter and morphology in solution spinning is easily obtained by adjusting polymer concentration or solvent composition.

There were some attempts in the literature to develop nanofiber structures using melt electrospinning. The setup for melt spinning is similar to the solution electrospinning apparatus except a heating mechanism added to melt inbound polymer, that keeps the melt heated until drawn out of the needle, and onto the target. The two important variables in this process are temperature and the polymer molecular weight. Lyons et al. used several polypropylene polymers of varying molecular weights, and observed that increase in molecular weight necessitated voltage increase to overcome the surface tension of the melt [65]. It was also observed that melt spinning required an electric field nearly 10 times stronger than that is normally required for solution electrospinning. Melt electrospinning of poly(ethylene glycol-*b*-caprolactone) (PEG-*b*-PCL) at temperatures between 60°C–90°C resulted in defect free solid fibers (200–1200 nm) at melt flow rates lower than 0.05 mL/h. In contrast higher flow rates resulted molten fibers because of the insufficient cooling [17]. Ogata et al. used a laser melting device to produce PLA nanofibers with an average fiber diameter less than 1 μm [66]. Coaxial melt electrospinning was also recently developed to encapsulate solid materials for microencapsulation and controlled delivery applications [67]. The melt spinning method is feasible with only the polymers that stay stable at elevated temperatures. More investigations in this direction are required to extend this method for various biodegradable polymers. Despite the unique challenges, melt electrospinning provides an opportunity to derive nanofiber scaffolds with no traces of toxic organics for various tissue regeneration applications.

7.3.3.2 Electrospinning with Different Collector Geometries

Collector geometry dictates fiber assembly in the electrospinning process. Various collector geometries were proposed to achieve fiber orientation, fiber patterning, and controlled deposition. Collector geometries including rotating drum, disc collector, parallel electrodes, array of counter electrodes, and ring electrodes were used to achieve control on the deposition via electrode or collection geometry [68]. In electrospinning jet velocity is measured to be in the range of 2–186 m/s. A drum or disc rotating close to the jet velocity yields significant alignment along the axis of rotation. Rotational speeds much lower or higher than the jet velocity results in mix or random orientation. It is observed that the oriented fiber scaffolds show much better mechanical properties along the axis of rotation when compared to the scaffolds with randomly oriented fibers [69]. Counter electrode (deposited on quartz) collection geometry was used to obtain layer-by-layer aligned arrays by alternatively grounding the opposite pair of electrodes [70]. Further ring electrodes were used to achieve fiber deposition minimized to a particular area, which is referred as controlled or constricted electrospinning [37]. The scaffolds derived using different collector geometries find suitable applications in tissue engineering.

7.3.3.3 Coaxial Electrospinning

A simple coaxial electrospinning setup consists of two coaxial needles connected to two individual syringes. Except the feeder assembly, other experimental setup remains same as the one used for conventional electrospinning. By feeding two different polymer solutions through the coaxial tip, it is possible to obtain core–shell polymer nanofibers. This method controls core and shell thickness and the final fiber diameter by controlling flow rates of inner and outer fluids. Drug or growth factor in combination with a biodegradable polymer as inner fluid result in functional tissue engineered scaffolds with a controlled and sustainable release pattern. Also, coaxial spinning can combine bioactivity (via shell polymer) and mechanical strength (via core polymer) to develop a scaffold that exhibit superior mechanical characteristics with excellent bioactivity [71]. Bio-electrospinning or cell spinning is one of the recent applications of co-electrospinning method. In addition to the above mentioned biological applications, coaxial spinning further used to create fibers incorporated with both active and passive components for various biotechnological applications [72,73].

7.3.3.4 Multispinneret Electrospinning

The low fluid throughput of a single jet (10 μL/min–10 mL/min) limits the industrial use of electrospinning process. One of the ways to increase the production rate is via a technique called multispinneret/emitter electrospinning [74]. In this method multiple syringes are arranged in square or line fashion to obtain uniform fiber spread with wide area coverage as shown in Figure 7.4a. Initial studies with two side by side syringes lead to the path deviation from the straight line due to the inter jet Columbic repulsion. However, the deviation observed was reduced by the introduction of more jets in a line or matrice configuration [75]. The multispinneret process was modeled to evaluate the right parameters such as the distance between the jets, multiple jet configuration, flow rate, and the distance between the nozzles and the collector to obtain uniform nonwoven nanofiber mats with increased production rate. In addition to high throughput deposition, multijet method was further used for multilayering (Figure 7.4b) or mixing electrospinning (Figure 7.4c) [76]. In these methods two or more jets were used (focused at the same area or spun on a rotating mandrel) with different polymer solutions or operated at different conditions to achieve multilayered or mixed fiber

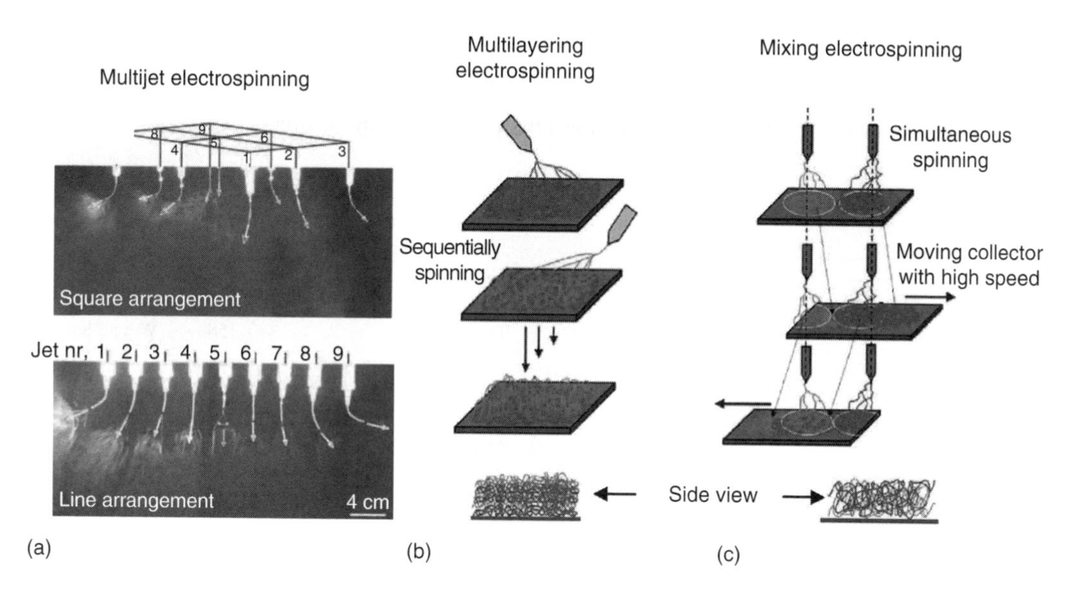

FIGURE 7.4 (See color insert following page 206.) Various methods of electrospinning: (a) Multijet electrospinning where jets are arranged in square and line configuration. (From Theron, S.A., *Polymer*, 46, 2889, 2005. With permission.) (b) Multilayered electrospinning. (c) Mixing electrospinning. Multijet electrospinning provided opportunities to improve spinning process for obtaining large area uniform thick nanofiber mats, with production rate that match the industrial fiber production process. Native tissues are often multicomponent with layered or mixed arrangement. Multilayering or mixing electrospinning are two novel approaches to build scaffolds that duplicate the native tissue structure. (From Kidoaki, S., Kwon, I.K., and Matsuda, T., *Biomaterials*, 26, 37, 2005. With permission.)

matrices. Novel mesoscopic spatial designs composed of biopolymer layers (with varied fiber size) fabricated by multispinneret electrospinning mimic natural tissue in structure and may find applications in functional tissue engineering.

7.4 ELECTROSPUN SCAFFOLDS IN TISSUE REGENERATION

Electrospun nanostructures due to their ability to mimic the ECM in topography and chemistry are increasingly becoming popular as scaffolds for tissue engineering use [77]. Electrospinning was effectively used to create scaffolds with required porosity, fiber size, and the orientation that best fit the required scaffold application [78]. This technique also has the ability to create scaffolds with varied mechanical properties from elastic to plastic based on the polymer selection (synthetic, natural, synthetic–natural, and polymer–ceramic composites) and the conditions used for spinning [69]. Cytocompatability studies with cells including endothelial, smooth muscle, neural, fibroblasts, osteoblasts, chondrocytes, and various stem cell types clearly demonstrated the potentiality of these scaffolds for various tissue regeneration applications [5,33,79]. The observed fiber size dependent cell attachment, proliferation, migration, and differentiation further exhibit the tunability of these scaffolds to evoke a particular cellular response. A list of biodegradable polymer nanofiber scaffolds, spinning parameters, scaffold properties, and their possible applications are presented in Table 7.1.

The ability to obtain oriented fiber structures led to the development of scaffolds for soft tissue (dermal, vascular, neural, and cartilage tissue) regeneration [5]. Electrospun flexible membranes with required porosity and barrier properties are appealing for wound dressing [44,80]. These membranes offer controlled water evaporation, excellent oxygen permeability, exudates removal, and also secure the wounds from exogenous microorganisms [81]. In addition to wound dressing, nanofiber mats generated by electrospinning (Figure 7.5a) mimic the ECM and can be used as

TABLE 7.1
Various Biodegradable Polymers Electrospun into Nanofiber Scaffolds: Spinning Parameters, Scaffold Properties, and Tissue Engineering Applications

	Polymer	Solvent and Polymer Concentration (wt/v)	Fiber Size (nm)	Tissue Engineering Application
Synthetic	PLAGA 85/15 [15]	THF:DMF(1:1), 5%	500–800	Smooth tissue
		THF + DMF	550–970	Bone, cartilage
	PLAGA 50/50 [44]	THF:DMF(3:1), 20%–30%	340–1500	Wound healing
	PCL [99–101]	Chloroform, 10%	400 ± 200	Bone
		THF:DMF (1:1), 14%	700	Cartilage
		$CHCl_3$:CH_3OH (1:1), 10%	250	Cardiac tissue
	PLLA-*b*-PCL [102]	HFIP, 3%	700–800	Vascular
	PNmPh [18]	Chloroform, 8%	1200	Bone
Natural	Col I [82]	HFIP, 8%	460	Wound healing
	Col II [103]	HFIP, 4%	60–270	Cartilage
	Elastin [51]	HFIP, 20%	500–1500	Soft tissue
	Gelatin [51]	HFIP, 8.3%	200–500	Soft tissue
	Fibrinogen [104]	HFIP:MEM (9:1), 8.3%	80–700	Wound dressing
Polymer–polymer blends	Col I–Ela–PDLA [84]	HFIP, 15%	720 ± 350	Vascular
	Polydioxanone [105]	HFIP, 30%	400–1200	Vascular
	Col I–P(LLA–PCL) [106]	HFIP, 5%	100–200	Vascular
	Col I–elastin [61]	10 mM HCl, 1%–5%	220–600	Vascular
	PCL–gelatin [60]	TFE, 10%	1000	Vascular
	PCL–Col I [98]	HFP + $CHCl_3$ + CH_3-OH, 8%	170	Dermal
	PLAGA–dextran [107,108]	DMSO:DMF (1:1), 60%	1000	Wound healing
	Polyaniline-gelatin [23]	HFIP, 8%	50–800	Cardiac, neural
	PLAGA–Gel–Ela [59]	HFIP, 38%	380 ± 80	Soft tissue
Polymer–ceramic composites	PCL–nHA [85]	Chloroform, 20% nHA	450–650	Bone
	PLA–CNT [109]	$CHCl_3$ + DMF, 1% MWCNT	550–860	
	PNEA–nHA [25]	DMF + THF, 50%	100–800	
	Gelatin–nHA [62]	HFP, 16%–53% HA	200–400	
	PCL–nHA–collagen [24]	HFP, 20% nHA	180 ± 50	

PLAGA 85/15, poly(D,L-85lactide-*co*-15glycolide); PLAGA 50/50, poly(D,L-50lactide-*co*-50glycolide); PCL, polycaprolactone; PLLA-*b*-PCL, poly(L-lactide-*co*-caprolactone); PNmPh, poly[bis(ethyl alanato)phosphazene]; Col I, collagen I; Col III, collagen III; Ela, elastin; Gel, gelatin; PLA, poly(DL-lactide); PNEA, poly[bis(ethyl alanato)phosphazene]; nHA, nano hydroxyapatite; THF, tetrahydrofuran; DMF, dimethyl formamide; DMSO, dimethyl sulfoxide; HCl, hydrochloric acid; MWCNT, multiwalled carbon nanotubes.

scaffolds for dermal tissue engineering [22,82]. The early break through in tissue engineering was in the form of synthetic skin substitutes. Since then, there is a great deal of interest and opportunity for improving current skin grafts, and this can be achieved with electrospun nonwoven interconnected membranes.

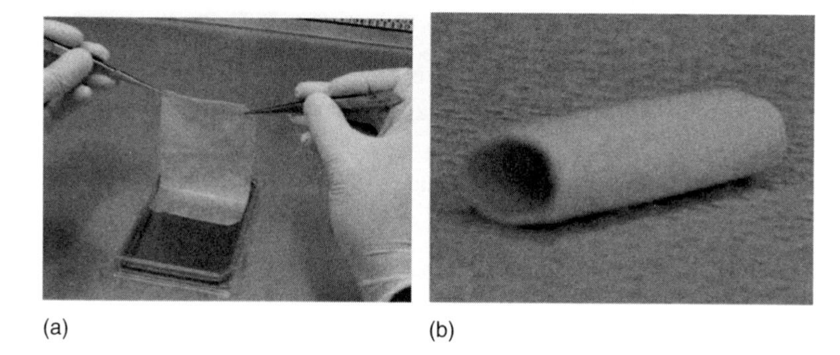

(a) (b)

FIGURE 7.5 Examples of biodegradable electrospun nanofiber scaffolds. (a) Nanofiber scaffolds in the form of membranes have potential applications as wound dressing, would healing, and dermal tissue engineering. (From Venugopal, J.R., *Artif. Organs*, 30, 440, 2005. With permission.) (b) Small diameter grafts for blood vessel tissue engineering. Though electrospinning has been used to derive scaffolds for various tissue engineering applications, the technique so far has established its reputation in producing nanofiber membranes for wound healing, dermal tissue regeneration and small diameter tubes for vascular tissue engineering. In authors perspective, these are the two scaffold types that have generated lot of excitement and may soon find applications in tissue engineering. (From Stitzel, J. et al., *Biomaterials*, 27, 1088, 2006. With permission.)

Electrospinning is a facile technique to generate small diameter tubes (less than 6 mm) as conduits for vascular tissue engineering (Figure 7.5b). Nanofiber tube conduits with appropriate elastic properties and bioactivity can serve as grafts for vessel development [83,84]. Because of the tube scaffold nanotopography, either preconditioning with cells or *in vivo* use as conduit results in collagen and elastin fiber deposition that gradually replace the biodegradable tube conduit and form tissue engineered vessel. This strategy can eliminate the use of prosthetic vascular grafts, which remain patent only 15%–30% after 5 years of use, for treating coronary artery and peripheral vascular diseases. The oriented electrospun fiber scaffolds were also considered for neural and cartilage tissue engineering use. There were also numerous efforts to load HA nanoparticles to achieve composite nanofiber scaffolds that best mimic the bone in structure and composition for bone tissue repair and regeneration [24,85].

Nanofiber scaffolds comprising cells and growth factors are ideal combinations for tissue engineering. Recently, co-electrospinning was used for bioelectrospinning or cell spinning. Since living organisms are involved in this process solvent selection poses stringent requirements. In addition to spinnability, solvent should be biocompatible for cell survival during spinning. Also the lingering question has been how living cells behave under strong electric fields. In an attempt to answer, Wagner et al. electrosprayed smooth muscle cells (SMCs) with a biodegradable, elastomeric poly(ester urethane) [86]. This study has successfully micro-integrated SMCs into a biodegradable matrix without observing any changes in the cell viability. In a recent study, Jayasinghe et al. demonstrated cell electrospinning using biosuspension with 10^6 EHDJ/1321N1 (immortalized human brain astrocytoma) cells/mL flow in the inner needle and a medical grade PDMS (high viscosity and low electrical conductivity) flow through the outer needle [87]. Although the study showed cell clumps along the fiber length, post-processed cells showed no significant difference in their cell viability compared to the unprocessed cells. Cell spinning with biodegradable polymers loaded with various bioactive agents is still challenging and can provide the opportunity to fabricate highly cellularized and mechanically functional tissue constructs for soft tissue engineering applications.

7.5 ELECTROSPUN SCAFFOLDS FOR DRUG AND FACTOR DELIVERY

Scaffolds with the ability to deliver drugs or growth factors are of high value in tissue engineering. Sustained local delivery of antibiotics, antifungal, and antimicrobial agents could protect from any

TABLE 7.2
Various Biodegradable Polymers and Bioactive Agents Used to Fabricate Electrospun Nanofiber Scaffolds for Drug, Protein, DNA, and Growth Factor Delivery

Scaffold	Drug/Protein/ DNA/Growth Factor	Release Characteristics	Agent Delivery
PLAGA 50/50 [44]	Cefazolin		Antibacterial
PDLA, PEVA, 50PDLA/50PEVA [88]	Tetracycline hydrochloride	SR for 5 days	
PLLA (cs) [110]	Tetracycline hydrochloride	SR for 30 days	
PCL (cs) [92]	Gentamicin sulphate	SR for 7 days	
PLAGA/PEG-*b*-PLA [111]	Mefoxin, cefoxitin sodium	BR + SR for 7 days	
HPMC [112]	Itraconazole	SR for one day	Antifungal
Chitosan + PEO [113]	Potassium 5-nitro-8-quinolinate		Antimicrobial
PLGA [89]	Paclitaxel	SR for 60 days	Anticancer
PEG-*b*-PLLA [114]	Doxorubicin hydrochloride	BR + SR for10 h	
PLLA [115]	Doxorubicin hydrochloride	SR for 4 h	
PLLA [116]	Rifampin	SR for 6 h	Antituberculosis
PLGA, PLA–PEG [93]	Plasmid DNA	BR + SR for 20 days	DNA
PLAGA/HAp [94]	BMP-2 plasmid DNA	SR for 45–55 days	
PCL + PEG (cs) [117]	BSA, lysozyme	SR for 30 days	Protein
PCL/Dextran/PEG (cs) [117]	BSA	SR for 28 days	
Silk + PEO/Silk + nHA/ Silk + PEO + nHA [95]	BMP-2		Growth factor
PCLEEP (cs) [96]	β-NGF/BSA	BR + SR for 90 days	
PCL (cs) [118]	PDGF/BSA		

Note: Release characteristics presented here are obtained from *in vitro* studies. PLAGA 50/50, poly(D,L-50lactide-co-50glycolide); PLA, poly(DL-lactide); PEVA, poly(ethylene-*co*-vinyl acetate); PCL, polycaprolactone; PEG-*b*-PLA, poly(ethylene glycol-*co*-DL-lactide); HPMC, hydroxypropylmethylcellulose; PEO, polyethylene oxide; PEG-*b*-PLLA, poly(ethylene glycol-*co*-L-lactide); PCLEEP, poly(caprolactone-*co*-ethyl ethylene phosphate); cs, core–shell nanofibers; BMP-2, bone morphogenic protein 2; BSA, bovine serum albumin; β-NGF, β-nerve growth factor; PDGF, platelet-derived growth factor; BR, burst release; SR, sustained release.

possible infections that may transmit during scaffold implantation. Scaffold can be further used as a delivery vehicle to locally deliver anticancer and antituberculosis cytotoxic drugs. Also delivering a single or multiple growth factors as chemical cues to the surrounding cells is detrimental in successful tissue engineering. A list of biodegradable scaffolds loaded with drug, protein, DNA, and growth factors is shown in Table 7.2.

Electrospinning has been used to fabricate drug encapsulated nanofiber scaffolds for various tissue engineering applications. Kenawy et al. used poly(ethylene-*co*-vinylacetate) (PEVA), PLA, and a blend of 50PEVA–50PLA to load antibacterial agent tetracycline hydrochloride [88]. The feasibility of incorporating Cefazolin into bioresorbable PLAGA 50/50 nanofiber (≈470 nm) matrices was studied by Katti et al. for antibacterial loaded nanofiber systems for wound dressing applications [44]. PLAGA based nanofiber scaffolds were further used to encapsulate an anticancerous drug paclitaxel [89]. Though this method of encapsulating drugs in nanofibers has seen success, most of the scaffolds with water soluble drugs exhibited burst release pattern and released 90% of the drug in the first few hours [35,88]. This is because of the fact that drugs in general are hydrophilic, and cannot be homogenized in organic solvents that are commonly used for making polymer solutions. Due to the surficial effect at nanoscale, the drug particles tend to accumulate on the fiber surface that ultimately caused burst release *in vitro* and *in vivo* [90].

Co-electrospinning was recently adopted as an alternative approach for encapsulating drugs and bioactive agents into nanofiber scaffolds. As described in Section 7.3.3.3, this method can effectively control the amount of drug that is incorporated by (1) varying the drug concentration, and (2) varying the flow rate of the solution through the inner needle. Jiang et al. fabricated nanofiber scaffolds with water soluble model proteins BSA, lysozyme in PEG as core, and PCL as shell. This study showed sustained release pattern with released lysozyme maintaining its enzymatic activity. Further, release pattern was controlled by changing the flow rate through the inner needle. This resulted in different core–shell thicknesses that ultimately translated into varied release patterns [91]. Incorporating drugs alone in the core without any fiber forming additives was studied by Huang et al. The nanofiber PCL scaffolds with antibiotic gentamicin sulphate, antioxidant resveratrol as core showed sustained release pattern with no burst release [92]. In this technique, since the core and clad feeder assemblies are well isolated, drug/bioactive agents are less exposed to harsh organic solvents and able to retain their structure and the bioactivity.

Scaffolds that are able to deliver growth factors provide specific biochemical cues to the surrounding cells and accelerate the tissue formation. Electrospinning was successfully used to load growth factors or the equivalent plasmid DNA that encode desired growth factors. Luu et al. developed PLAGA and PLA–PEG nanofiber matrix loaded with plasmid DNA [93]. It was observed that the DNA released from the scaffold was capable of transfecting cells to release the intended protein. Similar approach was adopted to achieve BMP-2 production by releasing BMP-2 plasmid DNA from PLAGA/HA nanofibers. The nanofiber scaffolds loaded with DNA/chitosan nanopraticles showed higher cell attachment, cell viability, and the desired production of BMP-2 for successful bone tissue engineering [94]. Li et al. electrospun silk fibroin/nHA/BMP-2 scaffolds. The released BMP-2 was found to be bioactive and further induced *in vitro* bone formation from hMSCs [95]. Other studies have also confirmed the encapsulation and the delivery of β-nerve growth factor (NGF) from copolymer ε-caprolactone-ethyl ethylene phosphate (PCLEEO), and platelet-derived growth factor from ε-caprolactone electrospun nanofiber scaffolds [96,97].

7.6 CONCLUSIONS AND CHALLENGES

Electrospinning has emerged as an elegant and leading technique to create nanofiber scaffolds for tissue engineering. This technique has been successfully applied to create scaffolds from various biodegradable synthetic, natural polymers as well as their blends and composites. The fine control over the fiber diameter, fiber orientation, and the scaffold shape are the most appreciated features of this process over other scaffolding methodologies. Electrospun nanofiber scaffolds with appropriate mechanical and biochemical properties are ideal candidates for regenerating tissues such as skin, blood vessel, cartilage, nerve, and bone. Further, electrospinning has demonstrated its ability to encapsulate and deliver drug/growth factors in a controlled fashion for successful tissue regeneration.

Biodegradable electrospun matrices are increasingly becoming popular as scaffolds; however, there is still scope and need to improve this technique further to create bioactive scaffolds for future tissue engineering use. Some of the challenges include spinning natural polymers without loosing their bioactivity, achieving highly aligned scaffolds, and 3D structures with significant length, width, and thickness.

Electrospinning biopolymers with fluorinated solvents such as HFIP, 2,2,2-trifluoroethanol (TFE) often result in reduced or loss of bioactivity. Therefore, it is required to develop water-based solvents where in biopolymers like collagen, elastin, fibrinogen etc., retain their bioactivity after electro processing. Electrospinning technique is recognized for its ability to create oriented scaffolds for vascular, dermal, neural, and cartilage tissue regeneration. However, there is need to develop methods or modify the collection geometry to obtain large area meshes with the same degree of orientation throughout the scaffold thickness. The technique has its reputation in generating fiber meshes, but still lacks the ability to create 3D scaffolds with certain thickness for bone or vascular tissue scaffolding applications. Recently, multispinneret electrospinning has demonstrated some success in creating large area scaffolds with uniform thickness; however, the creation of 3D scaffolds with certain thickness is still an open challenge. For tissue engineering applications, it is also important that we develop sterilization methods that do not affect the nanostructure and the porosity of these electrospun scaffolds. Despite the challenges, electrospinning is continuing to be a popular technique for creating biodegradable scaffolds for various tissue engineering needs.

ACKNOWLEDGMENT

The authors acknowledge funding from the NIH (R01 EB004051 and R01 AR052536). Dr. Laurencin is the recipient of a presidential faculty fellow award from the National Science Foundation.

REFERENCES

1. Li, D. and Xia, Y.N. 2004. Electrospinning of nanofibers: Reinventing the wheel? *Adv. Mater.* 16:1151–1170.
2. Huang, Z., Zhang, Y., Kotaki, M., and Ramakrishna, S. 2003. A review on polymer nanofibers by electrospinning and their applications in nanocomposites. *Compos. Sci. Technol.* 63:2223–2253.
3. Nair, L.S., Bhattacharyya, S., and Laurencin, C.T. 2004. Development of novel tissue engineering scaffolds via electrospinning. *Expert. Opin. Biol. Ther.* 4:659–668.
4. Chew, S.Y., Wen, Y., Dzenis, Y., and Leong, K.W. 2006. The role of electrospinning in the emerging field of nanomedicine. *Curr. Pharm. Des.* 12:4751–4770.
5. Kumbar, S.G., James, R., Nukavarapu, S.P., and Laurencin, C.T. 2007. Electrospun nanofiber scaffolds: Engineering soft tissues. *Biomed. Mater.* 2:1–15.
6. Laurencin, C.T. and Ko, F.K. 2004. Hybrid nanofibril matrices for use as tissue engineering devices. US patent 6689166.
7. Laurencin, C.T., Nair, L.S., Bhattacharyya, S., Allcock, H.R., Bender, J.D., Brown, P.W., and Greish, Y.E. 2005. Polymeric nanofibers for tissue engineering and drug delivery. US patent 7235295.
8. Kumbar, S.G., Nair, L.S., Bhattacharyya, S., and Laurencin, C.T. 2006. Polymeric nanofibers as novel carriers for the delivery of therapeutic molecules. *J. Nanosci. Nanotechnol.* 6:2591–2607.
9. Mikos, A.G., Thorsen, A.J., Czerwonka, L.A., Bao, Y., Langer, R., Winslow, D.N., and Vacanti, J.P. 1994. Preparation and characterization of poly(L-lactic acid) foams. *Polymer* 35:1068–1077.
10. Mooney, D.J., Baldwin, D.F., Suh, N.P., Vacanti, J.P., and Langer, R. 1996. Novel approach to fabricate porous sponges of poly(D,L-lactic-co-glycolic acid) without the use of organic solvents. *Biomaterials* 17:1417–1422.
11. Hsu, Y.Y., Gresser, J.D., Trantolo, D.J., Lyons, C.M., Gangadharam, P.R., and Wise, D.L. 1997. Effect of polymer foam morphology and density on kinetics of *in vitro* controlled release of isoniazid from compressed foam matrices. *J. Biomed. Mater. Res.* 35:107–116.
12. Giordano, R.A., Wu, B.M., Borland, S.W., Cima, L.G., Sachs, E.M., and Cima, M.J. 1996. Mechanical properties of dense polylactic acid structures fabricated by three dimensional printing. *J. Biomater. Sci. Polym. Ed.* 8:63–75.
13. Srouji, S., Kizhner, T., Suss-Tobi, E., Livne, E., and Zussman, E. 2007. 3-D nanofibrous electrospun multilayered construct is an alternative ECM mimicking scaffold. *J. Mater. Sci. Mater. Med.* (in press). www.springerlink.com/content/m727571243802621/fulltext.pdf.
14. McManus, M.C., Boland, E.D., Koo, H.P., Barnes, C.P., Pawlowski, K.J., Wnek, G.E., Simpson, D.G., and Bowlin, G.L. 2006. Mechanical properties of electrospun fibrinogen structures. *Acta. Biomater.* 2:19–28.
15. Li, W.J., Laurencin, C.T., Caterson, E.J., Tuan, R.S., and Ko, F.K. 2002. Electrospun nanofibrous structure: A novel scaffold for tissue engineering. *J. Biomed. Mater. Res.* 60:613–621.
16. Shortkroff, S., Li, Y., Thornhill, T.S., and Rutledge, G.C. 2002. Cell growth on electrospun PCL scaffolds. *PMSE Prepr.* 87:457–458.

17. Dalton, P.D., Lleixa Calvet, J., Mourran, A., Klee, D., and Moller, M. 2006. Melt electrospinning of poly-(ethylene glycol-block-epsilon-caprolactone). *Biotechnol. J.* 1:998–1006.
18. Nair, L.S., Bhattacharyya, S., Bender, J.D., Greish, Y.E., Brown, P.W., Allcock, H.R., and Laurencin, C.T. 2004. Fabrication and optimization of methylphenoxy substituted polyphosphazene nanofibers for biomedical applications. *Biomacromolecules* 5:2212–2220.
19. Xu, C., Inai, R., Kotaki, M., and Ramakrishna, S. 2004. Electrospun nanofiber fabrication as synthetic extracellular matrix and its potential for vascular tissue engineering. *Tissue Eng.* 10:1160–1168.
20. Buschle-Diller, G., Hawkins, A., and Cooper, J. 2006. Electrospun nanofibers from biopolymers and their biomedical applications. In *Modified Fibers with Medical and Speciality Applications*, V. Edwards, G. Buschle-Diller, S. Goheen (Eds.), pp. 67–80. Springer, Netherlands.
21. Boland, E.D., Matthews, J.A., Pawlowski, K.J., Simpson, D.G., Wnek, G.E., and Bowlin, G.L. 2004. Electrospinning collagen and elastin: Preliminary vascular tissue engineering. *Front. Biosci.* 9:1422–1432.
22. Chong, E.J., Phan, T.T., Lim, I.J., Zhang, Y.Z., Bay, B.H., Ramakrishna, S., and Lim, C.T. 2007. Evaluation of electrospun PCL/gelatin nanofibrous scaffold for wound healing and layered dermal reconstitution. *Acta Biomater.* 3:321–330.
23. Li, M., Guo, Y., Wei, Y., MacDiarmid, A.G., and Lelkes, P.I. 2006. Electrospinning polyaniline-contained gelatin nanofibers for tissue engineering applications. *Biomaterials* 27:2705–2715.
24. Catledge, S.A., Clem, W.C., Shrikishen, N., Chowdhury, S., Stanishevsky, A.V., Koopman, M., and Vohra, Y.K. 2007. An electrospun triphasic nanofibrous scaffold for bone tissue engineering. *Biomed. Mater.* (*Bristol, U.K.*) 2:142–150.
25. Bhattacharyya, S., Nair, L.S., Singh, A., Krogman, N.R., Bender, J., Greish, Y.E., Brown, P.W., Allcock, H.R., and Laurencin, C.T. 2005. Development of biodegradable polyphosphazene–nanohydroxyapatite composite nanofibers via electrospinning. *Mater. Res. Soc. Symp. Proc.* 845:91–96.
26. Deng, X.L., Xu, M.M., Li, D., Sui, G., Hu, X.Y., and Yang, X.P. 2007. Electrospun PLLA/MWNTs/HA hybrid nanofiber scaffolds and their potential in dental tissue engineering. *Key Eng. Mater.* 330–332, 393–396.
27. Ma, P.X. 2004. Scaffolds for tissue fabrication. *Mater. Today* 7:30–40.
28. Silva, G., Czeisler, C., Niece, K., Beniash, E., Harrington, D., Kessler, J., and Stupp, S. 2004. Selective differentiation of neural progenitor cells by high-epitope density nanofibers. *Science* 303:1352–1355.
29. Xu, H. 2003. PhD dissertation, University of Akron.
30. Yarin, A.L., Koombhongse, S., and Reneker, D.H. 2001. Bending instability in electrospinning of nanofibers. *J. Appl. Phys.* 89:3018–3026.
31. Fridrikh, S.V., Yu, J.H., Brenner, M.P., and Rutledge, G.C. 2003. Controlling the fiber diameter during electrospinning. *Phys. Rev. Lett.* 90:144502–144505.
32. Kumbar, S.G., Bhattacharyya, S., Sethuraman, S., and Laurencin, C.T. 2007. A preliminary report on a novel electrospray technique for nanoparticle based biomedical implants coating: Precision electrospraying. *J. Biomed. Mater. Res. B. Appl. Biomater.* 81:91–103.
33. Nukavarapu, S.P., Kumbar, S.G., Nair, L.S., and Laurencin, C.T. 2007. Nanostructures for tissue engineering/regenerative medicine. In *Biomedical Nanostructures*, K. Gonsalves, C. Halberstadt, C.T. Laurencin and L.K. Nair (Eds.), pp. 371–401. John Wiley & Sons, Inc, Hoboken, New Jersey.
34. Shenoy, S., Bates, W.D., Frisch, H., and Wnek, G. 2005. Role of chain entanglements on fiber formation during electrospinning of polymer solutions: Good solvent, non-specific polymer–polymer interaction limit. *Polymer* 46:3372–3384.
35. Zong, X., Kim, K., Fang, D., Ran, S., Hsiao, B., and Chu, B. 2002. Structure and process relationship of electrospun bioabsorbable nanofiber membranes. *Polymer* 43:4403–4412.
36. Kim, T.G. and Park, T.G. 2006. Biomimicking extracellular matrix: Cell adhesive RGD peptide modified electrospun poly(D,L-lactic-co-glycolic acid) nanofiber mesh. *Tissue Eng.* 12:221–233.
37. Deitzel, J.M., Kleinmeyer, J., Harris, D., and BeckTan, N.C. 2001. The effect of processing variables on the morphology of electrospun nanofibers and textiles. *Polymer* 42:261–272.
38. Pitt, S., Chidchanok, M., Manit, N. 2005. Ultrafine electrospun polyamide-6 fibers: Effects of solvent system and emitting electrode polarity on morphology and average fiber diameter. *Macromol. Mater. Eng.* 290:933–942.
39. Casper, C.L., Stephens, J.S., Tassi, N.G., Chase, D.B., and Rabolt, J.F. 2004. Controlling surface morphology of electrospun polystyrene fibers: Effect of humidity and molecular weight in the electrospinning process. *Macromolecules* 37:573–578.

40. Megelski, S., Stephens, J.S., Chase, D.B., and Rabolt, J.F. 2002. Micro- and nanostructured surface morphology on electrospun polymer fibers. *Macromolecules* 35:8456–8466.
41. Boland, E.D., Wnek, G.E., Simpson, D.G., Pawlowski, K.J., and Bowlin, G.L. 2001. Tailoring tissue engineering scaffolds using electrostatic processing techniques: A study of poly(glycolic acid) electrospinning. *J. Macromol. Sci., Pure Appl. Chem.* A38:1231–1243.
42. Stitzel, J.D., Pawlowski, K.J., Wnek, G.E., Simpson, D.G., and Bowlin, G.L. 2001. Arterial smooth muscle cell proliferation on a novel biomimicking, biodegradable vascular graft scaffold. *J. Biomater. Appl.* 16:22–33.
43. Xin, X., Hussain, M., and Mao, J.J. 2006. Continuing differentiation of human mesenchymal stem cells and induced chondrogenic and osteogenic lineages in electrospun PLGA nanofiber scaffold. *Biomaterials* 28:316–325.
44. Katti, D.S., Robinson, K.W., Ko, F.K., and Laurencin, C.T. 2004. Bioresorbable nanofiber-based systems for wound healing and drug delivery: Optimization of fabrication parameters. *J. Biomed. Mater. Res. B. Appl. Biomater.* 70:286–296.
45. Boland, E.D., Coleman, B.D., Barnes, C.P., Simpson, D.G., Wnek, G.E., and Bowlin, G.L. 2005. Electrospinning polydioxanone for biomedical applications. *Acta Biomater.* 1:115–123.
46. Reneker, D.H., Kataphinan, W., Theron, A., Zussman, E., and Yarin, A.L. 2002. Nanofiber garlands of polycaprolactone by electrospinning. *Polymer* 43:6785–6794.
47. Kumbar, S.G., Bhattacharyya, S., Nukavarapu, S.P., Khan, Y.M., Nair, L.S., and Laurencin, C.T. 2006. *In vitro* and *in vivo* characterization of biodegradable poly(organophosphazenes) for biomedical applications. *J. Inorg. Organomet. Polym. Mater.* 16:365–385.
48. Bhattacharyya, S., Nair, L.S., Singh, A., Krogman, N.R., Greish, Y.E., Brown, P.W., Allcock, H.R., and Laurencin, C.T. April 2006. Electrospinning of poly[bis(ethyl alanato) phosphazene] nanofibers. *J. Biomed. Nanotechnol.* 2:36–45.
49. Sombatmankhong, K., Sanchavanakit, N., Pavasant, P., and Supaphol, P. 2007. Bone scaffolds from electrospun fiber mats of poly(3-hydroxybutyrate), poly(3-hydroxybutyrate-co-3-hydroxyvalerate) and their blend. *Polymer* 48:1419–1427.
50. Matthews, J.A., Wnek, G.E., Simpson, D.G., and Bowlin, G.L. 2002. Electrospinning of collagen nanofibers. *Biomacromolecules* 3:232–238.
51. Li, M., Mondrinos, M.J., Gandhi, M.R., Ko, F.K., Weiss, A.S., and Lelkes, P.I. 2005. Electrospun protein fibers as matrices for tissue engineering. *Biomaterials* 26:5999–6008.
52. Barnes, C.P., Smith, M.J., Bowlin, G.L., Sell, S.A., Tang, T., Matthews, J.A., Simpson, D.G., and Nimtz, J.C. 2006. Feasibility of electrospinning the globular proteins hemoglobin and myoglobin. *J. Eng. Fibers Fabrics* 1:16–29.
53. Bhattarai, N., Edmondson, D., Veiseh, O., Matsen, F.A., and Zhang, M. 2005. Electrospun chitosan-based nanofibers and their cellular compatibility. *Biomaterials* 26:6176–6184.
54. Bhattarai, N., Li, Z., Edmondson, D., and Zhang, M. 2006. Alginate-based nanofibrous scaffolds: Structural, mechanical, and biological properties. *Adv. Mater.* 18:1463–1467.
55. Yin, G. and Zhang, Y. 2006. Structure and properties of electrospun regenerated silk fibroin nanofibers. *Jingxi Huagong* 23:882–886.
56. Duan, B., Wu, L., Li, X., Yuan, X., Li, X., Zhang, Y., and Yao, K. 2007. Degradation of electrospun PLGA-chitosan/PVA membranes and their cytocompatibility *in vitro*. *J. Biomater. Sci. Polym. Ed.* 18:95–115.
57. Huang, L., Nagapudi, K., Apkarian, R.P., and Chaikof, E.L. 2001. Engineered collagen-PEO nanofibers and fabrics. *J. Biomater. Sci. Polym. Ed.* 12:979–993.
58. Reneker, D.H. and Fong, H. 2006. *Polymeric Nanofibers*. Oxoford University Press.
59. Li, M., Mondrinos, M.J., Chen, X., Gandhi, M.R., Ko, F.K., and Lelkes, P.I. 2006. Co-electrospun poly (lactide-co-glycolide), gelatin, and elastin blends for tissue engineering scaffolds. *J. Biomed. Mater. Res. A.* 79:963–973.
60. Zhang, Y., Ouyang, H., Lim, C.T., Ramakrishna, S., and Huang, Z.M. 2005. Electrospinning of gelatin fibers and gelatin/PCL composite fibrous scaffolds. *J. Biomed. Mater. Res. B. Appl. Biomater.* 72:156–165.
61. Buttafoco, L., Kolkman, N.G., Engbers-Buijtenhuijs, P., Poot, A.A., Dijkstra, P.J., Vermes, I., and Feijen, J. 2006. Electrospinning of collagen and elastin for tissue engineering applications. *Biomaterials* 27:724–734.

62. Kim, H.-W., Song J.-H., and Kim, H.-E. 2005. Nanofiber generation of gelatin–hydroxyapatite biomimetics for guided tissue regeneration. *Adv. Funct. Mater.* 15:1988–1994.
63. Bhattacharyya, S., Nair, L.S., Singh, A., Krogman, N.R., Bender, J., Greish, Y.E., Brown, P.W., Allcock, H.R., and Laurencin, C.T. 2005. Development of biodegradable polyphosphazene–nanohydroxyapatite composite nanofibers via electrospinning. *Mater. Res. Soc. Symp. Proc.* 845:91–96.
64. Seth, D. Mc., Kelly, L.S., Derrick, R.S., Wesley, A.R., Nancy, A.M., Laura, I.C., and Russell, E.G. 2007. Development, optimization, and characterization of electrospun poly(lactic acid) nanofibers containing multi-walled carbon nanotubes. *J. Appl. Polym. Sci.* 105:1668–1678.
65. Lyons, J., Li, C., and Ko, F. 2004. Melt-electrospinning part I: Processing parameters and geometric properties. *Polymer* 45:7597–7603.
66. Nobuo, O., Shinji, Y., Naoki, S., Gang, Lu., Toshiharu, I., Koji, N., and Takashi, O. 2007. Poly(lactide) nanofibers produced by a melt-electrospinning system with a laser melting device. *J. Appl. Polym. Sci.* 104:1640–1645.
67. McCann, J.T., Marquez, M., and Xia, Y. 2006. Melt coaxial electrospinning: A versatile method for the encapsulation of solid materials and fabrication of phase change nanofibers. *Nano. Lett.* 6:2868–2872.
68. Teo, W.E. and Ramakrishna, S. 2006. A review on electrospinning design and nanofibre assemblies. *Nanotechnology* 17:R89–R106.
69. Kwon, I.K., Kidoaki, S., and Matsuda, T. 2005. Electrospun nano- to microfiber fabrics made of biodegradable copolyesters: Structural characteristics, mechanical properties and cell adhesion potential. *Biomaterials* 26:3929–3939.
70. Li, D., Wang, Y., and Xia, Y. 2004. Electrospinning nanofibers as uniaxially aligned arrays and layer-by-layer stacked films. *Adv. Mater.* 16:361–366.
71. Kwon, I.K. and Matsuda, T. 2005. Co-electrospun nanofiber fabrics of poly(L-lactide-co-e-caprolactone) with type I collagen or heparin. *Biomacromolecules* 6:2096–2105.
72. Lam, H., Ye, H., Gogotsi, Y., and Ko, F. 2004. Structure and properties of electrospun single-walled carbon nanotubes reinforced nanocomposite fibrils by co-electrospinning. *Polym. Prepr.* 45:124–125.
73. Yang, H., Loh, L., Han, T., and Ko, F. 2003. Nanomagnetic particle filled piezoelectric polymer nanocomposite wires by co-electrospinning. *Polym. Prepr.* 44:163.
74. Bowman, J., Taylor, M., Sharma, V., Lynch, A., and Chadha, S. 2003. Multispinneret methodologies for high throughput electrospun nanofiber. *Mater. Res. Soc. Symp. Proc.* 752:15–19.
75. Theron, S.A., Yarin, A.L., Zussman, E., and Kroll, E. 2005. Multiple jets in electrospinning: Experiment and modeling. *Polymer* 46:2889–2899.
76. Kidoaki, S., Kwon, I.K., and Matsuda, T. 2005. Mesoscopic spatial designs of nano- and microfiber meshes for tissue-engineering matrix and scaffold based on newly devised multilayering and mixing electrospinning techniques. *Biomaterials* 26:37–46.
77. Pham, Q.P., Sharma, U., and Mikos, A.G. 2006. Electrospinning of polymeric nanofibers for tissue engineering applications: A review. *Tissue Eng.* 12:1197–1211.
78. Nam, J., Huang, Y., Agarwal, S., and Lannutti, J. 2007. Improved cellular infiltration in electrospun fiber via engineered porosity. *Tissue Eng.* 13:2249–2257.
79. Kumbar, S.G., Kofron, M.D., Nair, L.S., and Laurencin, C.T. 2007. Cell behavior toward nanostructured surfaces. In *Biomedical Nanostructures*, K. Gonsalves, C. Halberstadt, and C.T. Laurencin and L.S. Nair (Eds.), 257–291. John Wiley & Sons, Inc, Hoboken, New Jersey.
80. Simpson, D.G. 2006. Dermal templates and the wound-healing paradigm: The promise of tissue regeneration. *Expert. Rev. Med. Devices* 3:471–484.
81. Myung-Seob, K., Dong-Il, C., Hak-Yong, K., In-Shik, K., Narayan, B. 2003. Electrospun nanofibrous polyurethane membrane as wound dressing. *J. Biomed. Mater. Res. Part B: Appl. Biomater.* 67B:675–679.
82. Rho, K.S., Jeong, L., Lee, G., Seo, B.M., Park, Y.J., Hong, S.D., Roh, S., Cho, J.J., Park, W.H., and Min, B.M. 2006. Electrospinning of collagen nanofibers: Effects on the behavior of normal human keratinocytes and early-stage wound healing. *Biomaterials* 27:1452–1461.
83. Inoguchi, H., Tanaka, T., Maehara, Y., and Matsuda, T. 2007. The effect of gradually graded shear stress on the morphological integrity of a huvec-seeded compliant small-diameter vascular graft. *Biomaterials* 28:486–495.
84. Stitzel, J., Liu, J., Lee, S.J., Komura, M., Berry, J., Soker, S., Lim, G., Van Dyke, M., Czerw, R., Yoo, J.J., and Atala, A. 2006. Controlled fabrication of a biological vascular substitute. *Biomaterials* 27:1088–1094.

85. Thomas, V., Jagani, S., Johnson, K., Jose, M.V., Dean, D.R., Vohra, Y.K., and Nyairo, E. 2006. Electrospun bioactive nanocomposite scaffolds of polycaprolactone and nanohydroxyapatite for bone tissue engineering. *J. Nanosci. Nanotechnol.* 6:487–493.
86. Stankus, J.J., Guan, J., Fujimoto, K., and Wagner, W.R. 2006. Microintegrating smooth muscle cells into a biodegradable, elastomeric fiber matrix. *Biomaterials* 27:735–744.
87. Townsend-Nicholson, A. and Jayasinghe, S.N. 2006. Cell electrospinning: A unique biotechnique for encapsulating living organisms for generating active biological microthreads/scaffolds. *Biomacromolecules* 7:3364–3369.
88. Kenawy, E., Bowlin, G.L., Mansfield, K., Layman, J., Simpson, D.G., Sanders, E.H., and Wnek, G.E. 2002. Release of tetracycline hydrochloride from electrospun poly(ethylene-co-vinylacetate), poly(lactic acid), and a blend. *J. Controlled Release* 81:57–64.
89. Xie, J. and Wang, C.H. 2006. Electrospun micro- and nanofibers for sustained delivery of paclitaxel to treat C6 glioma *in vitro*. *Pharm. Res.* 23:1817–1826.
90. Zhang, Y., Lim, C.T., Ramakrishna, S., and Huang, Z. 2005. Recent development of polymer nanofibers for biomedical and biotechnological applications. *J. Mater. Sci. Mater. Med.* 16:933–946.
91. Jiang, H., Hu, Y., Li, Y., Zhao, P., Zhu, K., and Chen, W. 2005. A facile technique to prepare biodegradable coaxial electrospun nanofibers for controlled release of bioactive agents. *J. Control. Release* 108:237–243.
92. Huang, Z.M., He, C.L., Yang, A., Zhang, Y., Han, X.J., Yin, J., and Wu, Q. 2006. Encapsulating drugs in biodegradable ultrafine fibers through co-axial electrospinning. *J. Biomed. Mater. Res. A.* 77:169–179.
93. Luu, Y.K., Kim, K., Hsiao, B.S., Chu, B., and Hadjiargyrou, M. 2003. Development of a nanostructured DNA delivery scaffold via electrospinning of PLGA and PLA–PEG block copolymers. *J. Control. Release* 89:341–353.
94. Nie, H. and Wang, C.H. 2007. Fabrication and characterization of PLGA/HAp composite scaffolds for delivery of BMP-2 plasmid DNA. *J. Control. Release* 120:111–121.
95. Li, C., Vepari, C., Jin, H.J., Kim, H.J., and Kaplan, D.L. 2006. Electrospun silk-BMP-2 scaffolds for bone tissue engineering. *Biomaterials* 27:3115–3124.
96. Chew, S.Y., Wen, J., Yim, E.K.F., and Leong, K.W. 2005. Sustained release of proteins from electrospun biodegradable fibers. *Biomacromolecules* 6:2017–2024.
97. Liao, I.C., Chew, S.Y., and Leong, K.W. 2006. Aligned core–shell nanofibers delivering bioactive proteins. *Nanomed* 1:465–471.
98. Venugopal, J.R., Zhang, Y., and Ramakrishna, S. 2006. *In vitro* culture of human dermal fibroblasts on electrospun polycaprolactone collagen nanofibrous membrane. *Artif. Organs* 30:440–446.
99. Yoshimoto, H., Shin, Y.M., Terai, H., and Vacanti, J.P. 2003. A biodegradable nanofiber scaffold by electrospinning and its potential for bone tissue engineering. *Biomaterials* 24:2077–2082.
100. Li, W.W., Tuli, R., Okafor, C., Derfoul, A., Danielson, K.G.K.G., Hall, D.J.D.J., and Tuan, R.S.R.S. 2005. A three-dimensional nanofibrous scaffold for cartilage tissue engineering using human mesenchymal stem cells. *Biomaterials* 26:599–609.
101. Zong, X., Bien, H., Chung, C., Yin, L., Fang, D., Hsiao, B.S., Chu, B., and Entcheva, E. 2005. Electrospun fine-textured scaffolds for heart tissue constructs. *Biomaterials* 26:5330–5338.
102. Inoguchi, H., Kwon, I., Inoue, E., Takamizawa, K., Maehara, Y., and Matsuda, T. 2006. Mechanical responses of a compliant electrospun poly(l-lactide-co-ε-caprolactone) small-diameter vascular graft. *Biomaterials* 27:1470–1478.
103. Matthews, J.A., Boland, E.D., Wnek, G.E., Simpson, D.G., and Bowlin, G.L. 2003. Electrospinning of collagen type II: A feasibility study. *J. Bioact. Compat. Polym.* 18:125–134.
104. Wnek, G.E., Carr, M.E., Simpson, D.G., and Bowlin, G.L. 2003. Electrospinning of nanofiber fibrinogen structures. *Nano. Lett.* 3:213–216.
105. Sell, S.A., McClure, M.J., Barnes, C.P., Knapp, D.C., Walpoth, B.H., Simpson, D.G., and Bowlin, G.L. 2006. Electrospun polydioxanone–elastin blends: Potential for bioresorbable vascular grafts. *Biomed. Mater.* 1:72–80.
106. He, W., Yong, T., Teo, W.E., Ma, Z., and Ramakrishna, S. 2005. Fabrication and endothelialization of collagen-blended biodegradable polymer nanofibers: Potential vascular graft for blood vessel tissue engineering. *Tissue Eng.* 11:1574–1588.
107. Jiang, H., Fang, D., Hsiao, B.S., Chu, B., and Chen, W. 2004. Optimization and characterization of dextran membranes prepared by electrospinning. *Biomacromolecules* 5:326–333.

108. Pan, H., Jiang, H., and Chen, W. 2006. Interaction of dermal fibroblasts with electrospun composite polymer scaffolds prepared from dextran and poly lactide-co-glycolide. *Biomaterials* 27:3209–3220.
109. McCullen, S.D., Stano, K.L., Stevens, D.R., Roberts, W.A., Monteiro-Riviere, N.A., Clarke, L.I., and Gorga, R.E. 2007. Development, optimization, and characterization of electrospun poly(lactic acid) nanofibers containing multi-walled carbon nanotubes. *J. Appl. Polym. Sci.* 105:1668–1678.
110. He, C., Huang, Z., Han, X., Liu, L., Zhang, H., and Chen, L. 2006. Coaxial electrospun poly(l-lactic acid) ultrafine fibers for sustained drug delivery. *J. Macromol. Sci., Part B* 45:515–524.
111. Kim, K., Luu, Y.K., Chang, C., Fang, D., Hsiao, B.S., Chu, B., and Hadjiargyrou, M. 2004. Incorporation and controlled release of a hydrophilic antibiotic using poly(lactide-co-glycolide)-based electrospun nanofibrous scaffolds. *J. Control. Release* 98:47–56.
112. Verreck, G., Chun, I., Peeters, J., Rosenblatt, J., and Brewster, M.E. 2003. Preparation and characterization of nanofibers containing amorphous drug dispersions generated by electrostatic spinning. *Pharm. Res.* 20:810–817.
113. Spasova, M., Manolova, N., Paneva, D., and Rashkov, I. 2004. Preparation of chitosan-containing nanofibers by electrospinning of chitosan/poly(ethylene oxide) blend solutions. *e-Polymers* 56:1–2.
114. Xu, X., Yang, L., Xu, X., Wang, X., Chen, X., Liang, Q., Zeng, J., and Jing, X. 2005. Ultrafine medicated fibers electrospun from W/O emulsions. *J. Control. Release* 108:33–42.
115. Zeng, J., Yang, L., Liang, Q., Zhang, X., Guan, H., Xu, X., Chen, X., and Jing, X. 2005. Influence of the drug compatibility with polymer solution on the release kinetics of electrospun fiber formulation. *J. Control. Release* 105:43–51.
116. Zeng, J., Xu, X., Chen, X., Liang, Q., Bian, X., Yang, L., and Jing, X. 2003. Biodegradable electrospun fibers for drug delivery. *J. Control. Release* 92:227–231.
117. Jiang, H., Hu, Y., Zhao, P., Li, Y., and Zhu, K. 2006. Modulation of protein release from biodegradable core–shell structured fibers prepared by coaxial electrospinning. *J. Biomed. Mater. Res. B. Appl. Biomater.* 79:50–57.
118. Liao, I.C., Chew, S.Y., and Leong, K.W. 2006. Aligned core–shell nanofibers delivering bioactive proteins. *Nanomed* 1:465–471.
119. Formhals, A. 1934. US Patent 1975504.
120. Taylor, G.I. 1969. *Proc. Roy. Soc. London A* 313:453.
121. Baumgarten, P. 1971. Electrostatic spinning of acrylic microfibers. *J. Colloid Interface Sci.* 36:71–79.
122. Hayati, I., Bailey, A.I., and Tadros, T.F. 1987. Investigations into the mechanisms of electrohydrodynamic spraying of liquids: I. Effect of electric field and the environment on pendant drops and factors affecting the formation of stable jets and atomization. *J. Colloid Interface Sci.* 117:205–221.
123. Doshi, J. and Reneker, D. 1995. Electrospinning process and applications of electrospun fibers. *J. Electrostatics* 35:151–160.
124. Deitzel, J.M., Kosik, W., McKnight, S.H., Beck Tan, N.C., DeSimone, J.M., and Crette, S. 2001. Electrospinning of polymer nanofibers with specific surface chemistry. *Polymer* 43:1025–1029.
125. Sun, Z., Zussman, E., Yarin, A.L., Wendorff, J.H., and Greiner, A. 2003. Compound core–shell polymer nanofibers by co-electrospinning. *Adv. Mater.* 15:1929–1932.
126. Boland, E.D., Simpson, D.G., Wnek, G.E., and Bowlin, G.L. 2003. Electrospinning of biopolymers for tissue engineering. *Abstracts of Papers,226th ACS National Meeting, New York, United States, September* 7–11, 533.
127. Xu, C.Y., Inai, R., Kotaki, M., and Ramakrishna, S. 2004. Aligned biodegradable nanofibrous structure: A potential scaffold for blood vessel engineering. *Biomaterials* 25:877–886.

8 Structure and Mechanical Properties of Electrospun Nanofibers and Nanocomposites

Eunice Phay Shing Tan and Chwee Teck Lim

CONTENTS

8.1 INTRODUCTION

Although the technique of electrospinning originated from the 1930s (Formhals 1934), the rapid development of this technique to produce continuous nanoscale fibers only took place in the past decade. This is due to the development in various fields that require the controlled production of nanofibers such as tissue engineering, fuel cells, filter media, chemical and biosensors, etc. (Huang et al. 2003; Chronakis 2005; Subbiah et al. 2005; Zhang et al. 2005; Pham et al. 2006). The demand for electrospun nanofibers can be attributed to the ease and cost-effectiveness of mass production, as well as the high surface area-to-volume ratio and excellent mechanical properties that are inherent in electrospun nanofibers. In order to further enhance the unique properties of nanofibers, composite nanofibers have been produced by adding reinforcement materials or combining two or more polymers.

Understanding the mechanical properties and deformation mechanisms of electrospun nanofibers is essential for a greater understanding on the contributions of these nanofibers to the mechanical and performance-related characteristics of nanofiber composites. Although most mechanical tests have been performed on nanofiber mats, techniques for testing single nanofibers have been only developed in recent years. A summary of the mechanical testing techniques will be presented in the next section.

Electrospun nanofibers generally have superior mechanical properties such as higher elastic modulus and strength as compared to bulk materials mainly due to the high molecular orientation of polymer molecules. This molecular orientation is produced by the stretching of the polymer jet during electrospinning (Reneker and Chun 1996). The mechanical properties of electrospun nanofibers are highly dependent on the morphology and this, in turn, is dependent on the electrospinning processing parameters and method of nanofiber reinforcement. Thus, it is the aim of this chapter to explore how processing and the addition of another material to pure polymers affect the structure and mechanical properties of electrospun nanofibers. Although electrospun fibers have been made from other materials such as ceramics (Chronakis 2005), this chapter focuses on polymer nanofibers.

8.2 MECHANICAL TESTING OF SINGLE ELECTROSPUN NANOFIBERS

Due to the challenges involved in isolating and handling single nanofibers, a majority of the mechanical tests are performed on aligned or nonaligned electrospun nanofiber mats. The most common test is the uniaxial tensile test of nanofiber mats (Huang et al. 2001; Lee et al. 2003b; Pedicini and Farris 2003; Bolgen et al. 2005; Huang et al. 2005; Bhattarai et al. 2006; Chen et al. 2006; Choi et al. 2006; Junkasem et al. 2006; Li et al. 2006; Sombatmankhong et al. 2006; Sun et al. 2006; Zhang et al. 2006; Zussman et al. 2006; Chuangchote et al. 2007; Han et al. 2007; Hwang et al. 2007; McCullen et al. 2007). Mechanical properties obtained include Young's modulus, tensile strength, and tensile strain at break. Dynamic mechanical analysis has also been performed (Jose et al. 2007) to obtain the storage modulus of the nanofiber mat. In order to assess the effects of changing electrospinning processing parameters or the addition of filler materials, mechanical testing of nanofiber mats is sufficient. However, the deformation behavior observed includes interfiber interaction. Thus, there is still a need to characterize single nanofibers.

With the development of atomic force microscopy (AFM), micro-electro-mechanical systems (MEMS), and commercial mechanical testing systems capable of nano-Newton load measurements, it is now possible to conduct mechanical characterization of single electrospun nanofibers. A summary of the mechanical characterization techniques of single electrospun nanofibers is presented in Table 8.1. Interested readers can refer to our previous work (Tan and Lim 2006b) for a detailed discussion of the various mechanical characterization techniques.

8.3 EFFECT OF FIBER PROCESSING ON STRUCTURE AND MECHANICAL PROPERTIES

Various processing strategies have been employed to enhance the chemical, electrical, and mechanical properties of electrospun nanofibers to suit the intended applications. Such strategies include optimization of electrospinning parameters, application of post-electrospinning treatment, and production of composite nanofibers. In this section, the effect of these strategies on the structure and mechanical properties of electrospun nanofibers will be explored.

8.3.1 ELECTROSPINNING PARAMETERS

Electrospinning parameters have been classified into three groups, namely solution properties, controlled variables, and ambient parameters by Doshi and Reneker (1995). Solution properties include viscosity, conductivity, surface tension, polymer molecular weight, dipole moment, and dielectric constant. Controlled variables include flow rate, electric field strength, distance between

TABLE 8.1
Mechanical Characterization of Single Electrospun Nanofibers

Test Method	Instruments and Techniques	Materials	References
Tensile test	(a) AFM cantilever 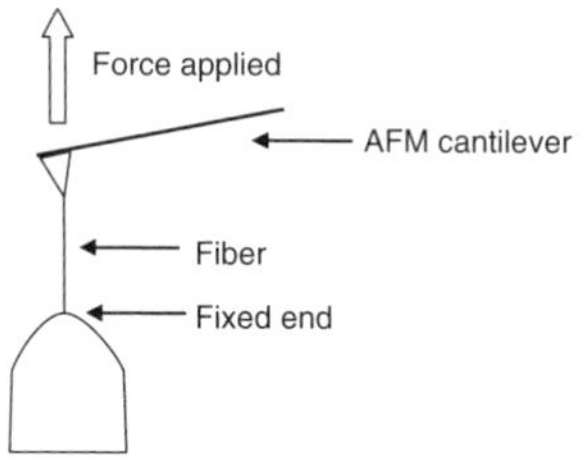	Polyethylene oxide (PEO)	(Warner et al. 1999; Tan et al. 2005)
		Nylon 6, 6	(Zussman et al. 2006)
		Pyrolyzed polyacrylonitrile (PAN)	(Zussman et al. 2005)
	(b) Commercial nano tensile tester 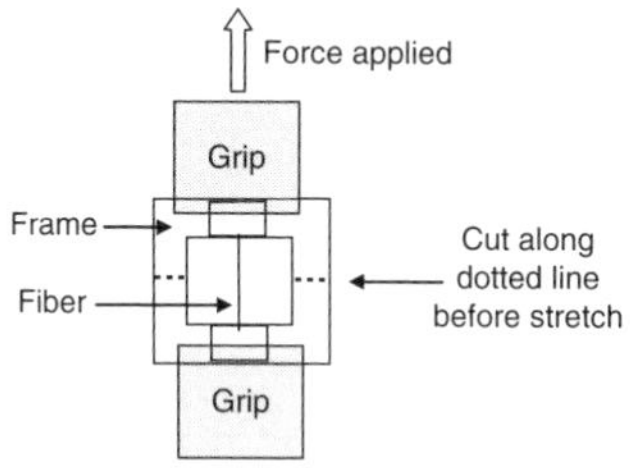	Polycaprolactone	(Tan et al. 2005)
		Poly (L-lactic acid) (PLLA)	(Inai et al. 2005; Tan and Lim 2006a)
		Nylon-6/montmorillonite (MMT) composite (particulate)[a]	(Ramakrishna et al. 2006)
	(c) MEMS devices	Pyrolyzed poly-furfuryl alcohol	(Samuel et al. 2007)
Bend test	(a) Three-point bend test using AFM 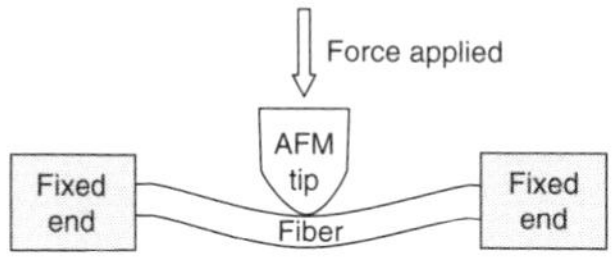	Electroactive polymers	(Shin et al. 2006b)
		Nylon-6/MMT composite (particulate)[a]	(Li et al. 2006)
		PEO	(Bellan et al. 2005)
		Piezoelectric	(Xu et al. 2006)
		PLLA	(Tan and Lim 2006a)
		Poly(vinyl alcohol)/ferritin composite (particulate)[a]	(Shin et al. 2006a)
	(b) Vertical deflection of free-end of fiber, with fixed end attached to AFM tip 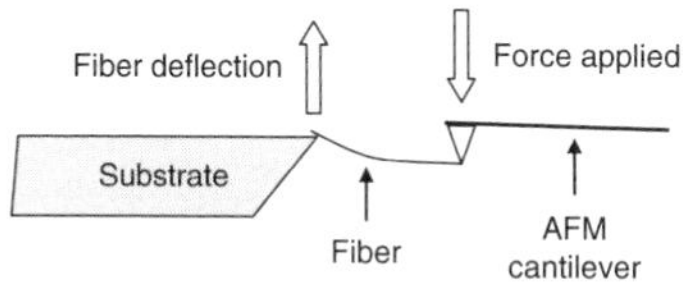	PAN	(Gu et al. 2005)

(continued)

TABLE 8.1 (continued)
Mechanical Characterization of Single Electrospun Nanofibers

Test Method	Instruments and Techniques	Materials	References
Nanoindentation	(a) Elastic–plastic indentation of fiber cross section using AFM 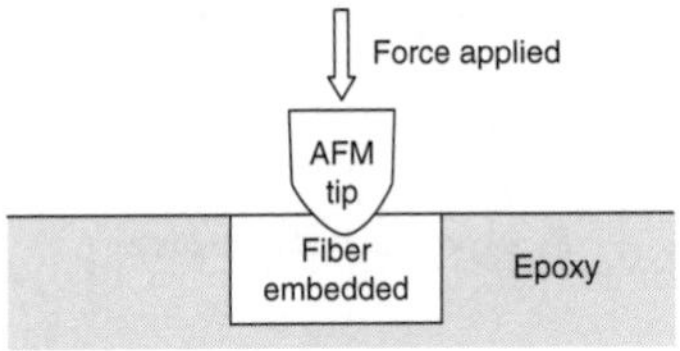	Silk/PEO composite (blend)[a]	(Wang et al. 2004)
	(b) Elastic indentation using AFM 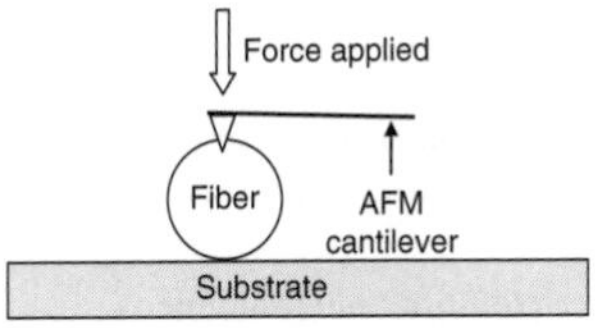	PAN/graphite composite (particulate)	(Mack et al. 2005)
		Poly(butylene terephthalate)/ carbon nanotube (CNT) composite (fiber)[a]	(Mathew et al. 2005)
Mechanical resonance	(a) Resonating microcantilever-nanofiber system 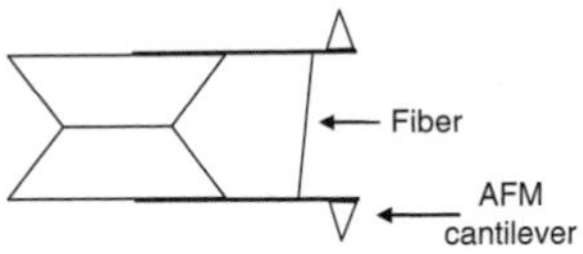	PAN	(Yuya et al. 2007)
	(b) Mechanical resonance with one end of fiber free 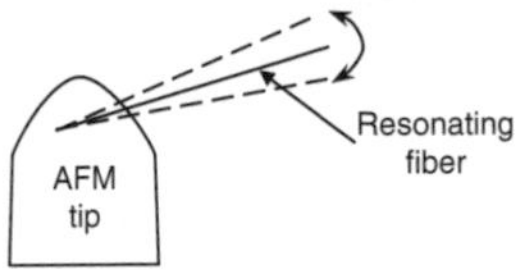	Pyrolyzed PAN	(Zussman et al. 2005)
Others	Shear modulation force microscopy 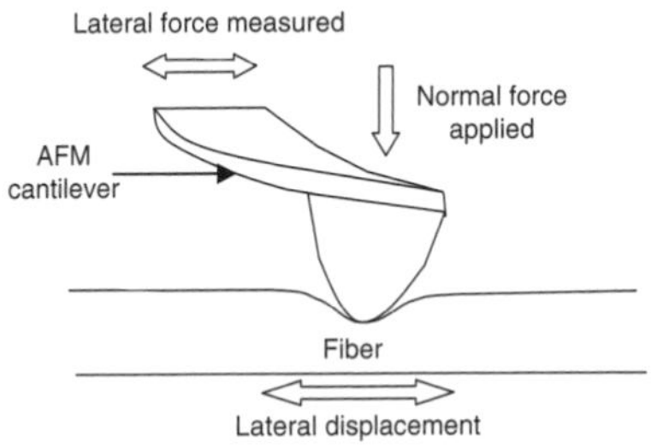	Polystyrene/MMT composite (particulate)[a]	(Ji et al. 2006)

[a] Denotes type of composite, i.e., particulate-reinforced, fiber-reinforced, or polymer blend with two distinct phases.

tip and collector, collector composition, and geometry. Ambient parameters include temperature, humidity, and air velocity (Pham et al. 2006). It can be seen that there is a variety of parameters that can be controlled. Pham et al. provided a review on the effect of process parameters on fiber morphology (Pham et al. 2006), with most studies investigating the effects of processing parameters on fiber diameter and uniformity of fibers. The processing parameters affect the morphology and mechanical properties of nanofibers at varying degrees but not all of these processing parameter–morphology–mechanical property relationships are explored. Commonly investigated parameters will be presented in this section.

Generally, any process that results in the production of uniform fibers with small diameters will result in stiff and strong fibers with reduced ductility. Larger fibers with uniform diameters are less stiff and strong but more ductile. This size effect may be explained by shear-induced molecular chain alignment during electrospinning. Large shear forces coupled with rapid solidification of the polymer jet stretched polymer chains along the fiber axis and prevented the molecular chains from relaxing back to their equilibrium conformations (Jaeger et al. 1996). As fibers with small diameters received a higher stress than those with larger diameters, the degree of molecular chain orientation in fibers with small diameters is higher than fibers with larger diameters. Thus, finer fibers tend to have higher stiffness and strength than fibers with larger diameters.

8.3.1.1 Fiber Take-Up Velocity

The most common processing parameter studied with respect to structure and mechanical properties of electrospun nanofibers is the rate of fiber collection (Inai et al. 2005; Thomas et al. 2006; Zussman et al. 2006; Jose et al. 2007). Collecting as-spun nanofibers on a rotating drum or wheel aligns the nanofibers in the direction of rotation. This method of collection also stretches the electrospinning jet further, thus resulting in the production of fibers with smaller diameters and higher degree of molecular orientation as compared with as-spun nanofibers. This effect is more pronounced when the collection rate or take-up velocity is increased. This can be seen from the x-ray diffraction results of nanofibers collected at two different take-up velocities in Figure 8.1.

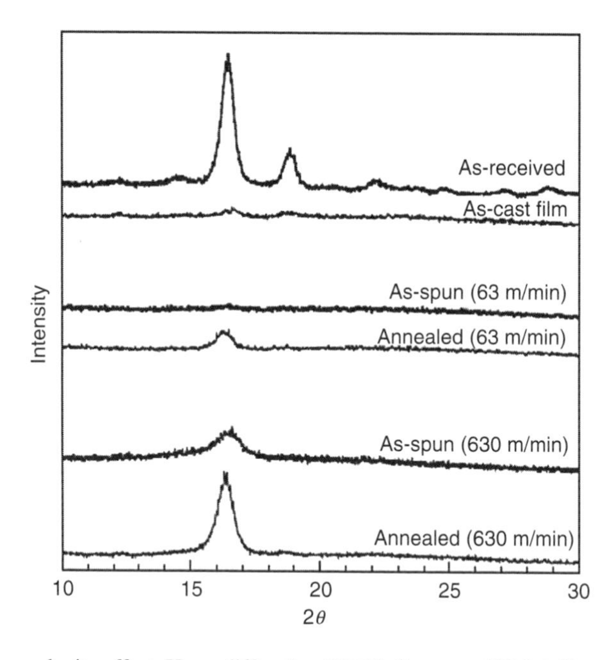

FIGURE 8.1 Take-up velocity effect: X-ray diffraction (XRD) diagram of PLLA fibers electrospun at take-up velocity of 63 and 630 m/min. (From Inai, R., Kotaki, M., and Ramakrishna, S., *Nanotechnology*, 16, 208, 2005. With permission.)

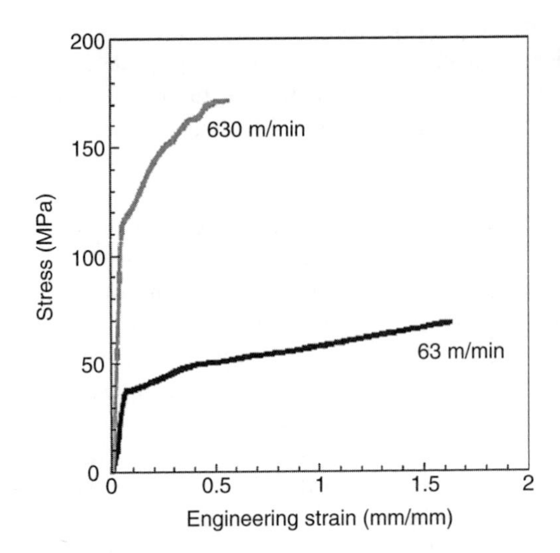

FIGURE 8.2 Tensile stress–strain curves of PLLA single nanofibers electrospun at take-up velocities of 63 and 630 m/min. (From Inai, R., Kotaki, M., and Ramakrishna, S., *Nanotechnology*, 16, 208, 2005. With permission.)

The nanofibers electrospun at 630 m/min exhibited a broad peak at 2θ of around 17°, whereas the nanofibers electrospun at 63 m/min exhibited no peak. The results indicated that the increased take-up velocity induced a highly ordered molecular structure. This feature was further confirmed by annealing the fibers. Due to the higher degree of stretching and molecular orientation, fibers produced from higher take-up velocities have higher Young's modulus and tensile strength but lower strain at break (Inai et al. 2005; Thomas et al. 2006; Zussman et al. 2006) as shown in Figure 8.2.

In the study by Jose et al. (Jose et al. 2007), higher take-up velocities were found to induce phase transformations of the nylon-6 crystals from the metastable γ phase to α phase, which are known to have better mechanical properties.

8.3.1.2 Polymer Concentration

Solution viscosity, as controlled by adjusting the polymer concentration, has been found to be one of the factors with the largest influence on fiber size and morphology. At low polymer concentrations, defects in the form of spherical- or spindle-shaped beads are formed along the fiber length. This is due to insufficient chain overlap for the formation of uniform fibers in dilute solutions (Pham et al. 2006). Fibers formed at low concentrations have smaller diameters. As the polymer concentration is increased, bead formation is reduced, fibers become more uniform and the fiber diameter is increased. When the solution is too concentrated or viscous, the droplet is dried at the tip before jets can be initiated. Figure 8.3 shows electrospun sodium alginate/PEO at two different polymer concentrations (Lu et al. 2006). At a lower polymer concentration, beads were seen whereas at a higher polymer concentration, smooth nanofibers were observed.

All the studies involving establishing the link between polymer concentration, morphology, and mechanical properties reported that nanofibers have inferior mechanical properties when beads are present in the nanofibers (Huang et al. 2004; Bolgen et al. 2005; Lu et al. 2006). Nanofibers with beads have lower Young's modulus, tensile strength, and elongation at break. This is due to stress concentrations induced by the beads as the nanofibers are stretched. It is interesting to note,

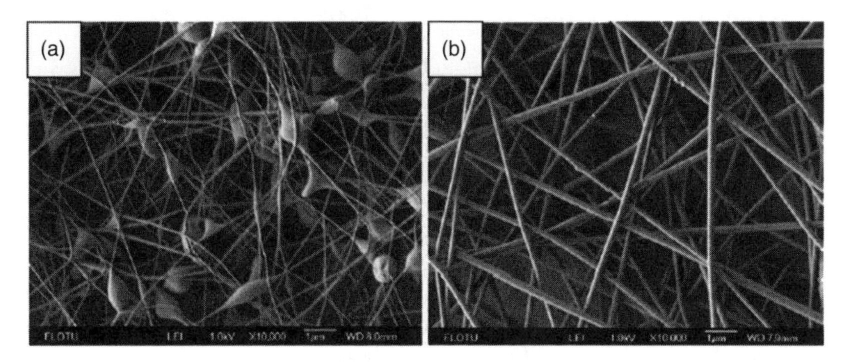

FIGURE 8.3 The morphology of electrospun fibers with different solution concentrations: (a) 1% and (b) 3%. (From Lu, J.-W., Zhu, Y.-L., Guo, Z.-X., Hu, P., and Yu, J., *Polymer*, 47, 8026, 2006. With permission.)

however, that once uniform nanofibers can be produced at a certain polymer concentration, an increase in concentration will result in a reduction in modulus and strength (Huang et al. 2004). This could be due to the fact that fibers produced at lower polymer concentrations have smaller diameters; hence size effects (Tan et al. 2005a,b; Chew et al. 2006; Ramakrishna et al. 2006) could play an important part. Another possible reason is the less viscous nature of dilute solutions which would imply that polymer chains have more mobility to align in the direction of the fiber axis during electrospinning. Thus, there could be a higher molecular orientation of polymer chains along the fiber axis in smaller fibers produced by more dilute solutions.

8.3.1.3 Other Processing Parameters

8.3.1.3.1 Solution Conductivity

In some cases, the solvent that the polymer is able to dissolve in has low conductivity; thus, only microsized fibers can be produced by electrospinning. A second solvent with a higher dielectric constant is usually added to increase the solution conductivity, and thus enable the production of nanosized fibers (Lee et al. 2003a). Bolgen et al. (2005) studied the effects of varying the percentage of a solvent with high dielectric constant. Increasing the percentage of this solvent resulted in a significant decrease in fiber diameter. Although it is expected that the smaller fibers produced implied that the Young's modulus and strength should increase while the elongation at break should decrease, Bolgen et al.'s results revealed that all the above mechanical properties decreased when the fiber diameter decreased. This could be due to the presence of beads in finer fibers.

8.3.1.3.2 Electric Field Strength

Higher applied voltage may give rise to stronger electric field and acceleration of charges in the electrospinning jet. This results in increased elongation of fibers and hence formation of finer fibers (Deitzel et al. 2001). In the study by Gu et al. (2005), higher applied voltage was found to induce a higher molecular orientation within the nanofibers. This resulted in a higher Young's modulus measured for single nanofibers.

8.3.2 Post-Electrospinning Treatments

Physical or chemical treatments may be applied to as-spun nanofibers in order to improve the stiffness, strength, or ductility of the final product. This may be important in biomedical applications, such as in cases where the nanofibers have to degrade more slowly in a wet environment at body temperature.

8.3.2.1 Cross-Linking

Cross-linking involves the formation of covalent bonds between the polymer chains and it is often applied to improve the thermomechanical performance of the resulting nanofiber mats (Zhang et al. 2006). Cross-linking by physical means such as UV irradiation involves adding cross-linking agent to the polymer solution and irradiating the electrospun product immediately after spinning (Choi et al. 2006). Choi et al. studied the effects of increasing crosslinker density and found that the Young's modulus and strength increase with an increase in crosslinker density (Choi et al. 2006). Increasing crosslinker density also resulted in fibers with better shape retention at room temperature for a polymer with very low glass transition temperature as shown in Figure 8.4. Cross-linking created intra- as well as interfiber covalent bonds. Thus, the improved tensile performance in this study can be explained by the increased number of permanently bonded junction points between fibers by co-curing between overlapping fibers, as well as by the increased stiffness of the fibers themselves by cross-linking of polymer molecules inside the fiber.

Cross-linking by chemical means involves exposing the nanofiber mats to a cross-linking agent by immersion in the cross-linking solution or by exposure to the vapor of the cross-linking agent in an enclosed environment (Zhang et al. 2006). Cross-linking by chemical means has been reported to be more efficient than physical means (Zhang et al. 2006). In the study by Zhang et al. (2006), cross-linking of gelatin nanofibers enhanced the Young's modulus and tensile strength by nearly ten times. As mentioned in the previous cross-linking example, this tremendous improvement in mechanical properties was due to the formation of inter- and intrafiber covalent bonds. Interestingly, the extensibility of the nanofiber mat was not reduced by the cross-linking process even though the stiffness and strength had increased.

8.3.2.2 Chemical Treatment

Chemical treatment involves immersing the nanofibers in a chemical which results in a change in the structure of the nanofibers. One common chemical treatment is the immersion of electrospun silk in methanol (Wang et al. 2004; Chen et al. 2006). As-spun silk fibers are often amorphous as there is not enough time for molecular arrangement and crystallization. In order to enhance the stiffness and strength of silk nanofibers, the as-spun fibers are immersed in methanol to induce molecular transformation from random coil to β-sheet crystals. Tensile tests of methanol-treated silk nanofiber mats reveal that the treatment resulted in increased strength and ductility. The observed reduction in Young's modulus was not conclusive as the tensile behavior was based on assemblies of fibers rather than structural character of fiber molecules (Chen et al. 2006). A nanoindentation study of single nanofibers revealed that this treatment indeed increased the Young's modulus (Wang et al. 2004) as

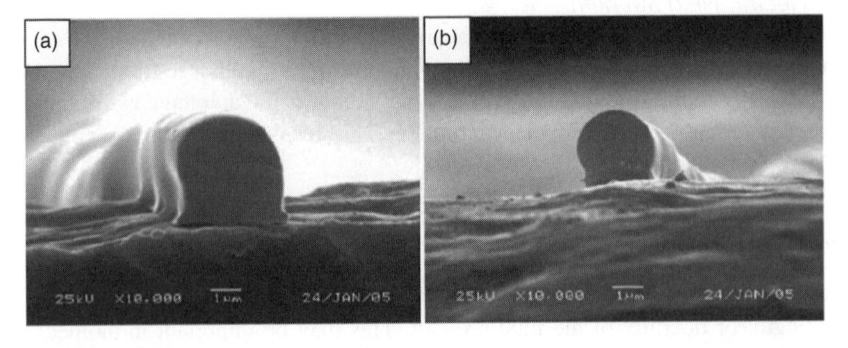

FIGURE 8.4 Scanning electron microscope (SEM) cross-sectional images of electrospun crosslinked polybutadiene fibers with different crosslinker percentages: (a) 0.5 wt% and (b) 3 wt%. (Reprinted from Choi, S.-S., Hong, J.-P., Seo, Y.S., Chung, S.M., and Nah, C., *J. Appl. Polym. Sci.*, 101, 2333, 2006. With permission.)

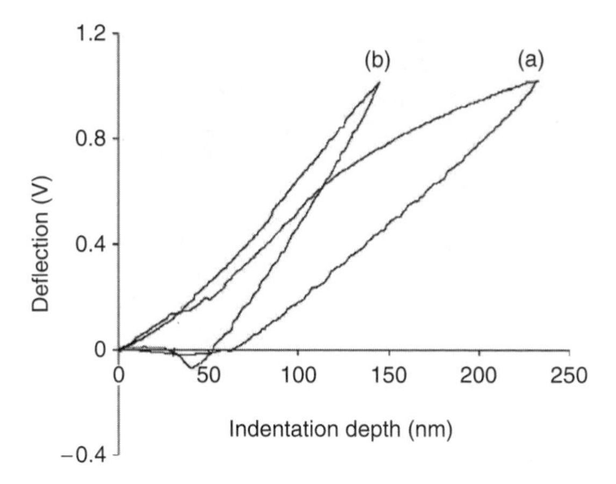

FIGURE 8.5 Representative loading and unloading curves of fibers: (a) as-spun fiber and (b) methanol-treated fiber. (Reprinted from Wang, M., Jin, H.-J., Kaplan, D.L., and Rutledge, G.C., *Macromolecules*, 37, 6856, 2004. With permission.)

shown in Figure 8.5 by the steeper slope of the unloading curve for methanol-treated fiber. It is interesting to note that the methanol treatment induced phase transformation from the surface of the nanofiber to the core. Figure 8.6 shows the AFM phase image of the cross section of methanol-treated silk nanofiber with an array of indentations. In AFM phase imaging, softer areas appear darker and stiffer areas appear brighter. The core appears darker than the shell. Thus, this indicates

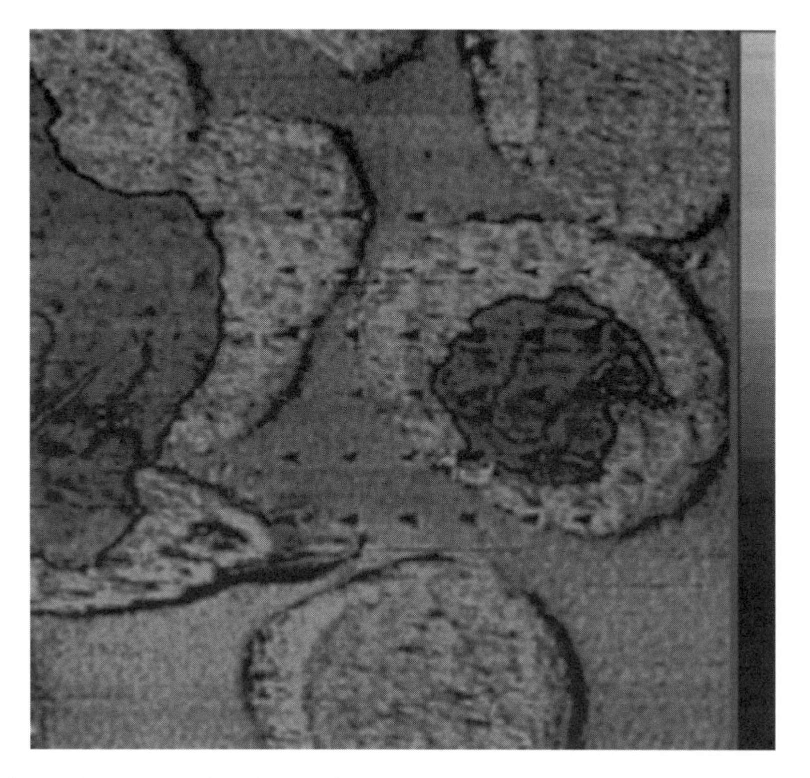

FIGURE 8.6 2 × 2 μm AFM phase image of the cross section of methanol-treated fiber. (From Wang, M., Jin, H.-J., Kaplan, D.L., and Rutledge, G.C., *Macromolecules*, 37, 6856, 2004. With permission.)

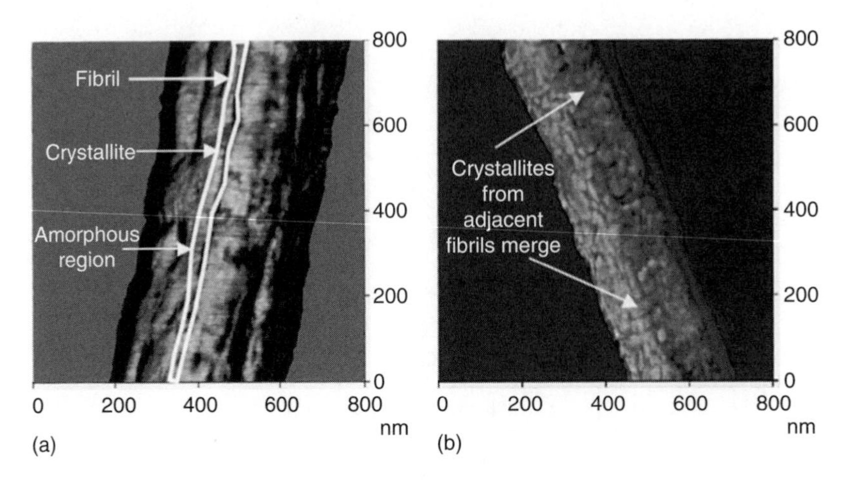

FIGURE 8.7 Phase image of electrospun PLLA nanofiber. (a) As-spun nanofiber and (b) annealed nanofiber. (Reproduced from Tan, E.P.S. and Lim, C.T., *Nanotechnology*, 17, 2649, 2006a. With permission.)

that the core, which is softer, is made of amorphous molecules whereas the shell, which is harder, is made up of β-sheet crystals.

8.3.2.3 Temperature

Heat treatment of polymers by annealing has been applied to increase the crystallinity of amorphous or semicrystalline polymers. This process serves to increase the stiffness and strength of the polymer. Annealing electrospun nanofibers above the glass transition temperature has been found to reduce fiber diameter by 10% and increase the Young's modulus by 150% in a study by Tan and Lim (2006a). This change in modulus is brought about by change in the morphology of the nanofibers as shown in Figure 8.7. Annealing promotes the growth of crystallites and merger of crystallites from adjacent fibrils, which serves to strengthen the bonds between fibrils within the nanofiber. This results in more resistance to deformation, as manifested by the higher modulus.

Although the effect of varying temperature is not a post-electrospinning treatment, it is illustrated here as some final electrospun products are used in applications at elevated temperatures and it is of interest to know the effects of increasing temperature on the structure and mechanical properties. Shear modulation force microscopy (SMFM) was used to study the effects of increasing temperature on electrospun polystyrene (PS) (Ji et al. 2006). Figure 8.8 shows the change in surface relative modulus as a function of temperature at different fiber diameters. As the temperature increased, the relative modulus decreased. The rate of decrease appears to be higher for smaller fibers. As the temperature approaches 377 K, which is the glass transition temperature, the modulus of all fibers approached 1, which corresponds to the bulk thin-film value. From these observations, it can be deduced that as temperature increases above the glass transition temperature, oriented molecular chains will relax to their equilibrium state, regardless of fiber diameter. As smaller fibers have a higher degree of molecular orientation (Section 8.3.1), the effect of molecular chain relaxation as temperature increases will be more pronounced, as is evident from Figure 8.8. This study has implications for electrospun fibers whereby the temperature at which it is put into service is around the glass transition.

8.3.3 Composite Nanofibers

Composite nanofibers are fibers that consist of two or more distinct components within each fiber. The types of electrospun composites include fiber- or particulate-reinforced fibers and polymer mixtures in the form of random blends or coaxial nanofibers as shown in Figure 8.9.

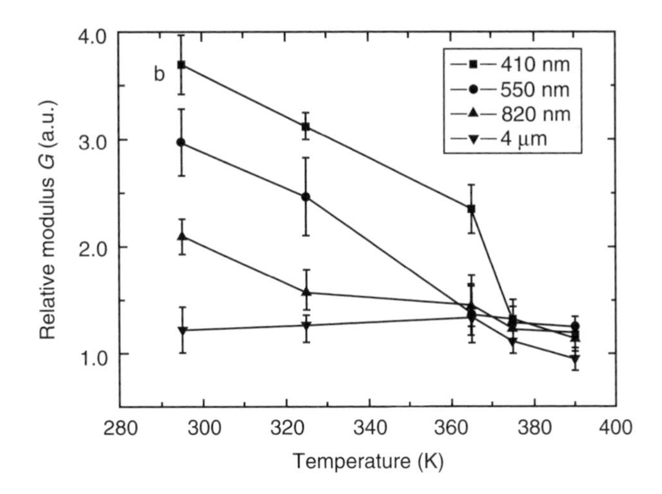

FIGURE 8.8 Relative modulus of electrospun polystyrene fibers as a function of the temperature at different fiber diameters. (Reprinted from Ji, Y., Li, B., Ge, S., Sokolov, J.C., and Rafailovich, M.S., *Langmuir*, 22, 1321, 2006. With permission.)

8.3.3.1 Filler-Reinforced Composites

Filler materials such as nanoparticles, nanoplatelets, carbon nanotubes (CNTs), or nanofibers are added into the polymer solution to improve the mechanical properties of the nanofibers in a similar way as bulk composites.

8.3.3.1.1 Fiber-Reinforced Nanofibers

Fiber reinforcements are one-dimensional reinforcement materials and usually occur in the form of short fibers. CNTs are the most common form of fiber reinforcement for electrospun polymeric

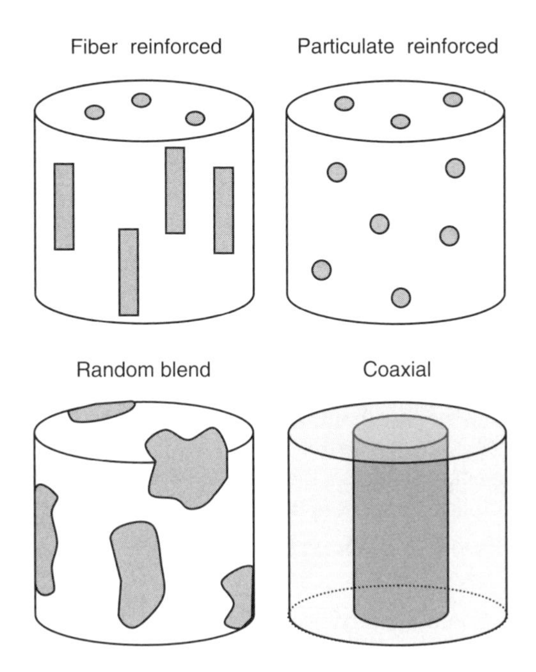

FIGURE 8.9 Types of electrospun composite nanofibers.

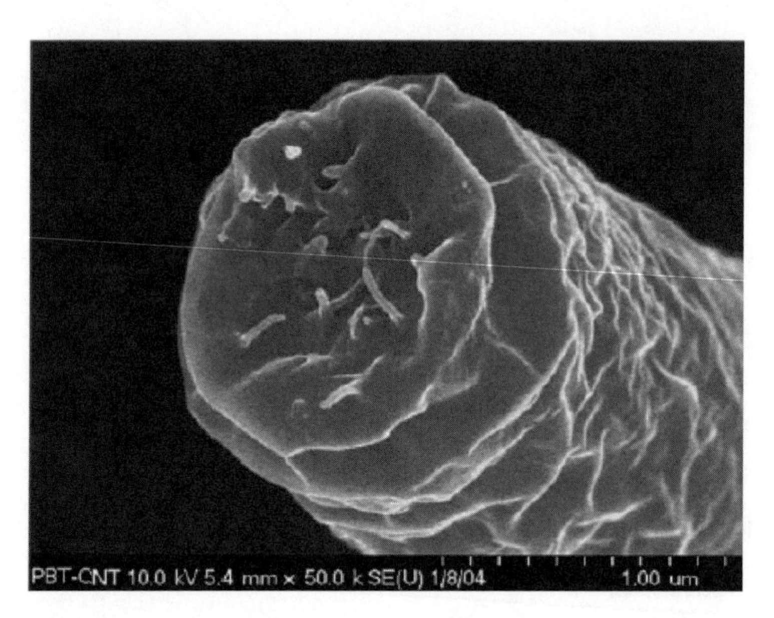

FIGURE 8.10 SEM cross-sectional image of a nanofiber showing embedded CNTs. (Reprinted from Mathew, G., Hong, J.P., Rhee, J.M., Lee, H.S., and Nah, C., *Polymer Testing*, 24, 712, 2005. With permission.)

nanofibers. The advantages of adding CNTs into the polymer matrix are not only limited to mechanical enhancements but also include the prospect of creating electrically conductive nanoscale products (Chronakis 2005). Generally, 1 wt% addition of CNTs to the nanofibers results in three to four times increase in Young's modulus (Ayutsede et al. 2006; Jose et al. 2007; McCullen et al. 2007). Higher percentages of CNTs up to 5 wt% have also been shown to increase the modulus of the resulting nanofibers (Mathew et al. 2005). This stiffness enhancement is due to the preferential alignment of CNTs along the fiber axis. CNTs have also been found to increase the crystallinity of the polymer matrix by acting as nucleating agents for the polymer molecules (Ayutsede et al. 2006), which contributes to the stiffening of the nanofibers. Figure 8.10 shows the cross section of CNTs embedded within a nanofiber. The strength and elongation at break, however, are usually adversely affected by the addition of CNTs. This is especially true when CNTs are not uniformly dispersed within the polymer matrix, are not aligned perfectly along the fiber axis, and form agglomerates. Poor interfacial bonding between CNTs and the polymer matrix can also lead to inefficient load transfer and thus lower strength. Nanotube agglomerates can act as critical flaws which decrease the ultimate strain and strength of the fibers (Ayutsede et al. 2006).

α-chitin whiskers are fiber reinforcements from biological sources which have been used to reinforce PVA nanofibers (Junkasem et al. 2006). Unlike CNT-reinforced composite nanofibers, the tensile strength of α-chitin-reinforced PVA nanofibers increased with a small percentage of reinforcement material. This is due to the interaction between PVA and chitin whiskers via hydrogen bonding. However, such interaction caused the nanocomposites to become more rigid with increasing whisker content, resulting in the observed decrease in the elongation at break. Once aggregation of chitin takes place, the strength of the nanofiber composite is reduced. Young's modulus has been found to increase (by four to eight times) with the addition of chitin until aggregates begin to form.

8.3.3.1.2 Particulate-Reinforced Nanofibers

In this chapter, we classify nanoplatelets/nanoflakes and nanoparticles as particulate reinforcements as they are two or three dimensional, respectively. The most common type of particulate-reinforced

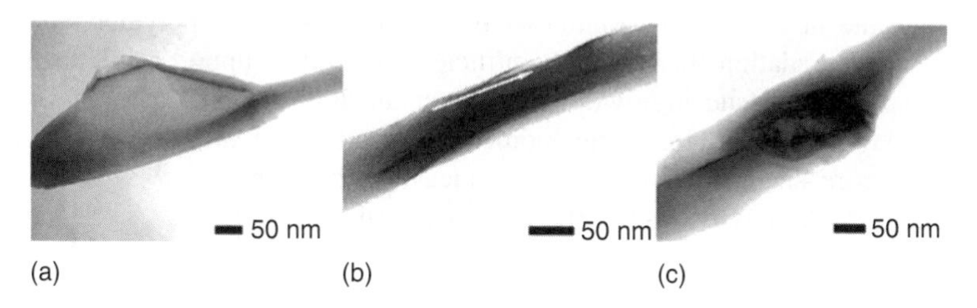

FIGURE 8.11 Transmission electron microscope (TEM) images of clay distribution inside electrospun polystyrene fibers at different clay weight percentages: (a) 1 wt%, (b) 4 wt%, and (c) 8 wt%. (Reprinted from Ji, Y., Li, B., Ge, S., Sokolov, J.C., and Rafailovich, M.S., *Langmuir*, 22, 1321, 2006. With permission.)

electrospun nanofibers is clay or montmorillonite (MMT) reinforced nanofibers (Ji et al. 2006; Li et al. 2006; Ramakrishna et al. 2006). The effectiveness of MMT reinforcement depends on how well the MMT layers are exfoliated inside the nanocomposite fibers. Li et al. (2006) reported that the addition of MMT increased the modulus of the nanofibers whereas Ramakrishna et al. reported a decrease in the modulus (Ramakrishna et al. 2006). Ji et al. (2006) reported the highest modulus at a certain weight fraction of MMT. The modulus depended on whether the MMT platelets are oriented along the fiber axis, form agglomerates, or distribute evenly (well exfoliated) within the nanofiber. Figure 8.11 shows the different distributions of MMT within the nanofiber at different MMT concentrations. At 4 wt% of MMT as shown in Figure 8.11b, the dark lines with a thickness of around 1 nm represent the MMT layers, which were aligned along the fiber axis as shown by the arrow. Other wt% of MMT resulted in inhomogeneous distribution of MMT. MMT platelets that are well exfoliated and oriented along the fiber direction result in an increase in crystallinity of the matrix polymer (nylon-6) as well as an increase in crystallite size (Li et al. 2006). This is because well exfoliated MMT platelets act as nucleating agents to facilitate nylon-6 crystallization. Polymer concentration and size effects are also observed in these studies.

Other platelet-like reinforcements include graphite nanoplatelets (Mack et al. 2005). Young's modulus has been found to increase with graphite content. The high aspect ratio of the nanoplatelets may provide an efficient means for stress transfer and reinforcement in nanofiber composites.

Round nanoparticles in the form of carbon black (Hwang et al. 2007) and ferritin (Shin et al. 2006a) have been used as reinforcement materials in polymeric nanofibers. The addition of nanoparticles has been shown to increase modulus and strength but reduce ductility (Hwang et al. 2007). The stiffening mechanism of adding nanoparticles could be due to the bonds formed between the particles and the matrix. For instance, in the study by Shin et al. (2006a), the ferritin nanoparticle contains a protein shell which consists of 24 peptide chains with 2 polar functional groups. These groups are capable of forming hydrogen bondings between the ferritin and PVA matrix. This enables efficient stress transfer from the matrix to the stiffer reinforcement material, which in turn enhances the overall mechanical properties. Unlike nanoplatelet reinforcements, there has been no mention of any changes in structure of the matrix material caused by nanoparticle reinforcements.

Thus, for reinforcement to improve the mechanical properties of the nanofiber effectively, at least two essential structural requirements have to be satisfied: uniform dispersion of reinforcement material to avoid localization of stress concentration; and good interfacial bonding between them and polymer matrix to achieve effective load transfer across the reinforcement–matrix interface (Kim et al. 2006). The strengthening mechanisms of reinforcement material within the polymer matrix during fiber deformation will be discussed in detail in Section 8.4.

8.3.3.2 Polymer Mixture Composites

There are a few reasons for adding another polymer to the polymer matrix. In tissue engineering applications, different types of collagen can be spun from two polymer solutions to create scaffolds that better mimic the in vivo ratios (Matthews et al. 2003). Another polymer may be added to control the rate of degradation that provides sufficient structural support while at the same time allows enough space for tissue ingrowth (Kidoaki et al. 2005). In the case of controlled drug delivery, there may be a need for some biomedical agent to be released at a later stage and nanofibers with a core–shell structure would be ideal for this application, with the agent as the core and a biodegradable polymer as the shell (Huang et al. 2005).

8.3.3.2.1 Randomly Blended Nanofibers

The two immiscible phases within the composite fibers can be visualized when one phase is leached out by a solvent that dissolves only one of the components. For instance, in the gelatin/PCL (polycaprolactone) blend, the two phases interpenetrate each other and coexist forming bicontinuous phase. This can be seen in Figure 8.12 after the water soluble gelatin has been leached out by phosphate buffered saline solution (Zhang 2006). The resulting mechanical properties of blended fibers as compared to single-phase fibers depend on how the additional phase affects the morphology of the primary phase and the quality of interface between the two phases.

For instance, gelatin/PCL nanofibers with a ratio of 1:1 has lower strength than either of the constituents, Young's modulus between that of gelatin and PCL, and higher elongation at break than either of the constituents. The blending of PCL with gelatin did not provide any strength reinforcement. This phenomenon is likely due to the immiscibility and microphase separation which will lead to easier slippage of chains under loading because of less entanglements and weak physical interactions among the chains of mixed polymers (Zhang 2006).

In the study by Chew et al., different types of drugs or proteins added to PCL-co-ethyl ethylene phosphate (PCLEEP) yielded different changes in mechanical properties (Chew et al. 2006). The

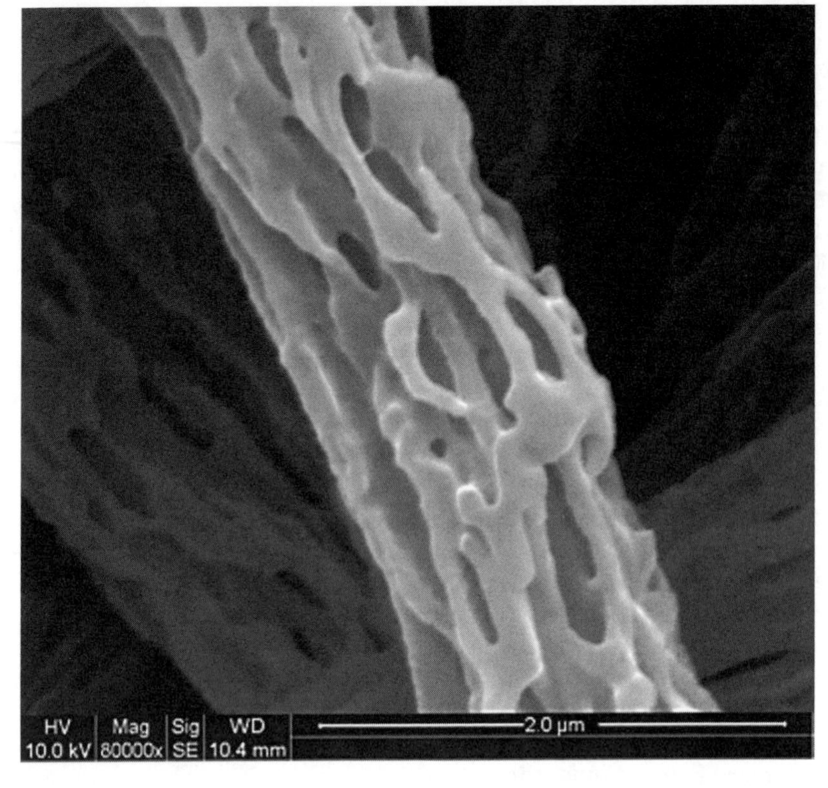

FIGURE 8.12 High-resolution SEM image of gelatin/PCL (polycaprolactone) randomly blended nanofiber after leaching of gelatin component. (From Zhang, Y.Z., Electrospinning of composite nanofibers for tissue scaffolding applications. Department of Mechanical Engineering. Singapore, National University of Singapore: 150. With permission.)

TABLE 8.2
Mechanical Properties and Morphology of Electrospun Sodium Alginate/PEO Mats with Different Sodium Alginate/PEO Ratios

Entry	Sodium Alginate/PEO Ratio	Tensile Strength (MPa)	Elongation to Break (%)	Morphology
1	3:1	0.9	8.5	114 nm, spindle-like beads
2	2:1	1.9	6.2	171 nm, spindle-like beads
3	1:1	4.5	3.4	228 nm, smooth fibers
4	1:2	4.0	3.3	266 nm, smooth fibers
5	0:1	1.2	2.9	308 nm, smooth fibers

Source: From Lu, J.-W., Zhu, Y.-L., Guo, Z.-X., Hu, P., and Yu. J., *Polymer*, 47, 8026, 2006. With permission.

addition of retinoic acid yielded nanofibers with smaller diameter and the nanocrystals of retinoic acid (RA) acted as reinforcing agents by restricting the movement of polymer chains as the fiber elongates under tensile force. As a result, the Young's modulus and strength were enhanced with the addition of RA. In contrast, the addition of bovine serum albumin (BSA) yielded fibers with larger diameter and phase separation between BSA and PCLEEP, which resulted in the restriction in polymer chain movement and elongation at the interface of these two phases during tensile deformation. As a result, both Young's modulus and strength were reduced with the addition of BSA.

The ratio of the polymers also affects the resulting mechanical properties. Table 8.2 shows the effect of varying sodium alginate/PEO ratio on the morphology and mechanical properties (Lu et al. 2006). It can be seen that the highest tensile strength can be obtained when the ratio is 1:1, which coincides with the production of smooth fibers with smallest diameter. The different trend observed for elongation at break could be due to the more dominant nature of PEO as its proportion increased.

8.3.3.2.2 Coaxial Nanofibers

Coaxial nanofibers are produced by using a conventional electrospinning setup but replacing the single capillary with two concentrically arranged capillaries that are connected to two separate reservoirs of polymer solutions (Sun et al. 2006). Figure 8.13 shows the TEM image of an electrospun coaxial nanofiber with PCL as the shell and gelatin as the core (Huang et al. 2005).

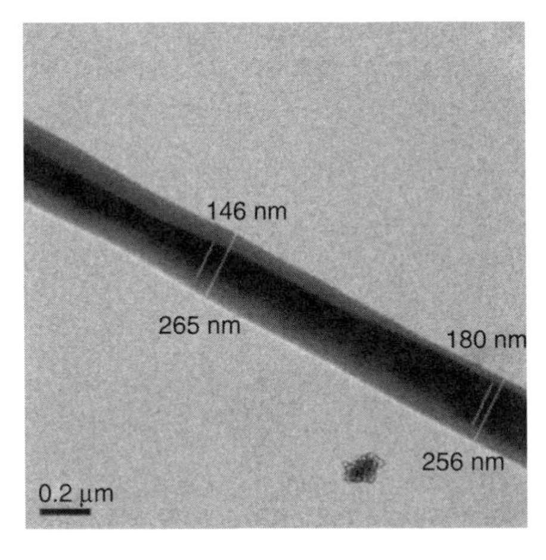

FIGURE 8.13 TEM image of coaxial nanofiber with polycaprolactone (PCL) as shell and gelatin as core. (Reprinted from Huang, Z.-M., Zhang, Y.Z., and Ramakrishna, S., *J. Polym. Sci. B: Polym. Phys.*, 43, 2852, 2005. With permission.)

The effects of adding a second polymer in the form of coaxial fibers on the mechanical properties for coaxial fibers are similar to randomly blended fibers. While some studies of coaxial fibers revealed inferior mechanical properties (Sun et al. 2006; Han et al. 2007), others observed improved mechanical properties (Huang et al. 2005). The lower tensile modulus and strength observed in some studies could be attributed to the weak physical interactions among the chains of mixed polymers and the imperfect morphology (Sun et al. 2006). Higher tensile modulus and strength are observed when the resulting fibers are the finest and without beads (Huang et al. 2005).

8.4 FIBER DEFORMATION AND FAILURE MECHANISMS

X-ray diffraction has been the most common equipment used to study the structure of polymeric nanofibers at the crystallite level. SEM has been used to observe the surface morphology while TEM has been used to observe the internal structure of larger particles within the nanofiber, as in the case of composites. However, there are only a handful of studies that report the deformation and failure mechanisms of electrospun nanofibers due to the difficulty in imaging the changes in morphology during deformation. The theories of how single-phase and composite nanofibers deform will be presented in this section.

8.4.1 Single-Phase Nanofibers

There is a need to understand the morphology of single nanofibers in order to understand the deformation and failure process of single nanofibers. Figure 8.7a shows the morphology of as-spun PLLA nanofiber and Figure 8.14 shows the corresponding schematic diagram of the fibrillar morphology (Cicero and Dorgan 2001). Although the schematic diagram was meant for melt spun PLLA fibers, it is still applicable for some electrospun nanofibers as both processes involve some form of drawing applied to the polymer nanofibers. The schematic diagram also matches closely with the morphology observed from the AFM phase image. The fibrillar morphology of electrospun nanofibers has also been observed by Zussman et al. in their study of failure modes of nanofibers (Zussman et al. 2003). However, due to the rapid process of electrospinning, some polymers may not have sufficient time to crystallize and thus have a different morphology from that

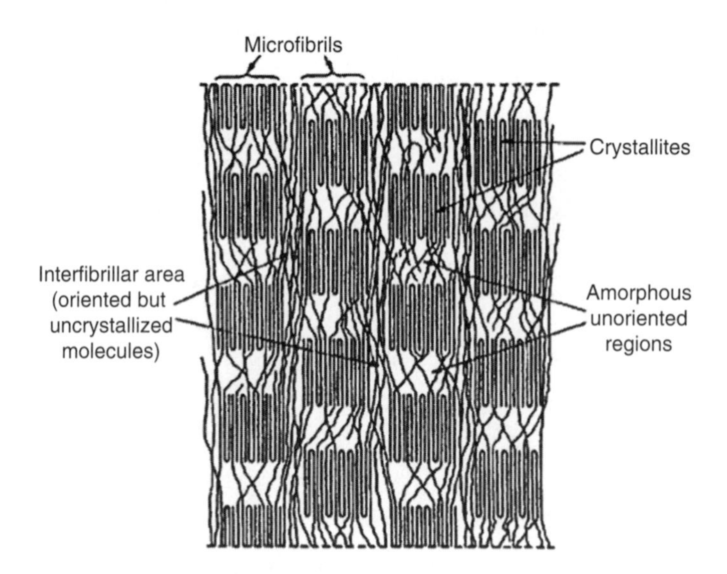

FIGURE 8.14 Schematic diagram of the fibrillar morphology with the vertical as the fiber axis. (From Cicero, J.A. and Dorgan, J.R., *J. Polym. Environ.*, 9, 1, 2001. With permission.)

suggested in Figure 8.14. PAN nanofibers produced by Ye and coworkers were found to have an amorphous morphology (Ye et al. 2004).

Based on the fibrillar morphology, the regions with least resistance to deformation are at the interfibrillar areas. During linear elastic deformation, any slack molecules will be straightened. During plastic deformation, the nanofiber would deform by shearing between the fibrils since the interfibrillar areas have the least resistance to deformation (Tan and Lim 2006a). SEM images of PEO nanofibers near failure show several fibrils connecting the two broken ends (Zussman et al. 2003), indicating that interfibrillar failure has taken place during large deformations. This appears similar to crazing often observed in cracking of polymers. Crazing has also been observed in PAN nanofibers (Ye et al. 2004). The formation of the crazing is related to the polymer chain entanglements. If the density of entanglements is high, the polymer appears brittle. At a low entanglement density, the polymer chains among entanglements may easily be stretched under tensile forces, leading to the formation of intrinsic fibrils and thus the crazing. The crazing phenomenon observed in the study by Ye et al. (2004) suggests that the polymer fibers produced by electrospinning contain entanglements with a low distribution density.

Besides interfibrillar deformation, necking (Inai et al. 2005; Zussman et al. 2006) or multiple necking (Zussman et al. 2003) has been observed at the site of fiber failure, indicating some form of ductile fracture. This is especially true for fibers with less oriented structures along the fiber axis. Inai et al. reported a neck-like fracture feature at the break point for electrospun fibers collected at a lower collection speed on a rotating wheel, which was absent from nanofibers collected at a higher collection speed (Inai et al. 2005). Multiple necking phenomena that are observed in nanofibers are unique to polymeric nanofibers as they are not observed in macroscopic polymer specimens. This could be due to the fact that nanofibers possess aspect ratios (length to diameter ratio) much higher than those of macroscopic polymer samples; they can accommodate many perturbation wavelengths, which leads to multiple necking (Zussman et al. 2003).

8.4.2 Composite Nanofibers

To date, all deformation studies available are limited to filler-reinforced composite nanofibers. Ye et al. (2004) and Kim et al. (2005a, 2005b, 2006) have proposed models for deformation mechanisms for different types of reinforcement systems. Lee et al. (2003b) proposed a general three-step mechanical deformation process. The process involves:

Stage 1: Stress concentration due to the presence of fillers or nanopores that have different elastic properties from the matrix.

Stage 2: Induced shear yielding of the matrix (instead of crazing) due to spatial confinement of the matrix by the fillers or pores.

Stage 3: Necking occurs when the strain reaches a certain critical value due to overlapping stress fields induced by stress concentrators. The neck propagates along the fiber until failure occurs.

Figure 8.15 shows the deformation of poly(methyl methacrylate) (PMMA)/SiO_2 composite nanofiber. Figure 8.16 shows the schematic diagrams for the three stages of deformation for PMMA/SiO_2 composite and polycarbonate with CNT and nanopores on the surface. Ye et al. (2004), on the other hand, suggested crazing as the main mode of deformation, with the reinforcement material (CNT) hindering the crazing extension, reducing stress concentration, and dissipating energy by CNT pullout. No polymer was found to attach to the single-walled CNTs (SWNT), indicating that the interfacial bonding between the CNT and matrix was not strong enough to keep polymer attached when SWNTs are pulled out.

The difference in deformation mechanism suggested could be due to the difference in volume fraction of filler materials used. Ye et al. used 1 wt% of CNT whereas Kim et al. used 4 wt% of

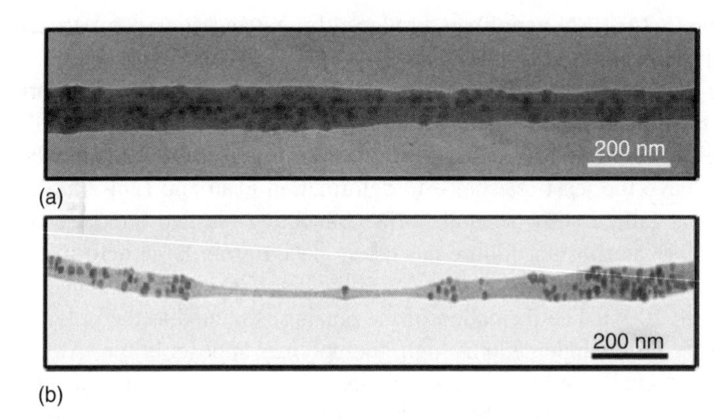

FIGURE 8.15 TEM images of a sequence of mechanical deformation processes: (a) before deformation of an electrospun PMMA/SiO_2 nanocomposite, and (b) a deformed state over a critical strain under uniaxial tensile load. (Reproduced from Kim, G.M., Lach, R., Michler, G.H., Pötschke, P., and Albrecht, K., *Nanotechnology*, 17, 963, 2006. With permission.)

CNT. When volume fraction of filler is increased, the interfiller distance is reduced (assuming uniform dispersion), thus reducing the available space that crazing ligaments can form. It has been found that by reducing the ligament thickness below a critical value, the excessive critical hydrostatic stresses can be effectively eliminated, which corresponds to a local plane-strain-to-plane-stress transition in the matrix (Margolina and Wu 1988). When this happens, craze formation is completely suppressed as the surface energy density for a void to form exceeds the volume strain energy density in the material during deformation of the sample. Thus, only shear yielding has been observed in Kim et al.'s studies. Another reason for the observed difference could be due to the different methods used to induce observable deformations in the nanofibers. Kim et al. induced in situ deformation of single nanofibers in TEM by electron beam-induced thermal stresses whereas Ye et al. simply picked a small amount of fibers out of electrospun yarns with tweezers and placed

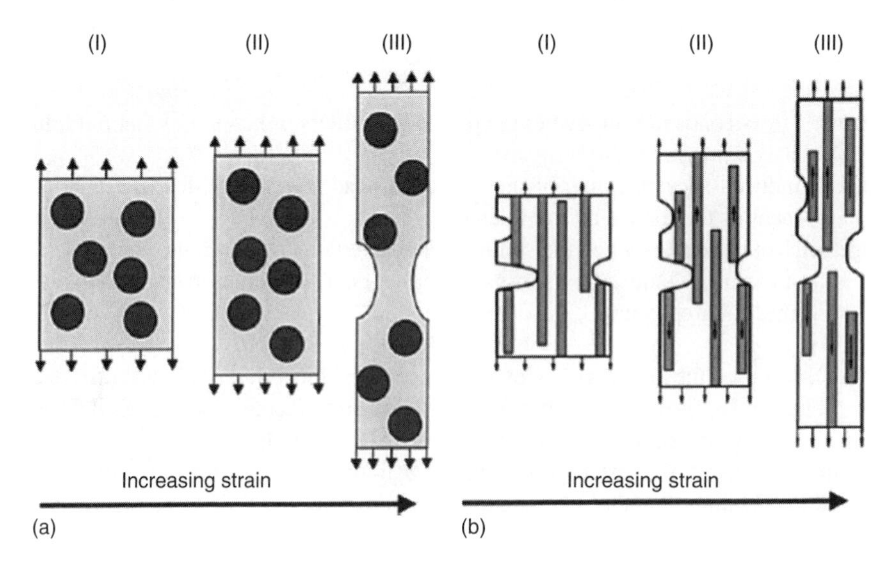

FIGURE 8.16 Schematic illustration of mechanical deformation processes of electrospun composite nanofibers under uniaxial tensile load for (a) PMMA with 10 wt% SiO_2, and (b) polycarbonate with 4 wt% multiwalled CNTs. (Reproduced from Kim, G.M., Lach, R., Michler, G.H., Pötschke, P., and Albrecht, K., *Nanotechnology*, 17, 963, 2006. With permission.)

them on copper grids for TEM observation. The former method probably induced a slower strain rate as well as increased the sample temperature during deformation. These two factors would increase the mobility of the polymer chains, and thus promote shear yielding during deformation.

When the filler does not form agglomerates as in the cases presented here, the toughness of the nanofiber can actually be improved as it requires more energy for void formation, either by CNT interference with crazing or the suppression of craze formation.

8.5 SUMMARY AND FUTURE OUTLOOK

Due to the rapid technological advancements in imaging and nano-Newton load sensing devices and systems in the past decade, the structure, mechanical properties, and deformation mechanisms of single electrospun nanofibers have been explored in greater detail. Utilizing electrospinning parameters that generate uniform nanofibers with small diameters have been found to yield a product with high stiffness and strength but reduced ductility, whereas any process that induces bead formation in electrospun nanofibers will affect the mechanical properties adversely. Post-electrospinning treatments that result in restricted movement between polymer chains such as cross-linking, annealing, and crystal phase change will result in stiffer nanofibers. Composite nanofibers in the form of filler-reinforced nanofibers and polymer mixture composites have been produced. Reinforcement effectiveness depends on the quality of interfacial bonding between reinforcement and matrix, uniformity of filler dispersion, and degree of alignment with respect to fiber axis (applicable for one- and two-dimensional fillers). Polymer mixture composites only induce stiffness and strength enhancements when the resulting fibers are smooth and have small diameters. Generally, there is phase separation in such composites and the bonding between the two polymers is weak; thus, the stiffness is adversely affected. Single-phase nanofibers often fail by interfibrillar failure or crazing, with necks often seen along the fiber length and at the site of failure. By adding reinforcement materials to the polymer nanofibers, the stiffness and strength can be increased as the fillers interfere with void formation and increase the energy required to initiate voids. Reinforcements also change the failure mode from crazing to shear yielding.

Although greater insight has been gained in recent years as to how the structure relates to the mechanical characteristics of single nanofibers, more research has to be done in the area of in situ imaging of single nanofiber deformation and relating the deformation observed to the stress–strain characteristics of the nanofibers. Thus, the method of simultaneously imaging and measuring the load-deformation behavior of single nanofibers has to be developed for this to be possible.

REFERENCES

Ayutsede, J., M. Gandhi, S. Sukigara, et al. 2006. Carbon nanotube reinforced *Bombyx mori* silk nanofibers by the electrospinning process. *Biomacromolecules* 7: 208–214.

Bellan, L.M., J. Kameoka, and H.G. Craighead. 2005. Measurement of the Young's moduli of individual polyethylene oxide and glass nanofibres. *Nanotechnology* 16: 1095–1099.

Bhattarai, N., Z. Li, D. Edmondson, and M. Zhang. 2006. Alginate-based nanofibrous scaffolds: Structural, mechanical, and biological properties. *Advanced Materials* 18: 1463–1467.

Bolgen, N., Y.Z. Menceloglu, K. Acatay, I. Vargel, and E. Piskin. 2005. In vitro and in vivo degradation of non-woven materials made of poly(ε-caprolactone) nanofibers prepared by electrospinning under different conditions. *Journal of Biomaterials Science. Polymer Edition* 16: 1537–1555.

Chen, C., C. Cao, X. Ma, Y. Tang, and H. Zhu. 2006. Preparation of non-woven mats from all-aqueous silk fibroin solution with electrospinning method. *Polymer* 47: 6322–6327.

Chew, S.Y., T.C. Hufnagel, C.T. Lim, and K.W. Leong. 2006. Mechanical properties of single electrospun drug-encapsulated nanofibres. *Nanotechnology* 17: 3880–3891.

Choi, S.-S., J.-P. Hong, Y.S. Seo, S.M. Chung, and C. Nah. 2006. Fabrication and characterization of electrospun polybutadiene fibers crosslinked by UV irradiation. *Journal of Applied Polymer Science* 101: 2333–2337.

Chronakis, I.S. 2005. Novel nanocomposites and nanoceramics based on polymer nanofibers using electrospinning process—A review. *Journal of Materials Processing Technology* 167: 283–293.

Chuangchote, S., A. Sirivat, and P. Supaphol. 2007. Mechanical and electro-rheological properties of electrospun poly(vinyl alcohol) nanofibre mats filled with carbon black nanoparticles. *Nanotechnology* 18: 145705.

Cicero, J.A. and J.R. Dorgan. 2001. Physical properties and fiber morphology of poly(lactic acid) obtained from continuous two-step melt spinning. *Journal of Polymers and the Environment* 9: 1–10.

Deitzel, J.M., J. Kleinmeyer, D. Harris, and N.C.B. Tan. 2001. The effect of processing variables on the morphology of electrospun nanofibers and textiles. *Polymer* 42: 261–272.

Doshi, J. and D.H. Reneker. 1995. Electrospinning process and applications of electrospun fibers. *Journal of Electrostatics* 35: 151–160.

Formhals, A. 1934. Process and apparatus for preparing artificial threads. [1,975,504]. US Patent.

Gu, S.-Y., Q.-L. Wu, J. Ren, and G.J. Vansco. 2005. Mechanical properties of a single electrospun fiber and its structures. *Macromolecular Rapid Communications* 26: 716–720.

Han, X., Z.M. Huang, C. He, and L. Liu. 2007. Preparation and characterization of core–shell structured nanofibers by coaxial electrospinning. *High Performance Polymers* 19: 147–159.

Huang, L., K. Nagapudi, R.P. Apkarian, and E.L. Chaikof. 2001. Engineered collagen-PEO nanofibers and fabrics. *Journal of Biomaterials Science. Polymer Edition* 12: 979–993.

Huang, Z.-M., Y.Z. Zhang, M. Kotaki, and S. Ramakrishna. 2003. A review on polymer nanofibers by electrospinning and their applications in nanocomposites. *Composites Science and Technology* 63: 2223–2253.

Huang, Z.-M., Y.Z. Zhang, and S. Ramakrishna. 2005. Double-layered composite nanofibers and their mechanical performance. *Journal of Polymer Science: Part B: Polymer Physics* 43: 2852–2861.

Huang, Z.-M., Y.Z. Zhang, S. Ramakrishna, and C.T. Lim. 2004. Electrospinning and mechanical characterization of gelatin nanofibers. *Polymer* 45: 5361–5368.

Hwang, J., J. Muth, and T. Ghosh. 2007. Electrical and mechanical properties of carbon-black-filled, electrospun nanocomposite fiber webs. *Journal of Applied Polymer Science* 104: 2410–2417.

Inai, R., M. Kotaki, and S. Ramakrishna. 2005. Structure and properties of electrospun PLLA single nanofibres. *Nanotechnology* 16: 208–213.

Jaeger, R., H. Schonherr, and G.J. Vancso. 1996. Chain packing in electro-spun poly(ethylene oxide) visualized by atomic force microscopy. *Macromolecules* 29: 7634–7636.

Ji, Y., B. Li, S. Ge, J.C. Sokolov, and M.H. Rafailovich. 2006. Structure and nanomechanical characterization of electrospun PS/Clay nanocomposite fibers. *Langmuir* 22: 1321–1328.

Jose, M.V., B.W. Steinert, V. Thomas, et al. 2007. Morphology and mechanical properties of Nylon 6/MWNT nanofibers. *Polymer* 48: 1096–1104.

Junkasem, J., R. Rujiravanit, and P. Supaphol. 2006. Fabrication of α-chitin whisker-reinforced poly(vinyl alcohol) nanocomposite nanofibres by electrospinning. *Nanotechnology* 17: 4519–4528.

Kidoaki, S., I.K. Kwon, and T. Matsuda. 2005. Mesoscopic spatial designs of nano- and microfiber meshes for tissue-engineering matrix and scaffold based on newly devised multilayering and mixing electrospinning techniques. *Biomaterials* 26: 37–46.

Kim, G.-M., R. Lach, G.H. Michler, and Y.-W. Chang. 2005a. The mechanical deformation process of electrospun polymer nanocomposite fibers. *Macromolecular Rapid Communications* 26: 728–733.

Kim, G.-M., G.H. Michler, and P. Pötschke. 2005b. Deformation processes of ultrahigh porous multiwalled carbon nanotubes/polycarbonate composite fibers prepared by electrospinning. *Polymer* 46: 7346–7351.

Kim, G.M., R. Lach, G.H. Michler, P. Pötschke, and K. Albrecht. 2006. Relationships between phase morphology and deformation mechanisms in polymer nanocomposite nanofibres prepared by an electrospinning process. *Nanotechnology* 17: 963–972.

Lee, K.H., H.Y. Kim, M.S. Khil, Y.M. Ra, and D.R. Lee. 2003a. Characterization of nano-structured poly(ε-caprolactone) nonwoven mats via electrospinning. *Polymer* 44: 1287–1294.

Lee, K.H., H.Y. Kim, Y.J. Ryu, K.W. Kim, and S.W. Choi. 2003b. Mechanical behavior of electrospun fiber mats of poly(vinyl chloride)/polyurethane polyblends. *Journal of Polymer Science: Part B: Polymer Physics* 41: 1256–1262.

Li, L., L.M. Bellan, H.G. Craighead, and M.W. Frey. 2006. Formation and properties of nylon-6 and nylon-6/montmorillonite composite nanofibers. *Polymer* 47: 6208–6217.

Lu, J.-W., Y.-L. Zhu, Z.-X. Guo, P. Hu, and J. Yu. 2006. Electrospinning of sodium alginate with poly (ethylene oxide). *Polymer* 47: 8026–8031.

Mack, J.J., L.M. Viculis, A. Ali, et al. 2005. Graphite nanoplatelet reinforcement of electrospun polyacrylonitrile nanofibers. *Advanced Materials* 17: 77–80.

Margolina, A. and S. Wu. 1988. Percolation model for brittle-tough transition in nylon/rubber blends. *Polymer* 29: 2170–2173.

Mathew, G., J.P. Hong, J.M. Rhee, H.S. Lee, and C. Nah. 2005. Preparation and characterization of properties of electrospun poly(butylene terephthalate) nanofibers filled with carbon nanotubes. *Polymer Testing* 24: 712–717.

Matthews, J.A., G.E. Wnek, D.G. Simpson, and G.L. Bowlin. 2003. Electrospinning of collagen nanofibers. *Biomacromolecules* 3: 232–238.

McCullen, S.D., D.R. Stevens, W.A. Roberts, S.S. Ojha, L.I. Clarke, and R.E. Gorga. 2007. Morphological, electrical, and mechanical characterization of electrospun nanofiber mats containing multiwalled carbon nanotubes. *Macromolecules* 40: 997–1003.

Pedicini, A. and R.J. Farris. 2003. Mechanical behavior of electrospun polyurethane. *Polymer* 44: 6857–6862.

Pham, Q.P., U. Sharma, and A.G. Mikos. 2006. Electrospinning of polymeric nanofibers for tissue engineering applications: A review. *Tissue Engineering* 12: 1197–1211.

Ramakrishna, S., T.C. Lim, R. Inai, and K. Fujihara. 2006. Modified Halpin–Tsai equation for clay-reinforced polymer nanofiber. *Mechanics of Advanced Materials and Structures* 13: 77–81.

Reneker, D.H. and I. Chun. 1996. Nanometre diameter fibres of polymer, produced by electrospinning. *Nanotechnology* 7: 216–223.

Samuel, B.A., M.A. Haque, B. Yi, R. Rajagopalan, and H.C. Foley. 2007. Mechanical testing of pyrolysed poly-furfuryl alcohol nanofibres. *Nanotechnology* 18: 115704.

Shin, M.K., S.I. Kim, S.J. Kim, S.-K. Kim, and H. Lee. 2006a. Reinforcement of polymeric nanofibers by ferritin nanoparticles. *Applied Physics Letters* 88: 193901.

Shin, M.K., S.I. Kim, S.J. Kim, S.-K. Kim, H. Lee, and G.M. Spinks. 2006b. Size-dependent elastic modulus of single electroactive polymer nanofibers. *Applied Physics Letters* 89: 231929.

Sombatmankhong, K., O. Suwantong, S. Waleetorncheepsawat, and P. Supaphol. 2006. Electrospun fiber mats of poly(3-hydroxybutyrate), poly(3-hydroxybutyrate-co-3-hydroxyvalerate), and their blends. *Journal of Polymer Science: Part B: Polymer Physics* 44: 2923–2933.

Subbiah, T., G.S. Bhat, R.W. Tock, S. Parameswaran, and S.S. Ramkumar. 2005. Electrospinning of nanofibers. *Journal of Applied Polymer Science* 96: 557–569.

Sun, B., B. Duan, and X. Yuan. 2006. Preparation of core/shell PVP/PLA ultrafine fibers by coaxial electrospinning. *Journal of Applied Polymer Science* 102: 39–45.

Tan, E.P.S., C.N. Goh, C.H. Sow, and C.T. Lim. 2005a. Tensile test of a single nanofiber using an atomic force microscope tip. *Applied Physics Letters* 86: 073115.

Tan, E.P.S., S.Y. Ng, and C.T. Lim. 2005b. Tensile testing of a single ultrafine polymeric fiber. *Biomaterials* 26: 1453–1456.

Tan, E.P.S. and C.T. Lim. 2006a. Effects of annealing on the structural and mechanical properties of electrospun polymeric nanofibres. *Nanotechnology* 17: 2649–2654.

Tan, E.P.S. and C.T. Lim. 2006b. Mechanical characterization of nanofibers—A review. *Composites Science & Technology* 66: 1099–1108.

Thomas, V., M.V. Jose, S. Chowdhury, J.F. Sullivan, D.R. Dean, and Y.K. Vohra. 2006. Mechano-morphological studies of aligned nanofibrous scaffolds of polycaprolactone fabricated by electrospinning. *Journal of Biomaterials Science. Polymer Edition* 17: 969–984.

Wang, M., H.-J. Jin, D.L. Kaplan, and G.C. Rutledge. 2004. Mechanical properties of electrospun silk fibers. *Macromolecules* 37: 6856–6864.

Warner, S.B., A. Buer, S.C. Ugbolue, G.C. Rutledge, and M.Y. Shin. 1999. A fundamental investigation of the formation and properties of electrospun fibers. Annual Report (M98-D01). USA, National Textile Center: 1–10.

Xu, S., Y. Shi, and S.-G. Kim. 2006. Fabrication and mechanical property of nano piezoelectric fibres. *Nanotechnology* 17: 4497–4501.

Ye, H., H. Lam, N. Titchenal, Y. Gogotsi, and F. Ko. 2004. Reinforcement and rupture behavior of carbon nanotubes-polymer nanofibers. *Applied Physics Letters* 85: 1775–1777.

Yuya, P.A., Y. Wen, J.A. Turner, Y.A. Dzenis, and Z. Li. 2007. Determination of Young's modulus of individual electrospun nanofibers by microcantilever vibration method. *Applied Physics Letters* 90: 111909.

Zhang, Y.Z. 2006. Electrospinning of composite nanofibers for tissue scaffolding applications. Department of Mechanical Engineering. Singapore, National University of Singapore: 150.

Zhang, Y.Z., C.T. Lim, S. Ramakrishna, and Z.M. Huang. 2005. Recent development of polymer nanofibers for biomedical and biotechnological applications. *Journal of Materials Science—Materials in Medicine* 16: 933–946.

Zhang, Y.Z., J. Venugopal, Z.M. Huang, C.T. Lim, and S. Ramakrishna. 2006. Crosslinking of the electrospun gelatin nanofibers. *Polymer* 47: 2911–2917.

Zussman, E., M. Burman, A.L. Yarin, R. Khalfin, and Y. Cohen. 2006. Tensile deformation of electrospun nylon-6,6 nanofibers. *Journal of Polymer Science: Part B: Polymer Physics* 44: 1482–1489.

Zussman, E., X. Chen, W. Ding, et al. 2005. Mechanical and structural characterization of electrospun PAN-derived carbon nanofibers. *Carbon* 43: 2175–2185.

Zussman, E., D. Rittel, and A.L. Yarin. 2003. Failure modes of electrospun nanofibers. *Applied Physics Letters* 82: 3958–3960.

9 Electrospinning Techniques to Control Deposition and Structural Alignment of Nanofibrous Scaffolds for Cellular Orientation and Cytoskeletal Reorganization

Joel K. Wise, Michael Cho, Eyal Zussman, Constantine M. Megaridis, and Alexander L. Yarin

CONTENTS

9.1 INTRODUCTION

Cells and extracellular matrix (ECM) fibrils in most natural tissues are not random, but exhibit well-defined patterns and specific spatial orientation. Recent findings demonstrated that oriented biopolymer-based nanofibrous scaffolds have the potential for engineering blood vessels [1], neural tissue [2], and ligament tissue [3]. Furthermore, it has been shown that cell adhesion and proliferation [1] is significantly improved on oriented nanofibrous scaffolds. The contact guidance theory suggests that cells have the greatest probability of migrating in preferred orientations which are associated with chemical, structural, and/or mechanical properties of the substrate [4–6]. Consequently, it may be postulated that an oriented nanofibrous scaffold would guide cell alignment along the nanofibers. The cell arrangement onto an oriented nanofibrous scaffold could be due to the contact guidance and/or to cytoskeletal reorganization. Aligned cells could then be used to remodel and modulate the regenerated ECM and microenvironment [7].

The particular motivation for this work is to engineer a tissue-like construct that will mimic the superficial zone of articular cartilage. It is recognized that the articular cartilage is mainly comprised of at least four distinct zones, each with a specific cell and ECM organization or orientation [8–10]. These zones are known as superficial or tangential, intermediate or transitional, deep or radial, and calcified zone. The specific cell and ECM organization within each cartilage zone can be attributed to developmental history and to the mechanical forces to which each of these cartilage zones is subjected, thereby supporting the overall functionality of articular cartilage tissue [12,13]. Several published reports have focused on the zonal organization of chondrocytes within biocompatible polymers for tissue-engineered articular cartilage constructs [14–18]. Furthermore, recently published papers have used chondrocytes and human mesenchymal stem cells (hMSCs) and random three-dimensional polymer nanofibrous scaffolds for cartilage tissue engineering [19–23]. However, attempts to specifically engineer the superficial zone of the articular cartilage using hMSCs and oriented nanofibrous scaffolds have not been reported. The superficial zone of natural articular cartilage tissue displays flattened, ellipsoidal-like chondrocytes and nanoscale collagen fibrils, which are oriented parallel to the plane of the articular surface, with a significant degree of orientation in that plane [8–11]. Although it is only a few layers thin, the importance of this superficial zone is often emphasized because it provides high tensile properties that are maintained by the orientation of cells and collagen fibrils [12,13], and therefore are crucial for normal function at the articular cartilage surface. The first necessary step toward functional tissue engineering of the cartilage superficial zone requires quantitative evaluation of cell adhesion and orientation.

In order to function properly, cells seeded in a scaffold must be able to communicate with their environment via appropriate biochemical, bioelectrical, and topological signals. The topography and composition of the microenvironment (i.e., native ECM) affect cellular functions, such as adhesion, growth, motility, secretion, gene expression, and apoptosis. It is well known that the physical structure of the native ECM has nanoscale dimensions. The high surface area-to-volume ratio of the nanofibrous scaffolds and the nanoscale diameter of the fibers are likely to provide attractive conditions for cell attachment and growth. For example, because collagen is the most abundant ECM component, collagen-based scaffolds composed of fibrils in the 10–300 nm diameter range are not only biocompatible but also could provide one of the most ideal nanofibrous structures for tissue engineering [24]. Indeed, cells have been shown to attach to and organize around fibers with diameters smaller than those of the cells [24,25].

Stem cells have the unique property of self-renewal without differentiation until and unless appropriate biological and physical signals are provided. In the context of tissue engineering, the use of stem cells has numerous advantages. Unlike engineered tissue constructs using differentiated cells, stem cells (1) have higher proliferative capacity; (2) provide excellent regenerative capability that will likely lead to desired integrity and functionality of the engineered tissue; (3) make it possible to contemplate multifunctional tissue constructs (e.g., osteochondral tissue); and (4) reduce or eliminate tissue rejection and failure. It is now well established that hMSCs can be cultured and expanded in vitro, and can be induced to proliferate and differentiate into tissue-specific cell phenotypes such as chondrogenic, osteogenic, adipogenic, and myogenic cells using biological and physical stimuli [26–29], and therefore provide potential for musculoskeletal tissue repair and regeneration. While the vast therapeutic potential of hMSCs has been recognized for treatment of damaged or diseased tissue, the complexity of events associated with transformation of these precursor cells leaves many unanswered questions about morphological, structural, proteomic, and functional changes in stem cells. More detailed knowledge of hMSC behavior would allow better and more effective approaches to cell expansion in vitro, and regulation of their commitment to specific phenotype. For example, control of hMSC orientation, proliferation, and differentiation has been shown to modulate the functional aspects of engineered tissue constructs [30,31]. Engineering an environment conducive to hMSC orientation, adhesion, proliferation, and differentiation is therefore of paramount importance in tissue engineering.

Electrospinning [32–37] is a promising technique to create patterned fiber architectures for regulating cellular and molecular interactions of stem cell adhesion, proliferation, differentiation, and orientation. A variety of natural or synthetic polymers have been used in the framework of electrospinning to fabricate and design biocompatible nanofibrous scaffolds for tissue engineering applications. For example, polycaprolactone (PCL) is a biodegradable material that shows virtually no toxicity, controllable degradation, and is inexpensive [38]. PCL is a semicrystalline polymer that belongs to the family of α-hydroxy polyesters (e.g., PLA, PGA) and can significantly enhance the mechanical properties of the scaffold due to its high tensile modulus (400 MPa; [39,40]). These mechanical properties exceed the elastic modulus of cartilage tissue ($\sim$10 MPa) and are in the range of the elastic modulus of collagen fibers ($\sim$500 MPa), suggesting that PCL nanofibrous scaffolds show excellent promise for cartilage tissue engineering. Furthermore, PCL has recently been used to demonstrate its compatibility to seed and accommodate hMSCs and support multilineage differentiation [20,21,38].

In the present work, we first review different techniques used to align nanofibers during electrospinning. In the subsequent sections, we aim to regulate and quantify hMSC orientation on at least two different nanofibrous PCL scaffolds fabricated using electrospinning. Using sophisticated image analysis, hMSC orientation on random and oriented nanofibrous scaffolds was quantified during an 18 day culture period. In addition, long-term cell viability, cytoskeletal reorganization, and cell–nanofiber adhesions were explored and correlated with contact-induced hMSC orientation on nanofibers. Stem cell-based therapeutic application and tissue engineering may be enhanced and facilitated by physical approaches, such as the one presented herein, to control the microenvironment for directed cell growth and differentiation.

9.2 FABRICATION OF ALIGNED ELECTROSPUN NANOFIBER SCAFFOLDS

Techniques for in situ alignment of as-spun nanofibers using electrostatic repulsion forces date back to works [41,42] where the first electrostatic lens for electrospinning was introduced. The idea of the electrostatic lens in electrospinning gained momentum in other studies [43–46] where some additional arrangements using this principle were used. A sketch of the experimental setup employed in Refs. [41,42] is shown in Figure 9.1. The jet flowed downward from the surface of a pendant drop of polymer solution suspended at the end of a needle toward a rotating wheel collector placed at a distance of 120 mm below the droplet. The aluminum wheel (with a diameter of 200 mm) had a tapered edge with a half angle of 26.6° in order to create a strongly converging electrostatic field. An electric potential difference of around 8 kV was initially applied between the surface of the liquid drop and the rotating wheel. When the potential difference between the pendant droplet and the grounded wheel was increased, the droplet acquired a cone-like shape (Taylor cone). At a high enough potential difference, a stable jet emerged from the cone and moved downward toward the wheel. The jet flowed away from the droplet in a nearly straight line and then bent into a complex path that was contained within a nearly conical region (envelope cone, [32]), with its apex at the top. Then, at a certain point above the wheel the envelope cone started to shrink, resulting in an inverted envelope cone with its apex at the wheel's edge.

During the electrospinning process, the wheel is rotated at a constant speed to collect the developing nanofibers onto its sharp edge. As the as-spun nanofiber reaches the wheel's edge, it is wound around the wheel. A small table (5 $\times$ 4 mm) made of aluminum can also be attached to the wheel edge to facilitate the subsequent collection of the nanofibers. The table can be rotated about its Z-axis (cf. Figure 9.1) when the wheel rotation is temporarily ceased; hence, the direction of the

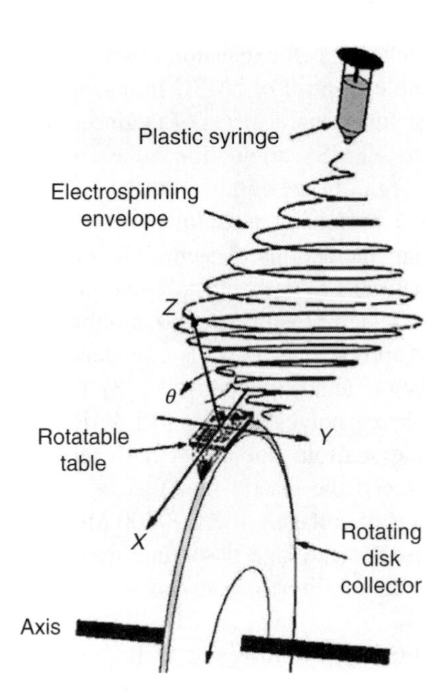

FIGURE 9.1 Schematic of the electrospinning process showing the double-cone envelope of the jet. The collector wheel can be equipped with a table that assists to collect nanofibers. The table can be rotated about the *Z*-axis when the wheel rotation is temporarily ceased to enable layer-by-layer collection at a desired angle between the nanofiber array layers. (From Zussman, E., Theron, A., and Yarin, A.L., *Appl. Phys. Lett.*, 82, 973, 2003. With permission.)

collected nanofibers can be controlled. The nanofibers are collected usually over a 10 s period. Typical two-dimensional nanofiber arrays fabricated from poly(ethylene oxide) (PEO) with this method are shown in Figure 9.2. The linear speed at the tip of the disk collector was 11 m/s in this case. The diameter of the nanofibers in Figure 9.2a is not uniform and varies from 250 to 400 nm, and from 200 to 500 nm in Figure 9.2b. The separation between the parallel nanofibers varies from 1 to 2 μm in both images.

A section of a long nanofiber deposited on the table remains charged for a brief period. Therefore, when a new loop of the nanofiber settles onto the table, it is repelled by the previously settled sections, which are still charged. This mechanism allows formation of two-dimensional arrays on the table similar to those shown in Figure 9.2. The linear rotation speed of the wheel plays a crucial role in stretching and aligning the nanofiber sections along the table after their touchdown (see Figure 9.3). At linear wheel rotation speeds of 5–10 m/s, sufficient nanofiber alignment can be achieved (Figure 9.3c and e). The angle (with respect to the direction of the wheel rotation) distributions shown in Figure 9.3b, d, and f reveal that at low speeds no preferred fiber orientation is attained (Figure 9.3b); at higher speeds, the angular distributions show a preferred fiber orientation and become narrower with increasing linear rotation speed (Figure 9.3d and f). The same mechanism essentially works when nanofibers are electrospun on rotating mandrels [46–48]. However, when there is no lateral restrictions imposed by the sharp edge of the wheel (the electrostatic lens), sideways excursions of the fibers are quite significant and a lower degree of alignment may be achieved. An intermediate variant corresponds to a metal ribbon (which is equivalent to an elongated narrow table) mounted on top of the sharpened wheel edge. Nanofiber strips accumulated on such a ribbon could also be well aligned. This is the reason that oriented nanofiber mats used in the subsequent sections of the present work were electrospun on a ribbon placed on top of the rotating sharpened wheel.

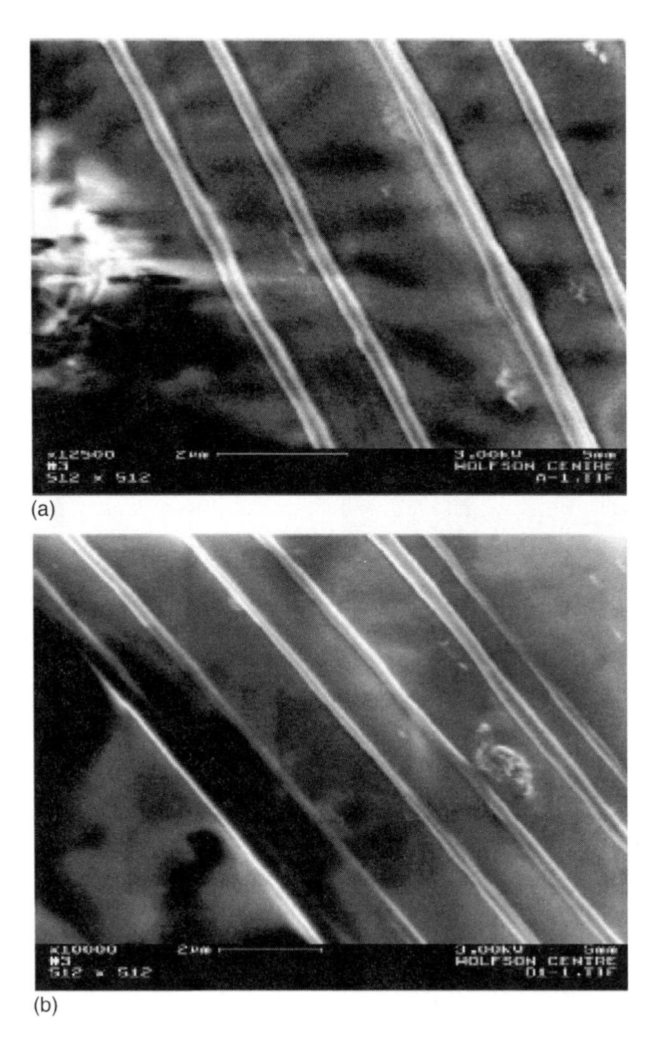

FIGURE 9.2 Scanning electron microscope (SEM) images of aligned PEO nanofibers that were collected on a carbon tape attached to the edge of the wheel collector. In (a), the diameter of the fibers varies from 250 to 400 nm. In (b), the diameter of the fibers varies from 200 to 500 nm. The pitch (center to center) varies from 1 to 2 μm in both images. (From Theron, A., Zussman, E., and Yarin, A.L., *Nanotechnology*, 12, 384, 2001. With permission.)

It is emphasized that stretching of nanofiber sections after the first touchdown on a rotating collector in addition enhances their tensile strength, which is desirable for a number of biomedical applications [46–48].

Typical three-dimensional nanofiber arrays (crossbars) are depicted in Figure 9.4. The collected PEO nanofibers show a high degree of alignment. The diameter of the nanofibers in this case is also nonuniform and varies in the range of 100–800 nm. When nanofibers were electrospun onto the wheel's sharp edge without the table, bundles (ropes) of nanofibers were obtained. A typical image of a nanofiber rope pealed off from the wheel edge is shown in Figure 9.5. The duration of the collection process in this case was 60 s. The rope was manually detached from the wheel edge [41]. Two SEM details of polymer fiber ropes produced in this manner are shown in Figure 9.6. The nanofibers are in contact, and remain nearly parallel for long distances.

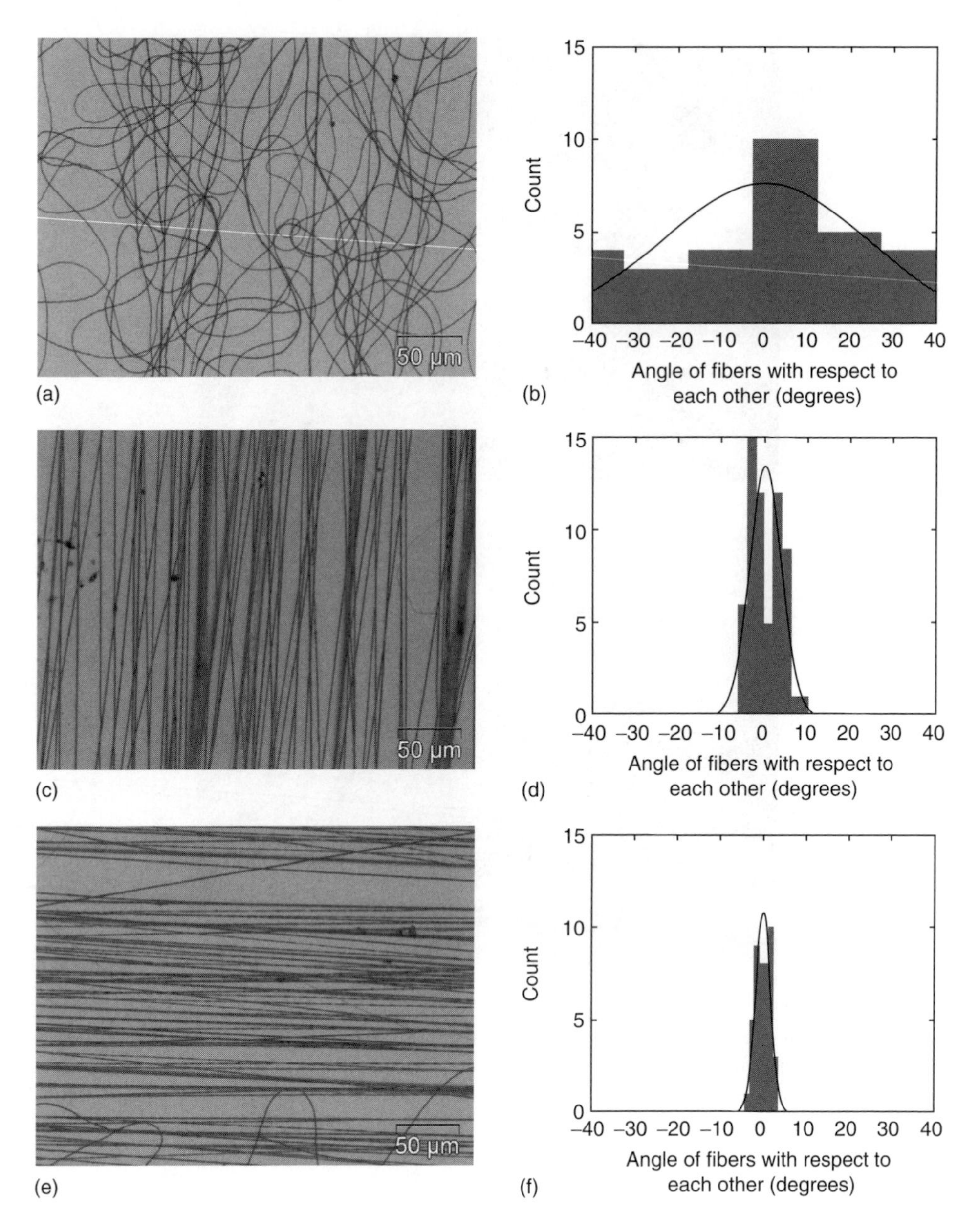

FIGURE 9.3 Nanofiber orientation at different linear wheel rotation speeds: (a) 2.9 m/s; (b) corresponding distribution of angular fiber orientation expressed in terms of angle with respect to the direction of the wheel rotation (standard deviation, SD = 23.44°); (c) 5.3 m/s; (d) corresponding distribution of angular fiber orientation (SD = 3.7°); (e) 7.7 m/s; (f) corresponding distribution of angular fiber orientation (SD = 1.8°). The wavy fiber section in (e) is attributed to the last part of the collected filament when the electric field has already been turned off. Nanofibers were electrospun from 4 wt% polyethylene oxide, PEO (MW = 1000 kDa) in ethanol/water.

9.3 MATERIALS AND METHODS

9.3.1 ELECTROSPINNING OF PCL NANOFIBROUS SCAFFOLDS

Nanofibers with diameters of several hundred nanometers were produced using the electrospinning method as described previously [32,33,41,42] (cf. Section 9.2). All electrospun polymer nanofiber

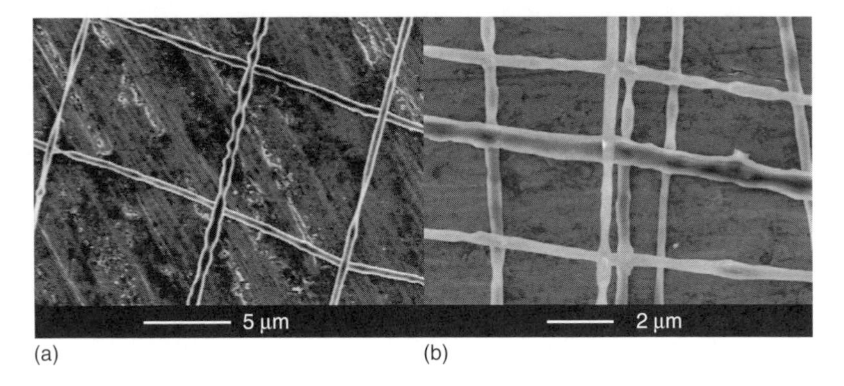

FIGURE 9.4 Typical SEM images of crossed arrays of PEO-based nanofibers collected on an aluminum table. The crossbar structures were obtained in a sequential assembly process with orthogonal table placement directions. (a) Two-layer assembly and (b) four-layer assembly. (From Zussman, E., Theron, A., and Yarin, A.L., *Appl. Phys. Lett.*, 82, 973, 2003. With permission.)

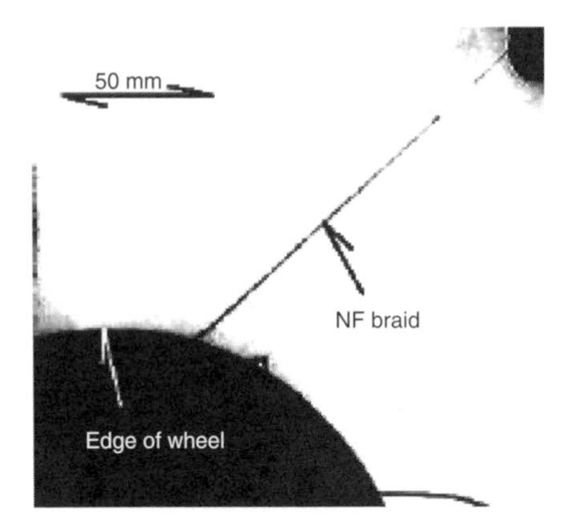

FIGURE 9.5 A bundle (rope) of aligned nanofibers manually pealed from the collector. Almost all the nanofibers were collected on the edge of the sharpened wheel collector. (From Theron, A., Zussman, E., and Yarin, A.L., *Nanotechnology*, 12, 384, 2001. With permission.)

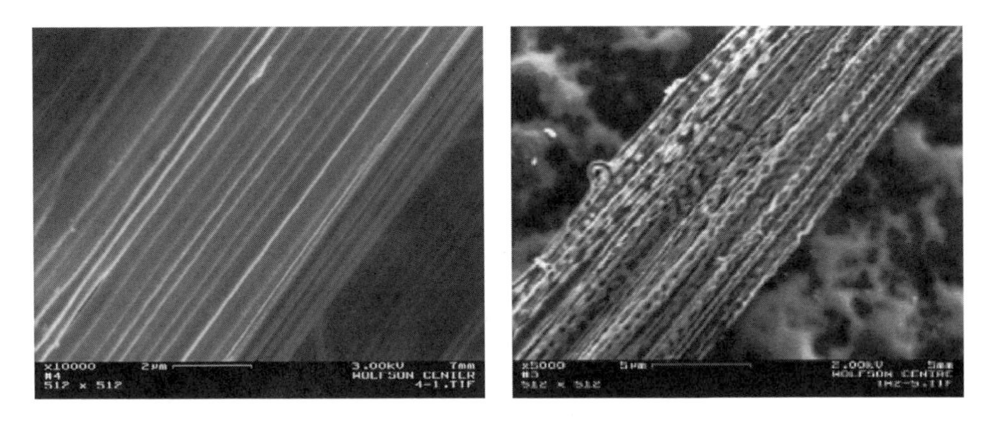

FIGURE 9.6 Typical SEM images of two bundles (ropes) of aligned PEO nanofibers. The density is about 100 nanofibers per square micrometer. (From Theron, A., Zussman, E., and Yarin, A.L., *Nanotechnology*, 12, 384, 2001. With permission.)

scaffolds were made from PCL with molecular weight of 80 kDa (Sigma–Aldrich) electrospun from a 10 wt% solution in methylene chloride/dimethylformamide with a ratio of 75/25 (vol.). Two different types of electrospun PCL nanofiber scaffolds were obtained and labeled as random or oriented. The random scaffold was produced by electrospinning polymer nanofibers onto a flat aluminum substrate, resulting in a nonwoven scaffold with no preferred nanofiber orientation. The oriented scaffold was produced by collecting the nanofibers on a narrow metal ribbon at the edge of a rotating wheel [41,42] (cf. Section 9.2), resulting in a nanofibrous scaffold with a clearly defined orientation. The polymer solution was electrospun from a 5 mL syringe attached to a hypodermic needle (inner diameter 0.1 mm) and using a flow rate of 1 mL/h. A copper electrode was placed in the polymer solution and the latter was electrospun toward the sharp edge of an electrically grounded collector (Figure 9.1). The strength of the electric field was 1.1 kV/cm. The linear speed of the edge of the wheel collector was 10 m/s. All experiments were performed at ambient temperature (~25°C) in air with 40% relative humidity. SEM was used to obtain images of each type of nanofibrous PCL scaffold. SEM observations of the PCL scaffolds were made with a Hitachi S-3000N variable pressure electron microscope. This instrument has a tungsten electron source and is capable of imaging specimens in variable pressures ranging from 1 to 270 Pa. Under variable pressure, nonconducting specimens can be imaged without coating of conductive film. The accelerating voltage can be varied over the range of 0.3–30 kV with a resolution of 5 nm at 25 kV. Nanofiber diameters were estimated using an image processor (MetaMorph, Molecular Device Corp., Downingtown, Pennsylvania).

9.3.2 Human Mesenchymal Stem Cell Culture and Seeding on PCL Nanofibrous Scaffolds

hMSCs were obtained from the Adult Mesenchymal Stem Cell Resource (Tulane University, New Orleans, Louisiana). Cells were cultured in complete growth media consisting of Dulbecco's modified eagle's medium (DMEM) supplemented with 15% fetal bovine serum (FBS), 2 mM L-glutamine, 1% antibiotics, antimycotics (final concentration: penicillin 100 units/mL, streptomycin 100 μg/mL, and amphotericin B 0.25 μg/mL). Cells at passage 6 were used in these experiments. From the bulk material of electrospun nanofibrous mats and strips, small scaffold disks with an area of approximately 0.3 cm^2 were cut out for cell culture and seeding experiments. The scaffolds were initially soaked in 70% ethanol for 1 h, dried, and sterilized under UV light for 6 h on each side, and pre-wetted by soaking in complete cell culture media with serum for 48 h to promote protein adsorption. To seed hMSCs onto the PCL nanofiber scaffolds, cells were pipetted directly onto the scaffolds at a density of 7.5×10^4 cells/cm^2. Culture media was added and, for long-term culturing, replaced every 2 to 3 days.

9.3.3 Laser Scanning Confocal Microscopy and Orientation Analysis

Samples were imaged on day 1 (24 h after initial stem cell seeding onto scaffolds), and then on days 4, 8, 12, 15, and 18. For cell orientation analysis, one set of samples was stained with CellTracker CMFDA (15 μM, 5-chloromethylfluorescein diacetate, Molecular Probes Invitrogen, Carlsbad, California) for fluorescence imaging and fixed in a 3.7% formaldehyde phosphate buffered saline (PBS) solution to ensure that cell orientations and morphologies were in unchanging states during imaging. Strong signal-to-noise ratios allowed for reliable quantitative image analysis of cell orientation. A multiphoton/laser scanning confocal microscope (Radiance 2000, Bio-Rad, Hercules, California) and a Nikon TE2000-S inverted microscope with a 20× Plan Apo objective (NA = 0.75) were used to produce images. While CMFDA-stained cells were fluorescently detected, the PCL nanofibers themselves were imaged in the reflection mode [49]. Alternating between fluorescent and reflection modes, the cells and nanofibers from the same area location on the sample could be imaged. Images were taken from at least five different and randomly chosen views on each sample at the different cell culture time points from day 1 to day 18. Using imaging analysis, the orientation of cells and fibers with respect to a reference (e.g., a horizontal axis) was quantified for each imaged view. Measurement of cell orientation was facilitated by the elongated shape of hMSCs, which displayed a clearly defined principal axis.

9.3.4 hMSC Viability and Fluorescence Visualization

Samples were stained using a cell viability assay (Molecular Probes). Briefly, calcein AM (2 μM) diffuses across the membrane of live cells and reacts with intracellular esterase to emit green fluorescence, while ethidium homodimer-1 (4 μM) can enter only dead cells with damaged cell membrane and emit red fluorescence upon binding to nucleic acids. For each sample, the total number of cells and percentage of live and dead cells were determined from at least three randomly selected fields of view.

Microfilaments were stained using rhodamine–phalloidin (170 nM, Molecular Probes). Samples from days 4 and 18 were selected for microfilament staining and were fixed in a 3.7% formaldehyde PBS solution. To permeabilize the cell membrane, samples were placed in a 0.5% Triton X-100 in PBS solution for 5 min at 4°C and washed three times with PBS. To block nonspecific binding sites, samples were pre-incubated with 1% BSA for 30 min and washed three times with PBS. Actin filaments were stained overnight with rhodamine–phalloidin in 1% BSA. Unique hMSC–nanofiber adhesion-like structures were found and further examined using high magnification microscope objectives.

9.3.5 Statistics

For each imaged view of cell-seeded oriented nanofibrous scaffold samples corresponding to days 1, 4, 8, 12, 15, and 18, the average cell orientation angle and the average fiber orientation angle with respect to a reference direction were calculated. The difference between these two quantities defined the average cell–nanofiber relative angle. In addition, the relative cell angle distributions with respect to the corresponding average fiber orientation angles were obtained for three views on each scaffold sample. These three distributions were statistically analyzed with Student's two-tailed t test and variances were determined not to be statistically significant ($p > 0.05$). Therefore, by combining the data from the three separate views, an overall distribution was formed for each day. The average cell–fiber relative angle was thus obtained along with the corresponding standard deviation for each day. Average relative cell–fiber angles of 10° or less suggest cell alignment parallel to the underlying oriented nanofibers. Regarding the random nanofibrous scaffolds, all values within the distributions of cell–nanofiber angles have equal probabilities and therefore no average angle values can be determined. Nonetheless, the distributions of cell–nanofiber angles in random nanofibrous scaffolds can be qualitatively compared with those in oriented nanofibrous scaffolds.

9.4 RESULTS AND DISCUSSION

9.4.1 Characterization of Electrospun PCL Nanofibrous Scaffolds

SEM imaging was performed to observe the structural morphology of nanofibers in each of the two types of scaffold (random or oriented). The random scaffold (Figure 9.7a) is shown to consist of nanofibers that have no preferred orientation. In contrast, the oriented scaffold (Figure 9.7b) exhibited nanofibers with a preferred orientation (north–south in this case). The majority of nanofibers in the oriented scaffold were aligned with respect to each other. The average nanofiber diameter was estimated as 820 ± 340 nm. Based on the SEM images, these scaffolds feature porosity that is suitable for cell adhesion, proliferation, and viability.

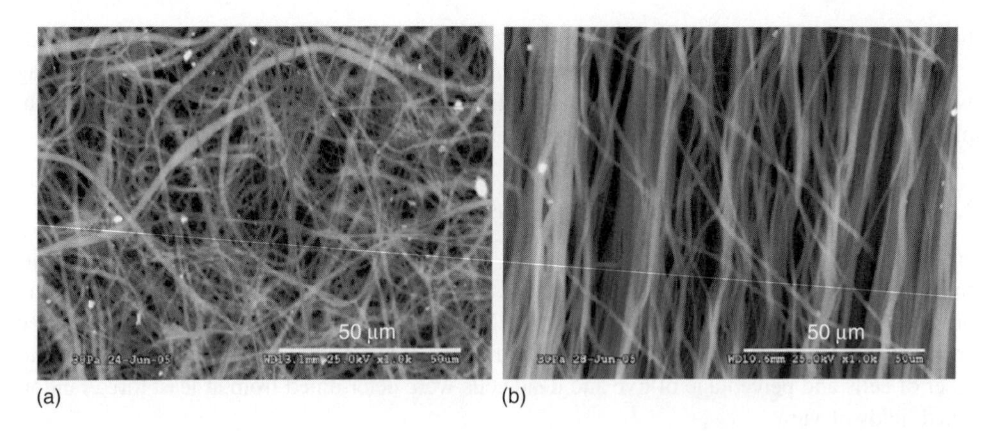

FIGURE 9.7 SEM images of PCL nanofibers in the (a) random and (b) oriented PCL scaffolds. The white specs seen in both frames are PCL fragments originating from the creation of a polymer crust at the nozzle exit of the electrospinning apparatus.

9.4.2 Orientation of Cells and Nanofibers for Random and Oriented Scaffolds

hMSCs and nanofibers were simultaneously imaged by laser scanning confocal microscopy (LSCM) to determine the effect of electrospun PCL nanofiber orientation on cell alignment. Figure 9.8 displays superimposed, pseudocolored images of hMSCs (green) and nanofibers (red) on (a1) random and (b1) oriented PCL nanofibrous scaffolds after 1 day cell culture. When images were superimposed, it became clearly noticeable that, just after 24 h incubation, there were significant differences in the orientation of hMSCs cultured on the random scaffold compared to that on the oriented scaffold. The orientation angles of hMSCs with respect to the horizontal axis were determined for the random and oriented scaffold samples. Histograms representing the distributions of hMSC orientation angles after 1 day cell culture are displayed in the green bar graphs of Figure 9.8. As verified by the narrow distribution of the cell angles shown in the histogram of Figure 9.8b2 (standard deviation 16°), hMSCs were induced to align on the oriented nanofibrous scaffold. Furthermore, the cell alignment on the oriented scaffolds closely matched with the underlying nanofiber orientation of the scaffolds (Figure 9.8b3), where the standard deviation of the distribution of nanofiber angles was 15°. On the other hand, cells seeded on the random nanofibrous scaffold for 24 h showed no preferred cell orientation (Figure 9.8a1), confirmed by a broad angular distribution in the histogram of Figure 9.8a2. Likewise, nanofibers in the random scaffolds also showed a broad angle distribution (Figure 9.8a3).

Orientations of hMSCs and nanofibers in each type of scaffold after a longer culture of 18 days were remarkably similar to those found after 1 day cell culture. The day 18 results are displayed in Figure 9.9. hMSCs cultured on the oriented nanofibrous scaffold (Figure 9.9b1) for 18 days continued to maintain an excellent degree of global cell alignment. This was again confirmed by the histograms for spatial cell alignment on the oriented scaffold (Figure 9.9b2) and the nanofiber orientation (Figure 9.9b3); both show narrow distributions of angles (standard deviations of 20° and 32°, respectively) and remarkable correspondence to each other. hMSCs cultured on the oriented scaffolds showed qualitatively a thinner and more elongated morphology than cells cultured on the random scaffolds, suggesting that cell shape may have been affected by the organized nanofiber structure. Similar to the results obtained after 1 day of incubation, hMSCs cultured on the random scaffold after 18 days (Figure 9.9a1) exhibited no significant cell orientation and maintained a prominently broader distribution of cell orientation angles ranging over 180° (Figure 9.9a2), which correlated with the random distribution of nanofiber orientation angles (Figure 9.9a3). It must be noted, however, that after 18 days of cell culture, hMSCs appeared to demonstrate a few areas of "localized" self-alignment in the random scaffold. As seen in the lower left section of Figure 9.9a1,

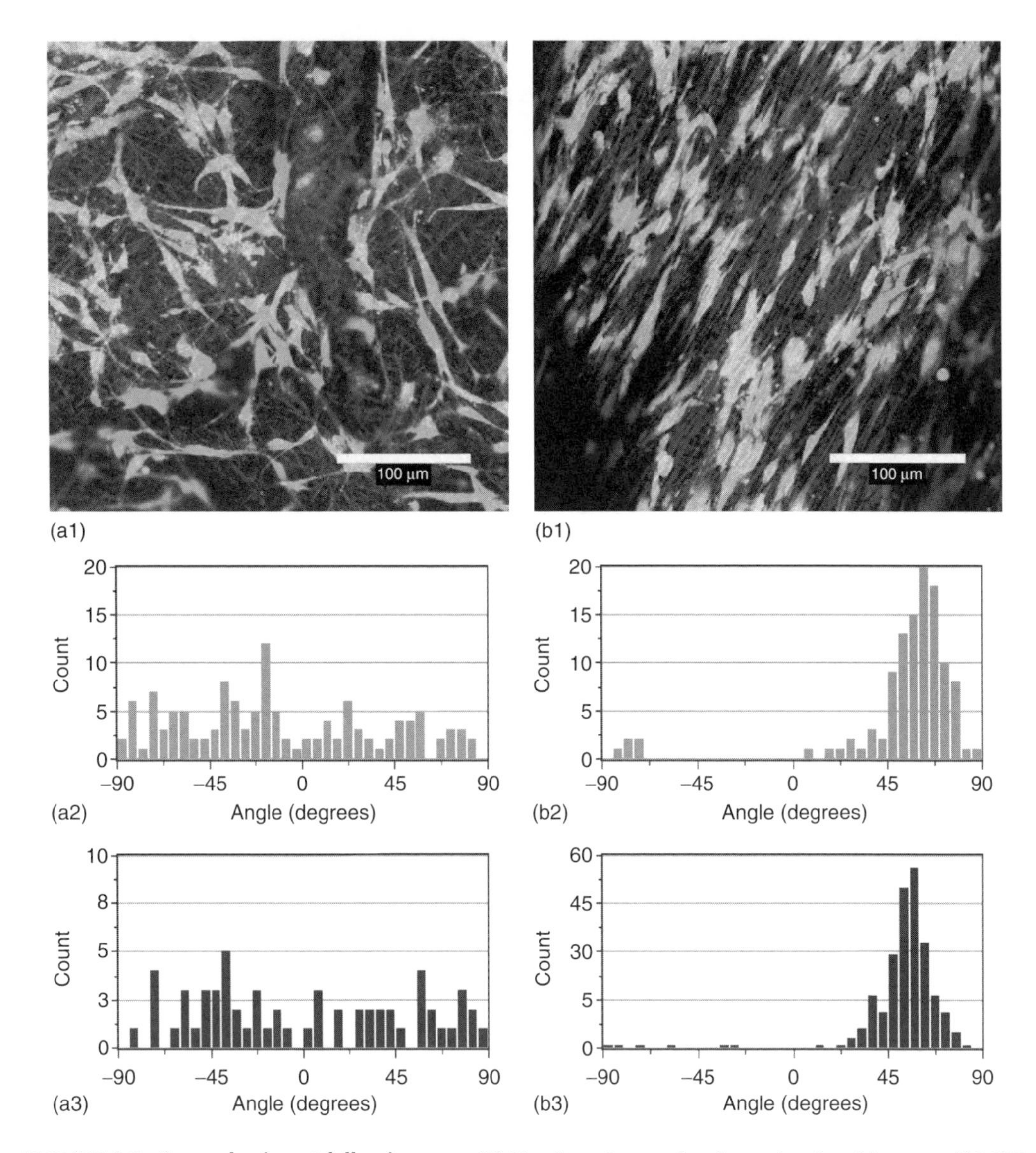

FIGURE 9.8 (See color insert following page 206.) Superimposed and pseudocolored images of hMSCs (green) and nanofibers (red) in the (a1) random and (b1) oriented PCL nanofibrous scaffolds after 24 h cell culture. Histograms representing distributions of cell orientation angles with respect to the horizontal axis for hMSCs on (a2) random and (b2) oriented scaffolds. Histograms representing distributions of nanofiber orientation angles with respect to the horizontal for (a3) random and (b3) oriented scaffolds. Histograms in (a2) and (a3) were constructed based on measurements of 127 cells and 234 nanofibers, respectively, while 124 cells and 249 nanofibers were used to construct the histograms in (b2) and (b3), respectively.

some cells seem to show a preferred orientation, as confirmed by the bell-shaped portion of the corresponding cell angle distribution in Figure 9.9a2. This finding is attributed to the proliferative and confluent cell culture rather than cell alignment guided by the underlying nanofibers. This localized self-assembly of hMSCs, which occurs despite the randomness of the underlying fibers, lacks the global organization features seen on the oriented scaffold (Figure 9.9b1). But the lack of control over the cell orientation—albeit at a localized scale—renders this local effect of little consequence to the overall goal of this work, namely, to align cells in a global and controllable fashion.

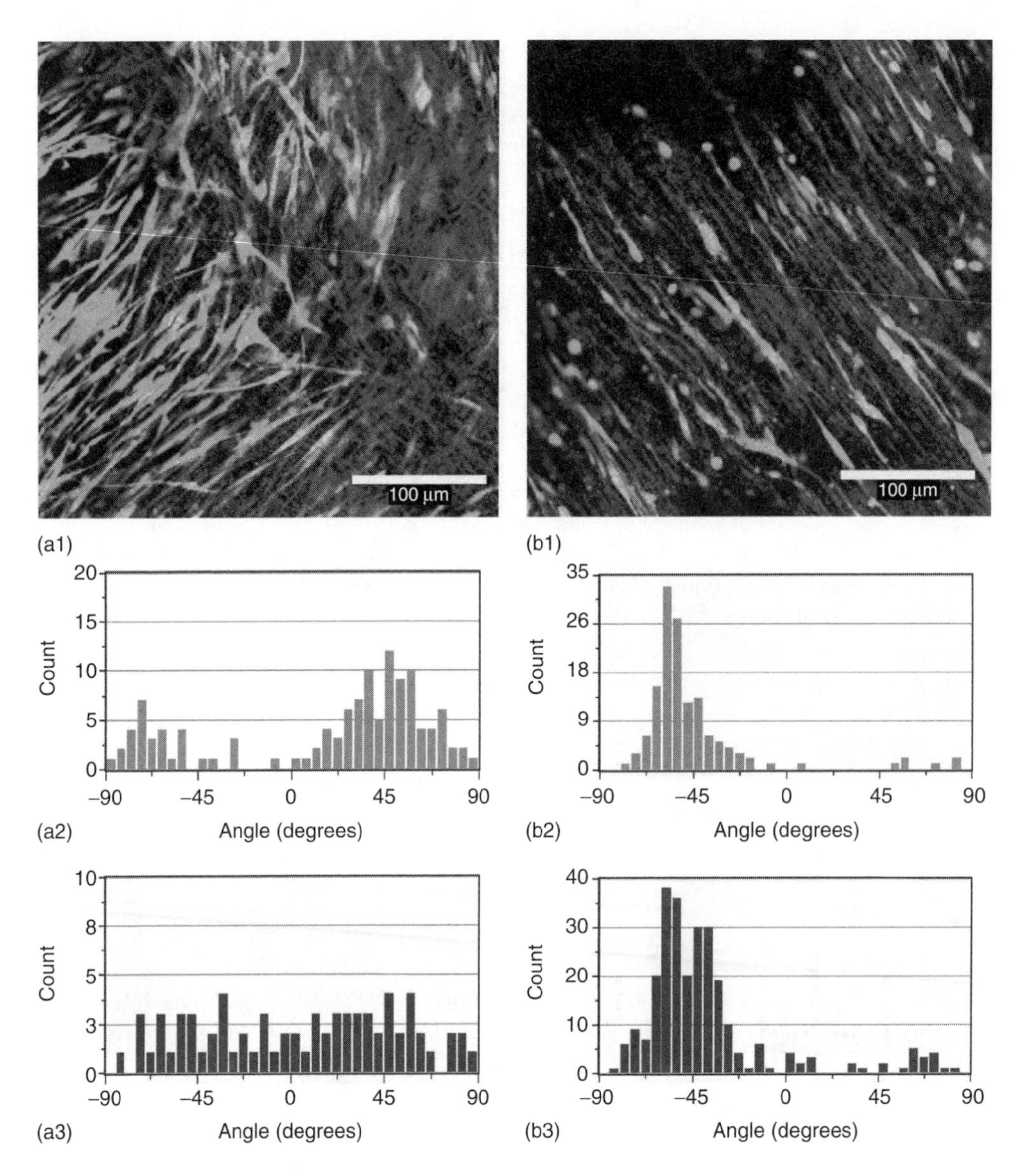

FIGURE 9.9 Superimposed and pseudocolored images of hMSCs (bright) and nanofibers (dark) in the (a1) random and (b1) oriented PCL nanofibrous scaffolds after 18 days of cell culture. Histograms representing distributions of cell orientation angles with respect to the horizontal axis for hMSCs on (a2) random and (b2) oriented scaffolds. Histograms representing distributions of nanofiber orientation angles with respect to the horizontal axis for (a3) random and (b3) oriented scaffolds. Histograms in (a2) and (a3) were constructed based on measurements of 132 cells and 194 nanofibers, respectively, while 138 cells and 258 nanofibers were used to construct the histograms in (b2) and (b3), respectively.

9.4.3 Consistent Long-Term hMSC Alignment on Oriented Nanofibrous Scaffolds

The efficacy of the oriented PCL nanofibrous scaffold in promoting cell alignment of seeded hMSCs was quantitatively analyzed at different time points. The relative cell angular distributions were determined for oriented scaffolds by subtracting the average nanofiber angle from the cell angle distributions, and determining the average relative cell–fiber angle at each time point. The results for samples cultured from day 1 until day 18 were used as a measure to assess the efficacy and any

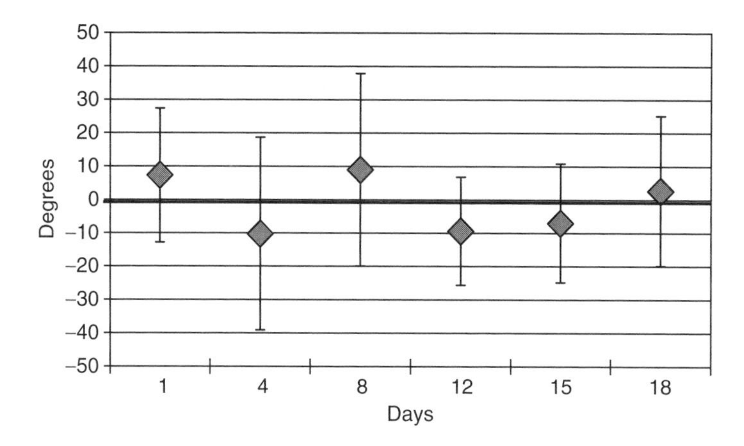

FIGURE 9.10 Temporal hMSC alignment on oriented nanofibrous scaffolds. *T*-test analysis (two-tailed, unpaired) showed that at least three views of each nanofibrous scaffold on each day were statistically not significant ($p > 0.05$), and an average relative cell orientation angle and corresponding standard deviation were determined for the oriented nanofibrous scaffold samples on days 1, 4, 8, 12, 15, and 18. Average relative cell–fiber angles on the oriented nanofibrous scaffolds on all days were in the range $\pm 10°$, indicating very good alignment with the underlying nanofibers. The error bars denote two standard deviations ($\pm\sigma$).

time-dependent changes in hMSC alignment. As shown in Figure 9.10, for all oriented nanofibrous scaffold samples from day 1 to day 18, relative cell angles were within $\pm 10°$. Therefore, accurate and consistent cell alignment on oriented nanofibrous scaffolds is confirmed for initial and long-term culture. From day 1 to day 18, the standard deviations of relative cell–fiber angular distributions for oriented nanofibrous scaffolds remained in the range of 15°–30°, with all average daily values within one standard deviation of the common reference angle of 0° (perfect alignment). This suggests that the initial hMSC attachment to the nanofibrous scaffolds was maintained for the entire 18 day incubation period.

9.4.4 hMSC Viability on PCL Nanofibrous Scaffolds

The short- and long-term viability of hMSCs cultured on the random and oriented scaffolds was tested by using a live/dead fluorescence assay. For each scaffold type, the total number of live and dead cells was counted from at least three different fields of view following 4 and 18 days of cell culture. The percentage of live cells from the total number of cells was calculated and the results are shown in Table 9.1. Furthermore, from the total number of cells observed, the average cell density was calculated by measuring the field of view area and extrapolating the data in units of cells per square centimeter. Most of the cells seeded on both types of scaffolds were live (>70%), but their viabilities were lower than that observed for cells cultured in the control experiment (e.g., on petri dishes). The apparent lower cell viability percentages could be attributed to the fact that we used undifferentiated but highly proliferative hMSCs. For example, even though the initial hMSC seeding density was modest (7.5×10^4 cells/cm^2), hMSCs quickly proliferated and became confluent. As shown in Table 9.1, the cell densities on both random and oriented nanofibrous scaffolds increased significantly from day 4 to day 18, whereas the percentage of live cells stayed relatively constant. Because PCL is a well-known biocompatible material, it is unlikely that it is responsible for the apparent lower hMSC viability. Rather, it is suggested that an optimal seeding cell density could have improved the cell viability. For example, the initial cell seeding density on a PCL nanofibrous scaffold has been demonstrated to be an important factor for short- and long-term hMSC viability, proliferation, and differentiation [20,21,38,50]. It is expected, however, that when

TABLE 9.1
hMSC Viability Expressed as Percentage of Live Cells and Total Number of Cells Examined

Day	Sample	% Live	Cells/cm^2 Viewed
4	Random	76	7.0×10^4
	Oriented	80	11.6×10^4
18	Random	73	32.9×10^4
	Oriented	76	20.2×10^4

hMSCs are induced to differentiate into a specific cell lineage, a decrease in hMSC proliferation would yield better cell viability. Additionally, although the electrospun nanofibrous scaffolds used in this work were air-dried prior to sterilization and subsequent presoaking in cell culture media, none of the scaffolds were placed in a vacuum chamber prior to sterilization. Performing this step in future studies should remove any residual organic solvent from the electrospinning process that may adversely affect cell viability.

9.4.5 Cytoskeletal Reorganization and hMSC–Nanofiber Adhesions

Cell elongation induced by the aligned PCL nanofibers is expected to reorganize the cytoskeletal structures that regulate the cell morphology, adhesion, and locomotion. Microfilaments of hMSCs cultured for 4 days on the random and oriented PCL nanofibrous scaffolds and on glass were stained brightly fluorescent with rhodamine–phalloidin and imaged using LSCM. hMSCs cultured on a glass substrate exhibited the typical and expected spread cell morphology (Figure 9.11a). Large actin stress fibers were visible and may be responsible for stronger hMSC adhesion on this flat (2D) substrate [51]. Many stress fibers appeared to come together and terminate at a common point on the cell membrane, probably forming a focal contact site. Compared to microfilaments found in hMSCs cultured on flat substrates, much different actin cytoskeletal organization was observed in hMSCs cultured on the random and oriented PCL nanofibrous scaffolds. For example, hMSCs seeded on the random nanofibrous scaffolds and cultured for 4 days (Figure 9.11b) displayed elongated cell morphology and densely concentrated actins. Interestingly, thin fibrous actin structures between hMSCs and the surrounding nanofibers were evident, being particularly noticeable in the circled region of Figure 9.11b with an arrow pointing to an adjacent PCL nanofiber. When hMSCs were seeded on the oriented PCL nanofibrous scaffold, cells were even more elongated and aligned next to a nanofiber bundle, which is again illustrated by the arrows in Figure 9.11c. Actins in these aligned cells were also densely concentrated, and some actin filament bundles were predominantly aligned in the same direction of the adjacent nanofibers. However, thin fibrous tether-like actin structures connecting the hMSC with the neighboring aligned nanofiber bundle were also observed, being particularly noticeable in the circled regions of Figure 9.11c. These findings could lead to the notion of unique hMSC binding characteristics to the nanofibers. Our results are consistent with a previous report that vascular smooth muscle cells cultured on aligned polymeric nanofibrous scaffolds also develop similar fibrous connections [1]. It should be noted, however, that the images reported therein providing evidence of possible cell–nanofiber adhesions were obtained using SEM. In contrast, our findings show that not only similar fibrous connections are found in the hMSC–nanofiber interactions, but also actin filaments are likely involved in the development of these unique adhesion-like structures. It is also interesting to note that the dense actin networks observed in the elongated hMSCs cultured on the random and oriented nanofibrous scaffolds appear similar to the cytoskeleton observed in mature articular chondrocytes, especially at the articular surface [52], suggesting that the use of oriented nanofibrous polymeric scaffolds could be advantageous to better

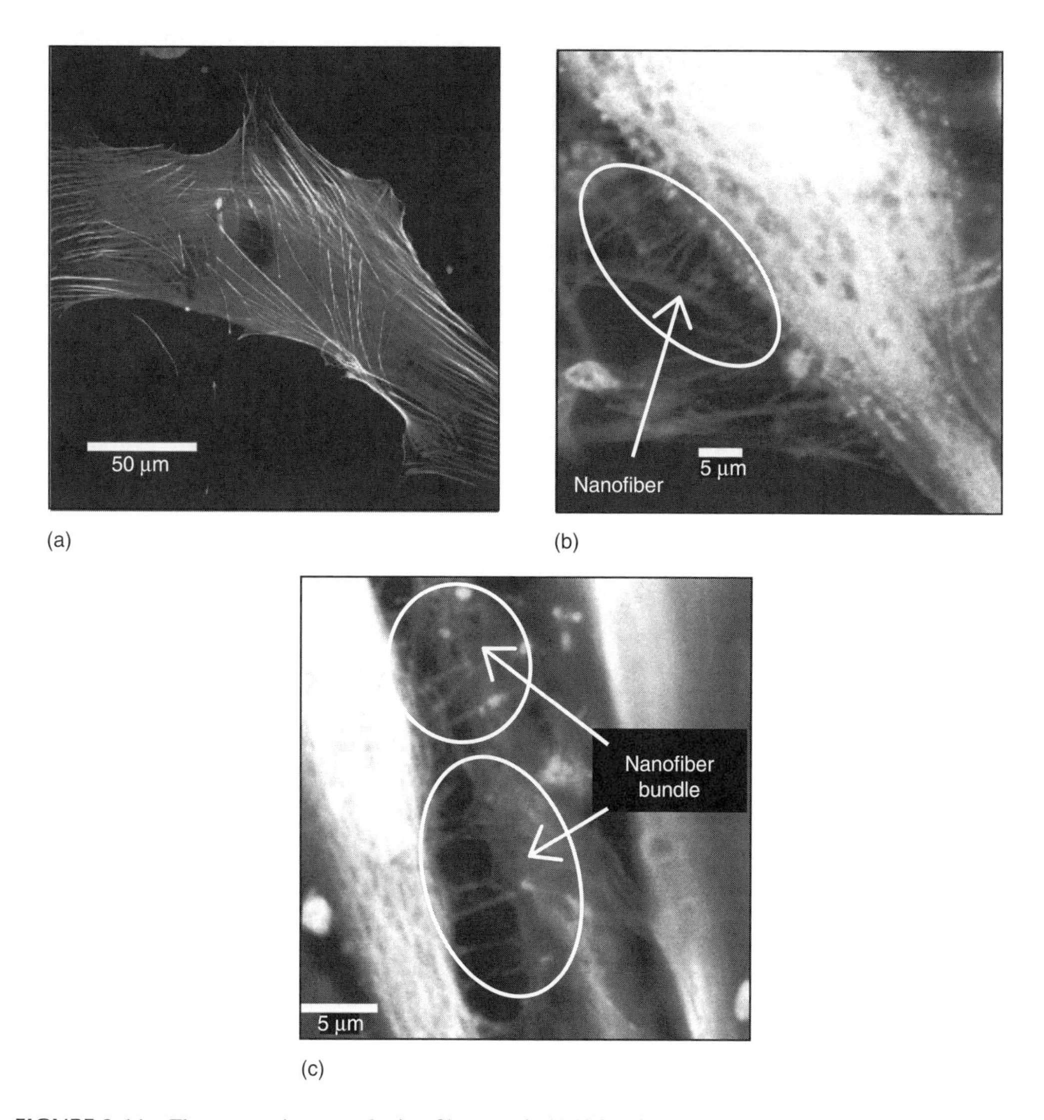

FIGURE 9.11 Fluorescent images of microfilaments in hMSCs after 4 days of cell culture on (a) a flat glass surface, and (b) in the random or (c) oriented PCL scaffolds. No fibrous tether-like adhesions were observed when cells were plated on a typical substrate, but the actin-containing tethers were apparently formed by hMSC to attach to the nanofibers. These unique structures will be further examined to test for the presence of the known adhesion proteins such as integrins and other intracellular proteins.

mimic articular cartilage tissue [19–21]. Because successful tissue engineering will demand that an engineered tissue construct mimic the micro- and nanostructures, spatial alignment, and functionality of the natural tissue, optimal use of aligned nanofibrous scaffolds combined with manipulation of hMSCs may produce one of the most compatible biomimetic scaffolds.

9.5 CONCLUSIONS

In the present study, techniques for nanofiber alignment in electrospinning were reviewed and employed in stem cell culturing. hMSCs were successfully cultured on two different types of electrospun PCL nanofibrous scaffolds. Cell orientation with respect to the supporting nanofiber network was quantified. The efficacy of hMSCs to align in the direction of oriented nanofibers was investigated. hMSCs cultured on an oriented PCL nanofibrous scaffold exhibited the most accurate and consistent cell alignment. Cell viability was tested to confirm that hMSCs were viable and proliferated on the PCL nanofibrous scaffolds during the 18 day cell culture period. In addition, hMSC actin cytoskeletal reorganization on nanofibrous scaffolds was observed, and the presence of microfilaments was revealed in fibrous tether-like adhesions that connected hMSCs to adjacent nanofibers. Further studies are under way to test chondrogenic differentiation of hMSCs on oriented PCL nanofibrous scaffolds, and the effects of varying nanofiber diameter and scaffold porosity. Taken together, our findings suggest that engineering an oriented ECM environment to regulate tissue alignment could be optimized by oriented electrospun nanofibers, and that specific tissue engineering applications such as creating the superficial zone of articular cartilage may be significantly improved by combining stem cells and nanomaterials.

ACKNOWLEDGMENTS

The authors acknowledge contributions of their students Andre Theron and Nicole Pearson. The authors are grateful to Dr. Igor Titushkin for his expertise and invaluable assistance with laser scanning confocal microscopy and image analysis. We also thank Kristina Jarosius of the UIC Research Resources Center for performing the SEM observations. This work was supported, in part, by NIH grants (GM60741, EB06067, MC), a grant from the Office of Naval Research (N00014-06-1-0100, MC), National Science Foundation under Grant NSF-NER-CBET 0708711, and a grant from the Volkswagen Foundation (AY, EZ). Some of the materials employed in this work were provided by the Tulane Center for Gene Therapy through a grant from NCRR of the NIH, Grant #P40RR017447.

REFERENCES

1. Xu C.Y., Inai R., Kotaki M. et al. 2004. Aligned biodegradable nanofibrous structure: A potential scaffold for blood vessel engineering. *Biomaterials* 25:877–886.
2. Yang F., Murugan R., Wang S. et al. 2005. Electrospinning of nano/micro scale poly(L-lactic acid) aligned fibers and their potential in neural tissue engineering. *Biomaterials* 26:2603–2610.
3. Lee C.H., Shin H.J., Cho I.H. et al. 2005. Nanofiber alignment and direction of mechanical strain affect the ECM production of human ACL fibroblast. *Biomaterials* 26:1261–1270.
4. Dunn G.A. 1982. Contact guidance of cultured tissue cells: A survey of potentially relevant properties of the substratum. In *Cell Behaviour*, ed. R. Bellairs, A. Curtis, G. Dunn, 247–280. Cambridge: Cambridge University Press.
5. Rajnicek A.M., Britland S., and McCaig C.D. 1997. Contact guidance of CNS neuritis on grooved quartz: Influence of groove dimensions, neuronal age and cell type. *J Cell Sci* 110:2905–2913.
6. Karlon W.J., Hsu P.-P., Li S. et al. 1999. Measurement of orientation and distribution of cellular alignment and cytoskeletal organization. *Ann Biomed Eng* 27:712–720.
7. Wang J.H.-C., Jia F., Gilbert T.W. et al. 2003. Cell orientation determines the alignment of cell-produced collagenous matrix. *J Biomech* 36:97–102.
8. Hunziker E.B. 1992. Articular cartilage structure in humans and experimental animals. In *Articular Cartilage and Osteoarthritis*, ed. K.F. Kuettner, R. Schleyerbach, J.G. Peyron, V.C. Hascall, 183–199. New York: Raven Press, Ltd.
9. Jackson D.W., Scheer M.J., and Simon T.M. 2001. Cartilage substitutes: Overview of basic science and treatment options. *J Am Acad Orthop Surg* 9:37–52.
10. Hunziker E.B., Quinn T.M., and Haeuselmann H.-J. 2002. Quantitative structural organization of normal adult human articular cartilage. *Osteoarth Cartilage* 10:564–572.
11. Eyre D. 2002. Collagen of articular cartilage. *Arthritis Res* 4:30–35.
12. Setton L.A., Elliott D.M., and Mow V.C. 1999. Altered mechanics of cartilage with osteoarthritis: Human osteoarthritis and an experimental model of joint degeneration. *Osteoarth Cartilage* 7:2–14.

13. Wong M. and Carter D.R. 2003. Articular cartilage functional histomorphology and mechanobiology: A research perspective. *Bone* 33:1–13.
14. Klein T.J., Schumacher B.L., Schmidt T.A. et al. 2003. Tissue engineering of stratified articular cartilage from chondrocyte subpopulations. *Osteoarth Cartilage* 11:595–602.
15. Kim T.-K., Sharma B., Williams C.G. et al. 2003. Experimental model for cartilage tissue engineering to regenerate the zonal organization of articular cartilage. *Osteoarth Cartilage* 11:653–664.
16. Sharma B. and Elisseeff J.H. 2004. Engineering structurally organized cartilage and bone tissues. *Ann Biomed Eng* 32:148–159.
17. Woodfield T.B.F., Van Blitterswijk C.A., De Wijn J. et al. 2005. Polymer scaffolds fabricated with pore-size gradients as a model for studying the zonal organization within tissue-engineering cartilage constructs. *Tissue Eng* 11:1297–1311.
18. Sharma B., Williams C.G., Kim T.K. et al. 2007. Designing zonal organization into tissue-engineered cartilage. *Tissue Eng* 13:405–414.
19. Li W.-J., Danielson K.G., Alexander P.G. et al. 2003. Biological response of chondrocytes cultured in three-dimensional nanofibrous poly(epsilon-caprolactone) scaffolds. *J Biomed Mater Res A* 67:1105–1114.
20. Li W.-J., Tuli R., Okafor C. et al. 2005. A three-dimensional nanofibrous scaffold for cartilage tissue engineering using human mesenchymal stem cells. *Biomaterials* 26:599–609.
21. Li W.-J., Tuli R., Huang X. et al. 2005. Multilineage differentiation of human mesenchymal stem cells in a three-dimensional nanofibrous scaffold. *Biomaterials* 26:5158–5166.
22. Kuo C.K., Li W.-J., Mauck R.L. et al. 2003. Cartilage tissue engineering: Its potential and uses. *Curr Opin Rheumatol* 18:64–73.
23. Xin X., Hussain M., and Mao J.J. 2007. Continuing differentiation of human mesenchymal stem cells and induced chondrogenic and osteogenic lineages in electrsopun PLGA nanofiber scaffold. *Biomaterials* 28:316–325.
24. Ma Z., Kotaki M., Inai R. et al. 2005. Potential of nanofiber matrix as tissue-engineering scaffolds. *Tissue Eng* 11:101–109.
25. Laurencin C.T., Ambrosio A.M., Borden M.D. et al. 1999. Tissue engineering: Orthopedic applications. *Ann Rev Biomed Eng* 1:19–46.
26. Caplan A.I. and Bruder S.P. 2001. Mesenchymal stem cells: Building blocks for molecular medicine in the 21st century. *Trends Molec Med* 7:259–264.
27. Tuan R.S., Boland G., and Tuli R. 2002. Adult mesenchymal stem cells and cell-based tissue engineering. *Arthritis Res Ther* 5:32–45.
28. Baksh D., Song L., and Tuan R.S. 2004. Adult mesenchymal stem cells: Characterization, differentiation, and application in cell and gene therapy. *J Cell Molec Med* 8:301–316.
29. Alhadlaq A. and Mao J.J. 2004. Mesenchymal stem cells: Isolation and therapeutics. *Stem Cells Dev* 13:436–448.
30. Yang S., Leong K.F., Du Z. et al. 2001. The design of scaffold for use in tissue engineering. Part 1. Traditional factors. *Tissue Eng* 7:679–689.
31. Karande T.S., Ong J.L., and Agrawal C.M. 2004. Diffusion in musculoskeletal tissue engineering scaffolds: Design issues related to porosity, permeability, architecture, and nutrient mixing. *Ann Biomed Eng* 32:1728–1743.
32. Reneker D.H., Yarin A.L., Fong H. et al. 2000. Bending instability of electrically charged liquid jets of polymer solutions in electrospinning. *J Appl Phys* 87:4531–4547.
33. Reneker D.H, Yarin A.L, Zussman E. et al. 2007. Electrospinning of nanofibers from polymer solutions and melts. In *Advances in Applied Mechanics*, Vol. 41, ed. H. Aref, E. van der Giessen, 43–195. Oxford: Elsevier.
34. Frenot A. and Chronakis I.S. 2003. Polymer nanofibers assembled by electrospinning. *Curr Opin Colloid Interface Sci* 8:64–75.
35. Huang Z.M., Zhang Y.Z., Kotaki M. et al. 2003. A review on polymer nanofibers by electrospinning and their applications in nanocomposites. *Composites Sci Technol* 63:2223–2253.
36. Dzenis Y. 2004. Spinning continuous fibers for nanotechnology. *Science* 304:1917–1919.
37. Ramakrishna S., Fujihara K., Teo W.E. et al. 2005. *An Introduction to Electrospinning of Nanofibers*. Singapore: World Scientific.

38. Shin M., Yoshimoto H., and Vacanti J.P. 2004. In vivo bone tissue engineering using mesenchymal stem cells on a novel electrospun nanofibrous scaffold. *Tissue Eng* 10:33–41.
39. Engelberg I. and Kohn J. 1991. Physico-mechanical properties of degradable polymers used in medical applications: A comparative study. *Biomaterials* 12:292–304.
40. Thomas V., Jose M.V., Chowdhury S. et al. 2006. Mechano-morphological studies of aligned nanofibrous scaffolds of polycaprolactone fabricated by electrospinning. *J Biomater Sci Polym Ed* 17:969–984.
41. Theron A., Zussman E., and Yarin A.L. 2001. Electrostatic field-assisted alignment of electrospun nanofibres. *Nanotechnology* 12:384–390.
42. Zussman E., Theron A., and Yarin A.L. 2003. Formation of nanofiber crossbars in electrospinning. *Appl Phys Lett* 82:973–975.
43. Deitzel J.M., Kleinmeyer J.D., Hirvonen J.K. et al. 2001. Controlled deposition of electrospun poly (ethylene oxide) fibers. *Polymer* 42:8163–8170.
44. Sundaray B., Subramanian V., Natarajan T.S. et al. 2004. Electrospinning of continuous aligned polymer fibers. *Appl Phys Lett* 84:1222–1224.
45. Li D., Wang Y., and Xia Y. 2004. Electrospinning nanofibers as uniaxially aligned arrays and layer-by-layer stacked films. *Adv Mat* 16:361–366.
46. Courtney T., Sacks M.S., Stankus J. et al. 2006. Design and analysis of tissue engineering scaffolds that mimic soft tissue mechanical anisotropy. *Biomaterials* 27:3631–3638.
47. Baker B.M. and Mauck R.L. 2007. The effect of nanofiber alignment on the maturation of engineered meniscus constructs. *Biomaterials* 28:1967–1977.
48. Li W.J., Mauck R.L., Cooper J.A. et al. 2007. Engineering controllable anisotropy in electrospun biodegradable nanofibrous scaffolds for musculoskeletal tissue engineering. *J Biomech* 40:1686–1693.
49. Hoheisel W., Jacobsen W., Luettge B. et al. 2001. Confocal microscopy: Applications in materials science. *Macromol Mater Eng* 286:663–668.
50. Yoshimoto H., Shin Y.M., Terai H. et al. 2003. A biodegradable nanofiber scaffold by electrospinning and its potential for bone tissue engineering. *Biomaterials* 24:2077–2082.
51. Curran J.M., Chen R., and Hunt J.A. 2005. Controlling the phenotype and function of mesenchymal stem cells in vitro by adhesion to silane-modified clean glass surfaces. *Biomaterials* 26:7057–7067.
52. Langelier E., Suetterlin R., Hoemann C.D. et al. 2000. The chondrocyte cytoskeleton in mature articular cartilage: Structure and distribution of actin, tubulin, and vimentin filaments. *J Histochem Cytochem* 48:1307–1320.

10 Nanolithographic Techniques in Tissue Engineering

Benjamin Moody and Gregory S. McCarty

CONTENTS

10.1 INTRODUCTION

The origins of lithography are perhaps surprising and to some, perhaps surprisingly interesting. Milestones in its development include a powerful Chinese dynasty (Tang Dynasty) [1], a successful Austrian playwright (Alois Senefelder) [2], a prolific German artist (Godefroy Engelmann) [3], an experimental French photographer (Nicephore Niepce) [4], and more recently, the multibillion dollar computer industry. Like many technologies, lithography had its beginnings out of the necessities of politics, was adopted by the creativity of artists, and has since become a process of unprecedented potential for the advancement of both commercial industries and multiple branches of science.

Truly it was the computer industry and the advent of the microchip that ushered lithography into the modern era but it will most likely be medicine and biology that carry it into the future. The biomedical applications of lithography and specifically of nanolithography have been growing exponentially, mainly as a focus for the creation of scaffolds that are intricately designed to replicate features and functions naturally inherent only in vivo scenarios. The term nanolithography is generally applied to an assortment of processes that are capable of "printing" features with less than 100 nm dimensions. Ideally, a single nanolithography technique would be able to (1) write with nanometer resolution at speeds of centimeter per second over wafer-scale areas, (2) impart different chemical/biological functionalities onto a variety of materials, under a variety of conditions, and

(3) be capable of massive parallelization for both writing and metrology [5]. There is to date no one single method that fits all the above criteria. There is, however, a host of uniquely individual methods that are each suited to at least one function. The specifics of each technique are tactically different but the guiding principle remains the same: at the very least, a nanolithographic process builds or transfers a desired pattern with submicron scale resolution onto a suitable surface. In general there are two ways of accomplishing this: methods known simply as "bottom–up" like block copolymer and self-assembly lithographies or those called top–down like optical, imprint, and scanning probe lithographies. While many of these approaches are still in the infant stages of scaffolding research in tissue engineering, all have the very real potential of playing important roles in biomedicine and tissue engineering.

10.2 BOTTOM–UP LITHOGRAPHY

Bottom–up refers to the process of building structures using molecules and molecular patterns as the direct building blocks of a structure, in other words, starting with molecules and building "up" to a useful pattern. This often involves short polymer chains (natural or synthetic) that are deposited on smooth or patterned surfaces resulting in thin but useful films. It is commonly used in one of the two ways, either as a means of patterning a surface with a certain geometry or as a way of creating a surface with a specific functionality. "Block copolymer lithography" and "self-assembling polymerization" subscribe to the first regime; "contact printing" usually subscribes to the latter.

10.2.1 Block Copolymer Lithography

A block copolymer is attained when two or three polymers are mixed, resulting in a polymer chain composed of repetitive units of monomers from the original polymers. For instance, if polymer A is composed of covalently bonded A monomers (A-A-A-A . . .) and polymer B is composed of covalently bonded B monomers (B-B-B-B . . .) then a diblock copolymer chain of polymer A and polymer B could look like A-A-A-A-B-B-B-B . . . or A-A-A-B-B-B-A-A-A-B-B-B . . . , etc. Different surface chemistries or other surface properties phase segregate the monomer units into predictable shapes. In other words, the copolymer chain spatially orients itself to a certain pattern and geometry depending on the polymer ratios (that in turn dictate the surface physics) that comprise the copolymer. After deposition on the surface, one polymer can be dissolved away by an appropriate solvent leaving only the other polymer with the geometry previously determined by the copolymer make-up (Figure 10.1). A drawback is that there is a limited set of possible patterns that can be formed using this technique and the arrays are generally simple and dense [1].

For surface applications that want simple, dense networks, block copolymer chemistry is proving useful. Combining block copolymer self-organization with a lift-off technique and reactive ion etching (discussed in Sections 10.3.2.1.1 and 10.4.1, respectively), Park et al. demonstrated a technique that uses a polystyrene-*b*-polyisoprene polymer blend as a sacrificial etch mask and is capable of depositing 20 nm gold dots with 40 nm periods onto silicon substrates [6]. Such dense gold arrays could conceivably be used for tissue engineering work in relation to cell adhesion [7] or DNA and self-assembling monolayer (SAM) applications [8].

10.2.2 Self-Assembling Polymerization

Self-assembling polymerization is a similar technique to block copolymer lithography in that it still uses copolymers to build up a three-dimensional structures but the actual construction is less related to conventional lithographies and more related to organic synthesis. Nonetheless, researches have begun using self-assembling polymers to manufacture microenvironments with structures on the nanometer scale. Polypeptides, when arranged as diblock polymers and treated with the appropriate initiator, have been shown to self-assemble into polymer films and hydrogels [9]. A 2002 article in

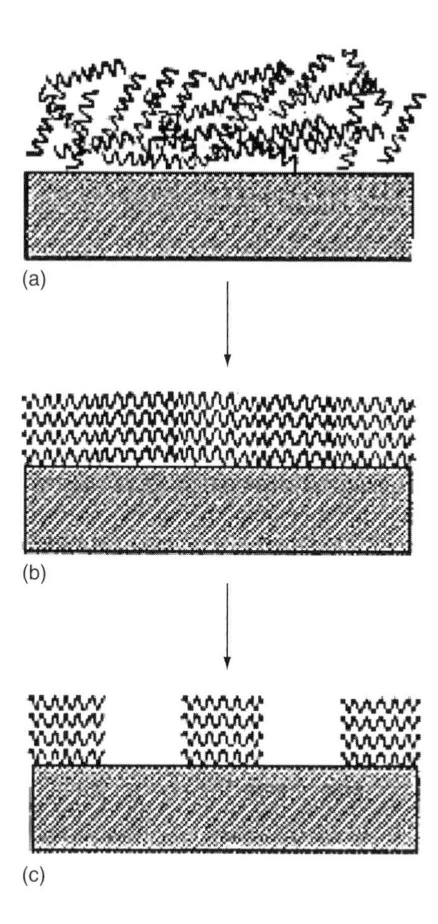

FIGURE 10.1 Schematic of block copolymer lithography. (Reprinted from Trawick, M., Angelescu, D., Chaikin, P., and Register, R., *Nanolithography and Patterning Techniques in Microelectronics*, Woodhead Publishing Limited, Cambridge, England, 2005. With permission.)

Nature describes the use of combinations of lysine, leucine, valine, and glutamic acid to form rigid hydrogels with porous structures that allow large molecular diffusions as might be necessary for scaffolds intent on cell nourishment or growth. Additionally, besides the physical advantages, hydrogels constructed from natural protein components offer the possibility of biofunctionality and biodegradation [9].

Under the right conditions, peptides have also been shown to self-assemble into nanofibers that orient and can be mineralized with hydroxyapatite in a similar manner as collagen does in bone [10]. Other researchers have used self-assembling nanofibers as three-dimensional scaffolds to promote chondrocyte production of a collagen-rich ECM [11] and to allow for a fuller differentiation of hepatocyte progenitor cells [12]. Holmes et al. used a self-assembling peptide scaffold culture of rat neurons. Not only did the scaffold promote neuronal attachment and neurite outgrowth but after several days some neurites began contacting adjacent neurons and other neurites, forming synaptically functional systems [13].

10.2.3 Contact Printing

Another approach that should be included as a bottom–up lithography is "micro-contact printing" (μCP). As the name suggests, much of the research in contact printing has been preformed on micron-sized features and has only recently begun to progress into the nanometer territory [14].

Nonetheless, the current uses of contact printing combined with future possibilities of enhanced nanometer resolutions necessitate a discussion here. Fundamentally, μCP has three steps: (1) create and pattern a "soft" elastomeric stamp, (2) "ink", or adsorb the chemical or biological entity of interest onto the stamp, and (3) use the stamp to print features onto target surfaces. Stamps are typically made of poly(dimethylsiloxane) (PDMS), a two-part elastomer commonly used for rapid prototyping applications. Producing a stamp from this material enables the same stamp to be used to print features onto multiple surfaces in excess of 100 times without losing performance. The ability to ink a wide variety of proteins has resulted in great interest in μCP for researchers involved in biomaterials, tissue engineering, and drug discovery [15] among others. Additional flexibility results from the capability of printing on a variety of surfaces including silicon, glass, and plastic [16]. μCP is widely used to print proteins onto surfaces for the purpose of using those proteins as anchors and manipulators to study in-depth cell functions such as life and death cycles [17], differentiation [18], and chemotaxis [19] to name a few.

However, because the stamp is generally patterned using micron-scale photolithography (discussed in Section 10.3.1.1), few applications at the nanoscale level are available. An exception to the rule is the preliminary work being done by Stellacci et al. at the Massachusetts Institute of Technology (MIT). Building upon the biological concept of how deoxyribonucleic acid (DNA) is copied onto ribonucleic acid (RNA) strands, the MIT research group uses single stranded DNA stamps to attach complimentary DNA to a target surface, usually gold. The technique can print DNA patches that boast resolutions down to 40 nm [20], breaking the 50 nm barrier that is common for other μCP techniques (Xia et al. 1995, 9576–9577). And though Stellacci et al. only report printing on gold surfaces, there are already plans to begin "printing on insulating substrates like glass or silicon dioxide, as well as on polymers" [20], as is possible with μCP techniques.

While block copolymer lithography and self-assembly methodologies boast superior resolutions and could someday be combined to give specific patterning control over 3D matrix formations, to date this has yet to be widely pursued. Conversely, contact printing is already widely used in tissue engineering but has yet to solidify its potential as a nanometer scale lithography.

10.3 TOP–DOWN LITHOGRAPHY

Though the uses of bottom–up approaches show promise, top–down lithographies have long been the true mainstays of surface engineers. The top–down approach contrasts the bottom–up in that the top–down adds patterns by stripping away a bulk material rather than simply depositing the patterned material. Thus, top–down lithographies generally begin by depositing a bulk material, carving out a certain pattern through chemical or physical means, and then building or transferring patterns on top of the base in order to complete the structure. Sometimes it is the substrate itself that is patterned in order to form a cell-inhabitable or functional surface [21]. Other times the pattern is built by stripping away a bulk resist by the addition of functional molecules onto the surface [22], [23]. Top–down lithographies can be subdivided into two separate categories—those that can effectively develop patterns over large areas on the scale of a standard silicon wafer, deemed wafer scale, and those known as direct write by which a pattern is written one line at a time. Thus, whereas wafer-scale methods are in general very temporally efficient, direct-write methods are temporally expensive.

10.3.1 Wafer-Scale Techniques

Wafer-scale techniques include optical and imprint lithographies. The great advantage of these methods is their ability to pattern large areas in comparatively little time. Through association and codevelopment with the computer and semiconductor industries, many are directly compatible with already established technologies making them convenient and cost-effective to use.

10.3.1.1 Optical

Optical lithographies have been the most widely used over the years as they technically require only a light source, a patterned photomask, and the resist that will be affected. In essence, a photomask is created with a transparent pattern on it. Light passes through the transparent section of the mask onto a photoresist thus illuminating the resist with a pattern identical to the one on the mask. Photoresists are specially designed polymer solutions that have their solubility altered by light exposure either making them more or less susceptible to removal by a solvent. "Negative" resists become less soluble upon light exposure and "positive" resists become more soluble. Therefore, after exposure and development, negative photoresists are dissolved away leaving only the exposed pattern. For positive photoresists, only the pattern dissolves away.

For the purposes of nanolithography, however, optical lithography becomes more difficult. Because light diffracts as it passes through an aperture, generally called Fraunhoffer diffraction [24], there is a fundamental limit to the resolution of optical lithographies that is proportional to the wavelength of light being used. The equation

$$d = k\left(\frac{\lambda}{\mathrm{NA}}\right)$$

describes the relationship between the size of the smallest possible feature (d), the light wavelength (λ), the numerical aperture of the light source (NA), and a constant (k) [25]. To reduce the scale of the diffraction limit or at least to limit its impact, several nonconventional techniques have recently been developed. Researchers initially began by decreasing the light wavelength into extreme ultraviolet and x-ray regions. Further, a few investigators have begun experimenting with the effects of metal-impregnated photoresists and masks in order to capitalize on light excited surface plasmons. Another attempt at avoiding the diffraction limit is near-field optical lithography which will be covered in Section 10.3.2.2.3.

The most obvious variable to be altered in the diffraction/resolution equation is wavelength. Theoretically, decreasing the wavelength of the incident light will decrease the feature size of the smallest feature. Many researchers have accordingly begun experimenting with x-ray and extreme ultraviolet (EUV) lithographies using the same source, mask, and resist blueprint but with x-rays or EUV light being used in place of conventional light. For comparison, visible light wavelengths range from roughly 400 to 700 nm whereas the wavelengths for x-rays used in lithography are typically anywhere from 0.5 to 4.5 nm depending on usage [26]. Seunarine et al. recently reported their group's use of "x-ray lithography" to form small polymethyl methacrylate (PMMA) pillars on a half-pipe quartz substrate [27]. Though previous findings state that regular nanopit arrays retard cell adhesion [28], Seunarine's group attempted to create nanopillar arrays on a tubular structure to foster a low cell adhesion environment reminiscent of human vasculature. By coating the quartz with 500 nm thick PMMA resist film and then exposing the resist at close proximity ($\sim$10 μm) with $\sim$2 keV (0.8 nm) x-rays, the group created pillars with diameters ranging from 178 to 213 nm. However, even at an x-ray's short wavelength, the proximity between the aperture and the sample is within the range to have the radiation affected by another type of diffraction called Fresnel diffraction [24]. So while small feature sizes are possible, distortions are caused that give the pillars a slightly hexagonal shape (instead of round) with small holes in the center of the pillar [27]. The group stopped short of reporting any cell adhesion studies on their coated half-pipe.

EUV lithography is another technique based on the same principle of using shortened wavelengths to reduce the diffraction limit. It boasts the ability to produce 20 nm resolution features with high quality over relatively large areas [25]. Though while EUV has been gaining interest in the microelectronics industry, it is far from mainstream in biomedicine, mostly on account of the special emission sources and optics it requires. Modern advances in synchrotrons and plasma sources, diffraction optics, and multilayer reflective EUV optics, however, have recently made

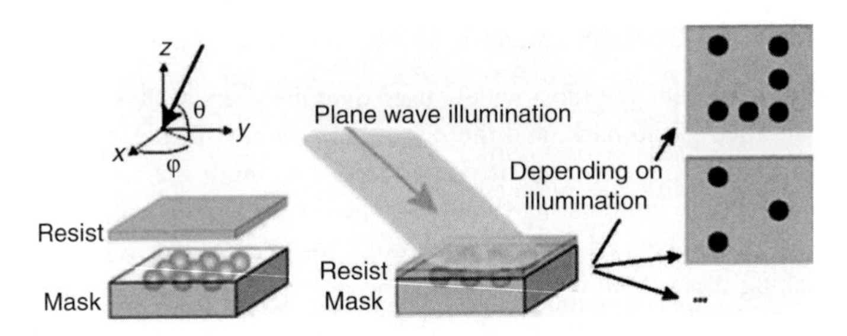

FIGURE 10.2 Plasmon lithography with silver nanoparticles. (Reprinted from Koenderink, A.F., Hernandez, J.V., Robicheaux, F. et al., *Nano Lett.*, 7, 745, 2007. With permission.)

these limitations less of a deterrent [25]. In fact, the 2004 International Strategy and Foresight Report on Nanoscience and Nanotechnology "strongly" supports EUV as a future nanotechnology, indicating that "more intensive research efforts are required at the academic and institutional level to ramp up the general state of all process elements" [29].

A further innovation in optical lithography is the use of "plasmon lithography." Researchers at the University of Texas have recently capitalized on the idea of surface plasmon lithography by using a quartz mask coated with a perforated metal (gold or titanium) in close proximity to a traditional photoresist that was spun onto a silicon substrate that was also coated with metal (titanium). A polarized 355 nm wavelength light was then used to create line patterns in the resists with widths down to 130 nm [30]. By explanation, the researchers suggest that the light is enhanced by excited surface plasmons as it passes through the subwavelength aperture and is then confined by coupling to excited surface plasmons on the substrate [30].

Additionally, in 2007, Koenderink et al., working on past experimental research by Hubert et al. [31], calculated that by implanting only the optical mask with silver nanoparticles it becomes possible to create different patterns on the resist by merely varying the incident light's wavelength, angle, or polarization. Then by moving the mask and sequentially exciting small subunits of the sample, they further claim a high level of design freedom can be achieved over a large area [32] (Figure 10.2).

As with block polymer nanolithography, while promising, optical nanolithography is also still in the very early stages of development and still suffers from some very important drawbacks. Analogous to what renowned Harvard chemist Dr. George Whitesides commented on photolithography in 1998:

> "It is expensive;... it cannot be easily adopted for patterning nonplanar surfaces;... it is poorly suited for introducing specific chemical functionalities; it is directly applicable to only a limited set of materials used as photoresists" [33].

As a result, most optical techniques still focus on exploring the different surface features possible to create while far fewer are yet at the point of discovering the consequences of applying optically nanopatterned surfaces to biological situations. However, biological and biomedical experimentation has begun on "nanoimprint lithography," a nonoptical wafer-scale technique.

10.3.1.2 Nanoimprint

Initially proposed in 1995, nanoimprint lithography (NIL) uses a nanostructured mold to press a pattern into a resist film on a substrate. Unlike optical lithography, NIL is not dependent on radiation thus its resolution is unaffected by such common lithography problems as diffraction, resist scattering, substrate backscattering, etc. [34] As seen in Figure 10.3 the NIL resist film is either (1) a polymer heated beyond its glass transition temperature that is compressed and then allowed to

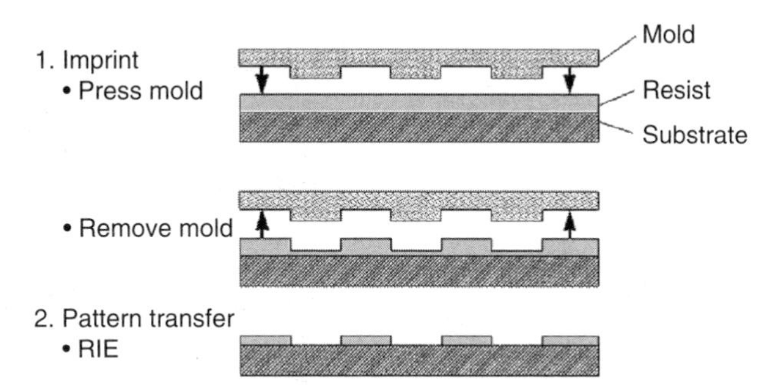

FIGURE 10.3 Overview of NIL process. (From Chou, S.Y., Krauss, P.R., Zhang, W. et al., *J. Vac. Sci. Technol. B*, 15, 2897, 1997. With permission.)

cool before the mold is removed or (2) the resist is a curable polymer that is pressed and then cured (via UV or curing agent) before being peeled back from the mold. Regardless, there is a final step, pattern transfer, which uses some form of etching (often reactive ion etching) to remove the resist that remains in the compressed areas (Figure 10.3). Molds are typically Si or SiO_2 and are initially nanostructured through direct writing techniques such as electron beam lithography [35], which will be further discussed in Section 10.3.2.1.1. Using NIL combined with conventional lift-off techniques, resist polymers like PMMA have been used to achieve patterned arrays of holes [35] or pillars [34] with diameters down to 6 nm. Compared to electron beam and optical lithographies, NIL can be inexpensive, with higher patterning speed, resolution, and reproducibility [21]. The drawbacks are the expense of the initial mold design and the high compression pressures that may be needed [36].

Besides resolution, another chief advantage of NIL over optical lithography is that experiments need not be limited to patterning inactive features onto a surface and then observing the cellular response to the geometry. Often in conjunction with contact printing, one can pattern proteins and other biofunctional molecules that are specifically designed for applications ranging from simple cell anchoring experiments [23] to phenotype expression and differentiation studies [37] (Figure 10.4).

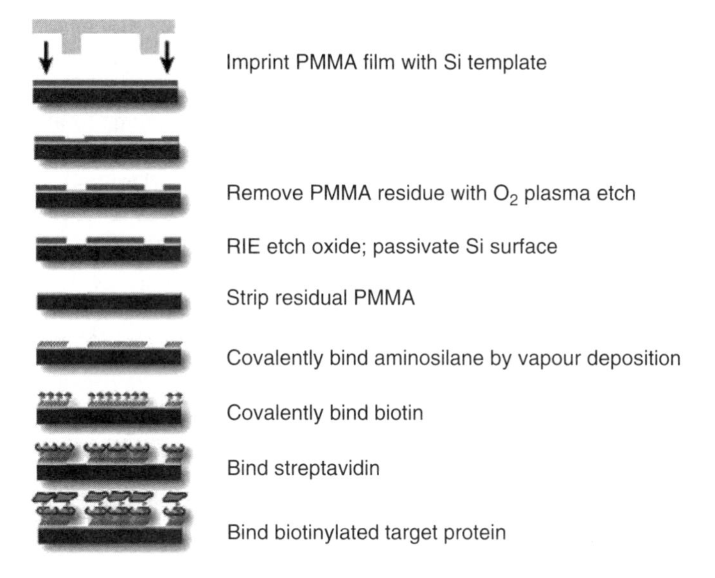

FIGURE 10.4 Protein immobilization on a patterned surface. (Reprinted from Hoff, D.J., Cheng, L., Meyhofer, E. et al., *Nano Lett.*, 4, 5, 853, 2004. With permission.)

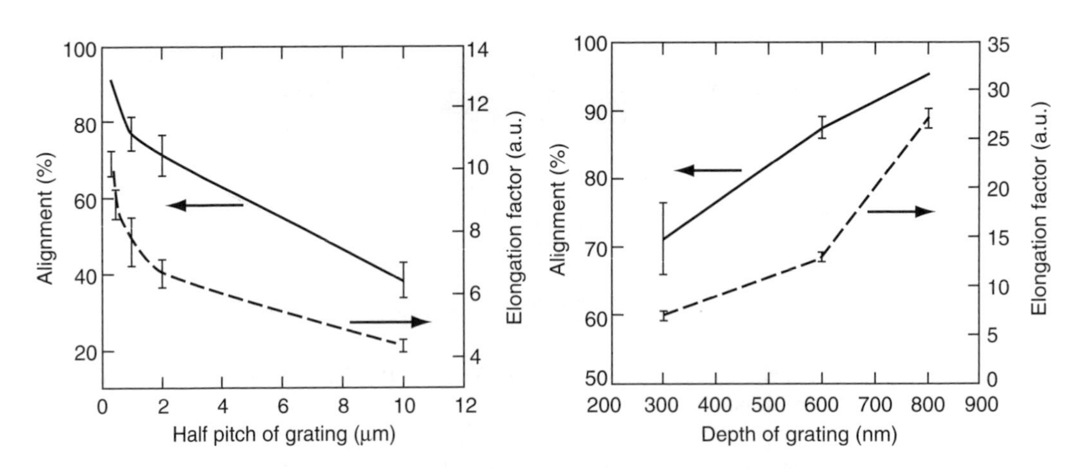

FIGURE 10.5 Effects of nanotopography on cell behavior. (From Hu, W., Yim, E.K.F., Reano, R.M. et al., *J. Vac. Sci. Technol. B*, 23, 2984, 2005. With permission.)

Gaubert et al. showed that ~60 nm gold nanopatterns could be deposited on a SiO_2 background and could then be functionalized with either fibronectin or polyethylene glycol (PEG) to promote or retard human umbilical vein endothelial cell growth in the respective areas [23].

One final benefit of NIL is its ability to print features onto a variety of substrates. Imprint patterning resolutions on high atomic weight substrates (thick gold substrates, for example) are not affected by the backscattering that troubles radiation-based lithographies [34]. It also becomes possible to pattern cheaper, less sophisticated substrates as shown by a joint project between the University of Michigan and John Hopkins School of Medicine who recently published a process to nanoimprint polystyrene tissue-culture dishes and used the patterned dishes to study reactions of bovine pulmonary artery smooth muscle cells to nanotopography. Figure 10.5 shows the cellular alignment and elongation on these patterned dishes as it relates to the topography [21] (Figure 10.5).

Reminiscent of NIL is a technique called capillary-assisted molding, or simply "capillary lithography". The two are similar in that they both often use electron beamed master templates to pattern a large area of polymer that is susceptible to heat, light, solvation, or pressure. Thermally induced capillary lithography uses a PDMS mold with recessed patterns to affect a thin polymer film. The mold is placed in contact with the film, the film is heated above its transition temperature, and capillary action draws the polymer up into the PDMS recesses. The film is allowed to cool and harden again and the PDMS mold is then removed leaving a positive pattern of the polymer. Inherent stability of the PDMS mold limits the density and resolution to about 100 nm [38]. "Solvent-induced capillary lithography" works similarly except that the PDMS mold with the recessed pattern is placed over a polymer film that is dissolved in some amount of solvent. Capillary action draws the solvent up into the recesses dragging the polymer with it. The solvent, however, is absorbed into the relatively diffusive PDMS mold leaving only the polymer behind. Eventually the solvent will diffuse through the mold and evaporate into the atmosphere. Density and resolution are still limited by the PDMS mold [38].

Researchers at a 2007 conference in China displayed their use of a solvent and UV based capillary lithography to promote the growth of cardiac myocytes on a glass substrate [39]. They used a modified polyurethane mold to wick up minute amounts of a PEG-DMA solution that had been drop-dispensed on the substrate. The PEG was then cured with UV light before removing the polyurethane mold leaving behind cone shaped nanopillars about 400 nm high and 150 nm wide at the base. Seeding rat primary cardiomyocytes then showed that the myocytes extended filopodia and

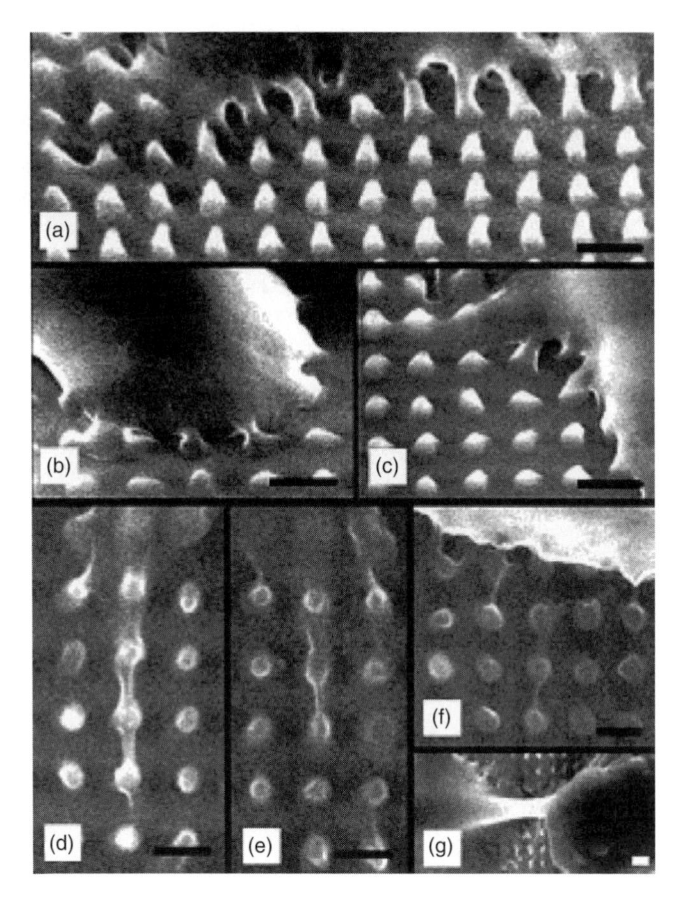

FIGURE 10.6 Response of rat cardiomyocyte cell to PEG nanopillars. (From Kim, D., Kim, P., Suh, K. et al., *Conf. Proc. IEEE Eng. Med. Biol. Soc.*, 4, 4091, 2005. With permission.)

lamellipodia primarily along the nanopillars, as seen in Figure 10.6. Though adhesion was far below that of the glass control, cell adhesion increased with the nanopillars when compared to a bare PEG substrate [39].

In contrast to forming a two-dimensional scaffold by coating polymers on a substrate, Pisignano et al. used capillary lithography to form small polyurethane fibers with diameters of ~300 nm. The process yields similar results to those of electrospinning but has the added ability of producing aligned and highly ordered fiber bundles [40] (Figure 10.7). No immediate biological studies were performed on the fibers though previous research has been done on the use of electrospun nanofibers in tissue engineering [41].

An additional variation of capillary lithography that has yet to find much use in biomedical engineering is "pressure assisted capillary lithography." As the name implies this lithography uses a pressure molded amorphous fluoropolymer powder mold instead of the PDMS and presses onto the polymer with 2 ~ 3 bar at 160°C. Feature sizes are still greater than 100 nm but have improved densities on the surface [38].

10.3.2 Direct-Write Techniques

Where wafer-scale techniques excel in speed and scale they are often lacking in design freedom because they use one large mold to affect many areas identically. Counterpart to these methods are direct-write techniques. Direct-write techniques are capable of producing very finely detailed

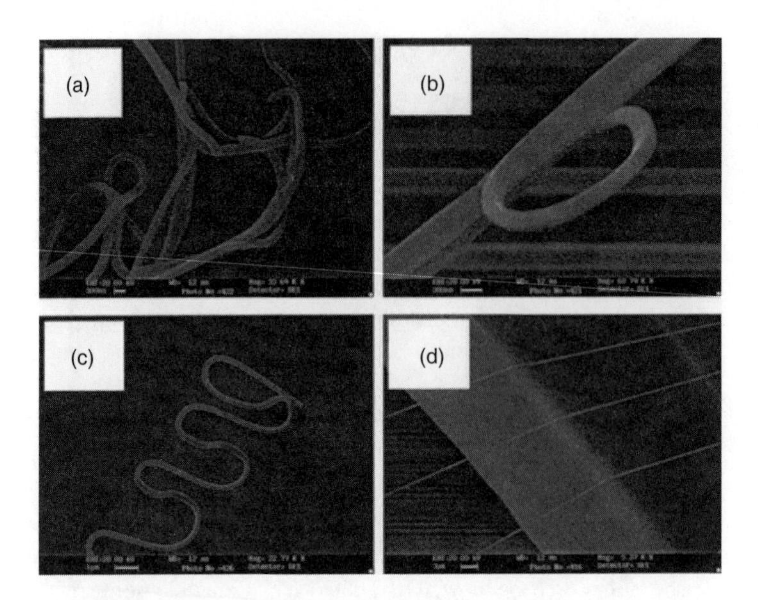

FIGURE 10.7 Nanofibers created by soft lithography. (Reprinted from Pisignano, D., Maruccio, G., Mele, E. et al., *Appl. Phys. Lett.*, 87, 123109, 2005. With permission.)

patterning with a great deal of freedom concerning the design but generally are very time consuming and only practical when used over small areas. The general scheme is for either a tightly focused particle beam or a small probe to scan across a surface, carefully affecting a pattern of predetermined design.

10.3.2.1 Accelerated Particle Beam

Accelerated particle beam techniques are a family of techniques in which charged particles are accelerated by electric fields and tightly focused into a beam by magnets. The particle beams can be directed to strike target locations with high accuracy and nanometer resolutions.

10.3.2.1.1 Electron Beam

The most common direct-write technique is "electron beam (e-beam) lithography" or EBL. EBL is a technology that has been around since the 1960s and naturally evolved from the development of the scanning electron microscope [26]. As the name implies, an electron source emits a tightly focused electron beam that is used to directly pattern resist much in the same way as photolithography uses light but with much smaller resolutions, conventionally down to 10 nm. For comparison, it is helpful to note that the wavelength of blue light is on the order of 480 nm, whereas the wavelength of an electron is only a fraction of a nanometer. The production scheme is a general lift-off procedure—the substrate is coated with e-beam resist, the electron beam etches a pattern into the resist, a metal is deposited over the surface, and finally the resist is removed leaving only the pattern of metal attached to the substrate where it was e-beamed. PMMA has long been the traditional resist [42] although researchers have since been able to directly etch SiO_2 [43] and AlF_3-doped lithium fluoride resists [44] to slightly finer resolutions using much higher electron exposures. However, if the advantages of using e-beam lithography are the enhanced resolution and the direct-write capabilities, the drawbacks are expensive equipment and long processing times [23].

Earlier biomedical studies have used e-beam lithography to create nano-pillared molds that are then used to hot emboss surfaces of polycaprolactone (PCL) with nanopits. Such studies have shown that human fibroblast cells, for instance, project more sensing filipodia onto surfaces with

35–120 nm diameter nanopits as compared to planar surfaces [45] and exhibit reduced adhesion [28]. It appears the pits are sensed by the filipodia who then guide the cell to be more motile rather than attaching [45]. E-beam patterning has also been combined with self-assembling monolayers [46] and thermally responsive polymers [47] to create surfaces that control cell adhesion and patterning during seeding. Even more extensively, e-beam lithography is used in conjunction with other lithographies such as nanoimprint and capillary lithography to drastically improve their natural resolutions. In fact, very few nanolithography techniques do not use EBL in some form, usually to create the pattern transfer masks.

10.3.2.1.2 Proton Beam

Proton beam writing (PBW) is a high energy ion beam direct lithography in which a small ion (often hydrogen or helium) with high energy is created and separated from its electron cloud using strong directional electric fields. When focused and directed at a target substrate, the beam bombards the target atom and affects mostly the atom's electron cloud thereby either affecting its solubility or sometimes its refractive index or crystallinity. As with previous resist techniques, the polymer resists are affected in such a way as to enhance the polymer cross-linking (negative resist) and making it impervious to dissolution or to break the polymer's covalent bonds through chain scission (positive resist) and thereby making it susceptible to solvents [48]. This process has been shown to produce features down to 50 nm in SU-8, 30 nm in PMMA [49], and 22 nm in hydrogen silsesquioxane (HSQ) [50]. Compared to EBL, PBW has a longer range with less lateral spreading and "can create structures with depths of tens of micrometers with a sidewall angle very close to 90° and very low roughness" [48]. It is therefore well suited for exposing thick resist materials like PMMA [51].

Though perhaps not as popular as e-beam lithography, PBW is making inroads into tissue engineering research. A 2007 article in *Nuclear Instruments and Methods in Physics Research B* cites the use of PBW to pattern SU-8 resist in order to study osteoblasts response to nanostructures [52] and several other papers show the promise of using PBW to pattern biomedically relevant surfaces such as PMMA, photosensitive glass, and GaAs [51].

10.3.2.1.3 Focused Ion Beam

Focused ion beam lithography (FIBL) uses ion bombardment physics similar to those of PBW but employs a larger atom or molecule. Common examples of focused ion beam particles are small gold clusters and C_{60} molecules. The focused beam of particles can remove material like "sputter etching" (discussed in Section 10.4.1) but is also generally considered a direct-write technique like EBL. FIBL uses electrostatic and magnetic fields to focus a low energy ion beam [48] commonly down to diameters of 50 nm [53] (Figure 10.8). The beam can be passed through a stencil onto a surface like traditional optical lithography but is more often used in free-form milling processes due to the complexities of ion beam optics [53]. Because of its slow write speeds, FIB is more often chosen for small-scale milling tasks [54,55] than for biomedical applications. However, the real promise for FIBL in the future of tissue and scaffold engineering remains in its ability to direct write on a variety of hard biocompatible surfaces (namely glass, silicon, and certain metals) without using polymer resists that are typically needed for patterning [55]. Optical and e-beam lithographies cannot yet claim to do this efficiently. Additionally, the technique can be precisely aimed and is well suited for repairing faulty masks [48].

A consequent drawback is often redeposition, i.e., the removed surface atoms are deposited elsewhere on the sample. Redeposition can be diminished with the use of a chemical decomposition gas. During milling, the displaced surface material either decomposes into a gas or reacts with an ambient gas, often oxygen, to form a stable gas, thus eliminating the problem of sputter drift and making for a cleaner pattern [48].

An expansion of FIBL that has yet to catch on as a tissue engineering tool is "ion beam deposition." Here a precursor gas is created by exposing a material with a low energy ion beam that decomposes the material into a solid and a gas (the solid being suspended in the gas). The gas is

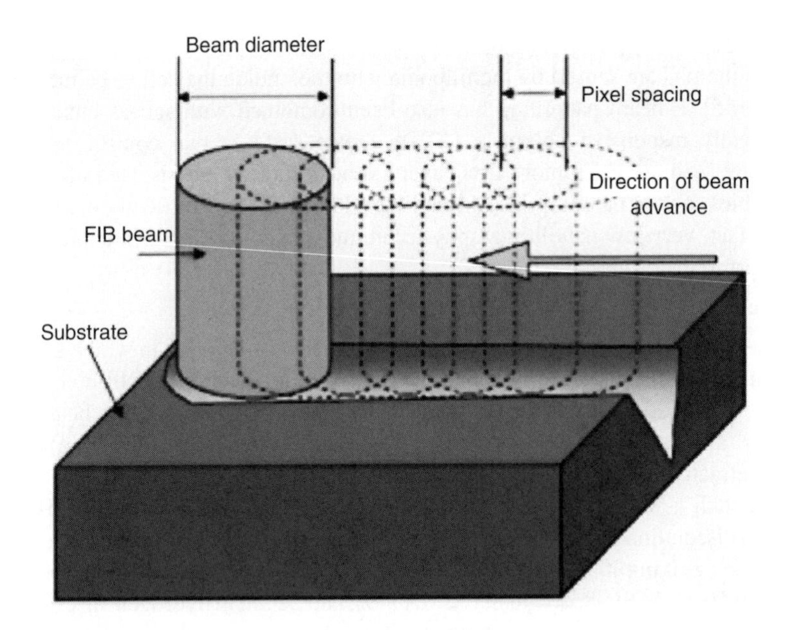

FIGURE 10.8 Schematic of FIB milling. (Reproduced from Tseng, A.A., *J. Micromech. Microeng*. 14, R15, 2004. With permission.)

introduced across the workpiece surface and a low energy, focused ion beam collides with the precursor at the surface, decomposing it into its solid form and depositing it onto the surface [48].

10.3.2.2 Scanning Probe

10.3.2.2.1 Atomic Force Microscopy

If common chemical etching can be considered a brute-force technique for inscribing a surface, "scanning probe nanolithography" (SPNL) is the delicate alternative. Using existing atomic force microscopy (AFM) technology, the surface is altered, usually scribed, with a small, sharp tip, and nearby chemicals react with the scribed portion (Figure 10.9). Diamond tips were initially used as

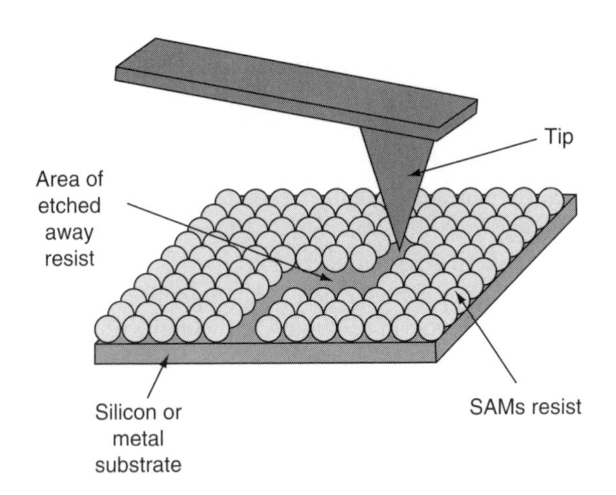

FIGURE 10.9 Schematic of scanning probe lithography with self-assembling monolayers (SAM) resist. (From Norman, J.J. and Desai, T.A., *Ann. Biomed. Eng*., 34, 89, 2006. With permission.)

the scribe but now AFM tips (SiN coated ~20 nm radii) are becoming common and can write features with widths down to 20 nm [56]. Often the surface is submerged in a chemical reagent so the reaction occurs as soon as the scribe exposes the new surface. Like other direct-write methods such as e-beam lithography, the major experimental benefits are the enhanced resolution and the freedom to scribe most imaginable two-dimensional patterns. Likewise, the drawbacks are low efficiencies and extended processing times though in recent years several new innovations are attempting to overcome this issue.

For instance, an idea involved in collagen patterning is to use normal biochemical methods for the deposition of self-assembling collagen but then to use AFM-based technology to pattern the protein. In 2004 Jian et al. showed that solubilized bovine dermal collagen would deposit and self-assemble when spread over a freshly cleaved mica surface. After 4–5 h the collagen array stabilized into rigid structures but within that time the protein could be manipulated by forcing an AFM tip across the surface. With the stylus scanning at 300 pN forces, the native collagen microfibrils could be realigned or even assembled in long fibrils of approximately 100 μm in length while sustaining natural height of ~3 nm. Additionally, even though the microfibrils are being mechanically forced, the authors state that "the material remains molecularly continuous," adding, "If the re-orientation of the microfibrils by the AFM stylus breaks bonds between collagen molecules, then the bonds evidently reform" [57].

A less forceful method is to resistively heat the AFM tip in lieu of applying more pressure. A heated AFM tip can modify a surface either chemically by affecting a thermochemically susceptible region (through oxidation, degradation, cross-linking, etc.) or physically, generally through deformation. In 2007 this method was shown to write down to 12 nm wide lines in a special copolymer and can create line densities of 2×10^7 lines per meter [5].

10.3.2.2.2 Dip-Pen

One increasingly popular idea that has come from SPNL is called "dip-pen nanolithography" (DPN). Here an AFM tip is "inked" or dipped into a solution of protein, peptide, or other chemical of interest and then scanned along a surface where it deposits the inked molecules through mass diffusion [58]. Of course collagen, being one of the more prominent ECM proteins, is attracting a fair amount of research via DPN. Wilson et al. in 2007 reported the positive printing of 30–50 nm lines of collagen on gold-coated muscovite green mica wafers. The real boon of this method is that it "preserved the triple-helical structure and biological activity of collagen and even fostered the formation of characteristic higher levels of structural organization" [58]. Furthermore, such patterning was successful for heights up to 300 nm and lengths of up to 100 μm.

Additionally, enzymes can be adhered to the SiN AFM tip to further incite highly localized chemical reactions. For example, it has been shown that an alkaline phosphatase–streptavidin conjugate, immobilized on an AFM tip, can be used to dephosphorylate BCIP in the presence of the cofactor NBT and thus precipitate the BCIP. Therefore, scanning such an AFM tip across a BCIP/NBT covered mica surface leaves linear traces of precipitate of 150–170 nm widths and 10 nm heights [59].

Because DPN writing speed is generally limited by mass diffusion rates of the molecules adhered to the tip (as they are transported to the substrate) or in thermochemical cases by heat diffusion [5], processing time and scalability are often cited as the major drawbacks to scanning probe lithography [22,23]. With this in consideration, researchers have begun exploring parallelization approaches to scanning probe lithographies to varying degrees of acceptance. While supporters contend that DPN efficiency can be bolstered dramatically by simply using multiple AFM tips in parallel [5,60], detractors maintain that other techniques such as NIL will likely remain "more flexible and faster" [23]. Nonetheless, IBM has experiments under way studying the efficacy of high-density two-dimensional chip arrays with AFM-like cantilever tips [61] in a project called the "Millipede" [61,62] (Figure 10.10). While conceptualized for data storage purposes [63], its use of

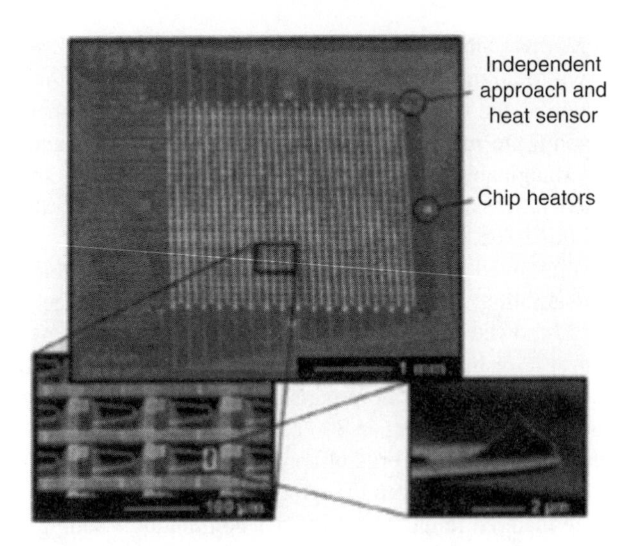

FIGURE 10.10 Two-dimensional chip arrays with AFM-like cantilever tips. (From Vettiger, P., Brugger, J., Despont, M. et al. 1999. *Microelectron. Eng.*, 46, 11, 1999. With permission.)

standard thermochemical techniques to scribe PMMA will no doubt find biomedical applications in the future.

Scanning probe lithography can also be used in cooperation with other patterning techniques. As previously mentioned, reactive ion etching is generally performed through a resistive mask that has been patterned via e-beam lithography. However, a recent publication by Zhang et al. reports the use of DPN to write a 16-mercaptohexadecanoic acid (MHA) etching mask with nanometer dimensions on a Au/Ti-coated Si substrate [60]. A wet chemical etch then etches the entire surface, exposing the written gold patterns. A subsequent reactive ion etch removes the gold leaving Si nanostructures in the pattern initially written by the AFM. Thusly the Si substrate was patterned with positive Si lines (312 nm wide and 16 nm high) and dots (385 nm diameter and 75 nm high) [60] (Figure 10.11).

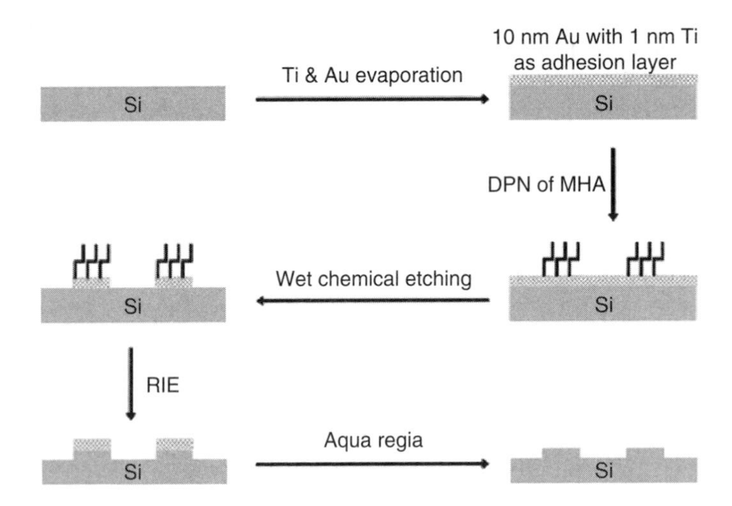

FIGURE 10.11 Use of DPN to create nanopatterned Si lines on a Si substrate. (Reproduced from Zhang, H., Amro, N.A., Disawal, S. et al., *Small*, 3, 81, 2007. With permission.)

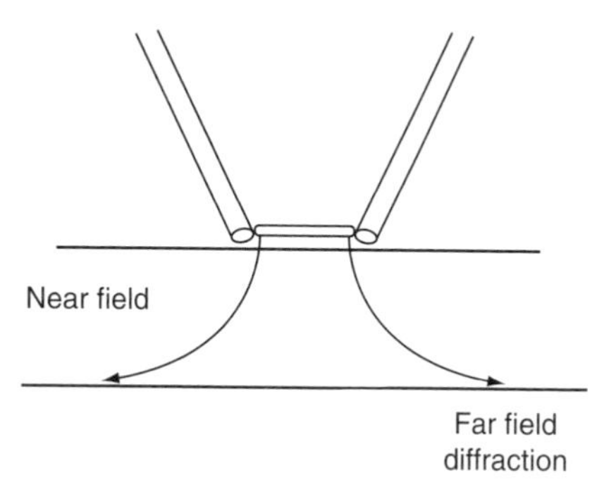

FIGURE 10.12 Near-field scanning optical lithography avoids the diffraction limit. (Reprinted from Legget, G.J., *Nanolithography and Patterning Techniques in Microelectronics*, Woodhead Publishing Limited, Cambridge England, 2005. With permission.)

10.3.2.2.3 Near-Field Scanning Probe

Another approach beyond wavelength manipulation to avoid the optical lithography diffraction limit previously discussed is to use a near-field approach. Placing the mask in contact with the resist limits the amount of space in which the light has to diffract thus increasing resolution. A maskless option called "near-field scanning optical lithography" projects light through a very narrow, aluminum-wrapped fiber-optic probe that scans across the surface. Depending on the size of the probe, resolutions of approximately 50 nm are possible (Figure 10.12). Because this technique requires an exact control over the distance between the light source and the resist, self-assembling monolayers (SAMs) are being studied with additional fervor. The thickness of SAMs can be tailored on a nanometer scale by sequentially adding 1–2 nm length chains. Interestingly, near-field photolithography with SAMs can achieve features between 20 and 40 nm even though the probe size is approximately 50 nm [64]. One explanation for this is that the light induces a highly contained plasmon on the gold surface which then affects the very local SAM chemistry leaving portions susceptible to removal by solvent [64].

10.4 MISCELLANEOUS

Miscellaneous nanolithographic techniques also exist that build upon other lithographies, sometimes both top–down and bottom–up. For example, nanoetching almost always involves some form of e-beam or FIB lithography to create a mask. Likewise, the resolution of shadow deposition is dependent on any number of combinations of accelerated particle beam, scanning probe, and chemical etching. The results and overall acceptance of these methods in nanolithography and biomedicine are therefore determined by a combination and balance between the advantages and disadvantages of their constitutive predecessors.

10.4.1 ETCHING

Recent curiosity over the relationship between cellular behavior and its nanostructured surroundings has also reinvigorated such classic chemistry techniques as chemical etching and sputter etching. Sputter etching uses a low energy ion species situation with heavier ions (O, Ar, Ga, or Xe) in a plasma form to etch away surface atoms from a "workpiece." Surface patterning is achieved by using typical stenciled resist masks to cover the workpiece. The plasma cloud is formed by applying a DC or radio-frequency electric field to a low pressure gas of atoms so the ions and ionized

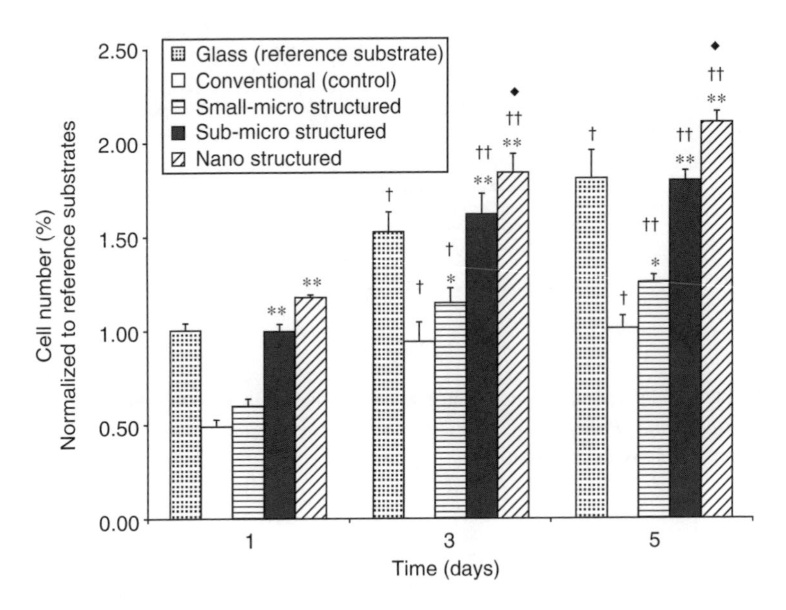

FIGURE 10.13 Effects of nanotopography on bladder SMC cell function. (From Thapa, A., Miller, D.C., Webster, T.J. et al., *Biomaterials*, 24, 2915, 2003. With permission.)

electrons then coexist in the same volume of constantly ionizing and recombining species. The submerged workpiece is therefore "randomly bombarded" by the positive ions which affect the exposed workpiece and any resist layer simultaneously. It is convenient to consider these low energy ion collisions as essentially atomic billiard balls; the generally larger positive ions barrel through the target atom's electron cloud, colliding with the nucleus and can eject surface target atoms if the energy of such a collision is greater than the target atom's lattice binding energy. Penetration depth is typically 1–100 nm and redeposition can again be a problem [48].

A variant of sputter etching is "reactive ion etching" (RIE), also called "dry etching." It is essentially like sputter etching except the plasma gas is reactive to the material to be etched relative to the resist. The same technique can also be used to react with removed particles in order to limit redeposition. The workpiece need not be submerged in plasma if a bias is placed across the system to draw positive ions out and the plasma density and energy can be adjusted by changing the excitation current or gas pressure. A variation called gas assisted etching uses an ion beam instead of plasma to resolve features. The setup is similar to the previously mentioned ion beam deposition scheme except that the precursor gas is replaced with a chemical etching gas [48].

A research group in Indiana led by Karen Haberstroh used etching in 2003 to manipulate the adherence of ovine bladder smooth muscle cells (SMCs) on PLGA and PU films. The group argues that through nanopatterning an increased cell–substrate interaction will foster new hopes for future artificial bladder replacement constructs. While the study does not demonstrate sputter or reactive ion etching but rather uses NaOH and HNO_3 chemical etches, a random nanotopography of features between 50 and 100 nm was nonetheless formed. The results in Figure 10.13 show that the bladder SMCs do indeed increase cell function on nanostructured PGLA as compared to a glass control substrate and untreated PGLA [65] (Figure 10.13).

The same year, Teixeira et al. in Wisconsin used the more specific reactive ion etching technique to etch silicon wafers into 70 nm wide grooves upon which they cultured human

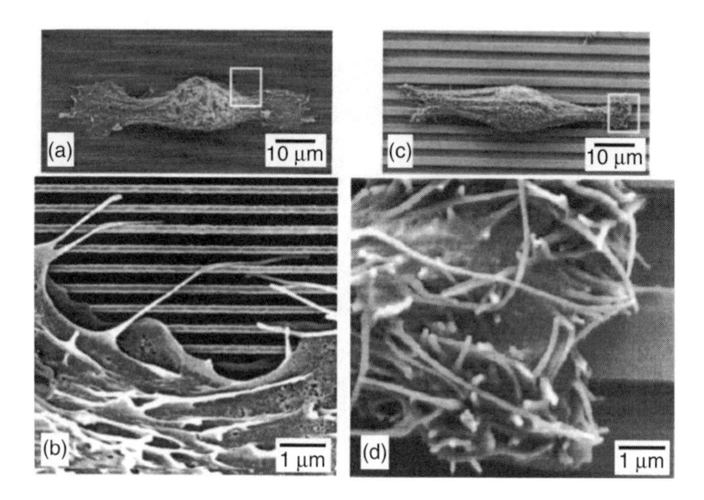

FIGURE 10.14 Cell alignment of nanopatterned surface. (From Teixeira, A.I., Abrams, G.A., Murphy, C.J. et al., *J. Vac. Sci. Technol. B*, 21, 683, 2003. With permission.)

corneal epithelial cells. They did this through a PMMA mask which was patterned by e-beam lithography and used SF_6 and $C_2H_2F_4$ as the reactive gas. As a result the cells aligned themselves along the grooves, abandoning the usually round shape that is common when cultured on smooth surfaces [66] (Figure 10.14). Likewise, the same year a group in Japan modified a silicon wafer surface with SF_6 and O_2 gases through a cobalt mask. Their end goal, however, was to deposit a SAM in the pattern behind the gas etch [67] which could then presumably be functionalized for biomedical purposes such as protein adsorption [68].

10.4.2 Shadow Deposition

Another technique that builds on e-beam resolution is "shadow deposition" through nanostencils. Here, the material to be deposited is either physically or chemically vaporized, often in a vacuum environment. The vapor travels perpendicular from the surface and can be screened through a stencil of specified pattern and is deposited on another surface behind the stencil. Stencils can be carved at submicron levels using e-beam, FIB, or scanning probe methods. One negative aspect is the possibility for the stencil to become subject to clogging and bending and problems associated with misalignment [69] (Figure 10.15).

In 2004, Pallandre et al. reported the use of a gas-phase silanization reaction to deposit a silane monolayer onto a silicon wafer through an e-beam-etched PMMA stencil. The resulting nanopattern was not only of high resolution (20–25 nm) but was then also susceptible to the attachment of a wide variety of chemical functionalities [70] which could then conceivably be used for biological attachments and manipulations. As proof of concept, Pallandre et al. followed the gas-silanization experiment later the same year with a report concerning the adsorption of a typical globular protein (P.69 pertactin) onto parallel nanostripes of alkylsilane. They found that not only could they localize protein adsorption but by varying the widths of the nanostripes they could also alter the protein's orientation. Nanostripes less than 50 nm wide would encourage the protein to attach with its longest axis perpendicular to the stripe and wider nanostripes promoted parallel or "flat" adsorption [71].

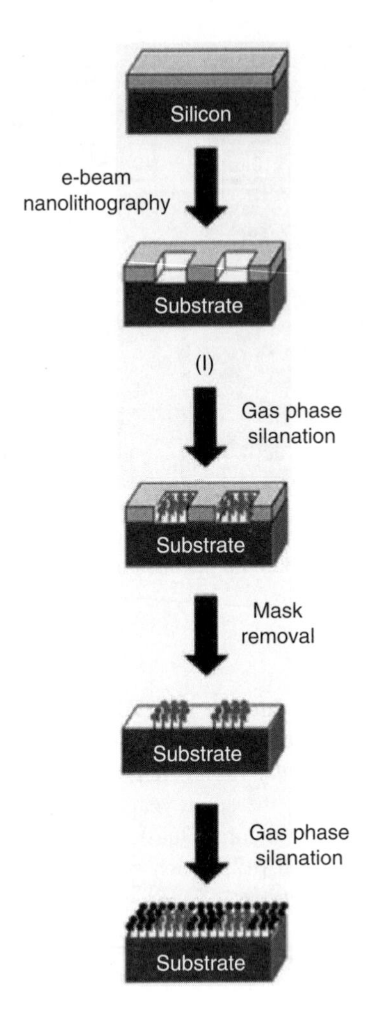

FIGURE 10.15 Gas-phase silanization reaction through e-beamed PMMA stencil. (Reprinted from Pallandre, A., Glinel, K., Jonas, A.M. et al., *Nano Lett.*, 4, 365, 2004. With permission.)

10.5 CONCLUSIONS

When discussing the uses of nanolithography in scaffolding and tissue engineering, it becomes simultaneously exciting to imagine the many possibilities that lie ahead and stifling to think of the great amount of research that has yet to be done. Though lithography has been around for centuries and even many of the advanced techniques have been around for several decades, the nanolithographic applications of such techniques are fairly recent developments. X-ray, e-beam, imprint, etching, and scanning probe lithographies have all been used in their own ways to pattern surfaces intent on being tissue engineering scaffolds. Each method has its own advantages and disadvantages and yet each method is still growing and expanding by the day with every new advancements in materials and technology. Perhaps now removed from its less scientific origins, nanolithography paradoxically both drives and is driven by the ever present goal of someday artificially creating surfaces capable of supporting whole functioning tissues.

REFERENCES

1. Trawick M, Angelescu D, Chaikin P, et al. 2005. Block copolymer nanolithography. In *Nanolithography and Patterning Techniques in Microelectronics*. 1st ed. Bucknall DG (Ed.), 1–38. Cambridge, England: Woodhead Publishing Limited.
2. Senefelder A. 1968. *A Complete Course of Lithography*. 2nd ed. New York: Da Capo Press.
3. Wenniger MA. 1983. *Lithography: A Complete Guide*. 1st ed. Englewood Cliffs, New Jersey: Prentice-Hall.
4. Fouque V. 1973. *The Truth Concerning the Invention of Photography: Nicephore Niepce, His Life, Letters, and Works*. 2nd ed. New York: Arno Press.
5. Szoszkiewicz R, Okada T, Jones SC, et al. 2007. High-speed, sub-15 nm feature size thermochemical nanolithography. *Nano letters* 7(4):1064–1069.
6. Park M, Chaikin PM, Register RA, et al. 2001. Large area dense nanoscale patterning of arbitrary surfaces. *Applied Physics Letters* 79(2):257–259.
7. Wei, XL, Mo, ZH, Li, B, and Wei, JM. 2007. Disruption of HepG2 cell adhesion by gold nanoparticle and paclitaxel disclosed by in situ QCM measurement. *Colloids & Surfaces B: Biointerfaces* 59(1):100–104.
8. Zhang, H, Amro, NA, Disawal, S, et al. 2007. High-throughput dip-pen-nanolithography-based fabrication of Si nanostructures. *Small* 3(1): 81.
9. Nowak AP, Breedveld V, Pakstis L, et al. 2002. Rapidly recovering hydrogel scaffolds from self-assembling diblock copolypeptide amphiphiles. *Nature* 417(6887):424.
10. Hartgerink JD, Beniash E, and Stupp SI. 2001. Self-assembly and mineralization of peptide-amphiphile nanofibers. *Science* 294(5547):1684.
11. Kisiday J, Jin M, Kurz, B, et al. 2002. Self-assembling peptide hydrogel fosters chondrocyte extracellular matrix production and cell division: Implications for cartilage tissue repair. *Proceedings of the National Academy of Sciences of the United States of America* 99(15):9996.
12. Semino CE, Merok JR, Crane GG, et al. 2003. Functional differentiation of hepatocyte-like spheroid structures from putative liver progenitor cells in three-dimensional peptide scaffolds. *Differentiation* 71(4–5):262.
13. Holmes T, de Lacalle S, Su X, et al. 2000. Extensive neurite outgrowth and active synapse formation on self-assembling peptide scaffolds. *Proceedings of the National Academy of Sciences of the United States of America* 97(12):6728.
14. Colin JR. 2005. MIT soft-lithography method harnesses DNA. *Electronic Engineering Times* (1376): 46–46.
15. Patel, N, Bhandari, R, Shakesheff, KM, et al. 2000. Printing patterns of biospecifically-adsorbed protein. *Journal of Biomaterials Science. Polymer Edition* 11(3):319–331.
16. Kane RS, Takayama S, Ostuni E, et al. 1999. Patterning proteins and cells using soft lithography. *Biomaterials* 20:2363–2376.
17. Chen CS, Mrksich M, Huang S, et al. 1997. Geometric control of cell life and death. *Science* 276(5317):1425–1428.
18. Dike LE, Chen CS, Mrksich M, et al. 1999. Geometric control of switching between growth, apoptosis, and differentiation during angiogenesis using micropatterned substrates. *In Vitro Cellular and Developmental Biology. Animal* 35(8):441–448.
19. Bailly M, Yan L, Whitesides GM, et al. 1998. Regulation of protrusion shape and adhesion to the substratum during chemotactic responses of mammalian carcinoma cells. *Experimental Cell Research* 241(2):285–299.
20. Colin JR. 2005. MIT soft-lithography method harnesses DNA. *Electronic Engineering Times* (1376):46–46.
21. Hu W, Yim EKF, Reano RM, et al. 2005. Effects of nanoimprinted patterns in tissue-culture polystyrene on cell behavior. *Journal of Vacuum Science Technology. B, Microelectronics Processing and Phenomena* 23(6):2984.
22. Hoff DJ, Cheng L, Meyhofer E, et al. 2004. Nanoscale protein patterning by imprint lithography. *Nano Letters* 4(5):853–857.
23. Gaubert HE and Frey W. 2007. Highly parallel fabrication of nanopatterned surfaces with nanoscale orthogonal biofunctionalization imprint lithography. *Nanotechnology* 18(13):135101.

24. Fowles, GR. 1968. *Introduction to Modern Optics*. 1st ed. 105–144. New York: Holt, Rinehart and Winston, Inc.
25. Solak HH. 2006. Topical review: Nanolithography with coherent extreme ultraviolet light. *Journal of Physics. D, Applied Physics* 39(10):R171–R188.
26. Hohn FJ. 1994. Nanolithography, the integrated system. In *Nanolithography: A Borderland between STM, EB, IB, and X-Ray Lithographies*. 1st ed. Gentili M, Giovannella C, and Selci S (Eds.), pp. 1–11. Boston: Kluwer Academic Publishers.
27. Seunarine K, Tormen M, Gadegaard N, et al. 2006. Progress towards tubes with regular nanopatterned inner surfaces. *Journal of Vacuum Science Technology. B, Microelectronics Processing and Phenomena* 24(6):3258.
28. Gallagher JO, McGhee KF, Wilkinson CDW, et al. 2002. Interaction of animal cells with ordered nanotopography. *IEEE Transactions on Nanobioscience* 1(1):24.
29. Luther W. 2004. International strategy and foresight report on nanoscience and nanotechnology. 1–47.
30. Shao DB and Chen SC. 2005. Surface-plasmon-assisted nanoscale photolithography by polarized light. *Applied Physics Letters* 86(25):253107.
31. Hubert C, Rumyantseva A, Lerondel G, et al. 2005. Near-field photochemical imaging of noble metal nanostructures. *Nano Letters* 5(4):615.
32. Koenderink AF, Hernandez JV, Robicheaux F, et al. 2007. Programmable nanolithography with plasmon nanoparticle arrays. *Nano Letters* 7(3):745.
33. Xia Y and Whitesides GM. 1998. Soft lithography. *Annual Review of Materials Science* 28:153.
34. Chou SY, Krauss PR, Zhang W, et al. 1997. Sub-10 nm imprint lithography and applications. *The Journal of Vacuum Science and Technology B* 15(6):2897.
35. Chou SY. 2003. Nanoimprint lithography. In *Alternative Lithography: Unleashing the Potentials of Nanotechnology*. 1st ed. Sotomayor Torres, Clivia M (Eds.), pp. 15–23. New York: Kluwer Academic/Plenum Publishers.
36. Choi S, Yoo PJ, Baek SJ, et al. 2004. An ultraviolet-curable mold for sub-100-nm lithography. *Journal of the American Chemical Society* 126:7744–7745.
37. Chen CS, Jiang X, and Whitesides GM. 2005. Microengineering the environment of mammalian cells in culture. *MRS Bulletin* 30(March):194–201.
38. Yoo PJ, Suh KY, Kim YS, et al. 2005. Patterning of polymer thin films. In *Nanolithography and Patterning Techniques in Microelectronics*. 1st ed. Bucknall DG (Ed.), 155–183. Cambridge England: Woodhead Publishing Limited.
39. Kim D, Kim P, Suh K, et al. 2005. Modulation of adhesion and growth of cardiac myocytes by surface nanotopography. *Conf Proc IEEE Eng Med Biol Soc* 4:4091.
40. Pisignano D, Maruccio G, Mele E, et al. 2005. Polymer nanofibers by soft lithography. *Applied physics letters* 87(12):123109.
41. Yoshimoto H, Shin YM, Terai H, et al. 2003. A biodegradable nanofiber scaffold by electrospinning and its potential for bone tissue engineering. *Biomaterials* 24(12):2077–2082.
42. Hatzakis M. 1994. Electron beam resists and pattern transfer methods. In *Nanolithography: A Borderland between STM, EB, IB, and X-Ray Lithographies*. 1st ed. Gentili M, Giovannella C, and Selci S (Eds.), 13–23. Boston: Kluwer Academic Publishers.
43. Pan X and Broers AN. 1994. Direct writing of nanoscale patterns in SiO2. In *Nanolithography: A Borderland between STM, EB, IB, and X-Ray Lithographies*. 1st ed. Gentili M, Giovannella C, Selci S (Eds.), 45–51. Boston: Kluwer Academic Publishers.
44. Langheinrich W and Beneking H. 1994. Sub-10nm Electron beam lithography: -AIF3-doped lithium fluoride as a resist. In *Nanolithography: A Borderland between STM, EB, IB, and X-Ray Lithographies*. 1st ed. Gentili M, Giovannella C, Selci S (Eds.), 53–56. Boston: Kluwer Academic Publishers.
45. Dalby MJ, Gadegaard N, Riehle MO, et al. 2004. Investigating filopodia sensing using arrays of defined nano-pits down to 35 nm diameter in size. *The International Journal of Biochemistry Cell Biology* 36(10):2005.
46. Geyer W, Stadler V, Eck W, et al. 2001. Electron induced chemical nanolithography with self-assembled monolayers. *Journal of Vacuum Science Technology. B, Microelectronics Processing and Phenomena* 19(6):2732.
47. Yamato M, Konno C, Utsumi M, et al. 2002. Thermally responsive polymer-grafted surfaces facilitate patterned cell seeding and co-culture. *Biomaterials* 23(2):561.

48. Grime GW. 2005. Ion beam patterning. In *Nanolithography and Patterning Techniques in Microelectronics*. 1st ed. DG Bucknall (Ed.), 184–217. Cambridge England: Woodhead Publishing Limited.
49. van Kan JA, Bettiol AA, and Watt F. 2003. Three-dimensional nanolithography using proton beam writing. *Applied Physics Letters* 83(8):1629.
50. van Kan, Jeroen A., Bettiol AA, et al. 2006. Proton beam writing of three-dimensional nanostructures in hydrogen silsesquioxane. *Nano Letters* 6(3):579.
51. Mistry P, Gomez-Morilla I, Grime GW, et al. 2006. Proton beam lithography at the University of Surrey's Ion Beam Centre. *Nuclear Instruments Methods in Physics Research. Section B, Beam Interactions with Materials and Atoms* 242(1–2):387–389.
52. Gorelick S, Rahkila P, Sajavaara T, et al. 2007. Growth of osteoblasts on lithographically modified surfaces. *Nuclear Instruments Methods in Physics Research. Section B, Beam Interactions with Materials and Atoms* 260(1):130.
53. Tseng AA. 2004. Recent developments in micromilling using focused ion beam technology. *Journal of Micromechanics and Microengineering* 14:R15.
54. Arshak K, Mihov M, Arshak A, et al. 2004. Novel dry-developed focused ion beam lithography scheme for nanostructure applications. *Microelectronic Engineering* 73/74:144–151.
55. Youn SW, Takahashi M, Goto H, et al. 2006. Microstructuring of glassy carbon mold for glass embossing—Comparison of focused ion beam, nano/femtosecond-pulsed laser and mechanical machining. *Microelectronic Engineering* 83(11–12):2482.
56. Linford M, Davis R, Magleby S, et al. 2005. Chemomechanical surface modification of materials for patterning. In *Nanolithography and Patterning Techniques in Microelectronics*. 1st ed. Bucknall DG (Ed.), 120–154. Cambridge England: Woodhead Publishing Limited.
57. Jiang F, Khairy K, Poole K, et al. 2004. Creating nanoscopic collagen matrices using atomic force microscopy. *Microscopy Research and Technique* 64(5–6):435.
58. Wilson DL, Martin R, Hong S, et al. 2001. Surface organization and nanopatterning of collagen by dip-pen nanolithography. *Proceedings of the National Academy of Sciences of the United States of America* 98(24):13660.
59. Riemenschneider L, Blank S, and Radmacher M. 2005. Enzyme-assisted nanolithography. *Nano letters* 5(9):1643.
60. Zhang H, Amro NA, Disawal S, et al. 2007. High-throughput dip-pen-nanolithography-based fabrication of Si nanostructures. *Small* 3(1):81.
61. Vettiger P, Brugger J, Despont M, et al. 1999. Ultrahigh density, high-data-rate NEMS-based AFM data storage system. *Microelectronic Engineering* 46(1–4):11.
62. Vettiger P, Despont M, Drechsler U, et al. 2000. Millipede' more than one thousand tips for future AFM data storage. *IBM Journal of Research and Development* 44(3):323.
63. Torres, Clivia M. Sotomayor. 2003. Alternative lithography: An introduction. In *Alternative Lithography: Unleashing the Potentials of Nanotechnology*. 1st ed. Sotomayor T, Clivia M (Eds.), 1–14. New York: Kluwer Academic/Plenum Publishers.
64. Legget GJ. 2005. Photolithography beyond the diffraction limit. In *Nanolithography and Patterning Techniques in Microelectronics*. 1st ed. Bucknall DG (Ed.), 238–266. Cambridge England: Woodhead Publishing Limited.
65. Thapa A, Miller DC, Webster TJ, et al. 2003. Nano-structured polymers enhance bladder smooth muscle cell function. *Biomaterials* 24(17):2915.
66. Teixeira AI, Abrams GA, Murphy CJ, et al. 2003. Cell behavior on lithographically defined nanostructured substrates. *Journal of Vacuum Science Technology. B, Microelectronics Processing and Phenomena* 21(2):683.
67. Wang C, More SD, Wang Z, et al. 2003. Patterning SiO thin films using synchrotron radiation stimulated etching with a Co contact mask. *Journal of Vacuum Science Technology. B, Microelectronics Processing and Phenomena* 21(2):818.
68. Sheller NB, Petrash S, and Foster MD. 1998. Atomic force microscopy and X-ray reflectivity studies of albumin adsorbed onto self-assembled monolayers of hexadecyltrichlorosilane. *Langmuir* 14 (16):4535.
69. Brugger J and Kim G. 2005. Nanofabrication by shadow deposition through nanostencils. In *Nanolithography and Patterning Techniques in Microelectronics*. 1st ed. Bucknall DG (Ed.), 218–237. Cambridge England: Woodhead Publishing Limited.

70. Pallandre A, Glinel K, Jonas AM, et al. 2004. Binary nanopatterned surfaces prepared from silane monolayers. *Nano Letters* 4(2):365–371.
71. Pallandre A, De Meersman B, Blondeau F, et al. 2005. Tuning the orientation of an antigen by adsorption onto nanostriped templates. *Journal of the American Chemical Society* 127:4320–4325.
72. Norman JJ and Desai TA. 2006. Methods for fabrication of nanoscale topography for tissue engineering scaffolds. *Annals of Biomedical Engineering* 34(1):89.

11 Designer Self-Assembling Peptide Scaffolds for Tissue Engineering and Regenerative Medicine

Akihiro Horii, Xiumei Wang, and Shuguang Zhang

CONTENTS

11.1 INTRODUCTION

Advancement of biology often requires development of new materials, methods, and tools that can in turn significantly accelerate scientific discoveries. The introduction of the petri dish over 100 years ago provided an indispensable tool for culturing cells in vitro, thus permitting detailed dissection of seemingly intractable biological and physiological systems into manageable units and well-defined studies. This simple dish has had profound impact on our understanding of complex biology, especially cell biology and neurobiology.

In this manner, regenerative medicine and tissue engineering require two complementary key ingredients. One is the biologically compatible scaffold that can be readily adopted by the body system without harm, and the other are suitable cells including various stem cells or primary cells that effectively replace the damaged tissues without adverse consequences. However, it would be advantageous if one could apply suitable and active biological scaffolds to stimulate and promote cell differentiation, in addition to regenerating tissues without introducing foreign cells.

The field of regenerative medicine is undergoing a rapid growth and it is impossible to cover it comprehensively in a few pages; thus in this review, we only focus on work concerning synthetic designer self-assembling peptide scaffolds developed in our laboratory and our colleagues since 1992. Readers interested in advances in biomaterials can consult other chapters in this book and broadly the literature.

11.2 SCAFFOLDS FOR TISSUE ENGINEERING

What are biomaterials? The definition varies. In our view, there is a distinction between biomaterials and biological materials. Biomaterials refer to materials that are used in medical applications in last few decades including several synthetic polymers, PLLA, i.e., poly-(D,L-lactide), PLGA, i.e., poly-(lactic-*co*-glycolic acid). On the other hand, biological materials refer to materials of truly biological origin including alginate, cellulose, lipid materials, chitin and peptide, and protein-based materials, for example, collagen, silk, spider silk, and bioadhesives.

The polymer biomaterial culture systems have significantly advanced our understanding of cell–material interactions and fostered a new field of tissue engineering [1]. However, high-porosity scaffolds comprising these biomaterials are often made of microfibers with diameter of ~5–50 μm or micropores of 10–50 μm. Since the size of most cells (~5–20 μm) are similar or smaller to these microstructures (~10–100 μm), upon attachment the cells still exhibit a 2-D topography with a curvature depending respectively on the microfiber diameters or on the pore size. To culture cells in a truly 3-D microenvironment, these dimensions must be significantly smaller than cells so that cells should be fully surrounded by the scaffolds, much like the extracellular environment.

Polymer biomaterials are often functionalized with an RGD peptide motif or other motifs to promote desired biological activities through chemical reactions or coating. Because of their micro-scale sizes, their mechanical strength usually prevents further material structural adaptations from the forces exerted by the cytoskeleton of cells during their adhesion, migration, and maturation processes. Thus, although these microfibers provide an artificial extracellular environment, they are still far from the natural nanoscale extracellular matrix (ECM). In the recent years a well-established technique, electrospinning, has been adopted to spin nanofibers of these same materials [2]; however, the harshness of the overall process and the presence of harmful chemical solvent prevent any possible addition of cells while the scaffold is forming. Attempts have been made to seed scaffold in dynamic conditions or just on their surfaces, but even if a considerable cell migration occurred a uniform and correct seeding of cells in real 3-D matrix is still an hurdle that has to be overcome.

For the encapsulation of labile bioactive substances and living cells, physically cross-linked nanofiber scaffolds are of great interest, especially if the scaffold formation occurs under mild physiological conditions without any organic solvent processes. There is a need for process scaffolds in aqueous environment with a desired pH range, temperature, or specific catalysts that could control their microstructure properties.

11.3 IDEAL BIOLOGICAL SCAFFOLD

There are a number of strategies to fabricate biological materials. However, the ideal biological scaffold should meet several requirements. (1) The building blocks should be derived from biological sources, (2) basic units should be amenable to design and modification to achieve specific needs, (3) exhibit a controlled rate of material biodegradation, (4) exhibit no cytotoxicity, (5) promote cell-substrate interactions, (6) elicit none or little immune responses nor cause inflammation, (7) afford economically scaleable and reproducible material production, purification, and processing, (8) be readily transportable, (9) be chemically compatible with aqueous solutions and physiological conditions, (10) and integrate in the body without harm.

11.4 SELF-ASSEMBLING PEPTIDE SCAFFOLDS

The self-assembling peptide scaffold belongs to a class of biologically inspired materials. The first member of the family, EAK16-II (AEAEAKAKAEAEAKAK), was discovered from a segment in a yeast protein, Zuotin [3]. The scaffolds consist of alternating amino acids that contain 50% charged residues [3,4]. These peptides are characterized by their periodic repeats of alternating ionic hydrophilic and hydrophobic amino acids that spontaneously form β-sheet structures. These β-sheets have distinct polar and nonpolar surfaces. They are isobuoyant in aqueous solution and readily transportable to different environments. Upon exposure to aqueous solutions with neutral pH, ions screen the charged peptide residues and alanines (forming the nonpolar surfaces of β-sheets) of different β-sheets pack together, thanks to their hydrophobic interactions in water, thus giving double-layered β-sheets nanofibers, a structure that is found in silk fibroin from silkworm and spiders. Thus the final self-assembly step creating the peptide scaffold takes place under physiological conditions. Individual fibers are ~10 nm in diameter. On the charged sides, both positive and negative charges are packed together through intermolecular ionic interactions in a checkerboard-like manner. In general, these self-assembling peptides form stable β-sheet structures in water, which are stable across a broad range of temperature, wide pH ranges in high concentration of denaturing agent urea and guanidium hydrochloride. The nanofiber density correlates with the concentration of peptide solution and the nanofibers retain extremely high hydration, >99% in water (5–10 mg/mL, w/v). A number of additional self-assembling peptides including RADA16-I (AcN-RADARADARADARADA-CNH2) and RADA16-II (AcN-RARADADARARADADA-CNH2), in which arginine and aspartate residues substitute lysine and glutamate, have been designed and characterized for salt-facilitated nanofiber scaffold formation.

Many self-assembling peptides that form scaffolds have been reported and the numbers are still expanding [5,6]. The formation of the scaffold and its mechanical properties are influenced by several factors, two of which are the level of hydrophobicity [7–12] and length of peptide sequence. That is, in addition to the ionic complementary interactions, the extent of the hydrophobic residues, Ala, Val, Ile, Leu, Tyr, Phe, Trp (or single letter code, A, V, I, L, Y, P, W), and consequently the number of repeats of the self-assembling motif, can significantly influence the mechanical properties of the scaffolds and the speed of their self-assembly. The higher the content of hydrophobicity and the longer the length of the peptide sequence, the easier it is for scaffold formation and the better for their mechanical properties [7,8,11,13].

11.5 IN VITRO TISSUE CULTURES

These new self-assembling peptide biological scaffolds have become increasingly important not only in studying spatial behaviors of cells, but also in developing approaches for a wide range of innovative medical technologies including regenerative medicine. One example is in the use of the peptide scaffolds to support neurite growth and maturation [14], neural stem cell differentiation, cardiac myocytes, bone, and cartilage cell cultures. The peptide scaffolds from RADA16-I and RADA16-II form nanofiber scaffold in physiological solutions that stimulated extensive rat neurite outgrowth and active synapses formation on the peptide surface was successfully achieved [14].

Navarro-Alvarez et al. [15] showed that nanofiber scaffold consists from RADA16-I can support isolated porcine hepatocytes culture and three-dimensional spheroidal formation for 2 weeks. This indicates that the hepatocytes maintained cell differentiation status. On the other hand, hepatocytes in collagen type I showed a spread shape which indicates dedifferentiation. Ammonia and drug-metabolizing capacities and albumin-producing abilities were maintained for 2 weeks in the hepatocytes cultured in RADA16-I scaffold, whereas there was a significant lose of those abilities in the hepatocytes cultured in collagen. This indicates that the self-assembling peptide scaffold can help maintaining cell function for longer term culture even better than natural derived matrix in some cell types.

A method to encapsulate chondrocytes within peptide scaffolds was developed using another self-assembling peptide KLD12 (AcN-KLDLKLDLKLDL-CNH2) for cartilage repair purposes [10]. During 4 weeks of culture in vitro, chondrocytes seeded within the peptide scaffold developed a cartilage-like ECM rich in proteoglycans and type II collagen indicative of a stable chondrocyte phenotype. Time dependent accumulation of this ECM was paralleled by increases in material stiffness indicative of deposition of mechanically functional tissue. The content of viable differentiated chondrocytes within the peptide scaffold increased at a rate that was 4-fold higher than that in parallel chondrocyte-seeded agarose culture, a well-defined reference chondrocyte culture system. These results demonstrate the potential of a self-assembling peptide scaffold as a scaffold for the synthesis and accumulation of a true cartilage-like ECM in a 3-D cell culture for cartilage tissue repair. These results demonstrated peptide scaffolds are feasible for additional explorations with other tissue types.

11.6 SELF-ASSEMBLING PEPTIDE SCAFFOLDS FOR REGENERATIVE MEDICINE

Misawa et al. [16] showed that self-assembling peptide nanofiber scaffold developed from RADA-I promoted bone regeneration. The self-assembling peptide scaffold was injected into bone defects of mice calvaria. Saline and Matrigel were injected as controls. After 4 weeks, x-ray radiograph and histological findings showed that there were stronger bone regenerations in peptide scaffold compared to the controls. Especially, the strength of the regenerated bone was >1.7-fold higher for RADA-I peptide scaffold than for Matrigel.

Ellis-Behnke and colleagues showed that self-assembling peptide material is a promising scaffold for neural regeneration medicine [17]. In vivo application to brain wounds was carried out using postnatal day-2 Syrian hamster pups. The optic tract within the superior colliculus (SC) was completely severed with a deep knife wound, extending at least 1mm below the surface. At surgery, 10 animals were treated by injection into the wound of 10–30 μL of 1% RADA16, 99% water, w/v. Control animals with the same brain lesion included 3 with isotonic saline injection (10 μL), and numerous additional cases, including 10 in which the dye Congo red was added into the peptide scaffold, and 27 earlier animals with knife cuts and no injection surviving 6–9 days. Animals were sacrificed at 1, 3, 6, 30, and 60 days for brain examinations. Histological specimen examinations revealed that only in the peptide scaffold-injected animals, but not in untreated animals, the brain tissue appears to have reconnected itself together in all survival times. Additionally, axons labeled from their retinal origin with a tracer molecule were found to have grown beyond the tissue bridge, reinnervating the SC caudal to the lesion. Most importantly, functional tests proved a significant restoration of visual function in all peptide scaffold treated animals.

In another work published by Richard Lee's group embryonic stem cell was suspended in RADA16-II peptide scaffold solutions and injected in the myocardium of 10 weeks old mice [18]. In that study it has been demonstrated that self-assembling peptides can be injected into the myocardium to create 3-D microenvironment. After 7, 14, and 28 days these microenvironments recruit both endogenous endothelial and smooth muscle cells, and exogenously injected cells survive in the microenvironments: self-assembling peptides can thus create injectable microenvironments that promote vascularization.

In addition Lee's group also developed an appealing drug delivery strategy by using a biotinylated version of RADA-II to demonstrate a slow release of IGF-1 in infarctuated rat myocardia [19]. The biotin sandwich strategy allowed binding of IGF-1 and did not prevent self-assembly of the peptides into nanofibers within the myocardium. In conjunction with cardiomyocytes transplantation the strategy showed that cell therapy with IGF-1 delivery by biotinylated nanofibers significantly improved systolic function after experimental myocardial infarction.

Remarkably, since the building blocks of this class of designer peptide scaffolds are made of pure natural L-amino acids, RADA16, unlike most of the other synthetic microfibers, has been shown not to elicit detectable immune response, nor inflammatory reactions in animals [16–19], and the degraded natural amino acid products can be reused by the body. Therefore, this class of scaffold may be useful as a bio-reabsorbable scaffold for tissue repair and neuroengineering to alleviate and treat a number of trauma and neuro-degeneration diseases as well other tissue injures, damage, and aging.

Alternative strategies are under evaluation to address the directionality and alignment of these scaffolds, by using microfluidic or magnetic approaches to drive the self-assembling process, since the nanofibrous scaffolds do not have a predetermined porosity, pore orientation, or a predetermined 3-D oriented architecture.

11.7 DESIGNER SELF-ASSEMBLING PEPTIDE SCAFFOLDS

Although self-assembling peptides are promising scaffolds, they show no specific cell interaction because their sequences are not naturally found in living systems. The next logical step is to search active and functional peptide motifs from a wealth of cell biology literature, thus the second generation of designer scaffolds will significantly enhance their interactions with cells and tissues [20–22].

The simplest method to incorporate the functional motifs is to directly synthesize it by extending the motifs on to the self-assembling peptides themselves (Figure 11.1) [22,23]. The functional motifs are on the C-termini since peptide synthesis start from C-termini to avoid deletion of the functional motifs during synthesis. Usually a spacer comprising two glycines residues is added to guarantee a flexible and correct exposure of the motifs to cell surface receptors. Different functional motifs in various ratios can be incorporated in the same scaffold. Upon exposure to solution with neutral pH the functionalized sequences self-assemble leaving the added motifs flagging on both sides of each nanofiber. Nanofibers take part to the overall scaffold thus giving microenvironments functionalized with specific biological stimuli (Figure 11.1).

The incorporation of functionalized peptide into the nanofiber was indicated by AFM [23]. 1% (w/v) peptide solution of RADA16-I, the functionalized peptide of RADA16-I with two-unit RGD binding sequence PRG (PRGDSGYRGDS) were examined. The mixed peptide solution of RADA16-I and the PRG functionalized peptide at a ratio of 1:1 was also examined using AFM tapping mode. Figure 11.2 shows AFM images of the peptide solutions. The nanofibers in aqueous solutions were observed in RADA16-I and all RADA16-I mixed solutions, although no fiber formation was observed in solely PRG functionalized peptide solution. These results were confirmed by visual inspection of increasing viscosity of the peptide scaffold solutions. We observed an increase in the fiber thickness in the mixed solution (29.5 ± 3.1 nm) from RADA16-I solution (16.3 ± 1.4 nm). The width of the peptide fiber thickness was correlated with the model shown in Figure 11.1. These results imply that PRG functionalized peptides were integrated in nanofiber when it is mixed.

The self-assembling peptide scaffolds with functional motifs can be commercially produced with a reasonable cost. Thus, this method can be readily adopted for wide spread uses including study of cell interaction with their local- and micro-environments; cell migrations in 3D, tumor, and

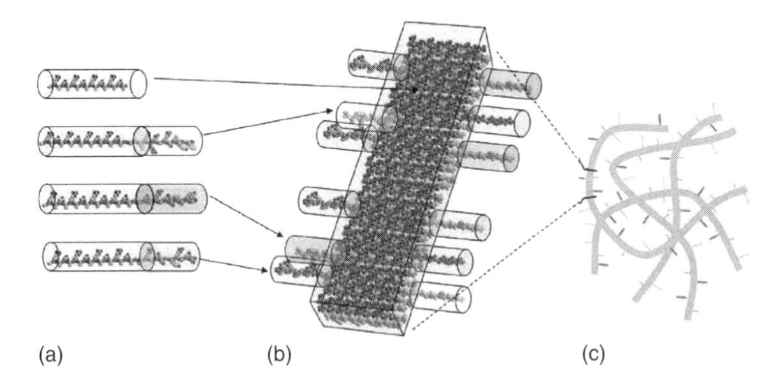

FIGURE 11.1 (See color insert following page 206.) Schematic illustration of the designer self-assembling peptide scaffold. (a) Direct extension of the self-assembling peptide sequence by adding different functional motifs. Longer cylinders on the left side represent the self-assembling backbone and the shorter cylinders on the right side represent various functional peptide motifs. (b) Molecular model of a self-assembling peptide nanofiber with functional motifs flagging from both sides of the double β-sheet nanofibers. Either mono or multiple functional (or labeled) peptide motifs can be mixed at the same time. The density of these motifs can be easily adjusted by simply mixing them in various ratios, 1:1–1,000,000 or more before the assembling step. (c) They then will be part of the self-assembled scaffold.

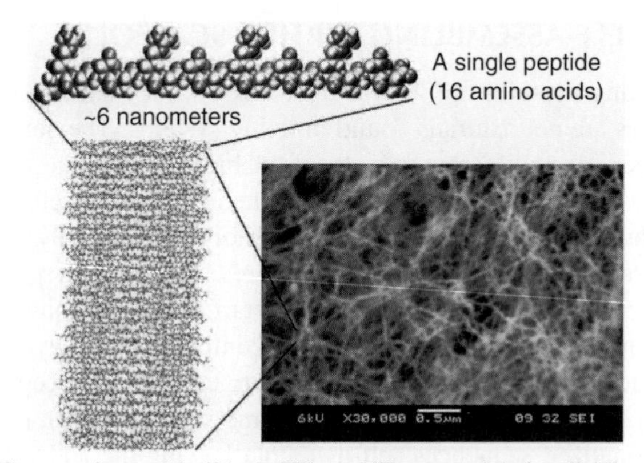

FIGURE 11.2 Designer self-assembling peptide nanofiber scaffold. A single peptide, ~6 nm, is shown. Thousands of peptides self-assemble to form a single nanofiber, trillions of peptides or billions of nanofibers form the scaffold, which contain ~99.5% water and 0.5% peptide materials.

cancer cells interaction with the normal cells; cell process and neurite extensions; cell-based drug test assays; and other diverse applications.

We have produced different designer peptides from a variety of functional motifs with different lengths [22,23]. We showed that the addition of motifs (up to 12 additional residues) to the self-assembling peptide RADA16-I did not inhibit self-assembling properties and nanofiber formations. Although their nanofiber structures appear to be indistinguishable from the RADA16-I scaffold (Figure 11.3), the appended functional motifs significantly influenced cell behaviors [22–26].

Using the designer self-assembling peptide nanofiber system, every ingredient of the scaffold can be defined and combined with various functionalities including the soluble factors. This is in sharp contrast with a 2-D petri dish where cells attach and spread only on the surface; cells reside in a 3-D environment where the extracellular matrix receptors on the cell membranes can bind to the functional ligands appended to the peptide scaffolds (Figure 11.4). It is possible that higher tissue architectures with multiple cell types, rather than monolayers, could be constructed using these designer 3-D self-assembling peptide nanofiber scaffolds.

Even if only a fraction of functionalized motifs on the 3-D scaffold are available for cell receptor binding, cells likely receive more external stimuli than when in contact with coated 2-D petri dishes

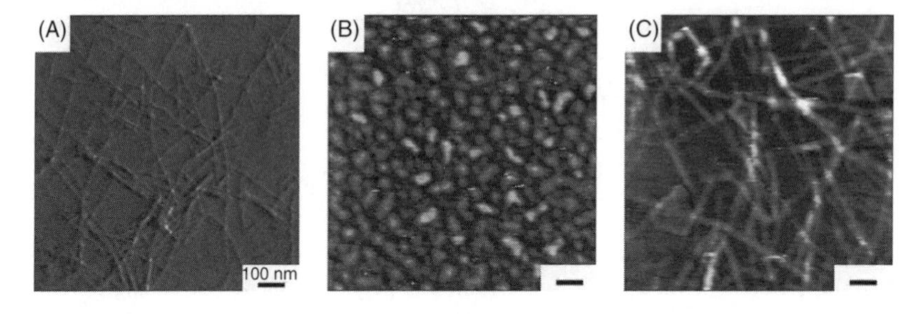

FIGURE 11.3 Tapping mode AFM images of 1% w/v peptides solution of (A) RADA16-I, (B) PRG functionalized peptide, (C) PRG functionalized peptide + RADA16-I (1:1). The bar represents 100 nm. The nanofiber formation is seen in (A) RADA16-I and (C) PRG + RADA16-I (1:1). An increase in the fiber thickness in (C) PRG + RADA16-I (1:1) (29.5 ± 3.1 nm) from (A) RADA16-I (16.3 ± 1.4 nm) was observed, which correlated to the width of the peptide fiber modeled in Figure 11.1.

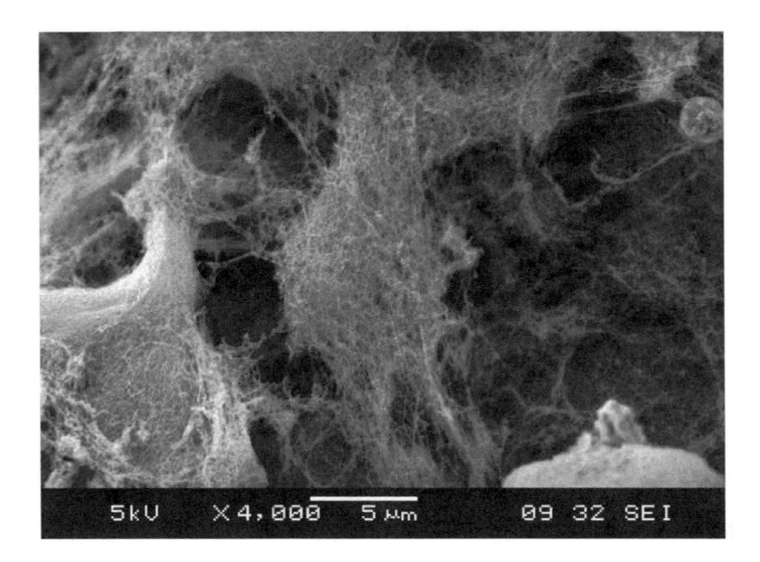

FIGURE 11.4 SEM image of a cell embedded in a 3-D designer self-assembling peptide nanofiber scaffold. The individual nanofiber scaffold completely wraps the cell body and membrane, thus bringing the chosen functional motifs all over the cell membrane.

or RGD- or other motif-coated polymer fibers in micron scale, which is substantially larger than the cell surface receptors and in most cases, larger than the cell themselves. There cells are not in real 3-D, rather, they are in 2-D wrapping around the polymers with a curvature depending on the diameter of the polymers. In a 2-D environment, where only one side of the cell body is in direct contact with the surface, receptor clustering at the attachment site may be induced; on the other hand, the receptors for growth factors, cytokines, nutrients, and signals are on the other sides that expose directly with the culture media. Thus cells may become partially polarized. In the 3-D environment, the functional motifs on the nanofiber scaffold surround the whole cell body in all dimensions and the factors may form a gradient in 3-D nanoporous microenvironment.

11.7.1 Designer Peptide Scaffolds for Cell Differentiation and Migration

The designer self-assembling peptide nanofiber scaffolds has been shown to be an excellent biological material for 3-D cell cultures and capable to stimulate cell migration into the scaffold as well for repairing tissue defects in animals by adding specific cell binding or chemotactic sequences. We developed several peptide nanofiber scaffolds designed specifically for osteoblasts [23]. We designed one of the pure self-assembling peptide scaffolds RADA16-I through direct coupling to short biologically active motifs. The motifs included osteogenic growth peptide ALK (ALKRQGRTLYGF) bone cell-secreted signal peptide, osteopontin cell adhesion motif DGR (DGRGDSVAYG), and two-unit RGD binding sequence PRG (PRGDSGYRGDS). We made the new peptide scaffolds by mixing the pure RAD16 and designer peptide solutions. Compared to pure RAD16 scaffold, we found that these designer peptide scaffolds significantly promoted mouse pre-osteoblast MC3T3-E1 cell proliferation. Moreover, alkaline phosphatase (ALP) activity and osteocalcin secretion, which are early and late markers for osteoblastic differentiation, were also significantly increased. The maker results were confirmed by ALP staining (Figure 11.5). We demonstrated that the designer, self-assembling peptide scaffolds promoted the proliferation and osteogenic differentiation of MC3T3-E1. Under the identical culture medium condition confocal images unequivocally demonstrated that the designer PRG peptide scaffold stimulated cell migration into the 3-D scaffold (Figure 11.6) [23]. Our results suggest that these designer peptide scaffolds may be very useful for promoting bone tissue regeneration.

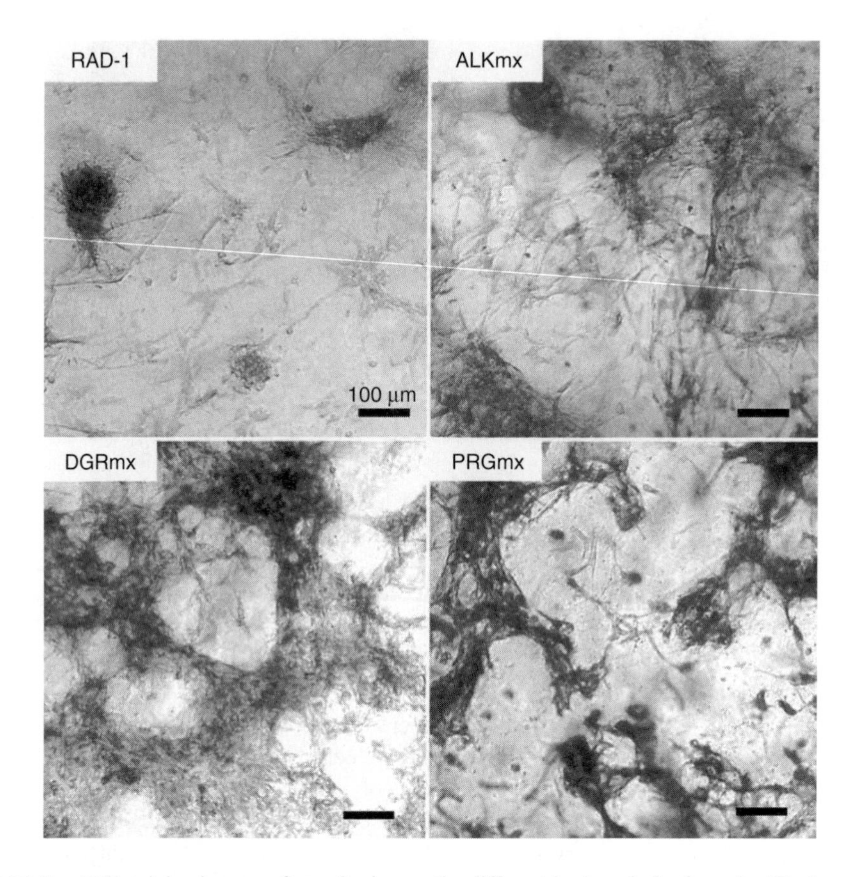

FIGURE 11.5 ALP staining images after culturing on the different hydrogels for 2 weeks. The bar represents 100 μm. RAD-I, RADA16-I 1% (w/v); ALKmx, ALK functionalized peptide 1% (w/v) + RADA16-I; DGRmx, DGR functionalized peptide 1% (w/v) + RAD; PRGmx, PRG functionalized peptide 1% (w/v) + RADA16-I (all mixture ratio is 1:1). The darkness correlates with the high ALP activity. RADA16-I shows low cell adhesion to the hydrogel and the cells are aggregated. The cell attachment increase of DGR and PRG was considered to be caused by RGD cell attachment sequence. ALK, DGR, and PRG showed higher ALP activities compared to RADA16-I, especially staining intensity of PRG.

In addition to laminin-derived self-assembling peptides previously studied by adding a long alky chain that promotes self-assembly [24], another study evaluates common fibronectin and collagens derived sequences as well [25].

11.8 WHY DESIGNER SELF-ASSEMBLING PEPTIDE SCAFFOLDS?

Why one should choose designer self-assembling peptide scaffolds while there are a large number of biomaterials on the market and some have already been approved by FDA? The advantage of using the designer peptide nanofiber scaffolds is several folds. (1) One can readily modify the designer peptides at the single amino acid level at will, inexpensively and quickly. This level of modification is impossible with Matrigel and other polymer scaffolds. (2) Unlike Matrigel, which contains unknown ingredients and quality that varies from batch to batch, the designer self-assembling peptide scaffolds belong to a class of synthetic biological scaffolds that contain pure components and every ingredient is completely defined. (3) Because these designer peptide scaffolds are pure with known motifs, it can be used to study controlled gene expression or cell signaling process. Thus these new designer nanofiber scaffolds proved to be promising tools to study cell signal pathways in a selective way not possible with any substrates including Matrigel and collagen gels

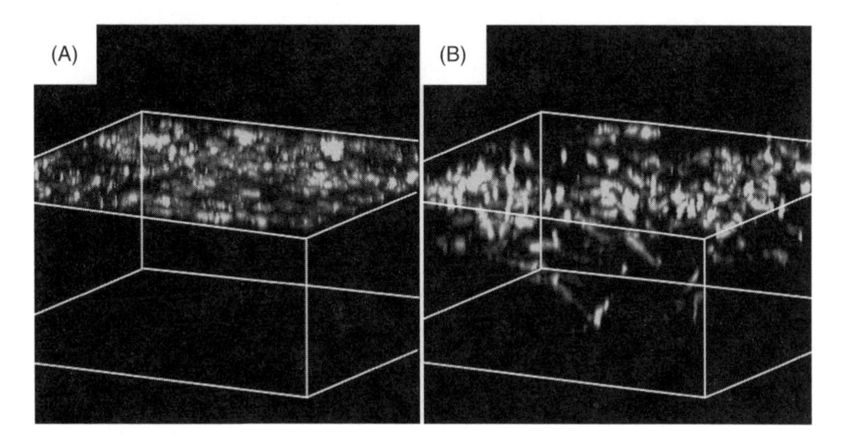

FIGURE 11.6 Reconstructed image of 3-D confocal microscope image of culturing on the different scaffolds consisting of different mix ratio of RADA16 1% (w/v) and PRG 1% (w/v) using calcein-AM staining. The bar represents 100 μm. (A) PRG 10% and (B) PRG 70%. In the case of 10% PRG scaffold, the cells were attached on the surface of the scaffold whereas the cells were migrated into the scaffold in the case of 70% PRG scaffold. There is a drastic cell migration into the scaffold with higher concentration of PRG motif.

that result in confusing cell signaling activation. (4) The initiation of the self-assembly process is through change of ionic strength at the physiological conditions without temperature influence. This is again unlike collagen gels, for which the gelation is through change of temperature that can sometimes induce unknown biological process including cold or heat shocks. (5) These scaffolds provide the opportunity to incorporate a number of different functional motifs and their combinations to study cell behavior in a well-defined ECM-analog microenvironment, not only without any chemical cross-link reactions but also fully bio-reabsorbable scaffolds.

The development of new biological materials, particularly those biologically inspired nanoscale scaffolds mimicking in vivo environment that serve as permissive substrates for cell growth, differentiation, and biological function is a most actively pursuit area which in turn could significantly advance regenerative medicine. These materials will be useful not only for further our understanding of cell biology in 3-D environment but also for advancing medical technology, tissue engineering, regenerative biology, and medicine.

REFERENCES

1. Lanza, R., Langer, R., and Vacanti, J. 2000. *Principles of Tissue Engineering*. Academic Press, San Diego.
2. Ashammakhi, N., Ndreu, A., Piras, A. et al. 2006. Biodegradable nanomats produced by electrospinning: Expanding multifunctionality and potential for tissue engineering. *J. Nanosci. Nanotechnol.* 6:2693–2711.
3. Zhang, S., Holmes, T., Lockshin, C., and Rich, A. 1993. Spontaneous assembly of a self-complementary oligopeptide to form a stable macroscopic membrane. *Proc. Natl. Acad. Sci. USA* 90:3334–3338.
4. Zhang, S., Holmes, T., DiPersio, M. et al. 1995. Self-complementary oligopeptide matrices support mammalian-cell attachment. *Biomaterials* 16:1385–1393.
5. Zhang, S. 2002. Emerging biological materials through molecular self-assembly. *Biotechnol. Adv.* 20:321–339.
6. Zhang, S. 2003. Fabrication of novel biomaterials through molecular self-assembly. *Nat. Biotechnol.* 21:1171–1178.
7. Caplan, M., Moore, P., Zhang, S., Kamm, R., and Lauffenburger, D. 2000. Self-assembly of a beta-sheet protein governed by relief of electrostatic repulsion relative to van der Waals attraction. *Biomacromolecules* 1:627–631.
8. Caplan, M., Schwartzfarb, E., Zhang, S., Kamm, R., and Lauffenburger, D. 2002. Control of self-assembling oligopeptide matrix formation through systematic variation of amino acid sequence. *Biomaterials* 23:219–227.

9. Marini, D., Hwang, W., Lauffenburger, D. et al. 2002. Left-handed helical ribbon intermediates in the self-assembly of a beta-sheet peptide. *Nano Lett.* 2:295–299.
10. Kisiday, J., Jin, M., Kurz, B. et al. 2002. Self-assembling peptide hydrogel fosters chondrocyte extracellular matrix production and cell division: Implications for cartilage tissue repair. *Proc. Natl. Acad. Sci. USA* 99:9996–10001.
11. Hwang, W., Marini, D., Kamm, R., and Zhang, S. 2003. Supramolecular structure of helical ribbons self-assembled from a beta-sheet peptide. *J. Chem. Physics* 118:389–397.
12. Yokoi, H., Kinoshita, T., and Zhang, S. 2005. Dynamic reassembly of peptide RADA16 nanofiber scaffold. *Proc. Natl. Acad. Sci. USA* 102:8414–8419.
13. Leon, E.J., Verma, N., Zhang, S., Lauffenburger, D., and Kamm, R. 1998. Mechanical properties of a self-assembling oligopeptide matrix. *J. Biomater. Sci., Polym. Ed.* 9:297–312.
14. Holmes, T.C., De Lacalle, S., Su, X. et al. 2000. Extensive neurite outgrowth and active synapse formation on self-assembling peptide scaffolds. *Proc. Natl. Acad. Sci. USA* 97:6728–6733.
15. Navarro-Alvarez, N., Soto-Gutierrez, A., Rivas-Carrio, J.D. et al. 2006. Self-assembling peptide nanofiber as a novel culture system for isolated porcine hepatocytes. *Cell Transplant.* 15:921–927.
16. Misawa, H. Kobayashi, N. Soto-Gutierrez, A. et al. 2006. Puramatrix facilitates bone regeneration in bone defects of calvaria in mice. *Cell Transplant.* 15:903–910.
17. Ellis-Behnke, R.G., Liang, Y.X., You, S.W. et al. 2006. Nano neuro knitting: Peptide nanofiber scaffold for brain repair and axon regeneration with functional return of vision. *Proc. Natl. Acad. Sci. USA* 103:5054–5059.
18. Davis, M.E., Motion, J.P., Narmoneva, D.A. et al. 2005. Injectable self-assembling peptide nanofibers create intramyocardial microenvironments for endothelial cells. *Circulation* 111:442–450.
19. Davis, M.E., Hsieh, P.C., Takahashi, T. et al. 2006. Local myocardial insulin-like growth factor 1 (IGF-1) delivery with biotinylated peptide nanofibers improves cell therapy for myocardial infarction. *Proc. Natl. Acad. Sci. USA* 103:8155–8160.
20. Ayad, S., Boot-Handford, R.P., Humphreise, M.J., Kadler, K.E., and Shuttleworth, C.A. 1998. *The Extracellular Matrix: Facts Book.* Academic Press, San Diego.
21. Kreis, T. and Vale, R. 1999. *Guide Book to the Extracellular Matrix, Anchor, and Adhesion Proteins.* Oxford: Oxford University Press.
22. Ricard-Blum, S., Dublet, B., and Van Der Rest, M. 2000. *Unconventional Collagens: Types VI, VII, VIII, IX, X, XIV, XVI & XIX.* Oxford: Oxford University Press.
23. Gelain, F., Bottai, D., Vescovi, A., and Zhang, S. 2006. Designer self-assembling peptide nanofiber scaffolds for adult mouse neural stem cell 3-dimensional cultures. *PLoS ONE* 1:e119.
24. Horii, A., Wang, X., Gelain, F., and Zhang, S. 2007. Biological designer self-assembling peptide nanofiber scaffolds significantly enhance osteoblast proliferation, differentiation and 3-D migration. *PLoS ONE* 2:e190.
25. Silva, G.A., Czeisler, C., Niece, K.L. et al. 2004. Selective differentiation of neural progenitor cells by high-epitope density nanofibers. *Science* 303:1352–1355.
26. Nowakowski, G.S., Dooner, M.S., Valinski, H.M. et al. 2004. A specific heptapeptide from a phage display peptide library homes to bone marrow and binds to primitive hematopoietic stem cells. *Stem Cells* 22:1030–1038.

Section IV

Applications of Nanostructured Scaffolds in Biology and Medicine

12 Nanotechnologies in Neural Tissue Engineering

Vivek Mukhatyar, Julie Yeh, and Ravi Bellamkonda

CONTENTS

12.1 INTRODUCTION

Biological systems are created from hierarchical structures assembled from nanoscale building blocks. These nanoscale components are assembled to form macroscale structures with specific shape and function. Nanotechnology provides a new tool to fabricate, measure, probe, and interface with these biological systems at the atomic and molecular levels, typically ranging from 1 to 100 nm. This capacity affords precise control over a number of physical and functional properties that are unique to specific applications [1], including mechanical, physical, chemical, and electrical properties of materials. Potentially, materials created using nanotechnology have size scales and properties engineered to interact in controlled ways with their biological counterparts, ultimately leading to enhanced biocompatibility and functionality [2].

In this chapter, we review ongoing efforts in nanotechnology and its impact on the field of neural tissue engineering. Motivation for applying nanotechnology to neural tissue engineering comes from the many unsolved problems in the field of neuroscience and neuropathology. One obvious implication of small scales stems from the fact that neural tissue is often located in confined spaces (vertebral column and cranial activity) where space is a premium and any implant can cause complications by exerting pressure. Molecular specificity is also important in interacting and modulating the nervous system, as multiple types of cells with different and specific functions are

TABLE 12.1
Impact of Nanotechnology of Neural Tissue Engineering

Nanotechnologies	Application
Nanoscale Tools to Probe the Nervous System	
Fabrication of nanoscale topographies and micro/nanoscale molecular patterns	Evaluate axonal response to spacial and temporal cues
Nanolithography to fabricate ECM structures from natural and synthetic materials	Study neuronal cell response to ECM-like structures
Quantum dots	Image signaling pathways and target cellular function; peptide conjugation to deliver peptides to molecular targets
Nanoscale Tools to Guide Axons	
Self-assembling nanofibers	Incorporate epitopes to encourage neurite growth
Electrospinning nanofibers from natural, synthetic, and biosynthetic materials	Provide topographical cues to Schwann cells and axons to guide nerve regeneration
Nanotechnology and Neural Interfacing	
Nanoelectrode arrays (carbon nanotubes)	Stimulate and record from individual axons and dendrites; improve signal-to-noise ratio
Layer-by-layer coating of laminin	Improve cell attachment and differentiation on neural electrodes
Nitrocellulose coating to incorporate drugs	Limit inflammatory response; provide sustained delivery of drugs to implantation site
Nanotechnology and Controlled Delivery of Drugs to the Nervous System	
Cerium oxide and ChABC nanoparticles	Reduce scarring and provide neuroprotection in the spinal cord
Nanoparticles	Able to cross blood–brain barrier (BBB); produce sustained release of drugs

densely packed together. Given the complex nature of neural tissue that involves both electrical and biochemical connectivity underlying its "healthy" function, nanotechnology that helps us better understand neural function will play an important role in providing the design criteria for neural tissue engineering.

This chapter discusses several nanotechnologies which collectively impact neural tissue engineering. A summary of these technologies is provided in Table 12.1. Even though some of these technologies are still in their infancy, their potential for clinical impact is clear and discussed in this chapter.

12.2 NANOTECHNOLOGIES USED TO BETTER UNDERSTAND NEURAL TISSUE TO RATIONALLY FASHION ITS REPAIR

12.2.1 Micro/Nanopatterning In Vitro to Better Understand Axon Guidance

Multitudes of physical and biochemical signals are required for optimal regeneration of the central nervous system (CNS) and the peripheral nervous system (PNS). Understanding the rules governing axonal guidance could inform strategies to design regenerative therapies such that optimal topographical and biochemical cues can be presented to regenerating axons. Micro/nanopatterning on two-dimensional substrates provides us with the degree of spatial control necessary to probe axonal response to well-controlled topographical and biochemical cues [3,4]. Some of the nanotechnology-based efforts described in this section are invaluable in informing the next generation of tissue engineering strategies aimed at repairing or regenerating the nervous system.

12.2.1.1 Microfabrication to Create Patterns to Observe the Effects of Topographical Cues

Neural tissue engineering scaffolds are often designed with specific topographical guidance cues which aid in the adhesion and migration of neuronal cells. Microfabrication techniques developed for use in the microelectronics industry have been adapted to achieve micro- and nanoscale patterning on experimental substrates. For instance, using photolithography, lanes with depths ranging from 50 nm to 6 μm can be etched and different chemistries laid on the fused-silica slide substrate, to understand the effects of hierarchical interactions on cellular behavior [5]. For instance, lanes patterned with alternating hydrophobic methylsilane and laminin can be applied to evaluate neurite orientation in response to patterning and surface chemistry simultaneously. These studies point to the conclusion that neurites preferentially adhere to adhesive laminin substrates and also respond to topographical gratings of differing depths. The neurites become highly aligned when the groove depths exceed 1 μm. Other fabrication techniques such as electron beam lithography also allow us to fabricate substrates with features down to a few nanometers [6,7], potentially allowing multiple cues to be sampled by the same axon or growth cones. This level of control enables further study into how information might be processed or integrated by single axons or growth cones. Thus, nanoscale patterning technologies provide an important tool in promoting our understanding of growth cone function with regard to their ability to detect and process topographical and biochemical cues.

12.2.1.2 Observing Axonal Interaction with ECM Proteins Using Dip-Pen Nanolithography

Dip-pen nanolithography (DPN) is another technique which has been applied to study how axons interact with their extracellular environment. DPN has been used to print collagen and collagen-like structures with line widths of 30–50 nm [8]. This technique provides a robust means of organizing collagen into its hierarchical structure, while maintaining its bioactivity. Briefly, lyophilized rat tail collagen can be thiolated in an acidic environment to maintain its bioactivity. Atomic force microscopy (AFM) is then used to perform the DPN, by dipping the AFM tip in the aqueous peptide solution and patterning in tapping mode, while changing the angle to achieve collagen-like structures. This technology has vast potential in nanopatterning because it can be used to fabricate a variety of polymers at size scales similar to their natural states. In terms of neural tissue engineering, this technology can be very useful in understanding neuronal responses to different topographical cues and peptide epitopes, as well as in designing scaffolds with nanoscale features to enhance cellular behavior.

Proteins, synthetic amino acids, and small peptides can be patterned on different substrates with extraordinary spatial control to modulate axonal growth. Nanopatterning can thus provide higher resolution to differentially provide chemoattractants and repellants at specific sites within the vicinity of a single growth cone for the purpose of probing basic cell behavior [4] and improving the design of constructs used in neural tissue engineering.

12.2.2 Quantum Dots

The use of nanotechnology in labeling or monitoring biological states and processes is promising because nanomaterials often possess unique fluorescence characteristics. Semiconductor quantum dot (QD) nanocrystals with sizes ranging from 2 to 8 nm in diameter have highly stable fluorescent properties.

The physical structure of QDs also provides several advantages to their use in biological systems. QDs usually have a heavy metal core composed of cadmium–selenium with an inert zinc sulfide shell and a variable outer coating which can be fabricated with different biomolecules with a specific application [9]. These nanoparticles are able to absorb light over a broad spectrum with narrow emission spectra. This property of QDs provides two advantages to their use in imaging neurobiological systems. First, the narrow emission spectrum reduces spectral overlap which helps in distinguishing multiple fluorophores. Second, the broad excitation spectrum allows for the use of single excitation wavelength to excite QDs of different colors [2,9,10]. The fluorescence of the QDs can easily be modulated by varying their size or chemistry. Compared to fluorescence markers used in immunohistochemistry, QDs experience minimal photobleaching and result in a higher signal-to-noise ratio [9–12]. These characteristics make them useful for visualizing subcellular trafficking in vitro and for tracking and characterizing cells in vivo.

12.2.2.1 Visualizing Ligand–Receptor Interaction Using Peptide Conjugated Quantum Dots

QDs are beneficial for visualizing specific activated signaling pathways. QDs can be conjugated to peptides that bind to specific receptors to observe the activation of the pathway. For example, peptide conjugated QDs can be used to activate signaling pathways in PC12 cells in vitro. The β-nerve growth factor (NGF) peptide was conjugated to QDs using streptavidin–biotin conjugates, in a high affinity binding interaction, that left intact the bioactivity of the peptide. Figure 12.1 demonstrates the ability of NGF–QDs to get internalized and transported to the processes of PC12 cells. The study demonstrated that the functional QDs were successful in activating TrkA receptors and initiating differentiation of PC12 cells [13].

In another study, QDs conjugated with anti-TrkA receptor antibody enabled the visualization of internalization of the ligand–receptor complex and the vesicular transportation within the cell [14]. The anti-TrkA QDs were able to specifically fluoresce in the presence of the TrkA receptor. The study also showed that these QDs were able to interact with the cellular motors for several days after the initial endocytosis. Alternately, NGF can be conjugated to QDs to track the retrograde transport of NGF in dorsal root ganglion (DRG) cultures. This enables quantification of the kinetics of endosomal transport within the axon after internalization of the QD–NGF complex [15]. While radiolabeled isotopes have been used to study vesicular transport, conjugated QDs provide a much safer way to track and measure such kinetics. Thus, QDs are a powerful tool to study the mechanism

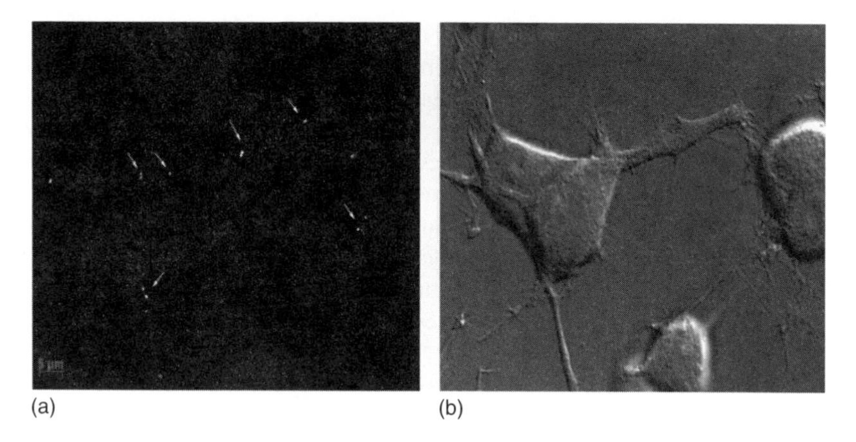

FIGURE 12.1 NGF-conjugated quantum dots (NGF–QDs) transported to the processes of the PC12 cells (b), pointed to by the arrows in (a). (Photograph courtesy of Dr. Tania Vu, Oregon Health and Science University.)

of uptake and intracellular trafficking of specific growth factors or extracellular matrix (ECM) proteins, and this in turn can be a powerful enabler of neural tissue engineering.

12.3 MICRO–NANO TOPOGRAPHIES THAT GUIDE AXONS IN VITRO AND IN VIVO

Another area of research where nanotechnology has had an influence in neural tissue engineering is in developing scaffolds which help in regeneration of damaged nerves. Due to the complex physiology of regenerating nerve, scaffolds for regeneration require features that aid in the proliferation, differentiation, and migration of neuronal and glial cells. Thus, three-dimensional scaffolds that mimic the complex physiological properties and chemical cues of the ECM, and enable guided cell and axonal migration in vitro and in vivo are necessary. These scaffolds also need to provide a viable environment for cell growth, not elicit any immune responses, have a high surface area, allow for the movement of nutrients, and provide strong support, while being able to degrade at a rate equal to or slower than the rate of regeneration [16]. Nanoscale control over molecular assembly and topography provides the ability to introduce some of the above features with spatial and temporal control, potentially impacting nerve guidance strategies significantly.

Three important characteristics of scaffolds that promote nerve regeneration include biocompatibility, degradability, and porosity. In addition, as discussed earlier, spatiotemporally controlled presentation of topographical and biochemical cues enabled by nanoscale patterning techniques can significantly influence regeneration, as described below.

12.3.1 Topographical Cues for Regenerating Nerves: Natural Fibers

Natural fibers can be formed from materials such as amphiphilic molecules, silk, and collagen. The inherent properties of fiber-based materials make them highly biocompatible, given their permeability and compliant nature. In particular, permeability allows for the diffusion of needed nutrients that enhance the adhesion and migration of cells. However, these same properties make the production of such fibers complex, and fiber orientation to guide cell migration is hard to control [16].

12.3.1.1 Self-Assembling Peptide Nanofibers

Natural fibers can be formed through the fabrication of peptide-based amphiphilic molecules. These molecules exist in solution and then self-assemble when introduced to suspensions of cells such as neural progenitor cells [17]. Forces from ionic bonding, hydrogen bonding, van der Waals interactions, and hydrophobic responses drive the formation of the peptide self-assembly process and generate a gel-like solid [17]. The hydrophilic heads of the amphiphilic molecules can be designed to include specific epitopes. For example, the incorporation of the sequence isoleucine–lysine–valine–alanine–valine (IKVAV) can help promote neurite growth [18]. Nanofibers with built-in IKVAV are 5–8 nm in diameter and range from hundreds of nanometers to a couple of micrometers in length [17]. When compared to laminin and poly(D-lysine) substrates, cells cultured on IKVAV nanofibers differentiated more quickly. Additionally, these nanofibers performed better than coatings of IKVAV soluble peptides because the nanofibers provide higher density of epitopes [17]. Thus, these fibers with the incorporation of specific epitopes into nanofibers are shown to aid in neuronal differentiation.

Peptide-amphiphile nanofibers can also be produced through soft lithography. In this process, a stamp of the fiber orientation is created and then pressed upon a solution of peptide-amphiphilic molecules [19]. This arrangement is then evaporated, sonicated, and dried to produce aligned nanofibers. The dimensions of the fibers depend on the percent weight of peptide amphiphiles. For 5 wt% peptide solution, the fibers produced had a width of 200–300 nm with a height of about 55.1 nm while the 1 wt% solution produced a width of 150 nm and a height of 23.1 nm [19]. However, they have only been shown to be compatible with media and further studies need to be done with cell cultures to better determine their effects.

12.3.1.2 Silk Fibers to Enhance Nerve Regeneration

Silk fibroin fibers represent another class of natural fibers that have been used to promote nerve regeneration. After the removal of sericin, which causes adverse immunological responses, the silk fibroin was extracted to produce fibers that were 15 μm in diameter. Their effects on DRGs and Schwann cells showed their biocompatibility and ability to promote cell growth [20,21]. Silk fibers from spiders have also been utilized to produce conduits. Extracted from the glands of spiders, these fibers are 10 μm in diameter and allow for better adhesion than polydioxanone monofilaments. In addition to adhesion, they are able to support cell migration and are biocompatible and biodegradable; however, their tedious extraction method is a major limitation [22].

12.3.1.3 Magnetically Aligned Nanofibers for Cell Alignment

Collagen is commonly used since it is a natural component of the ECM, providing structural strength through fibrous networks. Collagen can also be magnetically treated to produce a gel rod of aligned collagen fibrils. In strong magnetic fields a high degree of collagen alignment is produced. This was shown through a comparison of collagen gel produced in a magnetic field of 4.7 T and 9.4 T [23]. As a result, when seeded with DRGs, this alignment facilitated oriented neurite extension. When combined with Schwann cells, directed migration occurred and was further enhanced in the presence of 10% fetal bovine serum [23]. These data suggest that aligned collagen fibers are promising candidates as substrates for guided nerve regeneration.

12.3.2 Biosynthetic and Synthetic Fibers

Biosynthetic and synthetic fibers can be created through a variety of processes including extrusion and electrospinning. While naturally derived polymers including collagen, elastin, and gelatin would be useful starting materials for fabrication of oriented substrates for nerve regeneration, they present technical challenges from a fabrication perspective. One solution may be to mix natural materials with synthetic polymer solutions.

To "electrospin" a fiber, a difference in voltage is used to propel polymer fibers to a target (Figure 12.2). Oriented fibers can be fabricated by having the target spin at high speeds. Electrospun

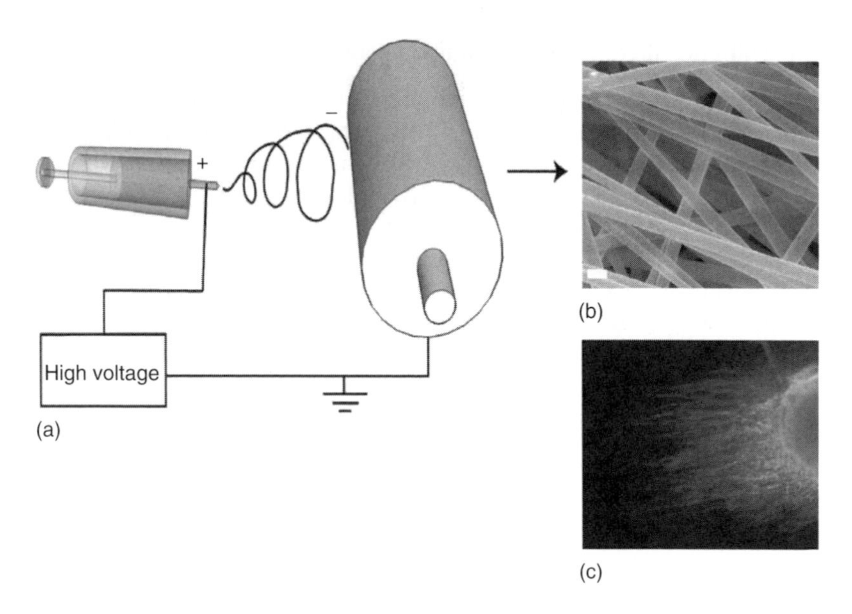

FIGURE 12.2 (See color insert following page 206.) (a) Schematic diagram of the electrospinning apparatus with rotating drum configuration to yield aligned biosynthetic or synthetic fibers. (b) Scanning electron micrographs of electrospun PAN-MA fibers. Scale bar = 500 nm. (c) Dorsal root ganglia from P3 pups cultured on aligned electrospun fibers.

fibers can have functional groups that can be exploited to aid in cell adhesion [16]. Furthermore, these aligned fibers guide glial alignment and promote directed growth (Figure 12.2c). Like natural fibers, most biosynthetic materials are hard to fabricate due to their inherent properties, while synthetic fibers are the least problematic to produce.

12.3.2.1 Fabrication through Electrospinning

Relative to extrusion and other techniques, electrospinning is a facile method of producing nanofibers. Fabrication starts with a polymer solution or melt that is pumped from a syringe. The polymer stream travels from a charged syringe to an oppositely charged collector. Before anchoring to the collector, electrostatic repulsion between the strands results in "splaying" which produces extrafine fibers [16]. To enhance the alignment of the fibers, collectors such as rotating drums or disks can be used. Electrospinning can be tailored to match the size of nerve fibers in vivo or to match mechanical properties of the nerve. These properties can be tailored through the manipulation of several factors, including the flow rate, voltage, distance between the syringe and the collector, design of the collector, and viscosity of the polymer solution or melt. The rotational speed of the collector can also be altered to change the orientation of the fibers. Nevertheless, the speed can only be increased as long as the collected polymer jet is still continuous. Examples of electrospun fibers for neural tissue engineering are given below.

12.3.2.2 Synthetic Electrospun Fibers for Neural Tissue Engineering

The most widespread synthetic polymers in neural tissue engineering are poly(α-hydroxy esters) which include poly(glycolic acid) (PGA), poly(lactic acid) (PLA), and a copolymer of the two poly (lactic-co-glycolic acid) (PLGA). These synthetic polymers are frequently used due to their advantageous biodegradable properties and their ease of electrospinning [24,25]. Another member, polycaprolactone (PCL), is also used when a slower rate of degradation is desired especially in some drug delivery applications.

12.3.2.3 Biosynthetic Fibers with Conductive Properties

Since natural materials are difficult to spin, they can be co-spun with synthetic polymers. For instance, a conductive polymer, polyaniline (PANi), can be combined with gelatin and then electrospun [26]. Several polymer solutions with varying amounts of PANi were produced and their fibers ranged from 924 to 48 nm in size [26]. The solutions with higher concentrations of PANi produced thinner fibers. Additionally, the proportion of gelatin in the solution should not be less than 5% due to the possibility of beads forming on the fibers. These fibers were then seeded with H9c2 rat cardiac myoblasts, and the cells were shown to proliferate [26]. Interestingly, when the width of the fiber was greater than 500 nm, cell alignment was induced. These properties of biocompatibility and conductivity show promise for future implementation in nanofiber scaffolds.

12.3.2.4 Biosynthetic Fibers to Aid in Schwann Cell Migration

To improve axonal regeneration, one study compared aligned PCL and a combination of PCL and collagen fibers. The mixed solution consisted of 25% collagen and 75% PCL constrained by the fact that increasing the proportion of collagen also increases the difficulty of spinning aligned fibers [27]. In this study, the PCL fibers were 889–259 nm in diameter while the collagen and PCL fibers were 705–377 nm in diameter [27]. After DRGs were seeded on the fibers, the collagen and PCL fibers were observed to produce more directed neurite outgrowth. With the seeding of Schwann cells, the mixed fibers were shown to have better adhesion and faster migration [27]. These effects are probably due to the presence of collagen which is naturally found in the ECM. Therefore, the incorporation of collagen into nanofibers can aid in Schwann cell migration and axonal regeneration.

12.3.2.5 Topography for Controlled Migration of Schwann Cells

Topographies are generated to mimic the existing physiological structures of the ECM in the hope that similar cues will promote cellular growth, adhesion, and differentiation. Schwann cells play a significant role in neurite extension and migration since they myelinate axons. They also secrete ECM molecules like laminin and produce numerous neurotrophic factors, such as NGF, BDNF, and NT-3, which aid in the nerve regeneration process. Utilizing micropatterned filaments allows Schwann cells to be properly aligned to form bands of Bungner, a critical phase in axonal regeneration. This alignment is what directs neurite outgrowth. These microstructured filaments can then be inserted into a conduit to enhance regeneration in rat nerve gaps. Conduits incorporated with oriented nanofiber scaffolds have also been able to recreate the bands of Bungner. These nanofiber scaffolds promote Schwann cell migration after peripheral nerve injuries and have shown functional outcomes comparable to autografts.

Generally, a multitude of fiber options exist. Nevertheless, the most important feature in producing these fibers is ability to control their topography. Consequently, this is also the most challenging aspect where nanotechnology has had an impact. Once the precise control over the physical structure is achieved, these materials can provide physical cues that can be paired with biochemical cues to promote regeneration.

12.4 NANOELECTRODE ARRAYS AND THEIR COATINGS TO UNDERSTAND ELECTRICAL AND BIOCHEMICAL FUNCTION AND IMPROVE TISSUE INTEGRATION

Nanotechnology has shown its presence for fundamental studies in neuroscience involving plasticity and for the treatment of CNS trauma, disease, and age-related degeneration. Multichannel electrodes have significant implications for our understanding of normal physiology, pathology, or treatment of several CNS disorders. The field of neural tissue engineering can greatly benefit from nanoscale electrode arrays (NEAs) because they provide an excellent analysis tool to evaluate functional neural networks so we can better design interfacing between biological and electrical systems. However, when electrodes are implanted into the brain tissue for long-term recording, their recording ability has been rather unreliable [28].

For example, formation of glial scar around the electrode arrays reduces the integration between the electrode and the nervous system [29,30]. The reactive response of the host system to the electrode is caused by various factors. The semiconductor materials which are usually used for electrode fabrication in combination with implantation injury elicit an immune response which leads to a fibrous capsule formation and causes an increase in the impedance. Damage to the microvasculature caused during implantations also contributes to the induction of an inflammatory response. These problems inhibit the function of electrode arrays and provide a major challenge in the field of neural tissue interfacing.

12.4.1 NEAs to Evaluate Neural Networks

The ability to successfully interface the brain to external electronics would have enormous implications for our society and facilitate a revolutionary change in the quality of life of persons with sensory and motor defects. Advances in the past half century in microelectronics and bio-MEMS/NEMS technology have allowed the study of the nervous system at a cellular level [31]. One of the offspring of these technologies is the development of NEAs designed to better understand electrical and biochemical function. Nanoelectrode arrays have been very useful to investigate the nervous system, especially to understand the effects of various stimuli on the morphological and functional behavior of single cells [32,33].

12.4.1.1 NEAs Provide Information on the Organization of the Nervous System

Neural electrodes have the ability to restore loss of function by stimulating neural pathways as well as to improve our understanding of the nervous system. Micro/nanoelectrode arrays have been widely used to record from the CNS and the PNS [28,34]. Arrays of nano/microneedles are inserted into the nervous tissue for recording electrical signals from the surrounding neurons. By designing these arrays to record from different neurons, in specific locations of the CNS and PNS, spatiotemporally preserved signal and control information can be extracted from the nervous system [35]. Nanoelectrodes provide faster mass transport which leads to faster electrochemical and chemical reactions and improves the signal-to-noise ratio while recording from biological cultures [36]. Since the materials typically used to fabricate these electrodes are not biologically favorable, there have been several technologies developed to enhance their interface. Some of these technologies are described below and have been successfully used to enhance neural interfacing with different electrodes.

12.4.1.2 Carbon Nanotube-Based Arrays

Nanotechnologies used in the fabrication of novel materials and coatings can overcome the challenges described above. NEAs by virtue of their small size can minimize scarring by minimizing the extent of microvascular damage. Carbon nanotubes (CNTs) are used for their unique characteristics, which can aid in enhancing better electrode integration [36–38]. Properties such as high mechanical strength, electrical conductivity, and potential to be chemically functionalized make them a very attractive choice of material for nanoelectrode fabrication. CNTs provide high surface-to-volume ratio, which leads to lower impedance and improves signal-to-noise ratio.

Plasma enhanced chemical vapor deposition on a silicon wafer in presence of catalysts is used to fabricate multiwalled graphene structures. These structures grow on silicon wafers with a bamboo-like morphology [37] and can be patterned to form desirable shapes (Figure 12.3). The average diameter of multiwalled CNTs is around 100 nm, which leads to less vascular damage around the

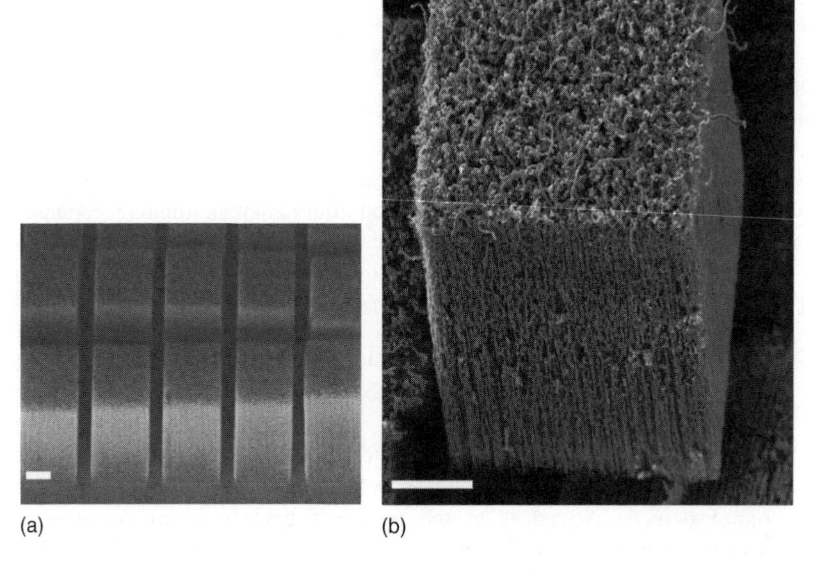

FIGURE 12.3 (a) CNT arrays grown on Si substrate for potential use as nanoscale neural electrodes. Scale bar = 10 μm. (b) Surface topography of CNTs. Scale bar = 10 μm. (Photograph courtesy of Dr. Jud Ready, Georgia Institute of Technology. With permission.)

implantation site of the electrodes. Besides their favorable mechanical and chemical properties, it is also possible to functionalize them with proteins and DNA to make them more biocompatible. One such study is investigating the use of streptavidin/biotin interaction for the adsorption of proteins onto CNTs [39]. In another study, electrochemically added nitro groups were reduced to amino groups to covalently link DNA to CNTs [40].

Specific applications to interface neural networks to CNTs have also been explored. PC12 cells grown on collagen coated CNTs grew in a monolayer compared to clustering observed on tissue culture plastic. Also, in comparison to metal electrodes, impedance was observed to be lower on CNT arrays [37]. Others have observed that CNTs with diameters <100 nm can reduce neural scar tissue formation because the low energy state of small diameter CNTs did not support astrocyte proliferation, a major player in the inflammation process [41]. Chemical functionalization to provide ionic charge on the surface of the CNTs provides a favorable substrate for neurite extension [42]. Even though the cytoxicity of CNTs is not completely realized, the numerous physical, electrical, and chemical properties of CNTs have made them a very attractive technology to use in the neural interfacing field.

12.4.2 Nanoscale Electrode Coatings

Coating electrodes with nanoscale thick layers for improving local reaction of neural prosthetics has also been a wide area of research in the field of nanotechnology and neural tissue engineering. Coatings for passivation, sustained release of factors which promote axonal growth and minimize the immune response around the electrode have been developed to improve cell attachment, electrical function, and enhance regeneration. Various techniques in nanotechnology allow for precise tailoring of thickness and release of drugs to minimize the inflammatory response.

12.4.2.1 Electrostatic Layer-by-Layer-Based Coatings

Electrostatic layer-by-layer (LbL) assembly is a widely used technique to deposit nanoscale thickness films which can be modulated to release specific drugs or act as inert coatings to evade the immune system. LbL assembly uses electrostatic interactions between charged substrates and an oppositely charged polyelectrolyte to fabricate films with thickness ranging from angstroms to a few nanometers. The process can be repeated several times to lay multilayer films with desired thickness. LbL assembly technology has been used to deliver biomolecules with the specific purpose to improve cell attachment and differentiation. In vivo studies with laminin-1 coatings on silicon microelectrode using LbL assembly have shown to modulate glial scar formation around implanted electrodes in the CNS [43].

12.4.2.2 iCVD Coatings

Initiated chemical vapor deposition (iCVD) is another technology used to deposit nanoscale coatings on neural prosthetic devices. This treatment entails the introduction of monomer vapor above the desired substrate and driving the free radical initiator species to induce vapor phase deposition of monomers. Since this technique uses vapor deposition, the resulting coatings conform to the shape of the substrate. This is very useful in the cases where material properties such as porosity are important in the function of the prosthetic device. Since the process is carried on in solvent-free environment, this technique can be also useful in coating polymer scaffolds, which are prone to solvent damage.

12.4.2.3 Using Nanoscale Nitrocellulose to Incorporate Anti-Inflammatory Drugs

The inflammatory reaction creates a barrier between the electrode and the surrounding neurons. The above methods are useful for creating stable coatings which passivate the electrode but fail to reduce the inflammatory response. To improve long-term recording in vivo, nitrocellulose-based coatings have been used to incorporate anti-inflammatory drugs such as α-melanocyte stimulating hormone (α-MSH) for sustained release over a period of weeks. Using spin coating techniques, micron thickness coatings were fabricated for incorporating bioactive α-MSH onto the silicon substrates. These coatings were successfully able to decrease nitric oxide formation due to the presence of activated microglia.

Another method is to coat electrodes with dexamethasone trapped in nitrocellulose films, which enables the release of drugs in a localized and sustained manner [44]. Nitrocellulose-based coatings of dexamethasone were used to cover probes, and coating adhesion tests showed that the coating survived the implantation procedures. Dexamethasone released from implanted electrodes decreased electrode impedance and thereby improved neural signaling. It also reduced the expression of chondroitin sulfate proteoglycans (CSPGs) and prevented neuronal loss at both week 1 and week 4 after implantation. Though the necessary long-term dosing of dexamethasone is unknown, the initial release of the drug may be able to stave off the effects of acute inflammation and result in better electrode–CNS interfaces.

Therefore, nanoscale coatings can significantly alter the fate of implanted electrodes and improve the recording stability of implanted electrodes. Stable in vivo recordings from the nervous system have significant implications not only for neuroprosthetic devices but also for promoting a more fundamental understanding of the functioning of the nervous system.

12.5 NANOPARTICLES FOR THE CONTROLLED RELEASE OF TROPHIC FACTORS IN THE NERVOUS SYSTEM

Nanoparticles are a means for targeted and sustained drug delivery in the nervous system. Their ability to provide precise control allows them to be used to mediate inflammatory responses or to deliver factors beneficial to neurite outgrowth. They can be fabricated from various substances including silica, polymers, lipids, and metallic and magnetic materials. To produce these nanoparticles, the polymer solution is dissolved in a solvent that is immiscible in water. This mixture is then emulsified and the solvent is allowed to evaporate, leaving behind the nanoparticles. Another method involves dissolving the polymer in a miscible solvent and allowing the solution to diffuse through an aqueous phase that causes the nanoparticles to precipitate.

12.5.1 Nanoparticles for the Release of Anti-Inflammatory Drugs

As discussed earlier, astroglial scarring around implanted electrodes can interfere with their ability to provide stable long-term recordings. An alternative to the nanoscale coating strategy described above has been to co-implant nanoparticles that release an anti-inflammatory agent in the immediate vicinity of the implanted electrodes. Dexamethasone-PGLA nanoparticles, approximately 400–600 nm in size, can be immersed in an alginate hydrogel and then co-implanted with electrodes to effectively reduce astroglial scarring. These particles were able to sustain continuous release of a drug for up to 3 weeks [45].

12.5.2 Nanoparticles for Release of Antiscarring Drugs

One major impediment to regeneration in the CNS is the formation of astroglial scar tissue around the lesion site [46]. CSPGs are a major inhibitory constituent of astroglial scar tissue, and their digestion has been shown to alleviate inhibition of scar, and promote regeneration after CNS injury [47]. One promising strategy to promote CNS regeneration has involved the digestion of scar tissue through the use of a bacterial enzyme chondroitinase ABC (ChABC) [46]. Although osmotic pumps have been used in the initial studies, their use in the injured cord is not ideal as the canula can cause inflammation and damage by its presence. To circumvent these issues, micro- and nanoparticles that release ChABC provide an attractive alternative. When ChABC nanospheres were used to deliver ChABC after spinal cord injury (SCI), CS-56 staining revealed a reduction in CSPGs, and a concomitant increase in GAP-43 staining at the site of injury [48]. Micro- and nanoparticle technology for digestion of CSPGs gains significance because there is a need to control the spatial distribution of CSPG digestion, due to their important role in stabilizing synapses in the cord via perineural nets [48].

12.5.3 Nanoparticles for Neuroprotection

Along the lines of spinal cord injuries, the administration of autocatalytic nanoceria particles has been investigated. Cerium oxide can be present in different valence states (Ce^{3+} and Ce^{4+}). This property results in antioxidant effects that promote neuroprotection. This effect was tested using a hydrogen peroxide-induced oxidative injury in which the nanoceria particles were able to provide a means of detoxification [49]. Other antioxidants are available through a B27 supplement which includes vitamin E, catalase, and gluthathione, but they are not long-lasting and need to be consistently replaced. In a comparison between nanoceria treated cultures and cultures with only the B27 supplement, the neuron survival of nanoceria cultures exceeded that of the controls [49]. Thus, nanoparticles that are anti-inflammatory and are scavengers of free radicals also represent an important arm in the reduction of secondary injury after SCI.

Numerous nanoparticles have been applied to drug delivery in the nervous system. Polybutylcyanoacrylate (PBCA) was able to move through the blood–brain barrier (BBB) once it was coated with polysorbate 80 [50]. However, its toxicity and inability to produce long-term effects make it an unlikely carrier for drugs. Methoxypoly(ethylene glycol)-polylactide or poly(lactide-co-glycolide) (mPEG-PLA/PLGA) nanoparticles are attractive candidates for drug delivery. They are biodegradable, biocompatible, and able to provide sustained drug release—properties that are utilized in other applications such as implants. Nevertheless, PEG-PLA nanoparticles still need further improvements and investigation before clinical potential can be reached [50].

Nanoparticles are promising due to their demonstrated ability to cross the BBB and provide a means of delivering needed drugs. Additionally, their degradation properties can be utilized to release treatments over a period of time. Their chemistry can also be tailored for targeted delivery to different parts of the nervous system and their size scales can be beneficial to evade the immune system.

12.6 FUTURE OUTLOOK

The use of nanotechnology in neural tissue engineering has shown great potential. The use of nanotechnology to evaluate and modulate cellular behavior has significantly advanced our understanding of the nervous system, and nanoscale materials with wide arrays of unique properties have been used to image and biologically record from the nervous system. Nanotechnology has given us precise control to interact with biological systems at a cellular level, and its benefits have been realized in this chapter.

While nanoscale control is extraordinarily promising, it is important to recognize that it is but one arm in a spectrum of potential interventions that will likely impact neural tissue engineering. This said, precise control over molecular features, which allows controlled assembly of nanoscale structures, will be potentially revolutionary in influencing the interface between materials and the nervous system. Given its complexity, the nervous system presents a challenging arena that will shape the evolution of nanotechnologies as they apply to biological systems.

ACKNOWLEDGMENTS

The authors would like to acknowledge Dr. Tania Vu from Oregon Health and Science University, for use of the QD image and Dr. Jud Ready from Georgia Institute of Technology, for use of the CNT array image. The authors would also like to acknowledge Dr. Anjana Jain and Isaac Clements for their expertise and input toward this manuscript. Funding to the Bellamkonda Laboratory is acknowledged from the National Institute of Health (NS44409, NS43486, NS45072, and DC006849) and the National Science Foundation (CBET 065176).

REFERENCES

1. Xuejun, W., S. Donglu, and N. Zhang, Applications of nanotechnology in tissue engineering, in *Handbook of Nanostructured Biomaterials and Their Applications in Nanobiotechnology*, H.S. Nalwa, Editor. 2005, American Scientific Publishers. pp. 1–23.
2. West, J.L. and N.J. Halas, Applications of nanotechnology to biotechnology commentary. *Curr Opin Biotechnol*, 2000, 11(2): 215–217.
3. Li, N. and A. Folch, Integration of topographical and biochemical cues by axons during growth on microfabricated 3-D substrates. *Exp Cell Res*, 2005, 311(2): 307–316.
4. Johansson, F., et al., Axonal outgrowth on nano-imprinted patterns. *Biomaterials*, 2006, 27(8): 1251–1258.
5. Britland, S., C. Perridge, M. Denyer, et al., Morphogenetic guidance cues can interact synergistically and hierarchically in steering nerve cell growth. *Exp Biol*, 1996, 1(2): 1–15.
6. Ressier, L., et al., Fabrication of planar cobalt electrodes separated by a sub-10 nm gap using high resolution electron beam lithography with negative PMMA. *Ultramicroscopy*, 2007, 107(10–11): 985–988.
7. Wilkinson, C.D., Making structures for cell engineering. *Eur Cell Mater*, 2004, 8: 21–25; discussion 25–26.
8. Wilson, D.L., et al., Surface organization and nanopatterning of collagen by dip-pen nanolithography. *Proc Natl Acad Sci U S A*, 2001, 98(24): 13660–13664.
9. Han, M., et al., Quantum-dot-tagged microbeads for multiplexed optical coding of biomolecules. *Nat Biotechnol*, 2001, 19(7): 631–635.
10. Jaiswal, J.K. and S.M. Simon, Potentials and pitfalls of fluorescent quantum dots for biological imaging. *Trends Cell Biol*, 2004, 14(9): 497–504.
11. Alivisatos, A.P., W. Gu, and C. Larabell, Quantum dots as cellular probes. *Annu Rev Biomed Eng*, 2005, 7: 55–76.
12. Silva, G.A., Neuroscience nanotechnology: Progress, opportunities and challenges. *Nat Rev Neurosci*, 2006, 7(1): 65–74.

13. Vu, T.Q., et al., Peptide-conjugated quantum dots activate neuronal receptors and initiate downstream signaling of neurite growth. *Nano Lett*, 2005, 5(4): 603–607.
14. Sundara Rajan, S. and T.Q. Vu, Quantum dots monitor TrkA receptor dynamics in the interior of neural PC12 cells. *Nano Lett*, 2006, 6(9): 2049–2059.
15. Cui, B., et al., One at a time, live tracking of NGF axonal transport using quantum dots. *Proc Natl Acad Sci U S A*, 2007, 104(34): 13666–13671.
16. Murugan, R. and S. Ramakrishna, Design strategies of tissue engineering scaffolds with controlled fiber orientation. *Tissue Eng*, 2007, 13(8): 1845–1866.
17. Silva, G.A., et al., Selective differentiation of neural progenitor cells by high-epitope density nanofibers. *Science*, 2004, 303(5662): 1352–1355.
18. Silva, G.A., Nanotechnology approaches for the regeneration and neuroprotection of the central nervous system. *Surg Neurol*, 2005, 63(4): 301–306.
19. Hung, A.M. and S.I. Stupp, Simultaneous self-assembly, orientation, and patterning of peptide-amphiphile nanofibers by soft lithography. *Nano Lett*, 2007, 7(5): 1165–1171.
20. Yang, Y., et al., Biocompatibility evaluation of silk fibroin with peripheral nerve tissues and cells in vitro. *Biomaterials*, 2007, 28(9): 1643–1652.
21. Yang, Y., et al., Development and evaluation of silk fibroin-based nerve grafts used for peripheral nerve regeneration. *Biomaterials*, 2007, 28(36): 5526–5535.
22. Allmeling, C., et al., Use of spider silk fibres as an innovative material in a biocompatible artificial nerve conduit. *J Cell Mol Med*, 2006, 10(3): 770–777.
23. Dubey, N., P.C. Letourneau, and R.T. Tranquillo, Guided neurite elongation and Schwann cell invasion into magnetically aligned collagen in simulated peripheral nerve regeneration. *Exp Neurol*, 1999, 158(2): 338–350.
24. Li, W.J., et al., Fabrication and characterization of six electrospun poly(alpha-hydroxy ester)-based fibrous scaffolds for tissue engineering applications. *Acta Biomater*, 2006, 2(4): 377–385.
25. Li, W.J., et al., Electrospun nanofibrous structure: A novel scaffold for tissue engineering. *J Biomed Mater Res*, 2002, 60(4): 613–621.
26. Li, M., et al., Electrospinning polyaniline-contained gelatin nanofibers for tissue engineering applications. *Biomaterials*, 2006, 27(13): 2705–2715.
27. Schnell, E., et al., Guidance of glial cell migration and axonal growth on electrospun nanofibers of poly-epsilon-caprolactone and a collagen/poly-epsilon-caprolactone blend. *Biomaterials*, 2007, 28(19): 3012–3025.
28. Polikov, V.S., P.A. Tresco, and W.M. Reichert, Response of brain tissue to chronically implanted neural electrodes. *J Neurosci Methods*, 2005, 148(1): 1–18.
29. Biran, R., D.C. Martin, and P.A. Tresco, Neuronal cell loss accompanies the brain tissue response to chronically implanted silicon microelectrode arrays. *Exp Neurol*, 2005, 195(1): 115–126.
30. Edell, D.J., et al., Factors influencing the biocompatibility of insertable silicon microshafts in cerebral cortex. *IEEE Trans Biomed Eng*, 1992, 39(6): 635–643.
31. Wise, K.D., D.J. Anderson, J.F. Hetke, et al., Wireless implantable microsystems: High-density electronic interfaces to the nervous system. *Proc IEEE*, 2004, 92(1): 76–97.
32. Irina Kleps, M.M., F. Craciunoiu, and M. Simion, Delevelopment of the micro- and nanoelectrodes for cells investigation. *Microelectron Eng*, 2007, 84: 1744–1748.
33. Franks, W., et al., Impedance characterization and modeling of electrodes for biomedical applications. *IEEE Trans Biomed Eng*, 2005, 52(7): 1295–1302.
34. Kipke, D.R., et al., Silicon-substrate intracortical microelectrode arrays for long-term recording of neuronal spike activity in cerebral cortex. *IEEE Trans Neural Syst Rehabil Eng*, 2003, 11(2): 151–155.
35. Normann, R.A., Technology insight: Future neuroprosthetic therapies for disorders of the nervous system. *Nat Clin Pract Neurol*, 2007, 3(8): 444–452.
36. Arrigan, D.W., Nanoelectrodes, nanoelectrode arrays and their applications. *Analyst*, 2004, 129(12): 1157–1165.
37. Nguyen-Vu, T.D., et al., Vertically aligned carbon nanofiber arrays: An advance toward electrical–neural interfaces. *Small*, 2006, 2(1): 89–94.
38. Nguyen-Vu, T.D., et al., Vertically aligned carbon nanofiber architecture as a multifunctional 3-D neural electrical interface. *IEEE Trans Biomed Eng*, 2007, 54(6) (Pt 1): 1121–1128.

39. Shim, M., et al., Functionalization of carbon nanotubes for biocompatibility and biomolecular recognition. *Nano Lett*, 2002, 2(4): 285–288.
40. Lee, C., et al., Electrically addressable biomolecular functionalization of carbon nanotubes and carbon nanofiber electrodes. *Nano Lett*, 2004, 4(9): 1713–1716.
41. McKenzie, J.L., et al., Decreased functions of astrocytes on carbon nanofiber materials. *Biomaterials*, 2004, 25(7–8): 1309–1317.
42. Ni, Y., et al., Chemically functionalized water soluble single-walled carbon nanotubes modulate neurite outgrowth. *J Nanosci Nanotechnol*, 2005, 5(10): 1707–1712.
43. He, W. and R.V. Bellamkonda, Nanoscale neuro-integrative coatings for neural implants. *Biomaterials*, 2005, 26(16): 2983–2990.
44. Zhong, Y. and R.V. Bellamkonda, Dexamethasone-coated neural probes elicit attenuated inflammatory response and neuronal loss compared to uncoated neural probes. *Brain Res*, 2007, 1148: 15–27.
45. Kim, D.H. and D.C. Martin, Sustained release of dexamethasone from hydrophilic matrices using PLGA nanoparticles for neural drug delivery. *Biomaterials*, 2006, 27(15): 3031–3037.
46. Li, H.P., et al., Regeneration of nigrostriatal dopaminergic axons by degradation of chondroitin sulfate is accompanied by elimination of the fibrotic scar and glia limitans in the lesion site. *J Neurosci Res*, 2007, 85(3): 536–547.
47. Bradbury, E.J., et al., Chondroitinase ABC promotes functional recovery after spinal cord injury. *Nature*, 2002, 416(6881): 636–640.
48. Osterhour, D.J., et al., Chondroitinase release from nanospheres induces axonal sprouting after spinal cord injury, in *Society of Neuroscience*. 2006: Atlanta.
49. Das, M., et al., Auto-catalytic ceria nanoparticles offer neuroprotection to adult rat spinal cord neurons. *Biomaterials*, 2007, 28(10): 1918–1925.
50. Olivier, J.C., Drug transport to brain with targeted nanoparticles. *NeuroRx*, 2005, 2(1): 108–119.

13 Nanotechnologies for Cardiovascular Tissue Engineering

Kristen A. Wieghaus and Edward A. Botchwey

CONTENTS

13.1 INTRODUCTION

The primary role of the circulatory system is the distribution of dissolved gases and other molecules necessary for nutrition, growth, and repair of tissues and organs of the body. Other functions of the circulatory system include fast chemical signaling to cells by circulating hormones or neurotransmitters, heat dissipation from the body's core to the surface, and mediation of inflammation and host defense responses.

The heart is composed of four chambers: the left and right atria and ventricles. The myocardium is composed of spontaneously contracting cardiac muscle fibers that allow heart contraction and blood pumping, and is covered with epi- and endocardial tissue for protection. Heart valves function to retain unidirectional blood flow between heart chambers and between chambers and connecting blood vessels. The aortic and pulmonary semilunar valves are trileaflets, while the atrioventricular valves (mitral and tricuspid) are bileaflet. Unlike most tissues, heart valves are not well vascularized, because the tissues are thin enough to maintain nourishment through direct diffusion from the heart's blood. Among the cardiac valves, the aortic valve is most commonly diseased and requires most frequent transplantation [1]. Two types of cells comprise the aortic valve: endothelial cells that cover its surface, and interstitial cells that have variable properties of fibroblast, smooth muscle, and myofibroblast cell types [2]. These cells are situated on a bed of extracellular matrix

(ECM) primarily composed of collagen, which provides strength and stiffness to the tissue, elastin, which allows for tissue extension during diastole and contraction in systole, and glycosaminoglycans, which accommodate shear to the cuspal layers and help cushion shock to valve tissue [1].

The vascular system is the first organ system to develop during vertebrate embryogenesis, supplying blood to the cells and tissues of the body. It consists of a complex network of arteries, veins, smaller microvessels, and capillaries that facilitate oxygen, nutrient, and waste exchange as well as the migration and trafficking of cells from the circulation into various tissues. The composition of blood vessels is quite complex [3]. In particular, arteries are composed of three distinct layers that surround a hollow lumen through which blood flows. The innermost, the tunica intima, is comprised of a single layer of endothelial cells (EC) supported by a basement membrane. The tunica media, the thickest of the three layers, includes several layers of vascular smooth muscle cells (vSMC) and surrounding ECM arranged in concentric layers to form a composite structure [4]. The outermost layer, called the adventitia, is formed from fibroblasts and type I collagen matrix, which imparts strength to the vessel wall [3].

As many significant age-associated diseases such as heart attack, peripheral limb ischemia, and chronic wounds arise from impaired or abnormal function of vasculature, the need for effective therapies to address cardiovascular deficiencies is an increasingly significant clinical problem. Cardiovascular diseases accounted for 2.4 million deaths in the United States in 2000 alone [5]. The largest cause of mortality in the Western world is atherosclerotic vascular disease [6], which presents as coronary artery disease or peripheral vascular disease. Arteriosclerosis is characterized by the thickening of the arterial wall (with corresponding decrease in lumen diameter), leading to eventual decrease in or loss of circulation beyond the arteriosclerotic area of the vessel. Additionally, peripheral vascular disease affects upward of 12 million Americans with few significant treatment options [3]. Heart failure is another leading cause of morbidity and mortality; because of the heart's limited ability for self-repair, surgical intervention is required for treatment of severe cardiac problems. Despite advances in surgical treatment, whole organ transplants remain the only option once more conservative treatments fail. Limited donor organ supply, significant potential for organ rejection, and other complications severely limit the success of these procedures.

Despite the large number of patients suffering from these diseases and enormous research expenditures supporting basic science underlying these disease pathologies, available treatments to address vascular insufficiency have relatively limited effectiveness. Among many considerations, this is clearly evidenced by as many as 25% of patients undergoing coronary bypass surgeries who suffer repeat heart attacks.

In recent years, increasing emphasis has been directed toward the study of the complex interplay of many soluble factors, insoluble adhesion molecules within the ECM, and hemodynamic factors that regulate vascular growth and remodeling in vivo. In this chapter, current strategies for cardiac and vascular tissue engineering are explored, focusing on the therapeutic manipulation of cardiac and vascular cell-specific signaling at the nanoscale. In Section 13.2, we highlight major paradigms of cardiovascular tissue engineering: the fabrication of cardiac tissue, cardiac valves, vascular stents, and small-diameter tissue-engineered bypass grafts, and the induction of neovascular structures into tissue-engineered scaffolds and implanted tissues. Subsequently, the nanoscale detail of the natural cardiovascular system and the importance of nanostructure for cardiac and vascular cell behavior are presented in Section 13.3. In Section 13.4, current nanotechnological strategies for improvement of cardiovascular tissue-engineered constructs are presented in detail. These areas of research can be organized into three types of approaches: modification of biomaterial surfaces with nanotopographical features, direct functionalization of surfaces with nanoscale motifs, and fabrication of biomimetic nanofibrous meshes and scaffolds. Finally, we conclude with challenges remaining in the field nanotechnology as it relates to cardiovascular tissue engineering.

13.2 VASCULAR TISSUE ENGINEERING PARADIGMS

13.2.1 Tissue-Engineered Cardiac Wall Tissue

The ability of the myocardium to regenerate itself after extensive infarction is quite limited; as a consequence, new modalities for regeneration of functional myocardial tissue must be investigated. Fabrication of a full tissue-engineered heart would require the assembly excitable and contractile atrial and ventricular myocardium, epicardium, and endocardium, valves, and vasculature [7]. Additionally, design requirements for heart tissue are far more demanding than for many other tissue and organ types, as cardiac muscle and its supporting tissues have complex and specialized properties that are integral for proper cardiac function. These include formation of an electrical syncytium, development of systolic force, withstanding of diastolic loading, and adaptation to both physiological stimuli, including for example, fight or flight response, and increased circulatory demand [7].

13.2.2 Tissue-Engineered Cardiac Valves

In addition to cardiac muscle, heart valves are specialized tissues that must be replaced due to disease. Currently, diseased valves are surgically removed and replaced with either a mechanically functioning prosthetic, xenograft, cadaveric allograft, or pulmonary-to-aortic autograft valves (also known as a Ross procedure) [1]. However, each type of valve implant is associated with significant limitations, including potential for thromboembolism, calcification, and structural deterioration [8–10].

Heart valves created through tissue-engineering techniques might be more advantageous than those currently in clinical use, as they would likely be nonthrombogenic, infection resistant, and induce more cell viability than other implant types [1]. However, the engineering of functional heart valve tissue presents a number of significant challenges. Heart valves are dynamic tissues composed of specialized cells and ECM that must be carefully mimicked for clinical success of tissue-engineered implants, as these tissues must respond and remodel to complex changes in local mechanical forces [11,12]. More than 40 million times per year, valve leaflet opening and closing induced repetitive stress and shape changes on valve tissue and supporting structures [13]; hence, the strength, flexibility, and durability of these tissues must be properly matched in tissue-engineered valves.

One unique advantage of tissue-engineered heart valves is the possibility for successful long-term clinical use in pediatric patients. The most urgent need for heart valve technology is for children and young adults, in which the outcomes of valve replacement procedures are not as favorable as in the adult population [14,15]. Additionally, the possibility for growth and remodeling of tissue-engineered valve implants may eliminate the need for repetitive surgeries typically necessary for young patients, as conventional prosthetic valves cannot enlarge as the patient grows [1].

13.2.3 Cardiac Stents

The most often used treatment option for severe atherosclerosis is vascular stenting, whereby stents are implanted into severely plaqued and stenosed arteries as a mechanism to restore normal blood flow to ischemic organs. These stents are often fabrication from titanium (Ti) metal, due to its mechanical properties. However, Ti and other conventional metals are not biocompatible, leading to acute and chronic complications such as thrombosis and restenosis [16] due to dysfunction of vascular smooth muscle cells in the lumen [17]. As such, various attempts have been made to modify Ti and other metals used as vascular stents [18–21]; however, few attempts have been made to alter the metallic surface itself in order to optimize interactions with surrounding tissue.

13.2.4 Tissue-Engineered Vessels and Vascular Grafts

Once blood flow is occluded due to coronary or peripheral artery diseases, vascular bypass is often the only treatment option to restore blood flow to distal tissues. An estimated 519,000 bypass procedures were performed in 2000 [6]. Typical autologous grafts include portions of the saphenous vain, the internal mammary artery, and venous homografts [3]. However, some 30% of atherosclerosis patients lack autologous bypass material sufficient for autografting due to disease or previous use [22,23], and hence, vascular grafts manufactured from alternative, biocompatible materials must be implanted. The first synthetic vascular prosthesis was applied clinically in 1954, demonstrating that prosthetic tubes could function as arterial replacements [24,25]. However, these tubes did not incorporate any cell types associated with vascular function.

As discussed previously, tissue engineering allows for creation of replacement structures with cells and biomaterial scaffolds without harvesting autologous grafts from the patient. The first generation of tissue-engineered small-diameter vascular prosthetics added autologous EC on the luminal surface of the graft in an attempt to create a function endothelium [26–29]; however, lack of endothelial confluence on the lumen of the polymer tubes limited the success of these experiments. Second-generation graft materials included the addition of more native elements of blood vessels, including EC, vSMC, ECM to the synthetic material support [30–33]. For example, simple versions of these grafts were composed of a synthetic adventitial layer, a collagen medial component seeded with vSMC, and a neointimal lining of EC [30]. The most modern vascular prosthetics consist completely of natural materials [34,35], although synthetic polymers are still commonly employed in these applications.

Tissue-engineered grafts may provide a critically important alternative to autologous vein grafts. Materials including poly(ethylene terephthalate) (Dacron) and polytetrafluoroethylene (PTFE) have been fabricated successfully into vascular grafts to replace large arteries; however, smaller diameter arteries (<6 mm diameter) have seen quite limited success [36]. Complications such as acute thrombosis caused by lack of endothelium, intimal hyperplasia, and restenosis due to compliance mismatch and inflammatory response, and susceptibility to infection [36–42] decrease patency of small-diameter graft implants to 30% after 5 years [36]. To circumvent these complications, new classes of biomaterials have been investigated for use as small-diameter vascular graft materials. Poly(orthoesters) including poly(lactic acid), poly(glycolic acid) and their copolymers have been fabricated into tubular scaffolds that support endothelial and vascular smooth muscle cell growth in vitro [43,44]. Novel polymer materials combined with emerging technologies to optimize cell growth and survival are currently being investigated to provide patients with synthetic small-diameter vessels for grafting procedures.

13.2.5 Tissue-Engineered Scaffolds That Must Support Neovascularization

In addition to the fabrication of small-diameter vascular grafts, vascular tissue engineering also encompasses the promotion of neovascularization (growth of new capillaries, arterioles, and venules) to a region of tissue during wound repair or after implantation of a tissue-engineered scaffold. Proper microvascular structure is essential for normal tissue function [45]; however, after peripheral artery disease, myocardial ischemia, islet cell transplantation, plastic surgery [46], and other tissue insults, the native microvasculature is often compromised. The formation of new blood vessels via angiogenesis is critical for many clinical applications [47–52], particularly those wound repair and tissue growth paradigms in which significant nutritive and metabolic demands are created. Therefore, one particularly important challenge for the advance of tissue engineering applications is the development of effective therapies to promote neovascularization of engineered tissues [46,53–56], as cell migration and tissue ingrowth into these structures is limited [57]. Because loss or failure of tissue and organs is a frequent and expensive problem in human health care, it is postulated that development of methods to regenerate human tissue will have significant influence on medical specialties in the future [58].

13.3 NANOSCALE FEATURES OF NATIVE CARDIOVASCULAR TISSUES

Living tissues possess a high degree of nanosurface roughness; indeed, proteins, and other biomolecules with which cells are accustomed to interacting, are nanosized structures [59,60]. In fact, all aspects of the ECM, including fibers, pores, ridges, and grooves, are all on the order of a few to more than 100 nm [61]. Varying types of ECM are found in all cardiovascular tissues, including myocardium, heart valves, and arterial and venular vascular systems. Despite the diversity of ECM structures in the human body caused by different macromolecules and their organization, nanoscaled dimensions are conserved throughout [62]. The ECM does more than just provide structural support for cells; the ECM is critical for modulation of cell motility, proliferation, morphology, metabolism, differentiation, and matrix deposition [63]. For example, it has been suggested that endothelial cells are more thrombogenic on materials that are "smooth" on the nanoscale; that is, any material surface features are much larger than nanosized. This lack of thromboresistance implies that the cells are not fully differentiated, as they are deficient in anticoagulant function [64,65]. It was hypothesized that lack of substrate-mechanical forces with cells on a nanoscale causes dedifferentiation; indeed, appropriate forces are most likely generated in vivo by topographic features on the 100–1000 nm scale range, where cell–ECM interactions take place [66].

The basement membrane is one particular type of ECM that underlies all epithelial and endothelial cell sheets [61], including the vascular endothelium. This 40–120 nm mat separates these tissues from underlying connective tissues, and is composed mainly of type IV collagen and laminin nanofibers embedded in heparin sulfate proteoglycan hydrogels [62]. These characteristics are critical to recreate in blood vessel tissue-engineering scaffold designs because endothelial monolayers grow directly on basement membrane in native blood vessels [62]. Similarly, the matrix that underlies vSMC is comprised of nanosized biological cues and structural elements. The ECM includes fibers of collagen types I (approximately 44%) and III (44%) elastin (12%) [4,67,68] dispersed among various proteoglycans. See Figure 13.1 for a representative image of native ECM as compared with some emerging technologies for nanofeatured biomaterials.

The deployment of vascular stents might also meet with improved clinical success if they are fabricated with nanoscale architecture. Although they are not natural in origin, they should be

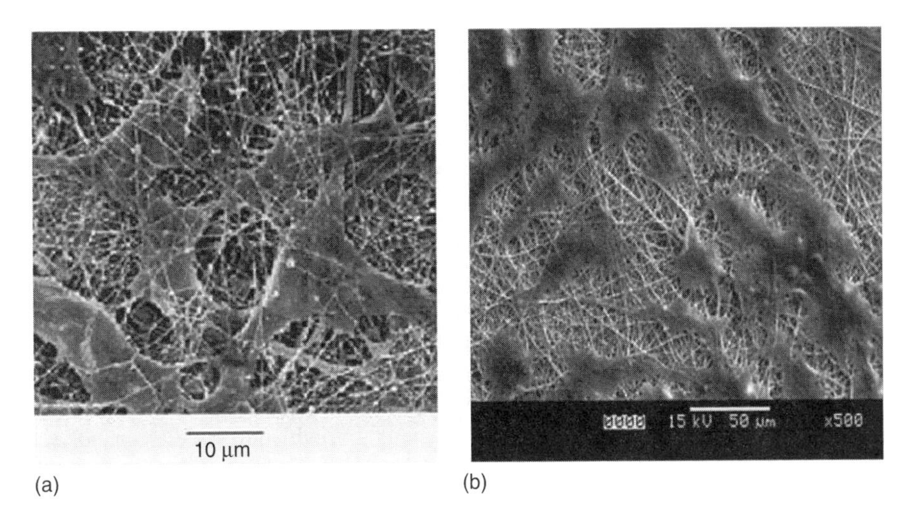

FIGURE 13.1 Similarity between native ECM protein structure and electrospun polymeric nanofiber matrix. (a) Fibroblasts cultured on collagen fibrils of rat cornea. (From Nishida, et al., *Invest. Ophthalmol. Vis. Sci.*, 29, 1888, 1988). (b) Endothelial cells cultures on electrospun poly(caprolactone) (PCL) nanofiber matrix. (From Ma, Z., et al., *Tissue Eng.*, 11, 101, 2005. With permission.)

designed to mimic the structure of natural vascular tissue, which possesses a high degree of nanostructure. Despite this fact, most currently utilized vascular stents are smooth on the nanoscale, and limit endothelial cell growth on the constructs [17].

13.4 NANOTECHNOLOGICAL STRATEGIES FOR IMPROVED CARDIOVASCULAR TISSUE ENGINEERING

The complex and specialized properties and features of the myocardium and valvular tissues, including mechanical properties that allow for contraction of cardiac tissue and deformation of valves, provide significant obstacles for the creation of functional tissue-engineered implants for cardiac interventions. Similarly, the composition and structure of the vascular wall imparts unique challenges for the development of living tissue-engineered small-diameter vessel replacements that can withstand high pressures and pulsatile environments [3]. The intricate nature of vessels creates a tremendous engineering challenge to resupply implanted and growing tissues with a functional microvasculature after disease, injury, or surgery. It has been hypothesized that biomaterial surface modification on the nanoscale might improve cell and tissue response to implanted biomaterials [69–71]. Nanotopographical cues can alter essential cellular functions including adhesion, recruitment, movement, morphology, apoptosis, gene expression, and protein production; for these reasons [69,72,73], nanostructuring of tissue-engineered materials has become a critical area of biomaterials research. For cardiovascular tissue engineering strategies including myocardial tissue, heart valve implants, small-caliber vessel graft development, and tissue-engineered scaffolding structures that promote neovascularization in surrounding tissue, it is clear that the creation of nanosized features on biomaterials might more accurately mimic the ECM features natural to cardiovascular cell types in order to improve cell adhesion and growth.

13.4.1 Engineering of Nanotopographical Features on Biomaterials

As discussed previously, procedures involving vascular stent materials would benefit from nanostructure on the metallic implant. Choudhary et al. [17] fabricated nanostructured and nanosmooth Ti structures to study the viability and function of vascular cells on such constructs. Atomic force microscopy (AFM) was utilized to confirm the presence of nanoarchitecture on the nanostructured Ti compacts, demonstrating an approximately 2.5-fold increased surface roughness, as compared with conventional Ti compacts. Adhesion of both vascular smooth muscle cells and endothelial cells were significantly greater on nanostructured Ti compared to nanosmooth Ti. Longer-term cell viability and density experiments showed that nanostructured Ti surfaces facilitated growth of a confluent monolayer of both ECs and vSMCs days earlier than on conventional Ti. Importantly, both ECs and vSMCs exhibited more than 2-fold increases in collagen and elastin production on nanorough Ti, showing that nanoarchitecture increased ECM synthesis and cell function.

The design of most of vascular tissue engineering strategies involves the seeding and culturing of EC and vSMC to create more functional vessels before implantation. These materials are, however, comprised of microsized fibers and surface features that do not adequately mimic the natural nanodimensional topography of the vascular tissue being replaced [61,74]. The engineering of nanofeatures onto the surfaces of biomaterials may be necessary to more accurately represent the natural ECM of endothelial and vascular smooth muscle cells.

Miller et al. [75] cultured rat aortic endothelial and smooth muscle cells on poly(lactic-co-glycolic acid) (PLAGA) films treated with 10 N NaOH for 1 h, to create nanofeatures on the film surface. This treatment etched 476%–502% greater surface roughness onto the films than the untreated samples, depending on method of film fabrication. These treated films were then compared against untreated and microstructured (treated with more dilute NaOH for a shorter period of time) films to determine the effects of nanotopography on the adhesion and growth of EC and vSMC. Interestingly, EC adhesion was decreased significantly on nanofeatured films, although it significantly augmented vSMC adhesion to the polymer. In a similar manner, cell density experiments over time demonstrated that nanofeatured PLAGA supported significantly increasing vSMC density over 5 days in culture, while significantly deterring the growth of EC on the same surface. However, significant promotion of EC growth was established on nanofeatured PLAGA fabricated via a silicone elastomer casting method instead of a more traditional solvent-casting approach [75]. These studies demonstrated that by better mimicking the surface characteristics of natural ECM, EC, and vSMC adhesion and growth on biomaterials were improved. Further experiments investigated the mechanism by which increased surface nanotopography promotes vascular cell adhesion and growth, and demonstrated that nanostructured polymer samples adsorbed significantly more fibronectin and vitronectin than nonroughened surfaces [76]. As these two proteins are critical for mediation of cellular adhesion, it is clear that increased attraction of ECM proteins to nanofeatured materials may provide the basis for enhanced vascular cell performance.

Nanotopography can also be applied to biomaterial surfaces in more precise, detailed manners than is possible with base-etching of polymers via colloidal lithography and embossing master. Using colloidal lithographic techniques, nanocolloids are spaced and electrostatically assembled over a polymer surface and utilized as a mask to alter the substrate surface. The colloids are then removed by etching, resulting in a surface of cylindrical nanocolumns [70,71]. Embossing master techniques employ metal or epoxy substrates with a desired pattern, produced by electron beam writing technology. This master is then hot-pressed into a material to transfer the reverse pattern onto the surface [72]. Giavaresi et al. [73] utilized these techniques to create nanopits 100 nm deep and 120 nm in diameter on the surface of a 3-hydroxybutyrate-co-3-hydroxyvalerate (3HB-3HV) polyester containing 8% 3HV, and 160 nm height, 100 nm diameter poly(caprolactone) (PCL) nanocylinders on PCL bulk. The authors investigated the in vivo local tissue response to subcutaneous implantation of the nanofeatured materials, in comparison to untreated polymer controls. Notably, the nanostructured surfaces presented thicker inflammatory capsules surrounding the implants than their respective, unmodified polymer surfaces. PCL implants stimulated significantly more vascular density after 12 weeks implantation than 3HB-3HV implants, and nanostructured PCL implants stimulated more vascular density than their respective PCL controls. This set of experiments marks the first evaluation of local tissue response to nanotopographical features, demonstrating that nanotopographical surfaces may promote increased neovascularization in vivo. Therefore, nanofeaturing (and in particular, creating nanocolumns) biomaterial surfaces might be an attractive method for fabricating tissue-engineered scaffolds designed to augment new blood vessel growth into regenerating tissue.

As an alternate strategy to forming nanofeatures in the form of surface roughness or nanocylinders on the surface of a polymer bulk, Goodman et al. [65] cast the actual, native ECM onto the surface of a polymer. An inverse replica of the subendothelial ECM surface of a denuded and distended vessel was prepared by casting the luminal space with a polymeric resin. The textured ECM surface was then replicated in polyurethane, and a complex topography was also observed at the submicrometer level, with many measurable features of the order of 100 nm. Goodman et al. then demonstrated that EC exhibited superior adhesion and morphology on the ECM-cast polymer films than on nontextured controls. While this technique certainly captured the nanofeatures of the native ECM, the cast was destroyed in order to recover the polyurethane replica for further study. Therefore, each graft/scaffold would require its own ECM-cast, which is inefficient in terms of time and expense [75].

13.4.2 Functionalization of Biomaterial Surfaces with Nanosized Motifs

As an alternative to the engineering of surface roughness or features on a nanoscale dimension on biomaterial surfaces, direct surface functionalization with bioactive peptide and protein motifs have been proposed to improve vascular cell adhesion, growth, and function. Materials suitable for tissue engineering applications, including vascular schemes, must have mechanical, structural, and degradation properties within rigid ranges, and must also support signal transmission with the surrounding tissue [77]. One way to incorporate all of these design constraints is to utilize a coating approach, or to functionalize the surface of a bulk material.

Osteopontin (OPN) is one such protein proposed for such strategies. A multifunctional protein, OPN binds many cell types through RGD-mediated interactions with various integrins [78], can itself bind to components of the ECM including collagen [79] and fibronectin [80], and has been proven to enhance wound healing [81], endothelial cell survival [82], and angiogenesis [83]. As a motif bound to the surface of biomaterials in a physiologically relevant conformation and orientation, OPN might facilitate endothelial survival and growth, wound healing, and decreased inflammatory reaction to implanted biomaterial scaffolds [84]. Liu et al. [84] exploited the control of surface coating afforded by hydrocarbon-based, end-functionalized, self-assembled monolayers (SAMs) to tether OPN on biomaterial surfaces. The authors determined that a —NH_2 terminated SAM with adsorbed OPN maximized protein coating; protein aggregation and incomplete OPN coverage were demonstrated on hydrophobic —CH_3 terminated samples. Coverage of bovine aortic endothelial cells on the material surface paralleled the AFM images, suggesting that protein-tethering to an —NH_2 moiety might optimize OPN conformation/orientation to that which EC adhere. As the extent of cell spreading has been found to promote cell growth and function [85,86], OPN-treated scaffolds, then, might augment endothelial survival and angiogenic potential [84]. These scaffolds may enjoy future utility to increase neovascularization into tissue-engineered scaffolds and provide a coating for biomaterial surfaces designed for small-diameter vascular grafts.

Multilayer nanofilms as fabricated via a layer-by-layer (LbL) method can also be used to create bioactive thin films on materials in a simple and controllable fashion. This type of assembly involves the alternate adsorption of positively and negatively charged polyelectrolytes [87–93], and the subsequent film can be later decorated with bioactive molecules capable of transmitting signals to cells. Wittmer et al. [77] created a nanofilm of layered poly(L-lysine) (PLL) and dextran disulfate (DS), and terminated the film with fibronectin moieties to promote endothelial cell attachment and spreading on the material surface [94]. This method allowed for the creation of nanothickness films: the PLL–DS layer was approximately 10 nm, with a 22 nm fibronectin coating overlaid. Wittmer et al. then investigated the kinetics of human umbilical vein EC on their surfaces, demonstrating that cells on the fibronectin-coated surface spread 140% of that of untreated titanium surfaces. Hence, this paradigm may serve as a promising strategy to display nanosized layers of biological cues, such as fibronectin, onto vascular tissue engineering biomaterials to modulate cell interactions and thereby augment cell function.

In addition to proteins, physiologically relevant inorganics/ceramics have also been proposed as coatings for vascular tissue engineering materials. Furozono et al. [95] created a coating of nanocrystalline hydroxyapatite (HA) crystals on Dacron fabric through covalent linkages, and suggested its use as small-diameter graft material. As discussed in Section 13.2, Dacron is a material successfully utilized as bypass graft material for blood vessels greater than 6 mm in diameter; when Dacron (or any other conventional material) is incorporated into smaller caliber vessels, the bypass fails due to thrombosis or restenosis. Furozono et al. proposed that the nanosized HA coating might provide less pathogenic interactions with endothelial cells than the Dacron fabric alone. The authors demonstrate that their HA-coated Dacron fabric allows for significantly better adhesion of human umbilical vein EC than the fabric alone, showing that the coating might promote enhanced endothelial function than what is currently clinically available.

13.4.3 Electrospun Nanofibrous Scaffolds

Electrospinning has become a popular fabrication method for biomaterials, particularly in the tissue engineering community, due to increased interest in creating tailored scaffolds with nanoscale properties [96]. The method requires only a simple and inexpensive apparatus incorporating a syringe pump, high voltage source, and a collector. An example of a typical electrospinning setup is shown in Figure 13.2. To create a nanofibrous mesh, a polymer solution is loaded into and slowly

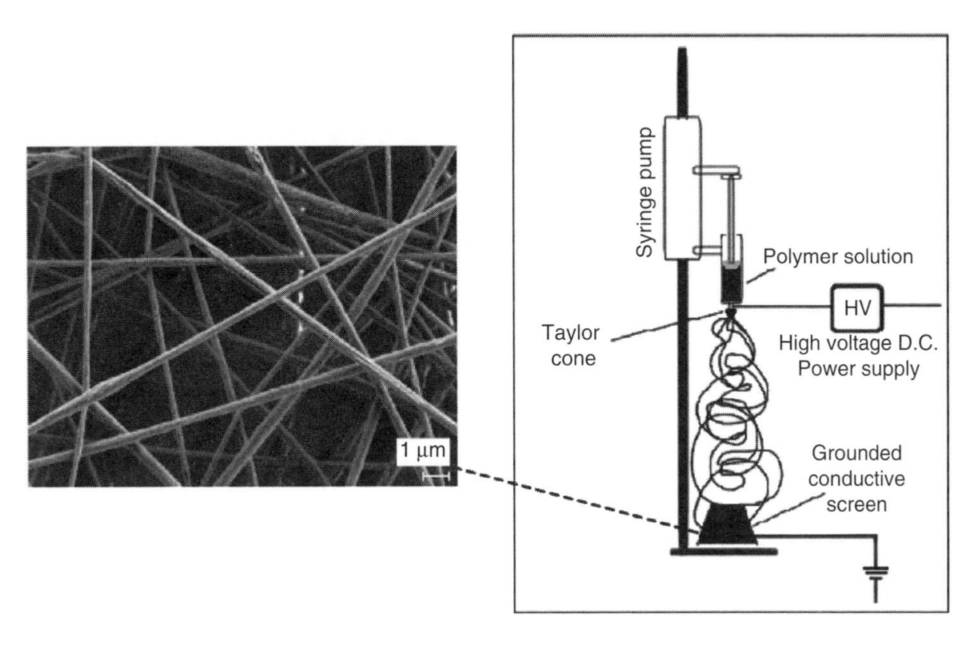

FIGURE 13.2 A typical electrospinning apparatus and its product: a nonwoven fiber mat. (From Yordem, O.S., *Mater. Des.*, 29, 34, 2008. With permission.)

released from the syringe, and a small amount of material is held at the needle tip by surface tension. A high voltage supplied by an electric field charges the solution, resulting in the creation of electrostatic force in the form of charge repulsion within the solution. As the surface tension is overcome, a jet of polymer solution is initiated. As the jet travels toward the collector, the solvent evaporates and fibers are captured [96].

This method of scaffold fabrication has been utilized to electrospin a number of synthetic and natural polymer materials [96] into fibers ranging in diameter from 3 nm to more than 5 μm [97]. The potential for nanoscale featuring on electrospun meshes has remained unmatched by any other biomaterials processing technique. High surface-to-volume ratio and porosity can be achieved in electrospun meshes for better cell incorporation and perfusion. Additionally, control over nanostructure allows for flexibility in tissue design [98]. It has been documented that fiber diameter of nanofibrous scaffold can exert appreciable influence on cell behavior [99], and therefore appropriate fiber ranges for particular cell types and tissue structures can be determined. Scaffolds formed from electrospun meshes can capitalize on the inherent nanoscale nature of native ECM and may thereby support the growth of cells and tissues that are not supported by macro- or microfeatured designs.

Owing to the many advantages of electrospun scaffolds, the method has been exploited by cardiovascular tissue engineers in an attempt to create cardiac tissue. Shin et al. [100] produced PCL, nanofibrous nonwoven meshes with an average fiber diameter of 250 nm, and coated them with collagen to promote cell attachment. The mesh was stretched across a wire ring, added as a passive stretching device to condition and mature cells seeded on the implants and permit contractions at the cells' natural frequency. After 3 days in culture, rat cardiomyocytes began to contract on the scaffold, and displayed continuous contractility throughout the 14 days of experimentation. From cross-sectional views of the meshes, the cardiomyocytes formed a tight arrangement and numerous intercellular contacts throughout the entire mesh (see Figure 13.3a). Additionally, staining for actin filaments (Figure 13.3b) demonstrate actin throughout the entire mesh thickness. The interior of the cell-scaffold structures also stain positively for cardiac troponin-I (Figure 13.3c). This work provides an approach for the fabrication of engineered myocardium with a synthetic polymer with nanoscale features.

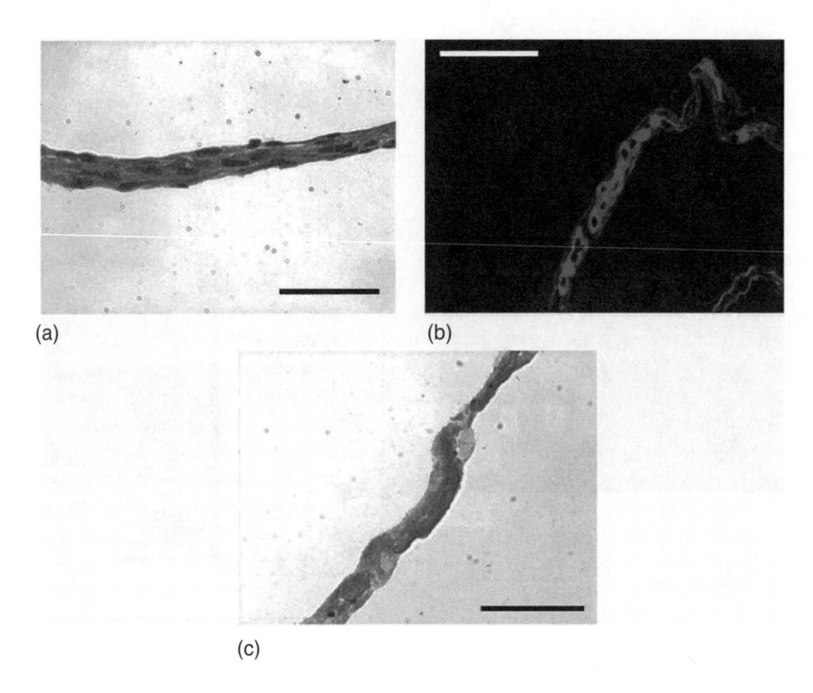

FIGURE 13.3 Staining of rat cardiomyocytes on a nanofibrous PCL electrospun mesh stretched over a wine ring to facilitate tissue contraction. (a) Hematoxylin and eosin staining showing that cells have attached through the entire mesh. (b) Immunohistochemical staining for F-actin; actin fibers traverse the entire thickness of the polymer scaffold. (c) Staining for cardiac troponin-I, found in the interior of the cell-scaffold construct. All images: 400X, scale bar–50 μm. (From Shin, M., et al., *Biomaterials*, 25, 3717, 2004. With permission.)

Electrospinning can be utilized to easily address issues of directional growth or anisotropy in cell/tissue organization. Anisotropy is functionally critical for cardiac tissue engineering [101–103], and therefore, electrospinning has become a popular method for scaffold fabrication for cardiac tissue engineering paradigms. For example, Zong et al. fabricated submicron nanostructured PLGA membranes with varying polymer ratios and compositions, including variations in hydrophobicity and degradation rate (polylactic acid, 75:25 PLGA, and a copolymer of 75:25 PLGA and polyethylene gycol-PLGA [PEG-PLGA] were utilized). To produce oriented, anisotropic fiber architecture from the randomly-oriented electrospun scaffolds, the meshes were uniaxially stretched to an extension ratio of 200% at a constant rate of 4 mm/min at 60°C, and then cooled to room temperature under tension. Figure 13.4 demonstrates results of the uniaxial stretching procedure, which produced uniformly oriented, anisotropic polymer scaffolds. Subsequent studies with rat cardiomyocytes showed that cells aligned with biomaterial fiber direction with elongated nuclei, and that post-processing fiber alignment provided guidance for cells to grow along the stretching direction. Hence, electrospinning techniques with subsequent processing procedures to create anisotropic meshes might advance cardiac tissue engineering, in terms of creating scaffolds with properties important for directional growth of cardiac tissue.

Electrospinning has also been proposed as a means for aortic valve implant fabrication, as cardiac valves have extremely complex geometries and mechanical requirements. Theoretically, meshes can be electrospun onto a collector of any conceivable geometry, and, as described above, methods for manipulating the mechanical properties of spun meshes are being developed. Van Lieshout et al. [104] compared the feasibility of electrospinning PCL for valve implants, as compared with knitting of the same polymer. Knitting refers to a technique used to create certain scaffolds and prostheses, including vascular stents, and that can also produce complex geometries. PCL was fabricated into the geometry of the aortic valve using both fiber-organizing techniques, and

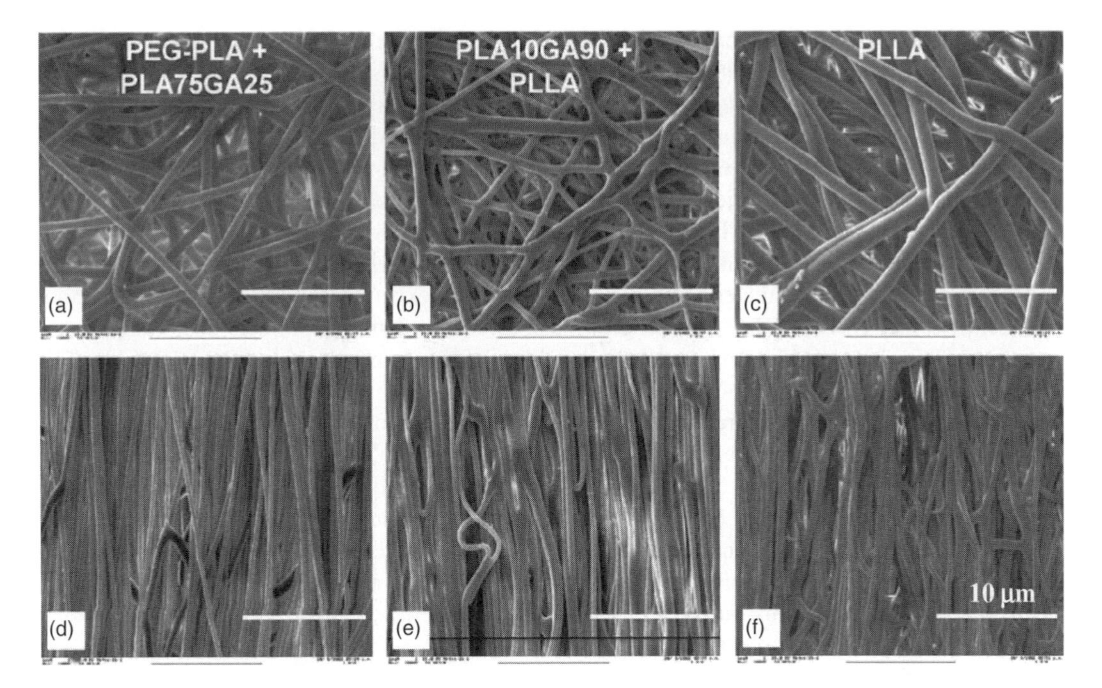

FIGURE 13.4 (a–c) SEM photographs of as-prepared electrospun membranes of (a) PEG–PLA + PLA75GA25, (b) PLA10GA90 + PLLA, and (c) PLLA. (d–f) SEM images of corresponding electrospun membranes after uniaxialstretching to fabricate oriented, anisotropic polymer meshes for cardiac tissue engineering. (From Zong, X., et al., *Biomaterials* 26, 5330, 2005. With permission.)

the resultant valve implants were subjected to pulse duplicator testing to investigate their function in a system replicating the flow of a human heart. Knitted valve implants performed in a superior manner to spun implants over 6 h during pulse duplication. The electrospun valve does not open symmetrically, detachment of one leaflet-sinus unit from the aortic wall is seen, and eventually, one of the leaflets ruptures completely. Nonetheless, both electrospun, knitted, and layered scaffolds containing both techniques are being investigated for use as aortic valve implants.

For vascular tissue engineering applications, investigators have studied the response of endothelial cells to various fiber size ranges to optimize their attachment and growth on nanoscaled surfaces. For example, Kwon et al. [105] seeded human umbilical vein EC on electrospun poly (lactide-co-caprolactone) (PLCL) matrices of various average fiber diameters and determined that cell growth was best on fibers on the highest end of the nanoscale (approaching 1 μm). Optimal endothelial cell growth will need to be optimized for each biomaterial investigated and each type of endothelial cell utilized in studies.

A myriad of biomaterials have been selected for electrospinning procedures for vascular tissue engineering. Poly(organophosphazenes) have been proposed as biomaterials suitable for electrospinning applications for vascular tissue engineering. These polymers have a high molecular weight backbone of alternating phosphorus and nitrogen atoms and two organic side groups. Properties including hydrolytic degradation into nontoxic products [106] and ability to support adhesion of various cell types, including endothelial cells [107], suggest that poly(organophosphazenes) may be suitable for vascular cell-polymer constructs. Carampin et al. [108] synthesized a new copolymer, poly[(ethyl phenylalanato)$_{1.4}$ (ethyl glycinato)$_{0.6}$ phosphazene] (PPhe–GlyP) and electrospun tubes with fibers of various size ranges. Selecting parameters that afforded an average fiber size of 850 ± 150 nm, the authors showed that rat neuromicrovascular EC proliferation significantly increased, as compared with tissue-culture controls. Over a culture period of 16 days, a confluent monolayer of EC spread through the inner diameter of the PPhe–GlyP tube without entering the inner layers; additionally, the tube did not collapse during this time frame. Hence, these nanofibrous tubular materials may show promise as small-diameter vessel substitutes.

In addition to biodegradable polymers, natural materials have also been proposed as scaffolds for vascular constructs. Collagen [109] and elastin [110], discussed previously as essential elements of vascular ECM, have been electrospun to produce nanodiameter fibers that mimic the molecular and structural properties of native membrane in a manner unmatched by synthetic materials. He et al. [111] blended PLCL with collagen to form nanofibers with average diameters between 100 and 200 nm. Mats of these nanofibers were much less stiff than traditional Dacron material, which is advantageous for tissue-engineering blood vessels that will experience pulsatile flow. The authors also demonstrated increased EC spreading on their blended polymer nanofibers than on synthetic nanofibers. Stitzel et al. [112] combined natural and synthetic polymers by electrospinning a blend of collagen (45%), elastin (15%), and PLAGA (40%, included for more superior electrospinning product than was possible with the natural materials alone). The nanofibrous tubes exhibited a compliance similar to that of natural vasculature (12%–14% diameter change vs. 9% for native vessels), and could withstand burst pressures nearly 10 times that of normal systolic blood pressure. The authors were able to culture confluent layers of EC and vSMC on the inner and outer surfaces of the tubular scaffold, respectively, with minimal inflammatory response. Novel polymer/protein composite scaffolds are beginning to exhibit both the mechanical properties and tissue compatibility that will be required for successful small-diameter bypass grafting materials.

In perhaps the most sophisticated electrospun vascular scaffold thus far, Boland et al. [3] developed a three-layered, 4 mm diameter vascular construct. These layers included separate medial and adventitial layers of electrospun collagen and elastin (average diameter 490 ± 220 nm) seeded with vSMC and fibroblasts, as well as an endothelium of seeded EC (see Figure 13.5). The authors

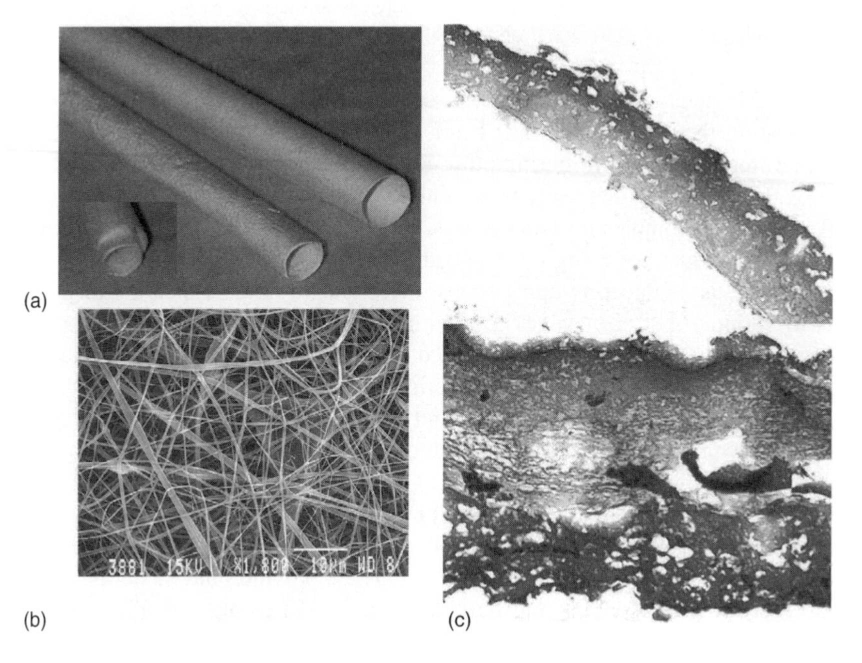

FIGURE 13.5 (See color insert following page 206.) (a) Collagen electrospun scaffolds fabricated in 2 and 4 mm tubular structures for small-diameter vascular graft engineering. (b) An electrospun blend of collagen type I, type III, and elastin (in a 40:40:20 ratio) showing fibers with an average diameter of 490 ± 220 nm and random fiber orientation. (c) Top: Histological section demonstrating three defined layers of vessel wall architecture: intimal (dark purple, left), medial (light purple, center) and adventitial (dark purple, right). Bottom: Three-layered prosthetic electrospun scaffold after seeding with endothelial cells, vascular smooth muscle cells, and fibroblasts, demonstrating similar definition of layering as native tissue. (From Boland, E.D., et al., *Front Biosci.*, 9, 1422, 2004. With permission.)

reported a morphologically mature layer of EC with a clearly defined basement membrane, with a dense layer of vSMC aligning in a circumferential fashion underneath. vSMC and fibroblasts in the adventitial layer were interspersed throughout the outer layers of collagen/elastin nanofibers. This multidimensional scaffold architecture allows for more complex cell–cell and cell–matrix interactions than are possible with only one layer of scaffold [3]. Creating scaffolds of alternating layers may prompt more physiologically relevant development of cellular and ECM organization.

13.5 CONCLUSIONS

It is clear that viable tissue-engineered cardiac structures and vessels will require full mimicry of native vascular structure [3]. Engineering designs for cardiovascular constructs are continually evolving and gaining complexity, as investigators attempt to include increasing numbers of relevant cell–cell and cell–matrix interactions into their designs. However, significant challenges remain for optimization and future clinical translation of nanotechnologies employed in vascular tissue engineering applications.

Advances have been made in cardiac tissue engineering, including electrospun structures that permit natural contractions of seeded cardiomyocytes and techniques that begin to provide nanoarchitectured anisotropy to cultured cells and tissues. However, truly successful cardiac grafts with mechanical and electrical properties of natural tissue have yet to be developed. As the only option for cardiac failure is currently full organ transplant, this area of research remains a priority for the cardiovascular research community. Additionally, better techniques for tissue engineering of cardiac valves would certainly advance the field of valve implantation. This area of research has yet to fully turn to nanofeatured paradigms for valve design.

One critical challenge for vascular tissue engineering that remains is to determine which size and type of nanofeatures will maximize EC and vSMC growth on vascular constructs. Additionally, the vascular tissue engineering community has yet to determine the most appropriate biomaterial and potential coating. Importantly, promising vascular grafts must parallel native vessels on critical mechanical properties, including compliance and burst pressure, or the grafts will ultimately fail in the clinic.

Additionally, many of the strategies proposed above have potential to be translated from techniques intended for small-diameter vascular grafts to designs for tissue-engineered constructs that promote implant microvascularization. Many of the nanotopographical features proposed to enhance endothelial growth on small-caliber bypass grafts could be exploited by tissue engineers and those interested in regenerative medicine to increase vascular density in new, growing tissue. It is clear that the clinical success of tissue engineering strategies will be contingent upon timely revascularization of the regenerating tissue. Advances in technologies that impart nanoarchitecture to promote endothelialization of tissue-engineered biomaterial implants may prove to be effective therapies for improved neovascularization.

REFERENCES

1. Mendelson, K. and F.J. Schoen. 2006. Heart valve tissue engineering: Concepts, approaches, progress, and challenges. *Ann Biomed Eng* 34:1799–1819.
2. Aikawa, E., P. Whittaker, M. Farber, et al. 2006. Human semilunar cardiac valve remodeling by activated cells from fetus to adult: Implications for postnatal adaptation, valve pathology, and tissue engineering. *Circulation* 113:1344–1352.
3. Boland, E.D., J.A. Matthews, K.J. Pawlowski, et al. 2004. Electrospinning collagen and elastin preliminary vascular tissue engineering. *Front Biosci* 9:1422–1432.
4. Rhodin, J. 1980. Architecture of the vessel wall. In *Handbook of Physiology: The Cardiovascular System*, D. Borh, A. Somlyo, and H. Sparks (Eds.), pp. 1–31. Massachusetts: American Physiological Society.

5. American Heart Association, 2003. Heart disease and stroke statistics–2003 update. http://www.americanheart.org/downloadable/heart/10590179711482003HDSStatsBookREV7-03.pdf
6. National Heart, Lung, and Blood Institute, 1981. Report of the Working Group on Arteriosclerosis, MD. U.S. Department of Health and Human Services, Public Health Service, National Institutes of Health; DHEW, NIH Publ. No. 82-2035:114–122.
7. Zimmermann, W.-H., M. Didie, S. Doker, et al. 2006. Heart muscle engineering: An update on cardiac muscle replacement therapy. *Cardiovasc Res* 71:419–429.
8. Hammermeister, K., G.K. Sethi, W.G. Henderson, et al. 2000. Outcomes 15 years after valve replacement with a mechanical versus a bioprosthetic valve: Final report of the Veterans Affairs randomized trial. *J Am Coll Cardiol* 36:1152–1158.
9. Jamieson, W.R., O. von Lipinski, R.T. Miyagishima, et al. 2005. Performance of bioprostheses and mechanical prostheses assessed by composites of valve-related complications to 15 years after mitral valve replacement. *J Thorac Cardiovasc Surg* 129:1301–1308.
10. Schoen, F.J. 2001. Pathology of heart valve substation with mechanical and tissue prostheses. In *Cardiovascular Pathology*, M.D. Silver, A.I. Gotlieb, and F.J. Schoen (Eds.), pp. 402–405. New York: Churchhill Livingstone.
11. Rabkin-Aikawa, E., M. Farber, M. Aikawa, et al. 2004. Dynamic and reversible changes of interstitial cell phenotype during remodeling of cardiac valves. *J Heart Valve Dis* 13:841–847.
12. Schoen, F.J. 1997. Aortic valve structure–function correlations: Role of elastic fibers no longer a stretch of the imagination. *J Heart Valve Dis* 6:1–6.
13. Schoen, F.J. and W.D. Edwards. 2001. Valvular heart disease: General principles and stenosis. In: *Cardiovascular Pathology*, M.D. Silver, A.I. Gotlieb, and F.J. Schoen (Eds.), pp. 402–405. New York: Churchhill Livingstone.
14. Erez, E.K., K.R. Kanter, E. Isom, et al. 2003. Mitral valve replacement in children. *J Heart Valve Dis* 12:25029.
15. Kanter, K.R., J.M. Bude, W.J. Parks, et al. 2002. One hundred pulmonary valve replacements in children after relief of right ventricular outflow tract obstruction. *Ann Thorac Surg* 73:1801–1806.
16. Robertson, S.W., V. Imbeni, H.R. Wenk, et al. 2005. Crystallographic texture for tube and plate of the superelastic/shape-memory alloy Nitinol used for endovascular stents. *J Biomed Mater Res* 72A: 190–199.
17. Choudhary, S., K.M. Haberstroh, and T.J. Webster TJ. 2007. Enhanced functions of vascular cells on nanostructured Ti for improved stent applications. *Tissue Eng* 13:1421–1430.
18. Schuler, P., D. Assefa, J. Ylanne, et al. 2003. Adhesion of monocytes to medical steel as used for vascular stents is mediated by the integrin receptor Mac-1. *Cell Commun Adhes* 10:17–26.
19. Cejna, M., R. Virmani, and R. Jones, 2002. Biocompatibility and performance of the Wallstent and the Wallgraft, Jostent, and Hemobahn stent-grafts in a sheep model. *J Vasc Interv Radiol* 12:823–830.
20. Chen, Y.J., T.X. Leng, X.B. Tian, et al. 2002. Antithrombogenic investigation of surface energy and optical bandgap and hemocompatibility mechanism of Ti $(Ta(+5)O_2)$ thin films. *Biomaterials* 23: 2545–2552.
21. Mason, M., K.P. Vercruysse, K.R. Kirker, et al. 2000. Attachment of hyaluronic acid to polypropylene, polystyrene, and polytetrafluoroethylene. *Biomaterials* 21:31–36.
22. Darling, R.C. and R.R. Linton. 1972. Durability of femoropopliteal reconstructions: Endarterectomy versus vein bypass grafts. *Am J Surg* 123:472–479.
23. Clayson, K.R., W.H. Edwards, T.R. Allen, et al. 1976. Arm veins for peripheral arterial reconstruction. *Arch Surg* 111:1276–1280.
24. Voorhees, A.B., A. Jaretski, and A.H. Blakemore. 1952. Use of tubes constructed from vinyon 'n' cloth in bridging arterial deficits. *Ann Surg* 135:332–338.
25. Blakemore, A. and A.B. Voorhees. 1954. The use of tube constructed of vinyon 'n' cloth in bridging arterial deficits: Experimental and clinical. *Ann Surg* 140:324–330.
26. Herring, M., A. Gardner, and J. Glover. 1978. A single-staged technique for seeding vascular grafts with autogenous endothelium. *Surgery* 84:498–504.
27. Anderson, J.S., T.M. Price, S.R. Hanson, et al. 1987. In vitro endothelialization of small-caliber vascular grafts. *Surgery* 101:577–586.
28. Foxall, T.L., K.R. Auger, A.D. Callow, et al. 1986. Adult human endothelial cell coverage of small-caliber Dacron and polytetrafluoroethylene vascular prostheses in vitro. *J Surg Res* 41:158–172.

29. Kempczinski, R.F., J.E. Rosenman, W.H. Pearce, et al. 1985. Endothelial cell seeding of a new PTFE vascular prosthesis. *J Vasc Surg* 2:424–429.
30. Weinberg, C. and E. Bell. 1986. A blood vessel model constructed from collagen and cultured vascular cells. *Science* 231:397–400.
31. Miwa, H. and T. Matsuda. 1994. An integrated approach to the design and engineering of hybrid arterial prostheses. *J Vasc Surg* 19:658–667.
32. Hirai, J. and T. Matsuda. 1996. Venous reconstruction using hybrid vascular tissue composed of vascular cells and collagen: Tissue regeneration process. *Cell Transplant* 5:93–105.
33. Ishibashi, K. and T. Matsuda. 1994. Reconstruction of a hybrid vascular graft hierarchically layered with three cell types. *ASAIO J* 40:M284–290.
34. L'Heureux, N., L. Germain, R. Labbe, et al. 1993. In vitro construction of a human blood vessel from cultured vascular cells: A morphologic study. *J Vasc Surg* 17:499–509.
35. L'Heureux, N., S. Paquet, R. Labbe, et al. 1998. A completely biological tissue-engineered human blood vessel. *FASEB J* 12:47–56.
36. Abbott, W.M., A. Callow, W. Moore, et al. 1993. Evaluation and performance standards for arterial prosthesis. *J Vasc Surg* 17:746–756.
37. Veith, F.J., S.K. Gupta, E. Ascer, et al. 1986. Six year prospective multicenter randomized comparison of autologous saphenous vein and expanded polytetrafluoroethylene grafts in infrainguinal arterial reconstructions. *J Vasc Surg* 3:104–114.
38. Quiñones-Baldrich, W.J., A. Prego, R. Ucelay-Gomez, et al. 1991. Failure of PTFE infrainguinal revascularization: Patterns, management alternatives, and outcome. *Ann Vasc Surg* 5:163–169.
39. Seeger, J.M. 2000. Management of patients with prosthetic vascular graft infection. *Am Surg* 66: 166–177.
40. Sarkar, S., H.J. Salacinski, G. Hamilton, et al. 2006. The mechanical properties of infrainguinal vascular bypass grafts: Their role in influencing patency. *Eur J Vasc Endovasc Surg* 31:627–636.
41. Murray-Wijelath, J., D.J. Lyman, and E.S. Wijelath. 2004. Vascular graft healing—III: FTIR analysis of ePTFE graft samples from implanted biografts. *J Biomed Mater Res B Appl Biomater* 70:223–232.
42. Hagerty, R.D., D.L. Salzmann, L.B. Kleinert, et al. 2000. Cellular proliferation and macrophage populations associated with implanted expanded polytetrafluoroethylene and polyethyleneterephthalate. *J Biomed Mater Res* 49:489–497.
43. Gao, K., L. Niklason, and R. Langer. 1998. Surface hydrolysis of poly(glycolic acid) meshes increases the seeding density of vascular smooth muscle cells. *J Biomed Mater Res* 3:417–423.
44. Solan, A., V. Prabhakar, and L. Niklason. 2001. Engineered vessels: Importance of the extracellular matrix. *Trans Proc* 33:66–68.
45. Jain, R.K. 2003. Molecular regulation of vessel maturation. *Nat Med* 6:685–693.
46. Brey, E.M., S. Uriel, H.P. Greisler, et al. 2005. Therapeutic neovascularization: Contributions from bioengineering. *Tissue Eng* 11:567–584.
47. Hockel, M., K. Schlenger, S. Doctrow, et al. 1993. Therapeutic angiogenesis. *Arch Surgery* 128:423–429.
48. Berglund, J.D. and Z.S. Galis. 2003. Designer blood vessels and therapeutic revascularization. *Br J Pharmacol* 140:627–636.
49. Bernas, G.C. 2003. Angiotherapeutics from natural products: From bench to clinics? *Clin Hemorheol Microcirc* 29:199–203.
50. Freedman, S.B. and J.M. Isner. 2002. Therapeutic angiogenesis for coronary artery disease. *Ann Intern Med* 136:54–71.
51. Hughes, G.C., S.S. Biswas, B. Tin, et al. 2004. Therapeutic angiogenesis in chronically ischemic porcine myocardium: Comparative effects of bFGF and VEGF. *Ann Thorac Surg* 77:812–818.
52. Lei, Y., H.K. Haider, J. Shujia, et al. 2004. Therapeutic angiogenesis: Devising new strategies based on past experiences. *Basic Res Cardiol* 99:121–132.
53. Madeddu, P. 2005. Therapeutic angiogenesis and vasculogenesis for tissue regeneration. *Exp Physiol* 90:315–326.
54. Levenberg, S., J. Rouwkema, M. Macdonald, et al. 2005. Engineering vascularized skeletal muscle tissue. *Nat Biotechnol* 23:879–884.
55. Li, W.W., K.E. Talcott, A.W. Zhai, et al. 2005. The role of therapeutic angiogenesis in tissue repair and regeneration. *Adv Skin Wound Care* 18:491–500.

56. Cassell, O.C.S., S.O.P. Hofer, W.A. Morrison, et al. 2002. Vascularization of tissue-engineered grafts: The regulation of angiogenesis in reconstructive surgery and in disease states. *Br J Plast Surg* 55:603–610.
57. Botchwey, E.A., M.A. Dupree, S.R. Pollack, et al. 2003. Tissue-engineered bone: Measurement of nutrient transport in three-dimensional matrices. *J Biomed Mater Res* 67A:357–367.
58. Laschke, M.W., Y. Harder, M. Amon, et al. 2006. Angiogenesis in tissue engineering: Breathing life into constructed tissue substrates. *Tissue Eng* 12:2093–2103.
59. Webster, T.J., E.L. Hellenmeyer, and R.L. Price. 2005. Increased osteoblast functions on theta + delta nanofiber alumina. *Biomaterials* 26:953.
60. Savaiano, J.K. and T.J. Webster. 2004. Altered responses of chondrocytes to nanophase PLGA/nanophase Titania composites. *Biomaterials* 25:1205.
61. Flemming, R.G., C.J. Murphy, G.A. Abrams, et al. 1999. Effects of synthetic micro- and nano-structured surfaces on cell behavior. *Biomaterials* 6:573–588.
62. Ma, Z., M. Kotaki, R. Inai, et al. 2005. Potential of a nanofiber matrix as tissue-engineering scaffolds. *Tissue Eng* 11:101–109.
63. Martins-Green, M. 1997. The dynamics of cell-ECM interactions with implications for tissue engineering. In *Principles of Tissue Engineering*, Lanza, R., R. Langer, and W. Chick (Eds.), pp. 23–46. Texas: Academic Press.
64. Williams, S.K. and B.E. Jarrell. 1987. Thrombosis on endothelializable prostheses. *Ann NY Acad Sci* 516:154–157.
65. Goodman, S.L., P.A. Sims, and R.M. Albrecht. 1996. Three-dimensional extracellular matrix textured biomaterials. *Biomaterials* 17:2087–2095.
66. Ingber, D.E. and J. Folkman. 1989. Mechanochemical switching between growth and differentiation during fibroblast growth factor-stimulated angiogenesis in vitro: Role of extracellular matrix. *J Cell Biol* 109:317–330.
67. Madri, J.A., B. Dreyer, F.A. Pitlick, et al. 1980. The collagenous components of the subendothelium. Correlation of structure and function. *Lab Invest* 43:303–315.
68. McCullagh, K.A. and G. Balian. 1975. Collagen characterisation and cell transformation in human atherosclerosis. *Nature* 258:73–75.
69. Lenhert, S., M.B. Meier, U. Meyer, et al. 2005. Osteoblast alignment, elongation and migration on grooved polystyrene surfaces patterned by Langmuir-Blodgett lithography. *Biomaterials* 26:563–570.
70. Dalby, M.J., M.O. Riehle, D.S. Sutherland, et al. 2004. Fibroblast response to a controlled nanoenvironment produced by colloidal lithography. *J Biomed Mater Res* 69A:314–322.
71. Dalby, M.J., M.O. Riehle, D.S. Sutherland, et al. 2004. Changes in fibroblast morphology in response to nano-columns produced by colloidal lithography. *Biomaterials* 25:5415–5422.
72. Rice, J.M., J.A. Hunt, J.A. Gallagher, et al. 2003. Quantitative assessment of the response of primary derived human osteoblasts and macrophages to a range of nanotopography surfaces in a single culture model in vitro. *Biomaterials* 24:4799–4818.
73. Giavaresi, G., M. Tschon, J.H. Daly, et al. 2006. In vitro and in vivo response to nanotopographically-modified surfaces of poly(3-hydroxybutyrate-co-3-hydroxyvalerate) and polycaprolactone. *J Biomater Sci Polymer Edn* 17:1405–1423.
74. Chu, C.F.L., A. Lu, M. Liszkowski et al. 1999. Enhanced growth of animals and human endothelial cells on biodegradable polymers. *Biochim Biophys Acta* 3:479–485.
75. Miller, D.C., A. Thapa, K.M. Haberstroh, et al. 2004. Endothelial and vascular smooth muscle cell function on poly(lactic-co-glycolic acid) with nano-structured features. *Biomaterials* 25:53–61.
76. Miller, D.C., K.M. Haberstroh, and T.J. Webster. 2004. Mechanisms controlling increased vascular cell adhesion to nano-structured polymer films. *Bioengineering Conference 2004: Proceedings of the IEEE 30th Annual Northeast*. 120–121.
77. Wittmer, C.R., J.A. Phelps, W.M. Saltzman, et al. 2007. Fibronectin terminated multilayer films: Protein adsorption can cell attachment studies. *Biomaterials* 28:851–860.
78. Giachelli, C. and S. Steitz. 2000. Osteopontin: A versatile regulator of inflammation and biomineralization. *Matrix Biol* 19:615–622.
79. Chen, Y., B.S. Bal, and J.P. Gorsik. 1992. Calcium and collagen binding properties of osteopontin, bone sialoprotein, and bone acidic glycoprotein-75 from bone. *J Biol Chem* 267:24871–24878.
80. Mukherjee, B.B., M. Nemir, S. Beninato, et al. 1995. Interaction of osteopontin with fibronectin and other extracellular matrix molecules. *Ann NY Acad Sci* 760:201–212.

81. Liaw, L., D.E. Birk, C.B. Ballas, et al. 1998. Altered wound healing in mice lacking a functional osteopontin gene (spp1). *J Clin Invest* 101:1468–1478.
82. Khan, S.A., C.A. Lopez-Chua, J. Zhang, et al. 2002. Soluble osteopontin inhibits apoptosis of adherent endothelial cells deprived of growth factors. *J Biol Chem* 85:728–736.
83. Asou, Y., S.R. Rittling, H. Yoshitake, et al. 2001. Osteopontin facilitates angiogenesis, accumulation of osteoblasts, and resorption in ectopic bone. *Endocrinology* 142:1325–1332.
84. Liu, L., S. Chen, C.M. Giachelli, et al. 2005. Controlling osteopontin orientation on surfaces to modulate endothelial cell adhesion. *J Biomed Mater Res* 74A:23–31.
85. Chen, C.S., M. Mrksich, S. Huang, et al. 1997. Geometric control of cell life and death. *Science* 276:1425–1428.
86. Singhvi, R., A. Kumar, G.P. Lopez, et al. 1994. Engineering cell shape and function. *Science* 264:696–698.
87. Decher, G., J.D. Hong, and J. Schmitt. 1992. Buildup of ultrathin multilayer films by a self-assembly process. 3. Consecutively alternating adsorption of anionic and cationic polyelectrolytes on charged surfaces. *Thin Solid Films* 210:831–835.
88. Decher, G. 1997. Fuzzy nanoassemblies: Toward layered polymeric multicomposites. *Science* 277: 1232–1237.
89. Decher, G., M. Eckle, J. Schmitt, et al. 1998. Layer-by-layer assembled multicomposite films. *Curr Opin Colloid Interface Sci* 3:32–39.
90. Hammond, P.T. 1999. Recent explorations in electrostatic multilayer thin film assembly. *Curr Opin Colloid Interface Sci* 4:430–442.
91. Bertrand, P., A. Jonas, A. Laschewsky, et al. 2000. Ultrathin polymer coatings by complexation of polyelectrolytes at interfaces: Suitable materials, structure and properties. *Macromol Rapid Commun* 21:319–348.
92. Decher, G. and J.B. Schlenoff. 2003. *Multilayer Thin Films*. Weinheim: Wiley-VCH.
93. Hammond, P.T. 2004. Form and function in multilayer assembly: New applications at the nanoscale. *Adv Mater* 16:1271–1293.
94. Potts, J.R. and I.D. Campbell. 1996. Structure and function of fibronectin modules. *Matrix Biol* 15:313–320.
95. Furuzono, T., M. Masuda, M. Okada, et al. 2006. Increase in cell adhesiveness on a poly(ethylene terephthalate) fabric sintered hydroxyapatite nanocrystal coating in the development of an artificial blood vessel. *ASAIO J* 52:315–319.
96. Pham, Q.P., U. Sharma, and A.G. Mikos. 2006. Electrospinning of polymeric nanofibres for tissue engineering applications: A review. *Tissue Eng* 5:1–15.
97. Subbiah, T., G.S. Bhat, R.W. Tock, et al. 2005. Electrospinning of nanofibers. *J Appl Polym Sci* 96:557.
98. Zong, X., H. Bien, C.-Y. Chung, et al. 2005. Electrospun fine-textured scaffolds for heart tissue constructs. *Biomaterials* 26:5330–5338.
99. Laurencin, C.T., A.M.A. Ambrosio, M.D. Borden, et al. 1999. Tissue engineering: Orthopedic applications. *Annu Rev Biomed Eng* 1:19–46.
100. Shin, M., O. Ishii, T. Sueda, et al. 2004. Contractile cardiac grafts using a novel nanofibrous mesh. *Biomaterials* 25:3717–3723.
101. Fast, V. and A. Kleber. 1994. Anisotropic conduction of monolayers of neonatal rat hearts cultured on collagen substrate. *Circ Res* 75:591–595.
102. Simpson, D.G., L. Terracio, M. Terracio, et al. 1994. Modulation of cardiac monocyte phenotype in vitro by the composition and orientation of the extracellular matrix. *J Cell Physiol* 161:89–105.
103. Bursac, N., N.N. Parker, S. Iravanian, et al. 2002. Cardiomyocyte cultures with controlled macroscopic anisotropy: A model for functional electrophysiological studies of cardiac muscle. *Circ Res* 91:45–54.
104. Van Lieshout, M.I., C.M. Vaz, M.C. Rutten, et al. 2006. Electrospinning versus knitting: Two scaffolds for tissue engineering of the aortic valve. *J Biomater Sci Polymer Edn*. 17:77–89.
105. Kwon, I.K., S. Kidoaki, and T. Matsuda. 2005. Electrospun nano- to micro-fiber fabrics mase of biodegradable copolyesters: Structural characteristics, mechanical properties and cell adhesion potential. *Biomaterials* 26:3929–3939.
106. Allcock, H.R., T.J. Fuller, D.P. Mack, et al. 1977. Synthesis of poly[(amino acid alkyl ester) polyphosphazenes. *Macromolecules* 10:824–830.
107. Nair, L.S., S. Bhattcharyya, J.D. Bender, et al. 2004. Fabrication and optimization of methylphenoxy substituted polyphosphazenes nanofibers for biomedical applications. *Biomacromolecules* 5:2212–2220.

108. Carampin, P., M.T. Conconi, S. Lora, et al. 2007. Electrospun polyphosphazene nanofibers for in vitro rat endothelial cells proliferation. *J Biomed Mater Res* 80A:661–668.
109. Matthews, J., G. Spinson, G. Wnek, et al. 2002. Electrospinning of collagen nanofibers. *Biomacromolecules* 3:232–238.
110. Bowlin, G., K. Pawlowski, J. Stitzel, et al. 2002. Electrospinning of polymer scaffolds for tissue engineering. In *Tissue Engineering and Biodegradable Equivalents: Scientific and Clinical Applications*, K. Lewandrowski, D. Wise, D. Trantolo, et al. (Eds.), pp. 165–178. New York: Marcel Dekker, Inc.
111. He, W., T. Yong, W.E. Two, et al. 2005. Fabrication and endothelialization of collagen-blended biodegradable polymer nanofibers: Potential vascular graft of blood tissue engineering. *Tissue Eng* 11:1574–1588.
112. Stitzel, J., J. Liu, S.J. Lee, et al. 2006. Controlled fabrication of a biological vascular substitute. *Biomaterials* 27:1088–1094.

14 Nanostructures for Musculoskeletal Tissue Engineering

Susan Liao, Casey K. Chan, and Seeram Ramakrishna

CONTENTS

14.1 INTRODUCTION

Diseases and injuries of musculoskeletal system cause pain, deformity, and loss of function. They limit activities of daily living and cause disability for more people than disorders of any other organ system. Musculoskeletal tissue engineering provides a promising approach to treat these diseases and injuries. In tissue engineering, scaffolds as temporary support for cells ingrowth and carrier for growth factors are gradually acknowledged as the key functions for successful tissue repair. More recently, the ability to engineer the nanostructures of scaffold materials to precisely control cell behaviors for tissue regeneration is becoming a reality. In this chapter, we highlight the current state of art of nanostructures in scaffolds to explore the potential breakthroughs for musculoskeletal tissue engineering. Firstly, the nanostructures of natural musculoskeletal tissues are discussed in Section 14.2. Diseases and injuries of the musculoskeletal system, the current treatments, and musculoskeletal tissue engineering are introduced in Section 14.3. Strategies of musculoskeletal tissue engineering and the effect of nanostructures for cell behaviors are discussed in Sections 14.4 and 14.5, respectively. Based on the preceding basic principles introduced in Sections 14.4 and 14.5, we illustrate the specific applications of nanostructures for musculoskeletal tissue engineering in Section 14.6. Finally, current challenges in the applications of nanostructures for musculoskeletal tissue engineering are discussed in Section 14.7. To take full advantage of nanostructures for tissue engineering, the fabrication of nanostructures must be incorporated seamlessly with the micro- as well as the macrostructure to achieve the optimal benefits of nanostructures and fulfill the intended purpose of tissue repair.

14.2 NANOSTRUCTURES IN NATURAL MUSCULOSKELETAL TISSUES

The stability and mobility of the body depend on the tissues that are formed from the musculoskeletal system—bone, cartilage, dense fibrous tissue (ligament and tendon), and muscle. These tissues differ in vascularity, innervation, mechanical and biological properties, and composition, but they originate from the same precursor cells, which are the mesenchymal stem cells (MSCs) [1]. To generate the musculoskeletal tissues by tissue engineering, we need to know the natural structure of these tissues, especially on a nanoscale level and how they develop in the body.

The matrices of the musculoskeletal tissues consist of elaborate, highly organized framework of organic macromolecule nanofibers embedded in an amorphous ground substance. The fibers consist of multiple types of collagen and elastin, whereas the ground substance consists primarily of water, proteoglycans, and noncollagenous proteins. In addition to an organic matrix, the bone has an inorganic matrix, nanophase hydroxyapatite (HA), for increasing the stiffness and compressive strength of bone tissue. The elegant assembly of collagen nanofiber and nanophase HA was discovered primarily by Landis and Weiner using transmission electron microscopy (TEM) [2–5]. The structure of bone includes five hierarchical levels from the nanolevel in the form of collagen molecular assembly to the microlevel as woven bone or lamellar bone and finally to the macrolevel as structural bone [6] (Figure 14.1). Most recently, fractured bone when observed under field emission scanning electron microscopy (FESEM) directly showed a mineralized fibrous nanostructure [7] (Figure 14.2). Because of lack of collagen fibril orientation, the high cell and water content, and the irregular mineralization, the mechanical properties of woven bone are weaker than lamellar bone. Nerve fibers in association with blood vessels have also been identified within the medullary canals of bone. These nerve fibers presumably control bone blood flow. For cells to be viable they must be within 300 μm from a blood vessel.

Except for bone, the other four basic tissues in musculoskeletal system are ligaments, cartilage, skeletal muscles, and tendons. They are attached to individual bones to provide the basic functions of stability and mobility for the musculoskeletal system. These tissues consist of collagen fibrils with varied patterns. For example, different alignment of the collagen nanofiber network in cartilage is clearly shown in Figure 14.3 [8]. Skeletal muscle is the most abundant tissue in the human body. These muscles also have a similar hierarchical structure as bone. At the lowest hierarchical scale, the myofibril is roughly a cylindrical nanofiber containing several nuclei, and has alternating light and dark bands called striations. A bundle of myofibrils form a muscle fiber and a bundle of muscle fibers in turn form a fascicle, which are usually visible to the naked eye. Finally the fascicles are histologically organized by the surrounding connective tissue into bulk muscle as the functional contractile unit. For sustained function, the skeletal muscles exist as composite structures composed of many muscle fibers, nerves, blood vessels, and connective tissue.

14.3 PROBLEM (DISEASE–DISORDER) OF MUSCULOSKELETAL TISSUES

Advances in technology and improvements in healthcare have resulted in longer life expectancy. An aging population in an affluent society where expectation of quality of life is higher could strain the

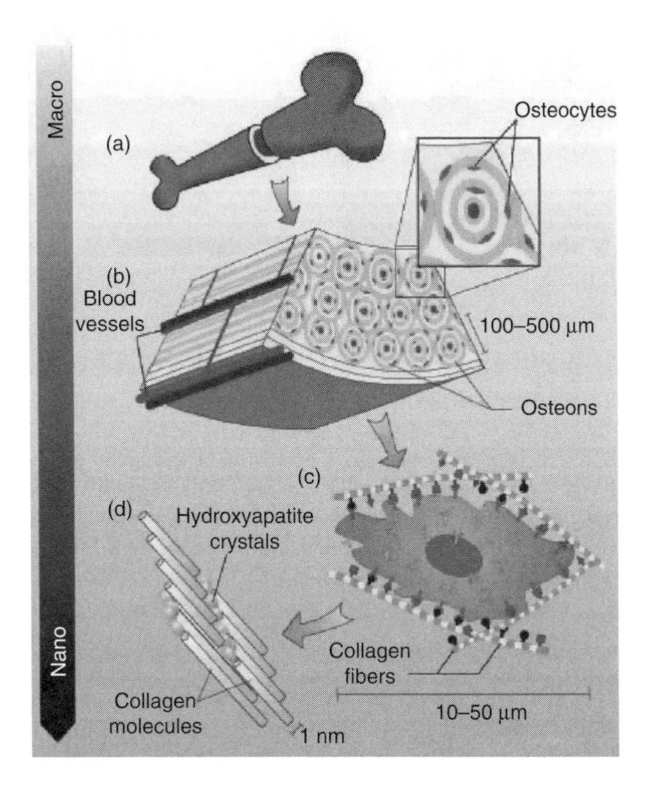

FIGURE 14.1 Schematic hierarchical organization of bone over different length scales. (a) Bone has a strong calcified outer compact layer, (b) which comprises many cylindrical Haversian systems, or osteons. (c) The resident cells are coated in a forest of cell membrane receptors that respond to specific binding sites and (d) the well-defined nanoarchitecture of the surrounding extracellular matrix. (From Stevens, M.M. and George, J.H., *Science*, 310, 1135, 2005. With permission.)

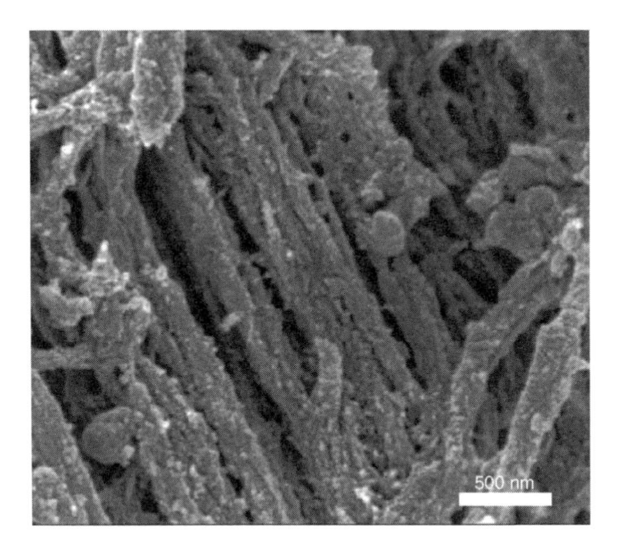

FIGURE 14.2 Scanning electron microscope (SEM) observation of fracture surface of human bone with mineralized collagen fibers in aligned bundle pattern. (From Fantner, G.E., Hassenkam, T., Kindt, J.H. et al., *Nat. Mater.*, 4, 612, 2005. With permission.)

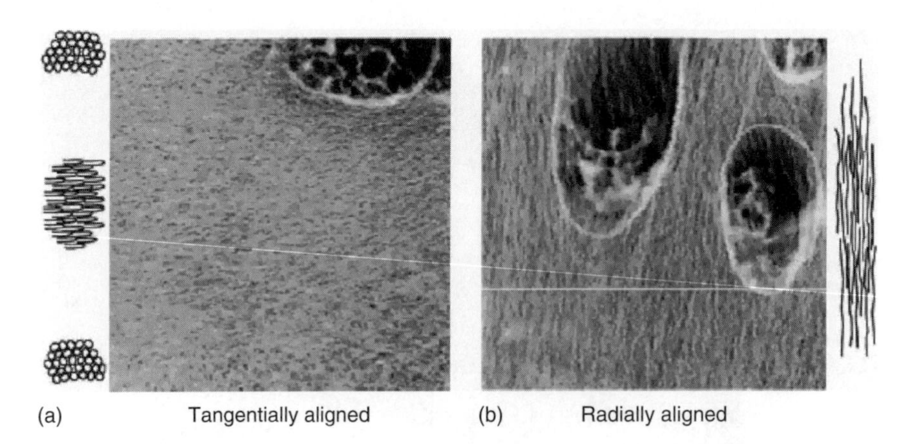

FIGURE 14.3 Network of collagen in cartilage by scanning transmission ion microscopy (STIM) observation: (a) tangentially aligned and (b) radially aligned structures. The structures in (a) can be interpreted as bunches of collagen tubuli in different tangential alignment (horizontal plane). In (b) the radially aligned (vertical) collagen fibers are readily visible. (From Reinert, T., Reibetanz, U., Schwertner, M. et al., *Nucl. Instr. Methods Phys. Res. B.*, 188, 1, 2002. With permission.)

healthcare system. Musculoskeletal problems rank second as the reason for which patients seek medical attention, and about 10% of U.S. citizens have such severe musculoskeletal problems that their activity is greatly restricted [1].

As the population ages, the number of sports and recreation-related injuries is expected to increase. This is due to a fast aging population that is increasingly shifting toward active lifestyles often leading to orthopedic injuries. In the elderly population, there will be an increasing number of age-related fractures such as hip fractures or pathological fractures due to cancer involvement of bone. Much degenerative arthritis of the knees and hips, and degenerative condition of the spine will require reconstructive surgical intervention to restore mobility and function.

14.3.1 Current Treatment Methods

Current treatments of musculoskeletal problems include use of medications, physical therapy, orthotics, prosthetics, and surgery. In general nonsurgical therapies are the first mode of treatment before surgical intervention is considered. Exercises that improve range of motion of musculoskeletal and muscle strength form part of the physical treatment of many musculoskeletal conditions. Orthotics is used in the treatment of scoliosis, lower back pain, foot pain, and some foot deformities. Operative treatments include manipulation, arthroscopic surgery, and open surgical reconstructions. Development of new synthetic materials and the better understanding of the musculoskeletal system have resulted in highly successful prosthetic joint replacement surgery for damaged joints. Total joint replacements for degenerative arthritis of the hips and knees are one of the most successful orthopedic surgeries with predictable favorable outcome in a majority of the patients. However, there are still many unsolved musculoskeletal problems such as healing of large bony defects, articular damage in the younger patients, and fusion of the spine. The application of nanostructures offers an advantage over traditional microbased scaffold to address some of these very challenging problems.

14.3.2 Tissue Engineering for the Treatment of Musculoskeletal Problems

Tissue engineering is an interdisciplinary biomedical field that combines the accumulated knowledge over the last 14 years of materials science, engineering principles, cell, and molecular and developmental biology. The ultimate goal of tissue engineering is to develop biological substitutes that will restore, maintain, or improve the functions of damaged tissues [9]. Although the construction of a functional tissue is an arduous task, it provides a promising way to overcome the current problems such as the shortage of suitable donor tissues [10]. Musculoskeletal tissues such as bone and cartilage are less complex when compared to solid organs such as kidney and lungs. Compared to other tissues, a number of practical applications of tissue engineering have been achieved in bone repair [11].

In 2003, the potential U.S. market for tissue-engineered musculoskeletal products totaled approximately $23.8 billion. By the year 2013, this figure is expected to increase to $39.0 billion. Cartilage and joint repair and replacement represent the largest potential in the musculoskeletal market with a 67.4% share of the potential market, followed by viscosupplementation with a 13.7% share, and fracture fixation and orthopedic/spinal repair with a 9.4% share [12].

14.4 STRATEGIES OF MUSCULOSKELETAL TISSUE ENGINEERING

14.4.1 General Principles for Tissue Engineering

The principles of tissue engineering encompass three basic components: scaffolds, cells, and signaling biomolecules (or growth factors), the so-called triad in tissue engineering. The principal components of extracellular matrix (ECM) are collagen biopolymers, mainly in the form of fibers and fibrils. The porous scaffold, preferably made of absorbable polymer, should mimic the ECM to support cell proliferation and organization. Other forms of polymer organization like gels, foams, and membranes have also been used as scaffolds for tissue engineering. The various forms can be combined in the laboratory to create imitations of biopolymer organization in specific tissues [13]. Scaffolds can be enriched with signaling biomolecules. The incorporated cells in the scaffolds will form ECM and also provide the signals needed for tissue building. Acellular collagen matrices, in the form of sponges with and without bone precursor minerals (calcium phosphate), have been used as vehicles for delivering a variety of bone morphogenetic proteins (BMPs). As an example of application of the triad of tissue engineering for bone tissue engineering, one strategy is to add calcium phosphate to the scaffold, seed the scaffold with bone marrow stromal cells (BMSCs), and supplement the scaffold with growth factors (BMP-2, BMP-7, TGF-β, etc). To further mimic natural bone ECM the scaffold can be fabricated with nanotexture and associated with nanocalcium phosphate.

14.4.2 Biomimetic Scaffolds

ECM in natural tissue supports cell attachment, proliferation, and differentiation. Ideally the scaffold should mimic natural ECM as much as possible. Microscaled scaffold has been studied in great detail in the last 20 years. It has been noted that the high porosity and pore size of 50–400 μm in the scaffold are necessary for cell ingrowth. To enhance cell attachment, proliferation, and differentiation it is necessary to mimic some of the nanostructure of natural ECM. The nanoscaled structure of ECM is a natural network of intricate nanofibers, which provides an instructive background to guide cell behavior [6,10,14,15]. The functional cues presented by the ECM can be chemical in nature, such as the presence of adhesive proteins, or purely physical, such as topography from nanometer to micrometer scale of the ECM. The detailed reactions between cells and nanoscaffolds will be discussed later in Section 14.5. The enhancement of cellular behavior by the superposition of nanotexture on microscaled scaffold will be discussed.

On a micrometer level, bone can be classified as spongy bone with many pores and high porosity, or as lamellar bone with compact cylindrical osteons. On the nanometer level, the collagen fibrils and nanophase HA are self-assembled into ordered and aligned pattern. The different tissues in the musculoskeletal system are seamlessly connected with each other, either hard to hard tissue, soft to hard tissue, or soft to soft tissue. The connection starts from the molecular level and is hierarchically matched. The superadhesive strength is believed to originate from electrostatic interaction. Hence, the nanostructure of scaffold is a crucial consideration when developing scaffold for composite tissues such as scaffold for an osteochondral graft.

14.4.3 Stem Cells

Currently, there are various cell lines available for the regeneration of specific tissue type. The use of mature cells for tissue regeneration such as the use of chondrocytes for cartilage regeneration has a successful history. Such use remains an important technique in tissue engineering. However, rapid advances in cell biology and in the identification of new cell types have improved capability of using a wider choice of cell lines to transform and expand cells into the appropriate tissues. Stem cells have been recognized as a promising alternative to mature cells owing to their potential to renew themselves through cell division and to differentiate into a wide range of specialized cell types. Initially, research on stem cells is about how an organism develops from a single cell and how healthy cells replace damaged cells in adult organisms. Progress in the understanding of self-renewal and directed differentiation has led scientists to propose the possibility of cell-based therapies to treat disease, including the use of stem cells in tissue engineering.

There are three types of stem cells: embryonic stem cells, cord blood stem cells, and adult stem cells. Adult stem cells are particularly suited for applications in tissue engineering for the musculo-skeletal system because of their availability and the lack of controversy. A convenient source of adult stem cell is from the bone marrow. Bone marrow, in addition to red blood cells, white blood cells, and platelets, contains two distinct stem cells: hematopoietic stem cells (HSCs), which are responsible for the formation of blood cells, and MSCs, which have the potential to differentiate into osteoblasts, chondrocytes, myocytes, and other cell lines. Hence, the potential advantage of using adult stem cells is that the patient's own cells could be easily harvested, isolated, expanded in culture, and then transplanted into the patient's body where tissue regeneration is required [16]. This autologous approach avoids immune rejection, which are problems when allografts or xenografts are used. One strategy is to seed autologous stem cells on biomimetic scaffolds to mimic the patient's natural tissues and thus the engineered tissue will retain some of the advantages of autografts. Such tissue engineering approaches overcome the morbidity associated with traditional reconstructive surgery where normal tissue is harvested from the patient's own body.

Certain adult stem cell types are pluripotent; that is, they can differentiate into cells derived from the three germ layers [17–19]. This ability to differentiate into multiple cell types is called plasticity or transdifferentiation. The following list offers examples of adult stem cell plasticity that have been reported during the past few years as shown in Figure 14.4. Hematopoietic stem cells in addition may differentiate into three major types of brain cells (neurons, oligodendrocytes, and astrocytes); skeletal muscle cells; cardiac muscle cells; and liver cells. BMSCs in addition to differentiating skeletal muscle cells may also differentiate into cardiac muscle cells [20]. The plasticity provides the basic possibility for multiple-tissue engineering using a certain type of stem cells.

Although standard cell culture techniques have been successful for mature cells' expansion, MSCs culture techniques for MSC-based tissue engineering especially for three-dimensional (3D) expansion require further developments. To maintain the phenotype and differentiated functions of MSCs, the simulated natural environment of the biomimetic ECM support has to provide the appropriate signals to the attached cells. Scaffolds with nanotexture can provide physical as well as spatial cues that are essential to mimic natural tissue growth. In addition to physical and spatial cues, the scaffold itself can be the carrier of signaling biomolecules.

14.4.4 Growth Factors

Growth factors have been in use clinically for a number of years and their efficacy to enhance tissue repair particularly for bone repair has been demonstrated to a certain degree. In the past, growth factors were either injected or applied directly to tissues. Delivering growth factors in this manner may require a higher dosage such as in the case of BMP-2 and BMP-7 (both being relatively small molecules they will rapidly diffuse away or become inactivated). Thus, some forms of carrier and delivery systems will be required for sustained release of BMPs in order for the growth factors to be

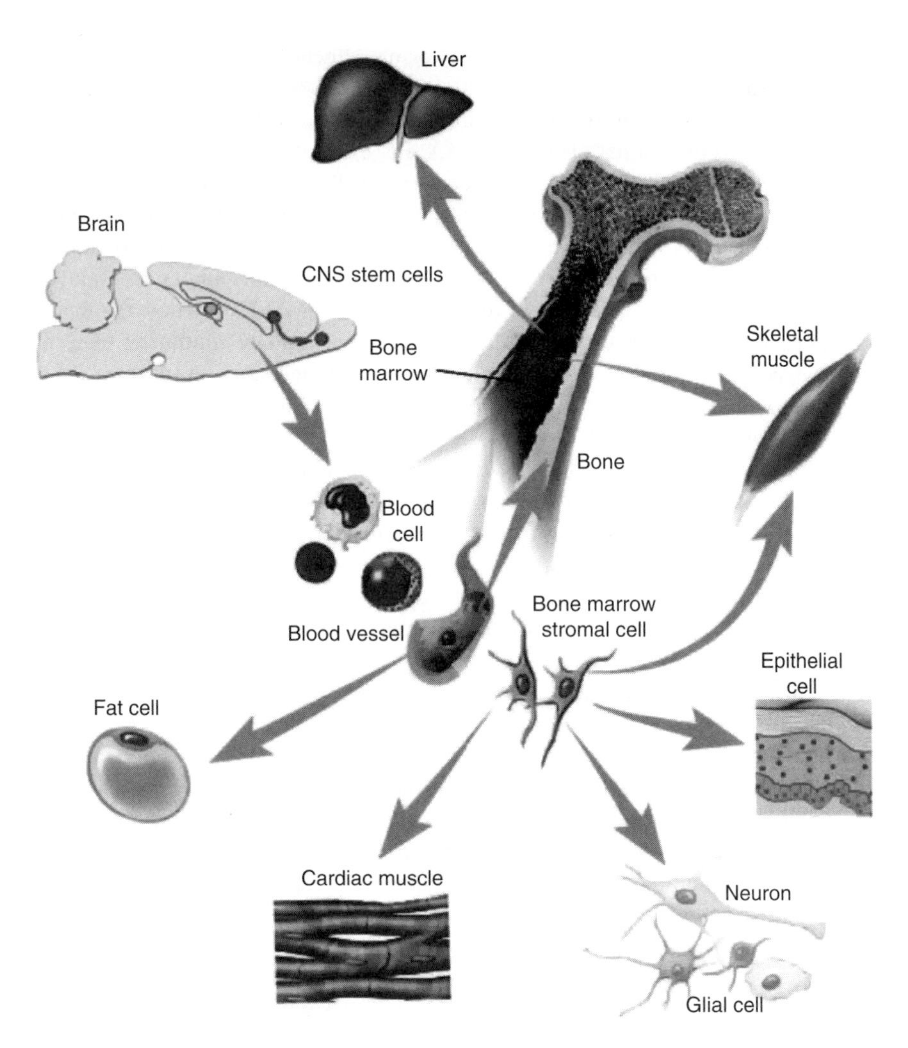

FIGURE 14.4 Plasticity of adult stem cells. (From NIH Stem Cell Report entitled: *Stem Cells: Scientific Progress and Future Research Directions*, Kirschstein, R. and Skirboll, L.R., Department of Health and Human Services, June 2001. http://stemcells.nih.gov/info/basics/basics4 [accessed Feb 06, 2007]. With permission.)

effective at the host bed during the healing period. One requirement for BMP carriers is that the bulk material after implantation be capable of resisting the imbibed BMPs from being squeezed out into the surrounding tissue and thus rapidly reabsorbed [21–23]. The early clinical experience to date suggests that current bone graft substitutes in combination with osteoinductive growth factors can replace the use of autogenous bone graft in certain clinical applications such as spinal fusion [24]. It is also speculated that the current recommended dosage can be reduced with better carrier and delivery systems. The challenge remains to optimize the dose of BMPs required and to determine the appropriate carrier for the BMPs [25].

The nanomaterials are perceived as suitable carriers because of their high surface area-to-volume ratio. Thus, the physically absorbed or chemically bonded signals on the surface can be more effectively controlled by the carrier materials. Thus, biomolecules, either physically absorbed or chemically bonded on the carrier surface, can be more effectively manipulated across a greater range of drug release kinetics. Nanophase materials have been developed as advanced carrier materials and these can be in the form of nanoparticles [26–28] or nanofibers [29] made of bioceramics and/or biopolymers. For example, nanophase HA has unique surface properties such as increased number of atoms, boundaries of the grains of the material and defects at the surface area, and altered electronic structure compared to conventional micron size HA. For example, nanophase HA (size 67 nm) has higher surface roughness of 17 versus 10 nm for conventional submicron size HA (179 nm) while the contact angles (a quantitative measure of the wetting of a solid by a liquid) are significantly lower for nanophase HA with 6.1° compared with 11.51° for conventional HA. Additionally, the diameter of individual pores in nanophase HA compact is five times (pore diameter 6.6 Å) smaller than the conventional grain size HA compacts (pore diameter 19.8–31.0 Å) [30]. Thus, the nanophase HA had 11% more proteins in the fetal bovine serum adsorbed per square centimeter than the conventional HA. The adsorption of specific proteins such as fibronectin, vitronectin, and denatured collagen from blood, bone marrow, and other tissues on the implant surface mediate the subsequent adhesion, differentiation, and growth of osteoblasts or other desirable cells.

Recently, it has been shown that biomolecules can be incorporated into the core part of core–shell structure of nanoparticles or nanofibers for a more sustained delivery [30–35]. However, nanoparticles are difficult to integrate into a 3D scaffold. Thus, nanofibrous scaffolds seem attractive. Of course, nanoparticles can be used as additives for nanofibrous composites [36]. In the following sections, we will focus on the use of nanofibrous scaffolds in tissue engineering, either as substrates for cells growth and/or as carriers for drug delivery.

14.5 NANOFIBROUS SCAFFOLD–CELL INTERACTIONS

By studying how functional cues influence cell positioning and alignment during the developmental process, tissue engineers can optimize their scaffold design for effective tissue regeneration. Nanoscale alterations in topography elicit diverse cell behavior, ranging from changes in cell adhesion, cell orientation, cell motility, surface antigen display, cytoskeletal condensation, activation of tyrosine kinases, and modulation of intracellular signaling pathways that regulate transcriptional activity and gene expression [37]. For example, osteoblasts have been found to adhere preferentially to carbon nanofibers in competition with chondrocytes, fibroblasts, and smooth muscle cells (SMCs) [38,39]. To regenerate tissue, engineered scaffolds serve as hosts to cells harvested from natural tissue. Cells that are bound to scaffolds with microscale architectures flatten and spread as if they are cultured on flat surfaces, but scaffolds with nanoscale architectures have larger surface areas to adsorb proteins, presenting many more binding sites to the cell membrane receptors (Figure 14.5). The adsorbed proteins may also change conformation, exposing additional cryptic binding sites. Our group has conducted the pioneering research on cell–polymer nanofiber interactions [40–48]. These observations indicate that electrospun nanofibrous scaffolds can promote positive cell–matrix interactions, encouraging the cells to anchor to the nanofibers illustrated in Figure 14.6a through c. In summary (1) nanofibers significantly promote the cell adhesion and proliferation; (2) the orientation of aligned nanofibers guides the orientation of cytoskeletal proteins; and (3) functionalized nanofibers can enhance the above-mentioned effects of nanofibers on cell behaviors. As an example of the effect of surface functionalization for P(LLA-CL), for nanofibers coated with collagen, the cell adhesion and proliferation improved (see Figure 14.6f and g). One-day cultured osteoblast on poly(ε-caprolactone)/$CaCO_3$ (PCL/$CaCO_3$) nanofibers adhered well with polygonal morphology [40]. SMCs cultured on aligned poly(L-lactide-co-ε-caprolactone) (PLLA-CL) nanofibers oriented along the nanofibers [41]. The SMCs attached and migrated along the axis of the aligned nanofibers and expressed a spindle-like contractile phenotype. The distribution and organization of smooth muscle cytoskeleton proteins inside SMCs were parallel to the direction of the nanofibers. The adhesion and proliferation rate of SMCs on the aligned nanofibrous scaffold was significantly improved than on the plane polymer films. After coating collagen on PLLA-CL nanofibers, cell adhesion and cell viability were both significantly increased [42].

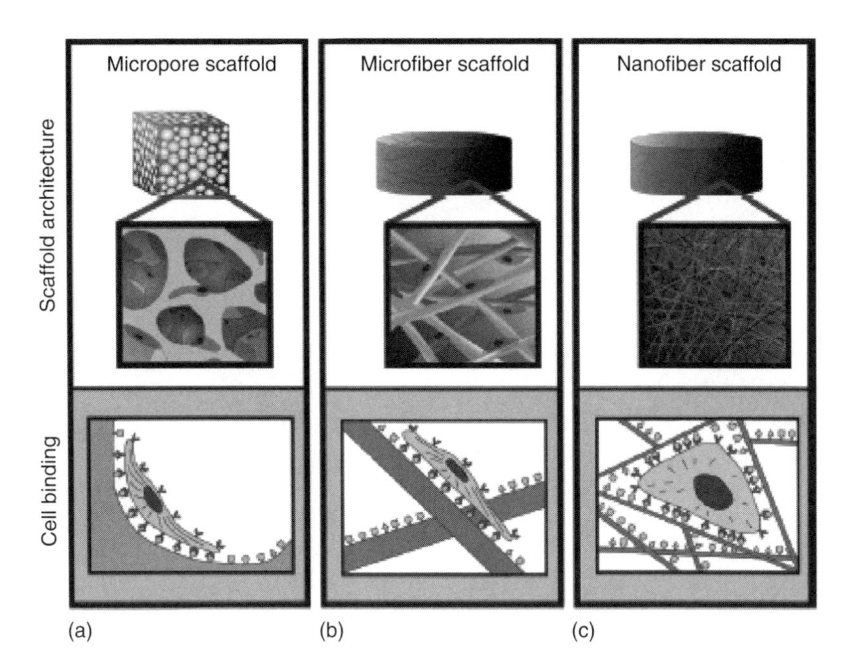

FIGURE 14.5 Scaffold architecture affects cell binding and spreading. (a, b) Cells binding to scaffolds with microscale architectures flatten and spread as if cultured on flat surfaces. (c) Scaffolds with nanoscale architectures have larger surface areas to adsorb proteins, presenting many more binding sites to cell membrane receptors. The adsorbed proteins may also change conformation, exposing additional cryptic binding sites. (From Stevens, M.M. and George, J.H., *Science*, 310, 1135, 2005. With permission.)

The biomineralization was also significantly increased (55%) in PCL/nHA/Col biocomposite nanofibrous scaffolds after 10 days of culture and appeared as minerals synthesized by the human fetal osteoblasts (hFOB). The unique nanoscale biocomposite system with nanofibrous morphology had inherent surface functionalization for hFOB adhesion, migration, proliferation, and mineralization [48]. The details of stem cells reaction on nanofibers are discussed for specific tissue regeneration in Section 14.6. The effect of nanofibers for stem cells' differentiation is a key question for further applications of nanofibers for tissue engineering. Stem cells can be induced to differentiate into different cell types by growth factors or other factors in the media, and we can incorporate such biomolecules into the nanofibers to direct differentiation to a desired cell type. The biomimetic morphology of nanofibers with different patterns may also help to direct the stem cells' differentiation. This is still questionable as researches on the above postulations are currently ongoing.

Besides electrospun nanofibers, nanofibers fabricated by other methods, such as self-assembly and phase separation, give similar results on cell reactions [49–53]. Basically, these nanofibers are randomly interweaved without long-term periodic pattern. Elia Beniash et al. [49] had developed a class of peptide amphiphile (PA) molecules possessing both hydrophilic and hydrophobic properties that will self-assemble into 3D nanofiber networks under physiological conditions in the presence of polyvalent metal ions. PA self-assembly entraps MC3T3-E1 cells in the nanofibrillar matrix, and the cells survive in culture for at least 3 weeks. The entrapment was able to support cell proliferation and motility. Biochemical analysis indicated that entrapped cells internalized the nanofibers and possibly utilized PA molecules in their metabolic pathways.

Nanofibrous poly(L-lactic acid) (PLLA) scaffolds with interconnected pores were developed by phase separation [54,55]. Scaffolds with walls of the interconnected pores having a nanofibrous texture adsorbed four times more serum proteins than scaffolds with the interconnected pores having

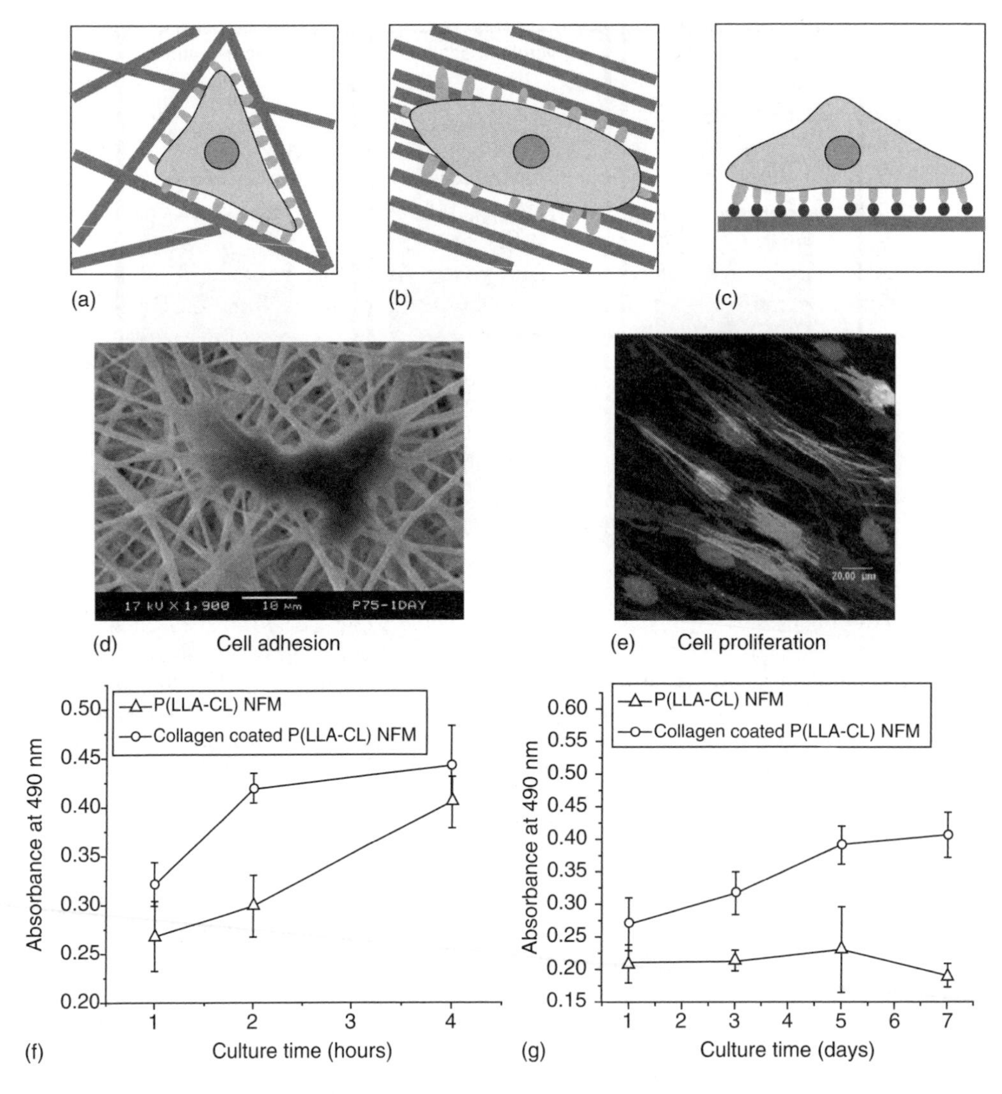

FIGURE 14.6 Schematic cell–nanofiber reactions on (a) random nanofibers, (b) aligned nanofibers, and (c) functionalized nanofibers. (d–g) are results from our group: (d) one-day cultured osteoblast on PCL/$CaCO_3$ nanofibers. (Reprinted from Fujihara, K., Kotaki, M., and Ramakrishna, S., *Biomaterials*, 26, 4139, 2005. With permission.) (e) Smooth muscle cells cultured on aligned PLLA-CL nanofibers. (From Xu, C.Y., Inai, R., Kotaki, M. et al., *Biomaterials*, 25, 877, 2004. With permission.) (f) Cell adhesion and (g) cell viability on collagen coated PLLA-CL nanofibers. (From He, W., Ma, Z., Yong, T. et al., *Biomaterials*, 26, 7606, 2005. With permission.)

solid walls. The nanofibrous architecture selectively enhanced protein adsorption including fibronectin and vitronectin, even though both scaffolds were made from the same PLLA material. Furthermore, nanofibrous scaffolds also allowed >1.7 times of osteoblastic cell attachment than scaffolds with solid pore walls. Osteoblasts cultured on nanofibrous scaffolds also exhibited higher alkaline phosphatase activity and an earlier and enhanced expression of the osteoblast phenotype versus solid-walled scaffolds (microscaffolds) fabricated by phase separation. Most notable were the increases in runx2 protein and in bone sialoprotein mRNA in cells cultured on nanofibrous scaffolds compared with those cultured on solid-walled scaffolds. At day 1 of culture, α_2 and β_1 integrins as well as α_v and β_3 integrins were highly expressed on the surface of cells seeded on nanofibrous scaffolds, and linked to this were higher levels of phospho-Paxillin and phospho-focal adhesion kinase (FAK) in cell lysates. Moreover, biomineralization was enhanced substantially on the nanofibous scaffolds compared to solid-walled scaffolds. Thus, nanofibrous architecture promotes osteoblasts differentiation and biomineralization.

14.6 APPLICATIONS OF NANOFIBROUS SCAFFOLDS FOR MUSCULOSKELETAL TISSUE ENGINEERING

Nanofibrous scaffolds have been successfully produced from many synthetic and natural polymers using the relatively simple technique of electrospinning. Fibers produced by electrospinning have diameters ranging from a few nanometers to micrometers [29,56,57]. The continuous nanofibers potentially allow for integrated manufacturing of 3D nanofiber matrices with high porosity, high spatial interconnectivity, and controlled alignment of fibers to direct cell orientation and migration of cells [58,59]. The currently used materials, cells, and growth factors for musculoskeletal tissue engineering are listed in Table 14.1. Principally, most of the currently used biopolymers can be used to fabricate nanofibers by electrospinning. Electrospinning supports the different assemblies of nanofibers and this technique can also be combined with other nanofabrication methods. This flexibility in fabrication in addition to other advantageous attributes makes nanofibrous composite a leading content for tissue engineering applications.

14.6.1 Bone

For bone tissue regeneration, biodegradable polymers are commonly combined with bioceramics either before or after the electrospinning process to mimic the component of natural bone tissue. Fujihara et al. [40] electrospun nanofibers as a PCL/$CaCO_3$ composite. This composite with nanoparticles of $CaCO_3$ encouraged osteoblast attachment in in vitro experiments. In another experiment, the electrospinning technique was used to fabricate PCL/nanophase HA/collagen biocomposite nanofibrous scaffolds to provide mechanical support and to direct the growth of hFOB for tissue engineering of bone. This biocomposite nanofibrous scaffold constructed from PCL, nanophase HA, and collagen type I is a highly porous (>80%) structure and provides a sufficiently open pore structure for cell occupancy while allowing free transport of nutrients and metabolic waste products and vascular ingrowth. The mineralization was significantly increased (55%) in PCL/nano-HA/collagen biocomposite nanofibrous scaffolds after 10 days of culture and appeared as minerals synthesized by osteoblastic cells [48].

Alternatively, HA can be incorporated with electrospun nanofibrous film by soaking in simulated body fluid (SBF), as in the study of Ito et al. [73]. The formed HA changed the hydrophobic poly(3-hydroxybutyrate-co-3-hydroxyvalerate) (PHBV) film to a more hydrophilic state, and as a result the degradation rate is increased. However, HA composition did not significantly affect the cell adhesion. On replacing the PHBV with silk/polyethylene oxide (PEO) containing poly (L-aspartate) (poly-Asp), and changing SBF to $CaCl_2$ and Na_2HPO_4 solution, the apatite growth occurred preferentially along the longitudinal direction of the fibers [74].

MSCs derived from the bone marrow of neonatal rats were used to seed PCL nanofibrous scaffolds by Yoshimoto et al. [75]. The cell–polymer constructs were cultured with osteogenic supplements under rotational oxygen-permeable bioreactor system. At 4 weeks, the surfaces of the cell–polymer constructs were covered with cell multilayers. In addition, both type I collagen secretion and cell mineralization were detected at the same time.

Gelatin/PCL composite electrospun nanofibers were fabricated by mixing 50% gelatin solution with 50% PCL solution; where gelatin concentration ranged from 2.5% w/v to 12.5% w/v [76]. Contact angle measurement and tensile tests indicated that the gelatin/PCL complex fibrous membrane exhibited increased wettability as well as improved elongation than that obtained from either gelatin or PCL alone. BMSCs not only favorably attach and grow well on the surface of these

TABLE 14.1
Materials, Cells, and Growth Factors for Musculoskeletal Tissue Engineering

Musculoskeletal Tissue	Materials for Scaffolds			Cells	Growth Factors
	Natural Polymer	Synthetic Polymer	Composite		
Bone	Collagen [56], Gelatin, Silk [57], HYA [60–63], ChS, Chitosan	PLA, PGA, PLGA, PCL, PLA-CL, PU	PLA/HA/collagen, PCL/HA/collagen, PCL/$CaCO_3$, PCL/HA, PCL/Gelatin/HA, Silk/HA [64]	Osteoblast, MSC	BMP-2, BMP-7, TGF-β
Cartilage			PLGA-gelatin/chondroitin/HYA [65]	Chondrocyte, MSC	TGF-β
Muscle				Skeletal myoblasts, Skeletal muscle stem cells [68,69], MSC	TGF-β, IGF-I, VEGF [70,71]
Ligament			PLLA/PLGA [66,67]	Fibroblasts, MSC	TGF-β, BMP-12, EGF, PDGF, b-FGF, VEGF, IGF-II [72]

Note: b-FGF, basic-fibroblast growth factors; BMP, bone morphologic protein; ChS, chondroitin sulfate; EGF, epidermal growth factor; HA, hydroxyapatite; HYA, hyaluronic acid; IGF, insulin-like growth factor; PCL, poly(ε-caprolactone); PDGF, platelet-derived growth factor; PGA, poly(glycolic acid); PLA, poly(L-lactic acid); PLGA, poly(L-lactide/glycolide); PLLA-CL, poly(L-lactide-co-ε-caprolactone); PU, polyurethane; TGF, transforming growth factor; VEGF, vascular endothelial growth factor.

scaffolds, but were also able to migrate inside the scaffold up to 114 μm within 1 week of culture. These cellular behaviors demonstrate better biocompatibility with gelatin/PCL composite nanofibers than with pure PCL nanofibrous material. MSCs on nanofiber matrix made of PCL or silk maintain the capability to differentiate into different lineages such as adipogenic, chondrogenic, and osteogenic lineages [77,78].

The bone growth factors such as BMP-2, BMP-7, etc. gradually became attractive clinical drugs for bone defects repair. The result of the previously mentioned mineralized collagen/PLA 3D scaffold with BMP-2 has already shown some favorable healing effect in vivo [79]. In principle, the nanofibrous materials absorb growth factors better because of their high surface area and nanotexture.

Silk fibroin nanofibrous scaffolds containing BMP-2 and/or nanoparticles of HA prepared via electrospinning were selected as matrix for in vitro bone formation from human bone marrow-derived mesenchymal stem cells (hMSCs) [80]. The scaffolds with BMP-2 supported higher calcium deposition, higher crystallinity of apatite, and enhanced transcript levels of bone-specific markers than the controls without BMP-2, suggesting that nanofibrous electrospun silk scaffolds can be an efficient delivery system for BMP-2. More importantly, the coexistence of BMP-2 and nanophase HA in the electrospun silk fibroin fibers resulted in the highest calcium deposition and upregulation of BMP-2 transcript levels. Fortunately, the mild aqueous process required to electrospin the fibers offers an important option for delivery of labile cytokines and other biomolecules.

14.6.2 Cartilage

Cartilage is a dense connective tissue that is to a certain extent pliable and resilient. These characteristics are due to the nature of its matrix, which is rich in proteoglycans consisting of a core protein attached by the repeating units of disaccharides termed glycosaminoglycans (GAGs) consisting mainly of hyaluronic acid (HYA), chondroitin sulfate, collagen II, and keratin sulfate. Articular cartilages are important in load-bearing and reducing friction of the articular surfaces. Due to the limited capacity of articular cartilage to repair itself, cartilage defect resulting from aging, joint injury, and developmental disorders can cause joint pain and loss of mobility [81]. As promising candidates, the electrospun HYA or HYA-based composite nanofibers have been fabricated successfully. Although, these are promising candidate scaffolds for cartilage regeneration, the efficacy of these scaffolds for cartilage repair and the chondrocyte response to these fibers is still unknown [60–63].

Electrospun collagen type II [82] nanofibers seeded with chondrocytes were investigated for their potential use in cartilage tissue engineering. Individual scaffold specimens were evaluated as uncross-linked, cross-linked, or cross-linked/seeded in the study of Shields et al. [83]. Scanning electron microscopy (SEM) of cross-linked scaffolds cultured with chondrocytes demonstrated the ability of the cells to infiltrate the surface and the interior of the scaffold. The investigators conclude that electrospun collagen type II scaffolds produced a suitable environment for chondrocyte growth, and this could be the foundation for the development of articular cartilage repair.

Aligned chitosan fibers were also studied for their biocompatibility with chondrocytes by Subramanian et al. [84]. As expected, the aligned electrospun chitosan provides better chondrocyte cell viability than cast chitosan film. This electrospun membrane has a significantly higher elastic modulus (2.25 MPa) than the cast film (1.19 MPa). As with all electrospun membrane, the electrospun chitosan nanofiber membrane still needs to be further processed into 3D scaffolds for cartilage tissue repair.

Synthetic biodegradable polymers were also studied for their suitability as chondrocytes scaffold. Shin and colleagues [85] investigated the potential of four different types of nanofiber-based PLGA scaffold (lactic acid/glycolic acid content ratio = 75:25, 50:50, or a blend of 75:25 and 50:50) for cartilage reconstruction. The mechanical properties of the nanofibrous scaffold were only slightly lower than those of human cartilage suggesting that the nanofibrous scaffold was sufficiently mechanically stable to withstand implantation and to support cartilage regeneration. Proliferation of porcine chondrocytes and ECM formation in nanofibrous scaffolds was observed to be superior to those in conventional cast membrane.

Nanofibrous PCL was evaluated for its ability to maintain chondrocytes in functional state using fetal bovine chondrocytes (FBCs) [86]. Gene expression analysis by reverse transcription–polymerase chain reaction (RT-PCR) showed that chondrocytes seeded on the PCL nanofibrous scaffold continuously maintained their chondrocytic phenotype by expressing cartilage-specific ECM genes, including collagen types II and IX, aggrecan, and cartilage oligomeric matrix protein. Expression of the collagen type II splice variant transcript, which is indicative of the mature chondrocyte phenotype, was significantly upregulated. In addition to promoting phenotypic differentiation, the nanofibrous scaffold also supported cellular proliferation as evidenced by a 21-fold increase in cell growth over 21 days when the cultures were maintained in serum-containing medium. Thus, PCL nanofibers may be a suitable candidate scaffold for cartilage tissue engineering. Moreover, nanofibrous composite consisting of natural polymer and synthetic polymer can combine the advantages of both materials for chondrocytes' affinity.

14.6.3 Muscle

To avoid the morbidity associated with sacrificing of normal function at the donor site, skeletal muscle tissue engineering offers compelling advantages. Research in the past has mainly focused on the use of microfibrous scaffolds in muscle tissue engineering. For smooth muscle tissue engineering, a substantial amount of research has looked at using nanofibers as potential tissue scaffolds. To understand the potential use of scaffolds for skeletal muscle regeneration, we reviewed some of the research work on microfibrous scaffolds and SMC–nanofibers interactions. For many applications related to muscle tissue engineering the ability to control scaffold geometry at the micro- and nanolevels are important. In particular, the ability to control the alignment of the fibers will influence the cell morphology to create an effective contractile muscle construct.

DegraPols is a degradable block polyesterurethane, consisting of crystallizable blocks of poly ((*R*)-3-hydroxybutyric acid)-diol and blocks of poly(ε-caprolactone-co-glycolide)-diol linked with a diisocyanate. Electrospun DegraPols microfibrous membranes were studied as a potential scaffold for skeletal muscle tissue engineering [87]. The fibers are about 10 μm in diameter with a fiber-to-fiber distance of about 10 μm. Due to the rotational direction of the collector during electrospinning processing, there is a slight preferential orientation of the fibers. The membranes exhibited no toxic residuals and have satisfactory mechanical properties (linear elastic behavior up to 10% deformation and Young's modulus in the order of MPa). Cell viability, adhesion, and differentiation on protein coated (Matrigels, fibronectin, and collagen) and uncoated DegraPols membrane were investigated using cell lines (C2C12 and L6) and primary human satellite cells (SCs). All cell types showed adhesion, fusion, and proliferation on the coated electrospun membranes. Positive staining for myosin heavy chain was seen in the C2C12 cell line indicating that differentiation of C2C12 multinucleated cells have occurred.

Neumann et al. [88] attempted to create skeletal muscle sheets with C2C12 skeletal myoblasts. The cells were seeded on arrays of parallel oriented polypropylene microfibers. The microfibers were coated with laminin and the fiber diameters ranged from 10 to 15 μm. There was no cell attachment in the control groups where laminin was not used, indicating the importance of surface modification by such attachment proteins. It was noted that the critical separation between the parallel fibers is 55 μm or less. Separation of fibers greater than 55 μm will result in either no cell sheets or incomplete sheets. They succeeded in obtaining layers of cultured tissue up to 50 μm thick formed by longitudinally aligned cells. More importantly they were able to maintain in vitro contractile cell sheet in a 1 by 2 cm^2 array for up to 70 days.

It has been argued that the non-biodegradable fibers will adversely affect the contractility of the muscle construct. The use of fibers made from biodegradable materials, such as PLLA, PLGA, or

PGA, already in clinical use are preferred. However, cells attach poorly to such synthetic polymers unless they have undergone surface modifications by treating with air plasma, grafting of ligands, or coating with adhesion proteins of which laminin and fibronectin are commonly used. It was also demonstrated that coating of ECM gel onto PLLA microfibers (60 μm diameter) enhanced the attachment and differentiation of the seeded myoblast into multinucleated myofibers whose presence was confirmed by the expression of muscle markers such as myosin and α-actin [89]. As opposed to the uncoated fibers, the surface of the ECM gel-coated microfibers appears fibrillar in nature and these submicron textures probably provided cells with numerous points of attachments.

Our laboratory has investigated the adhesion and proliferation of SMC on PLLA-CL nanofibrous scaffolds. Aligned PLLA-CL (75:25) copolymer nanofibrous scaffold was produced by a special setup of electrospinning [41]. PLLA-CL with L-lactide to ε-caprolactone ratio of 75 to 25 has been electrospun into nanofibers. The x-ray diffractometer and differential scanning colorimeter results suggested that the electrospun nanofibers developed highly oriented structure in CL-unit sequences during the electrospinning process [90]. The diameter of the generated nanofibers was around 500 nm with an aligned topography. Favorable interactions between these scaffolds with human coronary artery SMCs were demonstrated via MTS assay, phase contrast light microscopy, SEM, immunohistology assay, and laser scanning confocal microscopy. It was shown that the SMCs attached and migrated along the axis of the aligned nanofibers and expressed a spindle-like contractile phenotype; the distribution and organization of smooth muscle cytoskeleton proteins inside SMCs were parallel to the direction of the nanofibers. This synthetic aligned matrix with the advantages of synthetic biodegradable polymers, nanometer-scale dimension mimicking the natural ECM and a defined architecture replicating the natural tissue structure, may be advantageous in certain tissue engineering application in which SMCs play an important role such as in blood vessel engineering.

14.6.4 Ligament

A variety of synthetic materials have been used for ligament replacement (e.g., Versigraft, Dacron, Gore-Tex, Leeds-Keio polyester, polypropylene-based Kennedy ligament-augmentation device), but with very limited short-term success. Of 855 prosthetic ligaments tracked for 15 years, 40%–78% failed owing to wear debris, tissue reactions, and mechanical limitations. Experience with these synthetic ligament grafts replacement has been disappointing and the failure history is particularly instructive from the perspective of tissue engineering [91].

There is an unmet need for tissue-engineered ligament tissues for the repair of torn and ruptured ligaments and tendons. The functional requirement for ligament replacements is demanding because of the high physiological loads that the replacement ligaments may be subjected to. The tissue-engineered ligament must be sufficiently strong immediately after implantation to withstand physiological loads. For long-term survival of the replaced ligament the scaffolds must possess architecture that is capable of encouraging cell adhesion, proliferation, and differentiation. In order for the regenerated ligament to take over the load from the original scaffold material, the scaffold must support cell orientation and deposition of ECM in response to the physiological loads. Although natural polymer such as collagen and silk are potential candidates as scaffold material, they by themselves may not be strong enough to support physiological loads [92,93].

Silk matrix is noted to support the attachment, expansion, and differentiation of adult human progenitor bone marrow stromal cells based on SEM, DNA quantitation, and the expression of collagen types I, III, and tenascin-C markers. The results support the conclusion that properly prepared silkworm fiber matrices, aside from providing unique benefits in terms of mechanical properties as well as biocompatibility and slow degradability, are suitable biomaterial matrices for the support of adult stem cell differentiation toward ligament lineages. Table 14.2 lists the mechanical properties of musculoskeletal tissues and natural or synthetic polymers [16,42,66,67,83, 94–105]. Synthetic polymers are possible scaffold materials due to the similarity of their tensile strength with soft tissues. Bashur et al. [106] postulated that oriented micro-scale synthetic polymer fiber meshes formed by the electrospinning process can regulate cell morphology, and the microscale topographic features can induce cell orientation through a contact guidance phenomenon. To test this, fused fiber meshes of poly(D,L-lactic-co-glycolic acid) (PLGA) were electrospun onto rigid supports under conditions that produced mean fiber diameters of 0.14–3.6 μm, and angular standard deviations of 31°–60°. Analysis of the morphology of adherent NIH 3T3 fibroblasts indicated that projected cell area and aspect ratio increased systematically with both increasing fiber diameter and degree of fiber orientation. Importantly, cell morphology on 3.6 μm fibers

TABLE 14.2
Mechanical Properties of Tissues in Musculoskeletal System and Biopolymers as Scaffolds

Materials	Young's Modulus (GPa)	Tensile Strength (MPa)	Ultimate Tensile Load (N)
Hard Tissue [16,94]			
Cortical bone	7–30	50–150	
Cancellous bone	1–14	7.4	
Dentin	11–17	21–53	
Enamel	84–131	10	
Soft Tissue [94]			
Articular cartilage	$4.5–10.5 \times 10^{-3}$	27.5	
	0.110 [95]		1725–2160 [96]
Fibrocartilage	159.1×10^{-3}	10.4	
Ligament	303.0×10^{-3}	29.5	295, 298 [97]
Tendon	401.5×10^{-3}	46.5	700 [98]
	1.2–1.8 [97]	50–150 [98]	
Muscle		0.2 [99]	
Natural Polymers			
Twist/parallel silk matrix [93]			$2337 \pm 72/2214$
Type I collagen nanofibers [55]	$(52.3 \pm 5.2) \times 10^{-3}$	1.5 ± 0.2	1.17 ± 0.34
Type I collagen nanofibers	$(262 \pm 18) \times 10^{-3}$ [100]	7.4 ± 1.17	
		11.44 ± 1.2 (cross-linked) [101]	
Type II collagen nanofibers [101]	$(172.5 \pm 36.1) \times 10^{-3}$	3.3 ± 0.3	
Gelatin nanofibers [102]	$(499 \pm 207) \times 10^{-3}$	5.77 ± 0.96	
Synthetic Polymers [16]			
PLLA	2.7	50	
PDLA	1.9	29	
PCL	0.4	16	
PU	0.02	35	
Knitted PLLA-PLGA scaffold	0.283 [66]		29.4 [65]
PLA-CL nanofibers [41]	$(44 \pm 4) \times 10^{-3}$	6.3 ± 1.4	
poly(DTE carbonate) microfibers [103]	3.1	230	
PLLA microfibers [103]	4.9	299	
PLGA(10:90) microfibers [104]			
Single multifilament yarn		5.3 ± 1.8	2.4 ± 0.02
10-Yarn bundle		8.8 ± 1.1	25 ± 3
Rectangular braid		393 ± 29	606 ± 45
Circular braid		212 ± 25	907 ± 132
PLGA(10:90) nanofibers [105]		4.9–6	
PLGA(85:15) nanofibers [56]		22.67	

Note: Poly(DTE carbonate), poly(desamino tyrosyl-tyrosine ethyl ester carbonate).

was similar to that on spin-coated PLGA films. However, cell densities on electrospun meshes were not significantly different from spin-coated PLGA, indicating that fibroblast proliferation is not sensitive to fiber diameter or orientation. Another study compared the PLLA with poly (DTE carbonate) fibers [103]. The study was performed in three phases. In phase I, first-generation fibers were found to promote tissue ingrowth in a subcutaneous model. In phase II, second-generation fibers were fabricated from poly(DTE carbonate) and poly(L-lactic acid), with diameters of 79 and 72 μm, ultimate tensile strengths of 230 and 299 MPa, moduli of 3.1 and 4.9 GPa, and molecular weights of 65,000 and 170,000 Da, respectively. These fibers were evaluated on the basis of molecular weight retention, strength retention, and cytocompatibility. After 30 weeks of incubation in phosphate-buffered saline, poly(DTE carbonate) and poly(L-lactic acid) fibers had 87% and 7% strength retention, respectively. Fibroblasts attached and proliferated equally well on both scaffold types in vitro. Finally, in phase III, a prototype ACL reconstruction device was fabricated from poly(DTE carbonate) fibers with strength values comparable to those of the normal ACL (57 MPa). Collectively, these data suggested that poly(DTE carbonate) fibers are potentially useful for development of resorbable scaffolds for ACL reconstruction comparable to well-established PLGA.

Polyurethane nanofibrous scaffold was used to evaluate the effect of aligned nanofibers on the cellular response of human ligament fibroblasts (HLFs) and to evaluate the influence of HLFs alignment and strain direction on mechanotransduction [107]. The effects of fiber alignment and direction of mechanical stimuli on the ECM generation of HLFs were assessed. To align the nanofibers, a rotating target was used. The HLFs on the aligned nanofibers were spindle-shaped and oriented in the direction of the nanofibers. The degree of ECM production was evaluated by comparing the amount of collagen on aligned and randomly oriented structures. Significantly, more collagen was synthesized on aligned nanofiber sheets, although the proliferation did not differ significantly. This suggests that the spindle shape observable in intact ligaments is preferable for producing ECM. To evaluate the effect of strain direction on the ECM production, HLFs were seeded on parallel aligned, vertically aligned to the strain direction, and randomly oriented nanofiber sheets attached to flexcell plates. After a 48 h culture, 5% uniaxial strain was applied for 24 h at a frequency of the 12 cycles/min. The amounts of collagen produced were measured 2 days after halting the strain application. The HLFs were more sensitive to strain in the longitudinal direction. In conclusion, the aligned nanofibrous scaffold used in this study constitutes a promising base material for tissue-engineered ligament in that it provides more preferable biomimetic structure, along with proper mechanical environment.

Knitted scaffolds have been proven to favor deposition of collagenous connective tissue matrix, which is crucial for tendon/ligament reconstruction [67]. But cell seeding of such scaffolds often requires a gel system, which is unstable in a dynamic situation, especially in the knee joint. Sahoo et al. [108] developed a novel, biodegradable nano-microfibrous polymer scaffold by electrospinning PLGA nanofibers onto a knitted PLGA scaffold to provide a large biomimetic surface for cell attachment. Porcine bone marrow stromal cells were seeded onto either the novel scaffolds by pipetting a cell suspension (Group I) or the knitted PLGA scaffolds by immobilizing in fibrin gel (Group II). Cell attachment was comparable and cell proliferation was faster in Group I. Moreover, cellular function was more actively exhibited in Group I, as evident by the higher expression of collagen I, decorin, and biglycan genes. Thus, this nano-microfibrous scaffold, facilitating cell seeding and promoting cell proliferation, function, and differentiation, could be applied in tissue engineering of tendon/ligament regeneration.

14.7 CHALLENGES

Although it is increasingly clear from in vitro experiments that appropriate nanostructures can exert favorable cell reactions, the practical application of such knowledge is still far from being translated into patient care. There are three closely related problems of current nanoscaffolds:

1. Imprecise control of nanotexture to mimic the specific natural tissues.
2. Certain nanotextures are conducive to cell–substrate interaction; however, the precise control is still not well understood.
3. Inappropriate architextural transition from nanoscale to macroscale.

The ideal strategy is to simultaneously mimic the natural ECM chemically, texturally, and mechanically.

14.7.1 Nanoscaffolds with Tissue-Specific Patterns

Contemporary approaches to material design for tissue regeneration attempts to mimic particular conditions of the development of natural tissue to direct the growth of a single type of tissue. Therefore, approaches to regenerate complex, 3D organs are likely to require simultaneous presentation of multiple distinct stimuli within a single material construct. The progress of the tissue scaffold has moved beyond simply requiring a scaffold to being nontoxic or an inert structure with interconnected pores. The ability to influence cell behavior at the nanoscale level has been amply demonstrated. However, a knowledge gap with regard to the precise influence of nanoscale effects on cellular mechanism remains. Fabrication of scaffold with nanoscale features is still immature and far from what nature can do [6,109]. Artificially directed and precisely controlled nanofabrication of scaffold remains a fertile area for research.

14.7.2 Mechanical Property

Standard tensile testing method has been used to test the nanofibrous membrane. There is a lack of systematic approach to testing even for the most widely used PLGA polymer. In Table 14.2, we listed the mechanical properties of nanofibers of natural polymer and synthetic polymer. The tensile strength of nanofibers is still significantly lower than that of natural tissues. Zong et al. [105] obtained 6 and 4.9 MPa of tensile strength from 400 and 1000 nm diameter PLGA (LA: GA = 10:90) fibers, respectively, and from the study of Li et al. [57], it is 22.67 MPa from 500 ~ 800 nm diameter PLGA (LA:GA = 85:15) nanofibers. For aligned nanofiber membranes, the tensile property orthogonal directions should be considered.

The size effect on the mechanical property of single electrospun nanofiber is the subject of a number of scientific papers. Presently, there are three methods to measure tensile and bending properties of a single nanofiber: cantilever technique, atomic force microscopy (AFM)-based nanoindentation system, and nanotensile tester [110–114]. Using the AFM-based nanoindentation system, bending test of PLLA single fiber was conducted by a suspending fiber over the etched groove in a 4 μm wide and 2.5 μm deep silicon wafer [112]. Crosshead speed of cantilever tip was 1.8 μm/s and the applied maximum load to a single PLLA nanofiber was 15 nN. This method can potentially measure the mechanical properties of cultured cells on electrospun nanofiber and the effect of natural ECM fibers in situ. This nanotechnique may provide us a new tool to investigate the stress stimulus effect on cell and tissue growth. Research on the interaction between the cell and nanofibers under mechanical stress is still in its infancy.

14.7.3 Exploring Advanced Nanofiber Technologies

Electrospinning technology enables production of continuous nanofibers but it has the limitation of mainly producing nanofibers membrane. Attempts have been made to make yarns out of nanofibers and then to fabricate textiles. If 3D structure could be fabricated using such fabrics, then such a structure has an architecture with nano-, micro-, and macroscale properties.

In an attempt to obtain a continuous yarn made of aligned fibers, fiber bundles are fabricated by collecting the electrospun fibers on a rotating drum. The collected nanofibers were linked and twisted into yarns [115]. In the setup designed by Khil et al. [116] for yarn collection, a water bath was used to collect the electrospun fibers. The collected fibers were drawn out of the water as a bundle from one side of the water bath using filament guide bar and collected onto a roller in the form of a yarn.

In another setup, the parallel electrodes of straight bars are replaced with rings. Rotating one of the rings the aligned fibers are twisted into a bundle [112]. A significant drawback is that the length of the twisted fiber bundle is limited to the space between the rings.

14.8 CONCLUSIONS

Nature appears to be the ultimate specialist in the design and manufacture of novel materials with just the right sizes and properties from the nanoscale to microscale. Investigation and understanding of nature's mechanisms of nanomaterial formation can in turn provide inspirations for the new fabrication strategies of improved nanostructured biomaterials for tissue engineering. Despite the recent advances in the development of nanostructures for musculoskeletal tissue engineering, several challenges and critical issues still remain. Along with the development of nanotechnologies and a better understanding of cell–nanotexture interaction, the scaffold materials in the future will have nanostructure that will better mimic the natural ECM environment. The ultimate aim is to manipulate and incorporate not only nanotexture into biomaterials but also to incorporate the appropriate chemical and mechanical signals for better control of cell behavior for tissue regeneration.

ACKNOWLEDGMENTS

Financial support of the National University of Singapore and Lee Kuan Yew Postdoctoral Fellowship (LKY PDF) is gratefully acknowledged.

REFERENCES

1. Weistein, S.L. and Buckwalter, J.A. 1994. *Turek's Orthopaedics Principles and Their Application*, fifth edition. J.B. Lippincott Company, Philadelphia.
2. Landis, W.J., Song, M.J., Leith, A. et al. 1993. Mineral and organic matrix interaction in normally calcifying tendon visualized in three dimensions by high-voltage electron microscopic tomography and graphic image reconstruction. *J Struct Biol* 110(1): 39–54.
3. Landis, W.J., Hodgens, K.J., Arena, J. et al. 1996. Structural relations between collagen and mineral in bone as determined by high voltage electron microscopic tomography. *Microsc Res Tech* 33(2): 192–202.
4. Weiner, S. and Traub, W. 1986. Organization of hydroxyapatite crystals within collagen fibrils. *FEBS Lett* 206(2): 262–266.
5. Weiner, S. and Traub, W. 1989. Crystal size and organization in bone. *Connect Tissue Res* 21(1–4): 589–595.
6. Stevens, M.M. and George, J.H. 2005. Exploring and engineering the cell surface interface. *Science* 310(5751): 1135–1138.
7. Fantner, G.E., Hassenkam, T., Kindt, J.H. et al. 2005. Sacrificial bonds and hidden length dissipate energy as mineralized fibrils separate during bone fracture. *Nat Mater* 4(8): 612–616.
8. Reinert, T., Reibetanz, U., Schwertner, M. et al. 2002. The architecture of cartilage: Elemental maps and scanning transmission ion microscopy/tomography. *Nucl Instr Methods Phys Res B* 188: 1–8.
9. Langer, R. and Vacanti, J.P. 1993. Tissue engineering. *Science* 260(5110): 920–926.
10. Lutolf, M.P. and Hubbell, J.A. 2005. Synthetic biomaterials as instructive extrtacelluar microenvironments for morphogenesis in tissue engineering. *Nat Biotechnol* 23(1): 47–55.
11. Robert, F. 2000. Tissue engineers build new bone. *Science* 289(5483): 1498–1500.
12. Medtech@Insight, Tissue Engineering and Cell Transplantation: Technologies, Opportunities, and Evolving Markets. Industry Report July, 2004.
13. Lanza, R.P., Langer, R., and Vacanti, J. 2000. *Principles of Tissue Engineering*, second edition. Academic Press, USA.

14. Griffith, L.G. and Naughton, G. 2002. Tissue engineering—current challenges and expanding opportunities. *Science* 295(5557): 1009.
15. Langer, R. and Tirrell, D.A. 2004. Designing materials for biology and medicine. *Nature* 428(6982): 487–492.
16. Murugan, R. and Ramakrishna, S. 2006. Nanophase biomaterials for tissue engineering. In *Tissue, Cell and Organ Engineering*, ed. Challa Kumar. 226–256. Wiley-VCH Verlag Gmbh & Co. KgaA, Weinheim.
17. Pittenger, M.F., Mackay, A.M., Beck, S.C. et al. 1999. Multilineage potential of adult human mesenchymal stem cells. *Science* 284(5411): 143–147.
18. Seshi, B., Kumar, S., and Sellers, D. 2000. Human bone marrow stromal cell: Co-expression of markers specific for multiple mesenchymal cell lineages. *Blood Cells Mol Dis* 26(3): 234–246.
19. Caplan, A.I. and Bruder, S.P. 2001. Mesenchymal stem cells: Building blocks for molecular medicine in the 21st century. *Trends Mol Med* 7(6): 259–264.
20. NIH Stem Cell Report entitled: *Stem Cells: Scientific Progress and Future Research Directions*, Kirschstein, R. and Skirboll, L.R., Department of Health and Human Services, June 2001. http://stemcells.nih.gov/info/basics/basics4 [accessed Feb 06, 2007].
21. Kraiwattanapong, C., Boden, S.D., Louis-Ugbo, J. et al. 2005. Comparison of HEALOS/bone marrow to INFUSE(rhBMP-2/ACS) with a collagen-ceramic sponge bulking agent as graft substitutes for lumbar spine fusion. *Spine* 30(9): 1001–1007.
22. Martin, G.J. Jr., Boden, S.D., Marone, M.A. et al. 1999. Posterolateral intertransverse process spinal arthrodesis with rhBMP-2 in non-human primate: Important lessons learned regarding dose, carrier and safety. *J Spinal Diord* 12(3): 179–186.
23. Barnes, B., Boden, S.D., Louis-Ugbo, J. et al. 2005. Lower dose of rhBMP-2 achieves spine fusion when combined with an osteoconductive bulking agent in non-human primates. *Spine* 30(10): 1127–1133.
24. Schimandle, J.H. and Boden, S.D. 1997. Bone substitutes for lumbar fusion: Present and future. *Oper Tech Ortho* 7(1): 60–67.
25. Chan, C.K., Sampath Kumar, T.S., Liao, S. et al. 2006. Biomimetic nanocomposites for bone graft applications. *Nanomedicine* 1(2): 177–188.
26. Leroux, J.C., Allemann, E., Da Jaeghere, F. et al. 1996. Biodegradable nanoparticles—From sustained release formulations to improved site specific drug delivery. *J Control Release* 39: 339–350.
27. Soppimath, K., Aminabhavi, T., Kulkarni, A. et al. 2001. Biodegradable polymeric nanoparticles as drug delivery devices. *J Control Release* 70: 1–20.
28. Saltzman, W.M. and Olbricht, W.L. 2002. Building drug delivery into tissue engineering. *Drug Discov* 1: 177–186.
29. Ramakrishna, S., Fujihara, K., Teo, W.E. et al. 2005. *An Introduction to Electrospinning and Nanofibers*. World Scientific Publishing Co. Pte. Ltd.
30. Webster, T.J., Ergun, C., Doremus, R.H. et al. 2000. Specific proteins mediate enhanced osteoblast adhesion on nanophase ceramics. *J Biomater Med Res* 51: 475–483.
31. Soppimath, K.S., Tan, D.C.W., and Yang, Y.Y. 2005. pH-triggered thermally responsive polymer core–shell nanoparticles for drug delivery. *Ad Mater* 17(3): 318–323.
32. Lo, C.L., Lin, K.M., and Hsiue, G.H. 2005. Preparation and characterization of intelligent core–shell nanoparticles based on poly(D,L-lactide)-g-poly(N-isopropyl acrylamide-co-methacrylic acid). *J Control Release* 104(3): 477–488.
33. Feng, S.S. 2006. New-concept chemotherapy by nanoparticles of biodegradable polymers—where are we now? *Nanomedicine* 1(3): 297–309.
34. Zhang, Y., Huang, Z.M., Xu, X. et al. 2004. Preparation of core–shell structured PCL-r-Gelatin bi-component nanofibers by coaxial electrospinning. *Chem Mater* 16: 3406–3409.
35. Zhang, Y.Z., Wang, X., Feng, Y. et al. 2006. Coaxial electrospinning of fitcBSA encapsulated PCL nanofibers for sustained release. *Biomacromolecules* 7(4): 1049–1057.
36. Wei, G., Jin, Q., Giannobilea, W.V. et al. 2007. The enhancement of osteogenesis by nano-fibrous scaffolds incorporating rhBMP-7 nanospheres. *Biomaterials* 28(12): 2087–2096.
37. Curtis, A. and Wilkinson, C. 1999. New depths in cell behaviour: Reactions of cells to nanotopography. *Biochem Soc Symp* 65: 15–26.
38. Price, R.L., Ellison, K., Haberstroh, K.M. et al. 2004. Nanometer surface roughness increases select osteoblast adhesion on carbon nanofiber compacts. *J Biomed Mater Res* 70A: 129–138.

39. Price, R.L., Waid, M.C., Haberstroh, K.M. et al. 2003. Selective bone cell adhesion on formulations containing carbon nanofibers. *Biomaterials* 24(11): 1877–1887.
40. Fujihara, K., Kotaki, M., and Ramakrishna, S. 2005. Guided bone regeneration membrane made of polycaprolactone/calcium carbonate composite nano-fibers. *Biomaterials* 26(19): 4139–4147.
41. Xu, C.Y., Inai, R., Kotaki, M. et al. 2004. Aligned biodegradable nanofibrous structure: A potential scaffold for blood vessel engineering. *Biomaterials* 25(5): 877–886.
42. He, W., Ma, Z., Yong, T. et al. 2005. Fabrication of collagen-coated biodegradable polymer nanofiber mesh and its potential for endothelial cells growth. *Biomaterials* 26(36): 7606–7615.
43. Yang, F., Murugan, R., Wang, S. et al. 2005. Electrospinning of nano/micro scale poly(L-lactic acid) aligned fibers and their potential in neural tissue engineering. *Biomaterials* 26(15): 2603–2610.
44. He, W., Yong, T., Teo, W.E. et al. 2005. Fabrication and enthothelialization of collagen-blended biodegradable polymer nanofibers: Potential vascular graft for the blood vessel tissue engineering. *Tissue Eng* 11(9–10): 1575–1589.
45. Venugopal, J. and Ramakrishna, S. 2005. Fabrication of modified and functionalized polycaprolactone nanofiber scaffolds for vascular tissue engineering. *Nanotechnology* 16(10): 2138–2142.
46. Ma, Z.W., He, W., Yong, T. et al. 2005. Grafting of gelatin on electrospun poly(caprolactone) PCL nanofibers to improve endothelial cell's spreading & proliferation & to control cell orientation. *Tissue Eng* 11(7–8): 1149–1158.
47. He, W., Yong, T., Teo, W.E. et al. 2006. Biodegradable polymer nanofiber mesh to maintain functions of endothelial cells. *Tissue Eng* 12(9): 2457–2466.
48. Venugopal, J., Vadgama, P., Sampath Kumar, T.S. et al. 2007. Biocomposite nanofibres and osteoblasts for bone tissue engineering. *Nanotechnology* 18: 1–8.
49. Beniash, E., Hartgerink, J.D., Storrie, H. et al. 2005. Self-assembling peptide amphiphile nanofiber matrices for cell entrapment. *Acta Biomater* 1: 387–397.
50. Silva, G.A., Czeisler, C., Niece, K.L. et al. 2004. Selective differentiation of neural progenitor cells by high-epitope density nanofibers. *Science* 303: 1352–1355.
51. Holmes, T.C., de Lacalle, S., Su, X. et al. 2000. Extensive neurite outgrowth and active synapse formation on self-assembling peptide scaffolds. *Proc Natl Acad Sci U S A* 97(12): 6728–6733.
52. Semino, C.E., Kasahara, J., Hayashi, Y. et al. 2004. Entrapment of migrating hippocampal neural cells in three-dimensional peptide nanofiber scaffold. *Tissue Eng* 10(3–4): 643–655.
53. Rajangam, K., Behanna, H.A., Hui, M.J. et al. 2006. Heparin binding nanostructures to promote growth of blood vessels. *Nano Lett* 6(9): 2086–2090.
54. Woo, K.M., Jun, J.H., Chen, V.J. et al. 2007. Nano-fibrous scaffolding promotes osteoblast differentiation and biomineralization. *Biomaterials* 28(2): 335–343.
55. Woo, K.M., Chen, V.J., and Ma, P.X. 2003. Nano-fibrous scaffolding architecture selectively enhances protein adsorption contributing to cell attachment. *J Biomed Mater Res* 67A: 531–537.
56. Matthews, J.A., Wnek, G.E., Simpson, D.G. et al. 2002. Electrospinning of collagen nanofibers. *Biomacromolecules* 3: 232–238.
57. Li, W.J., Laurencin, C.T., Caterson, E.J. et al. 2002. Electrospun nanofibrous structure: A novel scaffold for tissue engineering. *J Biomed Mater Res* 60: 613–621.
58. Dzenis, Y. 2004. Spinning continuous fibers for nanotechnology. *Science* 304(5679): 1917–1919.
59. Ma, Z., Kotaki, M., Inai, R. et al. 2005. Potential of nanofiber matrix as tissue-engineering scaffolds. *Tissue Eng* 11(1–2): 101–109.
60. Wang, X., Um, I.C., Fang, D. et al. 2005. Formation of water-resistant hyaluronic acid nanofibers by blowing-assisted electro-spinning and non-toxic post treatments. *Polymer* 46: 4853–4867.
61. Um, I.C., Fang, D., Hsiao, B.S. et al. 2004. Electro-spinning and electro-blowing of hyaluronic acid. *Biomacromolecules* 5: 1428–1436.
62. Li, J., He, A., Han, C.C. et al. 2006. Electrospinning of hyaluronic acid (HA) and HA/gelatin blends. *Macromol Rapid Commun* 27: 114–120.
63. Ji, Y., Ghosh, K., Shu, X.Z. et al. 2006. Electrospun three-dimensional hyaluronic acid nanofibrous scaffolds. *Biomaterials* 27(20): 3782–3792.
64. Liao, S., Li, B., Ma, Z. et al. 2006. Biomimetic electrospun nanofibers for tissue regeneration. *Biomed Mater* 1: R45–R53.
65. Fan, H., Hu, Y., Zhang, C. et al. 2006. Cartilage regeneration using mesenchymal stem cells and a PLGA-gelatin/chondroitin/hyaluronate hybrid scaffold. *Biomaterials* 27(26): 4573–4580.

66. Ge, Z., Goh, J., and Lee, E.H. 2006. The effects of bone marrow-derived mesenchymal stem cells and fascia wrap application to anterior cruciate ligament tissue engineering. *Cell Transplant* 14(10): 763–773.
67. Ge, Z., Goh, J.C., Wang, L. et al. 2005. Characterization of knitted polymeric scaffolds for potential use in ligament tissue engineering. *J Biomater Sci Polym Ed* 16: 1179–1192.
68. Golding, J.P., Calderbank, E., Partridge, T.A. et al. 2007. Skeletal muscle stem cells express anti-apoptotic ErbB receptors during activation from quiescence. *Exp Cell Res* 313: 341–356.
69. Price, F.D., Kuroda, K., and Rudnicki, M.A. 2007. Stem cell based therapies to treat muscular dystrophy. *Biochim Biophys Acta* 1772: 272–283.
70. Payne, A.M., Messi, M.L., Zheng, Z. et al. 2007. Motor neuron targeting of IGF-1 attenuates age-related external Ca^{2+}-dependent skeletal muscle contraction in senescent mice. *Exp Gerontol*, 42(4): 309–319.
71. Becker, C., Lacchini, S., Muotri, A.R. et al. 2006. Skeletal muscle cells expressing VEGF induce capillary formation and reduce cardiac injury in rats. *Int J Cardiol* 113: 348–354.
72. Ge, Z., Yang, F., Goh, J.C.H. et al. 2006. Biomaterials and scaffolds for ligament tissue engineering. *J Biomed Mater Res* 77A: 639–652.
73. Ito, Y., Hasuda, H., Kamitakahara, M. et al. 2005. A composite of hydroxyapatite with electrospun biodegradable nanofibers as a tissue engineering material. *J Biosci Bioeng* 100(1): 43–49.
74. Li, C.M., Jin, H.J., Botsaris, G.D. et al. 2005. Silk apatite composites from electrospun fibers. *J Mater Res* 20: 3374–3384.
75. Yoshimoto, H., Shin, Y.M., Terai, H. et al. 2003. A biodegradable nanofiber scaffold by electrospinning and its potential for bone tissue engineering. *Biomaterials* 24(12): 2077–2082.
76. Zhang, Y.Z., Ouyang, H.W., Lim, C.T. et al. 2005. Electrospinning of gelatin fibers and gelatin/PCL composite fibrous scaffolds. *J Biomed Mater Res* 72B: 156–165.
77. Li, W.J., Tuli, R., Huang, X. et al. 2005. Multilineage differentiation of human mesenchymal stem cells in a three-dimensional nanofibrous scaffold. *Biomaterials* 26(25): 5158–5166.
78. Boudriot, U., Goetz, B., Dersch, R. et al. 2005. Role of electrospun nanofibers in stem cell technologies and tissue engineering. *Macromol Symp* 225: 9–16.
79. Liao, S.S., Cui, F.Z., Zhang, W. et al. 2004. Hierarchically biomimetic bone scaffold materials: Nano-HA/collagen/PLA composite. *J Biomed Mater Res* 69B: 158–165.
80. Li, C., Vepari, C., Jina, H.J. et al. 2006. Electrospun silk-BMP-2 scaffolds for bone tissue engineering. *Biomaterials* 27(16): 3115–3124.
81. Li, W.J., Tuli, R., Okafor, C. et al. 2005. A three-dimensional nanofibrous scaffold for cartilage tissue engineering using human mesenchymal stem cells. *Biomaterials* 26(6): 599–609.
82. Matthews, J.A., Boland, E.D., Wnek, G.E. et al. 2003. Electrospinning of collagen type II: A feasibility study. *J Bioact Comp Poly* 18: 125–134.
83. Shields, K.J., Beckman, M.J., Bowlin, G.L. et al. 2004. Mechanical properties and cellular proliferation of electrospun collagen type II. *Tissue Eng* 10(9–10): 1510–1517.
84. Subramanian, A., Vu, D., Larsen, G.F. et al. 2005. Preparation and evaluation of the electrospun chitosan/PEO fibers for potential applications in cartilage tissue engineering. *J Biomater Sci Polym Ed* 16: 861–873.
85. Shin, H.J., Lee, C.H., Cho, I.H. et al. 2006. Electrospun PLGA nanofiber scaffolds for articular cartilage reconstruction: Mechanical stability, degradation and cellular responses under mechanical stimulation in vitro. *J Biomater Sci Polym Ed* 17: 103–119.
86. Li, W.J., Danielson, K.G., Alexander, P.G. et al. 2003. Biological response of chondrocytes cultured in three dimensional nanofibrous poly(ε-caprolactone) scaffolds. *J Biomed Mater Res* 67A: 1105–1114.
87. Riboldi, S.A., Sampaolesi, M., Neuenschwander, P. et al. 2005. Electrospun degradable polyesterurethane membranes: Potential scaffolds for skeletal muscle tissue engineering. *Biomaterials* 26(22): 4606–4615.
88. Neumann, T., Hauschka, S.D., and Sanders, J.E. 2003. Tissue engineering of skeletal muscle using polymer fiber arrays. *Tissue Eng* 9(5): 995–1003.
89. Cronin, E.M., Thurmond, F.A., Bassel-Duby, R. et al. 2004. Protein-coated poly(L-lactic acid) fibers provide a substrate for differentiation of human skeletal muscle cells. *J Biomed Mater Res* 69A(3): 373–381.
90. Mo, X.M., Xu, C.Y., Kotaki, M. et al. 2004. Electrospun P(LLA-CL) nanofiber: A biomimetic extracellular matrix for smooth muscle cell and endothelial cell proliferation. *Biomaterials* 25(10): 1883–1890.
91. Vunjak-Novakovic, G., Altman, G., Horan, R. et al. 2004. Tissue engineering of ligaments. *Annu Rev Biomed Eng* 6: 131–156.

92. Chvapil, M., Speer, D.P., Holubec, H. et al. 1993. Collagen fibers as a temporary scaffold for replacement of ACL in goats. *J Biomed Mater Res* 27: 313–325.
93. Altman, G.H., Horan, R.L., Lu, H.H. et al. 2002. Silk matrix for tissue engineered anterior cruciate ligaments. *Biomaterials* 23(20): 4131–4141.
94. Black, J. and Hastings, G.W. 1998. *Handbook of Biomaterials Properties*. Chapman and Hall, London.
95. Noyes, F.R. and Grood, E.S. 1976. The strength of the anterior cruciate ligament in humans and Rhesus monkeys. *J Bone Joint Surg Am* 58: 1074–1082.
96. Dunn, M.G. 1998. Anterior cruciate ligament prostheses. In *Encyclopedia of Sports Medicine and Science*, ed. Fahey, T. Internet Society of Sport Science: http://sportsci.org, 10 March 1998.
97. LaPrade, R.F., Bollom, T.S., Wentorf, F.A., Wills, N.J., and Meister, K. 2005. Mechanical properties of the posterolateral structures of the knee. *The American Journal of Sports Medicine* 33: 1386–1391.
98. Mow, V.C. and Hayes, W.C., Eds. 1991. *Basic Orthopaedic Biomechanics*, Raven Press, New York.
99. Dee, R., Hurst, L.C., Gruber, M.A. et al. 1997. *Principles of Orthopaedic Practice*. The McGraw-Hill Co., Inc. New York.
100. Li, M., Mondrinos, M.J., Gandhi, M.R. et al. 2005. Electrospun protein fibers as matrices for tissue engineering. *Biomaterials* 26(30): 5999–6008.
101. Rho, K.S., Jeong, L., Lee, G. et al. 2006. Electrospinning of collagen nanofibers: Effects on the behavior of normal human keratinocytes and early-stage wound healing. *Biomaterials* 27(8): 1452–1461.
102. Li, M., Guo, Y., Wei, Y. et al. 2006. Electrospinning polyaniline-contained gelatin nanofibers for tissue engineering applications. *Biomaterials* 27(13): 2705–2715.
103. Bourke, S.L., Kohn, J., and Dunn, M.G. 2004. Preliminary development of a novel resorbable synthetic polymer fiber scaffold for anterior cruciate ligament reconstruction. *Tissue Eng* 10(1–2): 43–52.
104. Cooper, J.A., Lu, H.H., Ko, F.K. et al. 2005. Fiber-based tissue-engineered scaffold for ligament replacement: Design considerations and in vitro evaluation. *Biomaterials* 26(13): 1523–1532.
105. Zong, X., Ran, S., Fang, D. et al. 2003. Control of structure, morphology and property in electrospun poly(glycolide-co-lactide) non-woven membranes via post draw treatments. *Polymer* 44: 4959–4967.
106. Bashur, C.A., Dahlgren, L.A., and Goldstein, A.S. 2006. Effect of fiber diameter and orientation on fibroblast morphology and proliferation on electrospun poly(D,L-lactic-co-glycolic acid) meshes. *Biomaterials* 27(33): 5681–5688.
107. Lee, C.H., Shin, H.J., Cho, I.H. et al. 2005. Nanofiber alignment and direction of mechanical strain affect the ECM production of human ACL fibroblast. *Biomaterials* 26(11): 1261–1270.
108. Sahoo, S., Ouyang, H., Goh, C.H.J. et al. 2006. Characterization of a novel polymeric scaffold for potential application in tendon/ligament tissue engineering. *Tissue Eng* 12(1): 91–99.
109. Moroni, L. and Van Bitterswijk, C.A. 2006. Biomaterials converge and regenerate. *Nat Mater* 5(6): 437–438.
110. Dalton, P.D., Klee, D., and Moller, M. 2005. Electrospinning with dual collection rings. *Polym Commun* 46: 611–614.
111. Buer, A., Ugbolue, S.C., and Warner, S.B. 2001. Electrospinning and properties of some nanofibers. *Tex Res* 71(4): 323–328.
112. Tan, E.P.S. and Lim, C.T. 2004. Physical properties of a single polymeric nanofiber. *Appl Phys Lett* 84(9): 1603–1605.
113. Tan, E.P.S., Ng, S.Y., and Lim, C.T. 2005. Tensile testing of a single ultrafine polymeric fiber. *Biomaterials* 26(12): 1453–1456.
114. Inai, R., Kotaki, M., and Ramakrishna, S. 2005. Structure and property of electrospun PLLA single nanofibers. *Nanotechnology* 16: 208–213.
115. Fennessey, S.F. and Farris, R.J. 2004. Fabrication of aligned and molecularly oriented electrospun polyacrylonitrile nanofibers and the mechanical behavior of their twisted yarns. *Polymer* 45: 4217–4225.
116. Khil, M.S., Bhattarai, S.R., Kim, H.Y. et al. 2005. Novel fabricated matrix via electrospinning for tissue engineering. *J Biomed Mater Res* 72B: 117–124.

Index

A

B

C

D

E

F

G

O

P

T

V